Mathematics for Christian Living Series

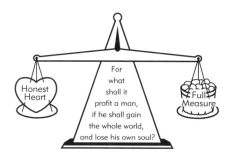

Mathematics for Christian Living Series

Applying Mathematics

Grade 8

Teacher's Manual
Part 1

Rod and Staff Publishers, Inc.
P.O. Box 3, Hwy. 172
Crockett, Kentucky 41413
Telephone: (606) 522-4348

Acknowledgements

We are indebted to God for the vision of the need for a *Mathematics for Christian Living Series* and for His enabling grace. Charitable contributions from many churches have helped to cover the expenses for research and development.

This revision was written by Brother Glenn Auker. The brethren Marvin Eicher, Jerry Kreider, and Luke Sensenig served as editors. Brother John Mark Shenk and Sisters Marian Baltozer, Amy Herr, and Christine Collins drew the illustrations. The work was evaluated by a panel of reviewers and tested by teachers in the classroom. Much effort was devoted to the production of the book. We are grateful for all who helped to make this book possible.

—*The Publishers*

Copyright, 2004

By Rod and Staff Publishers, Inc.
Crockett, Kentucky 41413

Printed in U.S.A.

ISBN 978-07399-0487-9
Catalog no. 13891.3

Materials for This Course

Books and Worksheets
Pupil's Textbook
2 Teacher's Manuals
Quizzes and Speed Tests
Chapter Tests

Tools Needed for Each Student
English ruler
Metric ruler
Protractor
Compass

To the Teacher

Basic Philosophy

Mathematics is a fundamental part of God's creation. Although a study of mathematics does not in itself lead to God and salvation, it can heighten one's awareness of the perfection in God's order. Because of that perfection, unchanging mathematical principles can be used to solve a host of everyday problems pertaining to quantity, length, size, proportion, and many other matters. The goal of this mathematics course is to develop in students the ability to solve mathematical problems so that they will be better able to serve the Lord.

May the Lord bless you as you teach this phase of the laws in His orderly world. Important as the mathematical concepts are that are presented in this course, they are superseded by the patience and wisdom that you will reveal as you present them. If you teach students to calculate well, yet influence them in a wrong direction, you will teach them to calculate for wrong purposes. But if you teach them to calculate well *and* teach godly virtues by word and example, you will help to train valuable servants for the Lord. May your teaching be fruitful to this end.

Plan of the Course

This book contains 170 lessons in twelve chapters. The last lesson of each chapter is a test which is bound in a separate, consumable booklet. This text is intended to be taught at the rate of one lesson per day. If you cannot have math class every school day, you will need to omit or combine some lessons in order to cover the material.

Pupil's Book

The lesson text presents explanations and illustrations of the new concepts. In teaching the lesson, briefly discuss the concepts in the explanation and then move promptly into solving several problems. Students learn most quickly by studying the examples, watching the teacher solve a few problems on the chalkboard, and then solving some problems themselves.

Class Practice is a set of problems to be used in teaching the lesson and providing class drill for the students. Solve one or two problems yourself, and then have the students solve the others either on paper or at the chalkboard. Check their work to make certain they are using the correct procedures.

Written Exercises are the homework problems, usually divided into Part A, Part B, and so on. The first several parts contain problems that develop the new concepts taught in the lesson. These are followed by a set of reading problems (usually six). Some of the reading problems relate to the new material, and some deal with review concepts.

Review Exercises give review of concepts taught in previous lessons. This textbook is designed to review more recent material more frequently and less recent material less frequently. Because of the review pattern and the pattern for introducing new material, it is important that the lessons are covered in consecutive order.

Challenge Exercises appear at the end of a few lessons. They are provided for the challenge of the more able and independent students, and need not be considered part of the basic assignment for all.

Pairing of problems. Most of the problems are arranged in pairs. That is, problems 1 and 2 are of the same kind, problems 3 and 4 are similar, and so on. If a lesson or set of problems is too long for the available time, the teacher can assign only the even-numbered or the odd-numbered problems with the confidence that all the main concepts will be drilled in those exercises. In general, each pupil should do at least the even- or odd-numbered problems in every part of the exercises.

Teacher's Manual

The Teacher's Manual provides a skeleton outline for each lesson. The order of the headings gives the suggested order to follow in presenting the lesson to the class.

Objectives lists the main concepts to be taught in the lesson. Watch especially for objectives that are marked with a star; they indicate concepts that are new in this math series. Plan to give special attention to these concepts as you teach.

Review gives concepts to be reviewed from previous lessons. Most often this review material corresponds to review exercises in the pupil's lesson. Use the examples given to prepare the students for the review portion of their lesson assignment. Spend more time with difficult concepts such as percentages, and less time with simpler concepts such as carrying in addition.

When reviewing, be sure to allow sufficient time for the main part of the lesson—the new material.

Introduction provides an illustration, a theme, or a train of logic with which to introduce the lesson. The Introduction is not intended to be a lesson of its own, but rather a springboard to the lesson itself.

Teaching Guide contains a set of numbered points to be taught, along with helps for presenting them to the class. Boldface is used for main lesson concepts, and regular type for further explanations and other points. After each main teaching point are several related exercises for illustration and practice.

Plan ahead so that you have sufficient time to develop this part of the lesson. Use both these exercises and the Class Practice problems in the pupil's text to reinforce the concepts being taught. In math, experience is usually the best teacher.

An Ounce of Prevention warns of wrong ideas or procedures that students need to avoid. Be sure to point these out to your students. Even though they may seem trivial, they can be real pitfalls.

Further Study presents material that is intended to broaden the teacher's base of knowledge. Although you may occasionally decide to present these facts to the class, this section is designed primarily for the teacher's benefit.

The *answer key* generally appears as colored text on the teacher's copy of the pupil pages. When more space is needed, answers are placed on the facing teacher's page.

Solutions for most reading problems are provided on facing pages for your reference.

When directions call for students to show their work, and for more advanced exercises such as square roots or algebraic equations, full solutions are given.

General Class Procedure

Following is the recommended procedure for teaching a typical lesson in this course.
1. Correct the homework from the previous lesson.
2. Review previous lessons, using the Review section of the teacher's guide.
3. Teach the material for the new lesson, using the Class Practice problems to make certain the lesson is well learned.
4. Assign the homework.
5. Administer any Quiz or Speed Test that is called for. If you have a multigrade classroom, it may fit your schedule better to give Quizzes or Speed Tests before math class while you are teaching another class.

Naturally, an experienced teacher will sometimes vary his approach to teaching the lesson. However, it is recommended that beginning teachers give careful consideration to the instructions in the Teaching Guide.

Quizzes and Speed Tests

The Quizzes and Speed Tests (found in a separate booklet) are an integral part of this course. Quizzes and Speed Tests are numbered according to the lessons with which they belong, and should be given as indicated in the Review section of the teacher's guide.

Quizzes

The purpose of the forty Quizzes is to review things that students have previously learned. These Quizzes are not timed, but are rather intended to reveal the understanding of concepts.

Speed Tests

The purpose of the eight Speed Tests is to stimulate and maintain the students' skill with basic computation. They drill basic arithmetic facts or concepts that should be so thoroughly mastered that they take little thought in the general math assignments. A good foundation in these basics is very important for accuracy.

Chapter Tests

A Chapter Test (found in a separate booklet) is to be given at the end of each chapter. Each test counts as a numbered lesson in the course.

Following are a few pointers for testing. These pointers also appear in the Teacher's Manual for the Chapter 1 Test (Lesson 16).

1. Only the test, scratch paper, pencils, and an eraser should be on each student's desk.
2. Steps should be taken to minimize the temptation of dishonesty and the likelihood of accidentally seeing other students' answers. Following are some suggestions.
    a. Desk tops should be level. If the desks are very close to each other, have the students keep their work directly in front of them on the desk.
    b. Students should not look around more than necessary during test time.
    c. No communication should be allowed.
    d. As a rule, students should remain seated during the whole test period. It is a good idea to sharpen a few extra pencils and have them on hand.
    e. Students should hand in their tests before going on to any other work.
3. Encourage the students to do their work carefully and to go back over it if they have time. Do not allow them to hand in their tests too soon. On the other hand, some students are so meticulous that they can hardly finish their tests. If you have this problem, set a time when you will collect all the tests. Once 90% of the tests are completed, the rest of them should generally be finished in the next five or ten minutes. Of course, there are exceptions for slower students.
4. A test is different from homework. Students should realize that they must rely on their own knowledge as they work. The teacher should not help them except to make sure that all instructions are clearly understood.

Evaluating Test Results

1. If you check the tests in class, have students check each others' work. Spot check the corrected tests.
2. Tests are valuable tools in determining what the students have grasped. Are there any places where the class is uniformly weak? If so, reinforcement is needed.
3. One effective way to discover the general performance of the class is to find the class median. This is done by arranging the scores in order from highest to lowest. The middle score is the class median. If there is an even number of students, find the average of the two middle scores.

Disposition of the Corrected Tests

1. As a rule, students should have the privilege to see their tests. Review any weak points and answer any questions about why an answer is wrong.
2. The teacher may use his discretion about whether the students should be allowed to keep their tests permanently. Some teachers prefer to collect them again so that students' younger siblings will have no chance of seeing the tests in later years.

Grading

Homework scores are the basic source of a student's math grade. Quizzes are fewer, but more significant in showing understanding of concepts. Chapter Tests are designed to show what the student has thoroughly mastered and have the greatest influence in determining the final grade.

It is suggested that a Quiz grade is given twice the value of a regular homework score, and that the Chapter Test(s) represent at least half the of the student's report card grade. Speed Tests, if graded, may be handled in the same way as homework scores.

To give a Quiz score double value, enter it twice in the list of addends when calculating an average, and also increase the divisor every time you make a double entry. For example, if you have a series of thirteen homework scores and four Quiz scores, the sum will be the total of those seventeen scores, plus the four Quiz scores repeated. The dividend for finding the average should be 13 plus 4 plus 4, or 21. A simple system for producing this result is to write the entry twice when entering a Quiz score in the gradebook. Then at grade-averaging time the double value is automatically worked in.

To have test scores represent half the value of the math grade, first find the average of all other scores. If there was more than one Chapter Test in the marking period, also average the test scores. Then find the average of those two results (add them and divide by two).

Table of Contents

Numbers given first are pupil page numbers.
Numbers at the far right are actual manual page numbers printed at the bottom of the pages.

To the Teacher 5

Chapter 1   Basic Mathematical Operations

1. Working With Large Numbers 12 16
2. Understanding Place Value and Rounding 15 23
3. Understanding Roman Numerals 18 28
4. Using the Commutative and Associative Laws 21 35
5. Adding Mentally 24 40
6. Reviewing Subtraction 27 47
7. Subtracting Mentally 30 52
8. Reviewing Multiplication 32 56
9. Multiplying Mentally 36 64
10. Using the Distributive Law 39 71
11. Reviewing Division 43 79
12. Dividing Mentally, Part 1 .... 47 87
13. Dividing Mentally, Part 2 .... 50 92
14. Reading Problems: Choosing the Correct Operation 52 96
15. Chapter 1 Review 56 104
16. Chapter 1 Test (in test booklet) 586

Chapter 2   English and Metric Measures

17. English Units of Linear Measure and Weight 60 112
18. English Units of Capacity 63 119
19. English Units of Area 66 124
20. Units of Time 69 131
21. Metric Units of Linear Measure 73 139
22. Metric Units of Weight 77 147
23. Metric Units of Capacity 80 152
24. Metric Units of Area 83 159
25. Adding and Subtracting Compound Measures 87 167
26. Multiplying and Dividing Compound Measures 90 172
27. Converting Between English and Metric Measures 93 179
28. Converting Between Fahrenheit and Celsius Temperatures 96 184
29. Distance, Rate, and Time 99 191
30. Understanding Bible Measures 102 196
31. Reading Problems: Choosing the Necessary Facts 106 204
32. Chapter 2 Review 109 211
33. Chapter 2 Test (in test booklet) 589

Chapter 3 Factoring and Fractions

34. Finding Prime Factors 114 200
35. Finding Greatest Common Factors and Lowest Common Multiples 117 227
36. Using Common Fractions to Express Parts 120 232
37. Adding and Subtracting Common Fractions .124 240
38. Multiplying Fractions 127 247
39. Dividing Fractions, Part 1 . . . . .131 255
40. Dividing Fractions, Part 2 . . . . .134 260
41. Finding a Number When a Fractional Part of It Is Known 137 267
42. Mental Division by Proper Fractions 139 271
43. Reading Problems Containing Fractions 141 275
44. Reading Problems: Deciding What Information Is Missing 145 283
45. Chapter 3 Review 148 288
46. Chapter 3 Test (in test booklet) 592

Chapter 4 Decimals, Ratios, and Proportions

47. Using Decimals to Express Fractional Parts .152 296
48. Decimals in Multiplication 156 304
49. Decimals in Division 159 311
50. Rounding Decimals and Expressing Fractions as Decimals 162 316
51. Multiplying by the Simpler Method 166 324
52. Ratios: Tools of Comparison 169 331
53. Direct Proportions 173 339
54. Inverse Proportions 178 348
55. Reading Problems: Using Direct and Inverse Proportions 183 359
56. Applying Direct Proportions to Maps 187 367
57. Applying Direct Proportions to Scale Drawings 191 375
58. Chapter 4 Review 196 384
59. Chapter 4 Test (in test booklet) 595

Chapter 5 Mastering Percents

60. Expressing Rates as Percents 200 392
61. Finding Percentages 204 400
62. Rates Greater Than 100% and Less Than 1% 207 407
63. Calculating Increase and Decrease (Method 1) 210 412
64. Calculating Increase and Decrease (Method 2) 213 419
65. Finding What Percent One Number Is of Another 216 424
66. Finding the Percent of Increase or Decrease 219 431
67. Calculating the Base 222 436

68.	Working With Commission	225	443
69.	Solving Percent Problems Mentally	228	448
70.	Reading Problems: Using Sketches	232	456
71.	Chapter 5 Review	236	464
72.	Chapter 5 Test (in test booklet)		598

Chapter 6   Statistics and Graphs

73.	The Arithmetic Mean	240	472
74.	The Median and the Mode	243	479
75.	The Histogram	247	487
76.	The Picture Graph	252	496
77.	The Bar Graph	257	507
78.	The Line Graph	261	515
79.	The Circle Graph	265	523
80.	The Rectangle Graph	269	531
81.	Chapter 6 Review	273	539
82.	Chapter 6 Test (in test booklet)		602
83.	Semester 1 Review	277	547
84.	Semester 1 Test (in test booklet)		606

Quizzes and Speed Tests 555

Chapter Tests 1–6 585

Index 573 611

Symbols 581 618

Formulas 582 619

Square Roots 584 621

Tables of Measure . . . . .Back endsheet

Contents of Teacher Book 2

Chapter 7	Lines and Planes in Geometry
Chapter 8	Geometric Solids and the Pythagorean Rule
Chapter 9	Mathematics and Finances
Chapter 10	Introduction to Algebra
Chapter 11	Signed Numbers, Tables, and Graphs
Chapter 12	Other Numeration Systems and Final Reviews

A rock is the proper foundation for a stable house.

Hearing and doing the words of Jesus is the proper foundation for a fulfilling life—"His divine power hath given unto us all things that pertain unto life and godliness, through the knowledge of him that hath called us to glory and virtue." Upon this foundation we build Christian virtues, the fruit of the Spirit—"Giving all diligence, add to your faith virtue; and to virtue knowledge; and to knowledge temperance; and to temperance patience; and to patience godliness; and to godliness brotherly kindness; and to brotherly kindness charity" (2 Peter 1:3, 5–8).

A solid grasp of the basic operations is the proper foundation for more advanced mathematical achievement. This means ability to answer the addition, subtraction, multiplication, and division facts quickly and accurately. It also means correctly performing the processes involved, such as carrying, borrowing, and the more complex procedures of multiplication and division.

Chapter 1
Basic Mathematical Operations

Addition, subtraction, multiplication, and division are called basic operations because all mathematics is based on them. As you work with more advanced concepts, such as exponents, square roots, literal numbers, and signed numbers, you will find that they are not entirely new. Rather, they are extensions that build on the four basic operations.

Even though you have been working with the four basic operations for many years, review is still valuable. It should firmly rivet in your mind the principles involved, and improve your accuracy so that you will avoid errors when doing more advanced mathematical computations. Practice makes perfect.

Focus on speed and accuracy as you practice the familiar operations in Chapter 1. This should improve your ability to learn more difficult concepts and also to apply mathematical principles in everyday life.

A wise man . . .
built his house upon a rock: . . .
and it fell not: for it was founded upon a rock.
Matthew 7:24, 25

12   Chapter 1   Basic Mathematical Operations

1. Working With Large Numbers

We can use words to write numbers. However, writing numbers is much easier if we use a system of symbols instead of words. This is called **arithmetic notation**.

Our system of notation probably originated with the Hindus in India. Later the Arab traders introduced it to the Europeans. Therefore, it became known as the Hindu-Arabic system, generally shortened to the **Arabic number system**.

The Arabic number system uses ten digits (0–9) and a system of place values. Large numbers are written with a series of digits. Long series of digits are divided into periods to make them easy to read. The first six periods are named in the example below.

Written with digits:

6, 043, 000, 000, 654, 321

Written with words:

"Six quadrillion, forty-three trillion, six hundred fifty-four thousand, three hundred twenty-one"

To write a number with digits, follow these rules.

1. Include commas in any number larger than three digits.
2. Use three zeroes to represent any period that has no named value. Be sure each period contains three digits, except the period farthest to the left.

To write a number with words, follow these rules.

1. Write the value of the number in any period that has non-zero digits. Add the period name according to its position in the place value system.
2. Insert commas between periods, the same as when using digits.
3. Hyphenate the numbers twenty-one through ninety-nine.

To read a number, follow these rules.

1. Identify the period farthest to the left in the number.
2. Read the number in the leftmost period and say the period name. Repeat this for all the periods.
3. Do not read periods containing only zeroes.
4. Do not use *and* in reading a number unless it includes a fraction or decimal.

LESSON 1

Objective

(New items are marked with a star.)

- To review reading and writing numbers *up to quadrillions. (Previous experience: Working with numbers up to trillions.)

Introduction

Beginning with thousands, write the smallest number in each period up to the quadrillions' period. As you write these numbers, ask the students for practical examples of uses for each number given.

1,000 (one thousand)—Population of a small town

1,000,000 (one million)—Population of a large city, such as Houston, Texas; Phoenix, Arizona; or Detroit, Michigan

1,000,000,000 (one billion)—Population of China
—Population of Europe, Central America, and South America combined

1,000,000,000,000 (one trillion)—Distance that light travels in about two months at 186,000 miles per second
—The United States produces about $6,000,000,000,000 worth of goods and services per year.
—The nearest star, Proxima Centauri, is about 25,000,000,000,000 miles from the earth.

1,000,000,000,000,000 (one quadrillion)—Distance that light travels in about $170\frac{1}{3}$ years
—The North Star (Polaris) is about 1.75 quadrillion miles away from the earth (1,750,000,000,000,000 mi.).

Note that the larger the number, the less we can comprehend it. However, very large numbers are used at times.

Teaching Guide

1. **To write a number with digits, follow these rules.**

   (1) Include commas in any number larger than three digits.

   (2) Use three zeroes to represent any period that has no named value. Be sure each period contains three digits, except the period farthest to the left.

   Write these numbers, using digits.
   a. Two thousand, seven   (2,007)
   b. Seventy-five trillion, sixteen million, five
      (75,000,016,000,005)
   c. Ninety billion, seven hundred fifty thousand
      (90,000,750,000)
   d. Twenty-five quadrillion, five hundred trillion
      (25,500,000,000,000,000)

2. **To write a number with words, follow these rules.**

   (1) Write the value of the number in any period that has non-zero digits. Add the period name according to its position in the place-value system.

   (2) Insert commas between periods, the same as when using digits.

   (3) Hyphenate the numbers twenty-one through ninety-nine.

   Identify the leftmost period in each number.
   a. 16,617   (thousands)
   b. 87,899,000,000   (billions)
   c. 781,000,000,000,000
      (trillions)
   d. 14,000,000,000,000,000
      (quadrillions)

T–13   Chapter 1   Basic Mathematical Operations

Write these numbers, using words.
- a. 450,000,500,000
 (four hundred fifty billion, five hundred thousand)
- b. 65,000,005,000,000
 (sixty-five trillion, five million)
- c. 6,500,400,300,200,100
 (six quadrillion, five hundred trillion, four hundred billion, three hundred million, two hundred thousand, one hundred)

3. **To read a number, follow these rules.**
   (1) Identify the period farthest to the left in the number.
   (2) Read the number in the leftmost period and say the period name. Repeat this for all the periods.
   (3) Do not read periods containing only zeroes.
   (4) Do not use *and* in reading a number unless it includes a fraction or decimal.

Read the following numbers.
- a. 12,300,015
 (twelve million, three hundred thousand, fifteen)
- b. 5,500,000,000
 (five billion, five hundred million)
- c. 75,000,000,000,000
 (seventy-five trillion)
- d. 6,500,000,000,000,000
 (six quadrillion, five hundred trillion)

Answers for CLASS PRACTICE a–f

- **a.** Two hundred seventy-five thousand, four hundred sixteen
- **b.** Seventy-five million, six hundred thousand, twenty-eight
- **c.** Six billion, four hundred twelve million
- **d.** Two hundred fifty billion, five hundred thousand
- **e.** Fifteen trillion, one hundred twenty-three billion
- **f.** Three quadrillion, six hundred trillion

Answers for Part B

9. Four hundred seventy-five thousand, eight hundred eight
10. Six hundred nine thousand, eight hundred ninety-seven
11. Fourteen million, five hundred fifty thousand
12. Nine hundred eighty billion, five hundred million
13. Fourteen trillion, six hundred billion
14. Seven trillion, five hundred billion
15. Forty-five quadrillion, six hundred twenty trillion
16. Eight hundred seventy-nine quadrillion, eight hundred trillion

CLASS PRACTICE

Read these numbers, using words. (Answers on facing page.)

a. 275,416

b. 75,600,028

c. 6,412,000,000

d. 250,000,500,000

e. 15,123,000,000,000

f. 3,600,000,000,000,000

Write these numbers, using digits.

g. Six hundred million, five hundred fifteen thousand, sixty-five 600,515,065

h. Seventy-seven billion, sixteen million 77,016,000,000

i. Seven trillion, five million 7,000,005,000,000

j. Fourteen quadrillion, five hundred trillion 14,500,000,000,000,000

WRITTEN EXERCISES

A. *Name the period that is underlined.*

1. 5,<u>478</u>,311   thousands
2. <u>25</u>,781,128,111   billions
3. 425,<u>171</u>,992,123   millions
4. 711,000,023,<u>111</u>   units (ones)
5. 16,<u>000</u>,000,000,000   billions
6. <u>271</u>,911,000,000,000   trillions
7. 26,<u>181</u>,882,717,578,000   trillions
8. <u>314</u>,192,000,000,000,000   quadrillions

B. *Write these numbers, using words.* (Answers on facing page.)

9. 475,808

10. 609,897

11. 14,550,000

12. 980,500,000,000

13. 14,600,000,000,000

14. 7,500,000,000,000

15. 45,620,000,000,000,000

16. 879,800,000,000,000,000

C. *Write these numbers, using digits.*

17. Forty-eight million, six hundred fifty-five thousand   48,655,000

18. Six hundred twenty-one million, two hundred two thousand   621,202,000

19. Seven hundred twenty-five billion, two hundred thousand   725,000,200,000

20. Six trillion, fifty-five million   6,000,055,000,000

21. Four quadrillion, seventy-seven trillion   4,077,000,000,000,000

22. Twenty-two quadrillion, ninety billion   22,000,090,000,000,000

14   Chapter 1   Basic Mathematical Operations

D. Solve these reading problems.

23. King Asa of Judah had an army of 580,000 men (2 Chronicles 14:8). Write this number, using words.   Five hundred eighty thousand

24. Matthew read in his science book that the average human body contains about fifty trillion cells. Write this number, using digits.   50,000,000,000,000

25. Scientists estimate that there are about 100,000,000,000 stars in the Milky Way galaxy. Write this number, using words.   One hundred billion

26. The star Arcturus, mentioned in Job 9:9 and 38:32, is about one hundred eighty-eight trillion, one hundred sixty billion miles from the earth. Write this number, using digits.   188,160,000,000,000

27. The star Rigel, the brightest star in the constellation Orion, is the seventh brightest star seen from earth. It is about 3,204,600,000,000,000 miles from the earth. Write this number, using words.   Three quadrillion, two hundred four trillion, six hundred billion

28. Cyrus, king of Persia, sent back to Israel the utensils that King Nebuchadnezzar had taken out of the temple. Among these vessels were 30 gold chargers, 1,000 silver chargers, 29 knives, 30 gold basins, and 410 silver basins. What was the number of all the chargers, knives, and basins combined?   1,499

REVIEW EXERCISES

E. Solve these problems. Write any remainder as a whole number.

29. 4,543   30. 75,200   31.   275   32. 226 R 23
    + 6,978   − 48,356   × 68   25)5,673
     11,521   26,844   18,700

O LORD our Lord,

how excellent is thy name in all the earth!

who hast set thy glory above the heavens.

Psalm 8:1

Lesson 1 T–14

Further Study

Would the students like the challenge of reading this number?

123,456,789,012,345,678,901,234,567,890,123,456,789,012,345,678,901,234,567,890,123,456

The answer is as follows (with digits to simplify reading): 123 vigintillion, 456 novemdecillion, 789 octodecillion, 12 septendecillion, 345 sexdecillion, 678 quindecillion, 901 quattuordecillion, 234 tredecillion, 567 duodecillion, 890 undecillion, 123 decillion, 456 nonillion, 789 octillion, 12 septillion, 345 sextillion, 678 quintillion, 901 quadrillion, 234 trillion, 567 billion, 890 million, 123 thousand, 456.

Although numbers are endless, there is no number large enough to describe God. "I am Alpha and Omega, the beginning and the ending, saith the Lord, which is, and which was, and which is to come, the Almighty" (Revelation 1:8).

Mathematicians have assigned names to periods far beyond the quadrillions. In the partial listing at the right, note the progression of the prefixes from *bi-* (two) to *deci-* (ten), and then continuing with *decillion* preceded by another set of prefixes representing 1 to 10.

Number of periods	Name
1	units (ones)
2	thousands
3	millions
4	billions
5	trillions
6	quadrillions
7	quintillions
8	sextillions
9	septillions
10	octillions
11	nonillions
12	decillions
13	undecillions
14	duodecillions
15	tredecillions
16	quattuordecillions
17	quindecillions
18	sexdecillions
19	septendecillions
20	octodecillions
21	novemdecillions
22	vigintillions
....	....
100	centillions

T–15   Chapter 1   Basic Mathematical Operations

LESSON 2

Objectives

- To review the concept of place value in the Arabic number system, and to identify the place value of each digit in a number.
- To review the rounding of numbers up to the nearest billion.

Review

1. *Write these numbers, using words.* (Lesson 1)

    a. 6,897,313
        (Six million, eight hundred ninety-seven thousand, three hundred thirteen)

    b. 7,800,000,000
        (Seven billion, eight hundred million)

    c. 22,000,200,000,000,000
        (Twenty-two quadrillion, two hundred billion)

    d. 6,000,500,400,000
        (Six trillion, five hundred million, four hundred thousand)

2. *Write these numbers, using digits.* (Lesson 1)

    a. Five million, six hundred thirty-six thousand, forty-four   (5,636,044)

    b. Seven billion, five hundred thousand   (7,000,500,000)

    c. Twenty trillion, two hundred billion   (20,200,000,000,000)

    d. Forty-four quadrillion   (44,000,000,000,000,000)

2. Understanding Place Value and Rounding

Place Value

The Arabic system of numerals is used throughout the world. Its place value system is both logical and convenient. The position or place of a digit determines its value. The place value principle allows any number to be written by using only ten digits.

The Arabic system is also called a **decimal system**. The value of a digit is multiplied by 10 for each place it is moved to the left, and its value is divided by 10 for each place it is moved to the right. Thus in the number 444, the middle 4 has a value of forty and is worth 10 times as much as the 4 at the right, which has a value of four. And the middle 4 has a value only one-tenth as great as the 4 at the left, which has a value of four hundred.

Below is a large number with each place value identified. The value of each digit is shown by multiplying the digit by its place value. Notice how the values of the digits decrease from left to right and increase from right to left. For example, the 4 in the hundred trillions' place is the same symbol as the 4 in the ten thousands' place, but its value is ten billion times as great.

Digit	Place value		Value
5	quadrillions	=	5,000,000,000,000,000
4	hundred trillions	=	400,000,000,000,000
3	ten trillions	=	30,000,000,000,000
2	trillions	=	2,000,000,000,000
1	hundred billions	=	100,000,000,000
0	ten billions	=	00,000,000,000
9	billions	=	9,000,000,000
8	hundred millions	=	800,000,000
7	ten millions	=	70,000,000
6	millions	=	6,000,000
5	hundred thousands	=	500,000
4	ten thousands	=	40,000
3	thousands	=	3,000
2	hundreds	=	200
1	tens	=	10
0	units (ones)	=	0

Total **5,432,109,876,543,210**

16   Chapter 1   Basic Mathematical Operations

Rounding Numbers

Large numbers are often rounded so that it is simpler to work with them. To round a number, follow these two steps.

1. Choose the place to which you will round the number.

2. Look at the digit to the right of the place being rounded. If the digit to the right has a value of 4 or less, the digit being rounded will remain unchanged. Replace the digits to its right with zeroes.

    If the digit to the right has a value of 5 or more, add 1 to the digit being rounded. Replace the digits to its right with zeroes. Notice in Example C that a number will sometimes round to zero.

Round each number to the place indicated.

Example A: 1<u>5</u>,501 to the nearest thousand   = 16,000

Example B: <u>2</u>,350,615,900 to the nearest billion = 2,000,000,000

Example C: 15 (<u>0</u>15) to the nearest hundred   = 0

CLASS PRACTICE

Give the place value of each underlined digit.

a. 3<u>2</u>,548,000   b. 4<u>8</u>7,300,000,000   c. <u>2</u>02,121,918,212
   millions   ten billions   hundred billions

Give the value of each underlined digit. The value of a zero is 0 regardless of its place.

d. <u>2</u>,717,471,000   2,000,000,000 (2 billion)   e. 7<u>0</u>,678,912,000,000   0

f. 4<u>3</u>,000,000,000,000,000
   3,000,000,000,000,000 (3 quadrillion)

Round each number to the place indicated.

g. 56,456 (ten thousands)   60,000   h. 41,111,577 (millions)   41,000,000

i. 17,890,000,000 (billions)   18,000,000,000   j. 45,679,000 (ten millions)   50,000,000

WRITTEN EXERCISES

A. Give the place value of each underlined digit.

1. 29,9<u>1</u>9   tens   2. <u>4</u>,515,000,000   billions

3. 2<u>8</u>8,600   ten thousands   4. 4,<u>2</u>19,668   hundred thousands

5. 17,<u>8</u>76,100,000   hundred millions   6. 17,318,<u>0</u>00,000,000   hundred millions

7. 2<u>7</u>,100,000,000,000   trillions   8. 4<u>2</u>2,500,000,000,000   ten trillions

9. <u>7</u>5,600,000,000,000,000   ten quadrillions   10. <u>2</u>92,000,000,000,000,000   hundred quadrillions

Lesson 2 T–16

Introduction

How it is possible to write any number by using only ten different digits? It is possible because the digits represent different values according to their position in a number.

Write the number 222,222,222,222 on the board. Select some of the 2s and have the students identify their values.

Compare any 2 with the 2 on its right. How many times larger is the 2 on the left compared to the 2 on its right? (ten times)

Compare any 2 with the 2 on its left. Write a fraction comparing the value of the 2 on the right compared to the 2 on its left. ($\frac{1}{10}$)

Teaching Guide

1. **The value of a digit is determined by its place in a number.** Discuss the value of the digits in the large number shown in the lesson example. Write the following numbers on the board, and have students identify the place and the value of each underlined digit. Note that the value of a zero is 0 regardless of its place.

   a. 66,<u>7</u>87   (hundreds; 700)
   b. 8<u>9</u>,900   (thousands; 9,000)
   c. 2<u>0</u>6,784   (ten thousands; 0)
   d. 1<u>7</u>,600,000   (millions; 7,000,000)
   e. <u>4</u>5,000,000,000,000
         (ten trillions; 40,000,000,000,000)

2. **To round a number, follow these two steps.**

   (1) Choose the place to which you will round the number.

   (2) Look at the digit to the right of the place being rounded. If the digit to the right has a value of 4 or less, the digit being rounded will remain unchanged. Replace the digits to its right with zeroes.

   If the digit to the right has a value of 5 or more, then increase the digit being rounded by one. Replace the digits to its right with zeroes. Notice in Example C that a number will sometimes round to zero.

Round these numbers as indicated.

   a. 7,789,600 to the nearest ten thousand   (7,790,000)
   b. 47 to the nearest hundred   (0)
   c. 445,350,000 to the nearest million
                              (445,000,000)
   d. 723,000,000 to the nearest ten million   (720,000,000)
   e. 478,000,000 to the nearest hundred million   (500,000,000)
   f. 7,688,900,000 to the nearest billion   (8,000,000,000)

T–17   Chapter 1   Basic Mathematical Operations

Further Study

The number system generally used throughout the world is the base ten system, meaning that all place values are based on the number ten. A digit has ten times the value of the same digit one place to its right. A digit has $\frac{1}{10}$ the value of the same digit one place to its left.

Not all numbers are base ten. Computers, for example, use both the base two (binary) and the base sixteen (hexadecimal) systems for their internal computing.

The binary system uses two in the way we use ten. Moving a digit to the left one place in a base two number increases its value two times. Below is an example of how to count from one to sixteen in binary (from 1_{two} to $10,000_{two}$).

The hexadecimal system uses sixteen in the way we use ten. Moving a digit to the left one place in a base sixteen number increases its value sixteen times. Moving a digit to the left two places increases its value 16 × 16 or 256 times. Because the hexadecimal system requires sixteen different digits, the letters *a–f* are often used in addition to the ten standard Arabic digits. The number $20_{sixteen}$ equals two sixteens, or 32_{ten}.

Note that it is not the numerals used that determine the base of the number. It is rather the number that is used to regroup for the next higher place value. The concept of base is taught in Chapter 12.

Decimal	Binary	(How to say the binary number)
1	1	
2	10	(one, zero, base two)
3	11	(one, one, base two)
4	100	(one, zero, zero, base two)
5	101	(one, zero, one, base two)
6	110	(one, one, zero, base two)
7	111	(one, one, one, base two)
8	1,000	(one, zero, zero, zero, base two)
9	1,001	(one, zero, zero, one, base two)
10	1,010	(one, zero, one, zero, base two)
11	1,011	(one, zero, one, one, base two)
12	1,100	(one, one, zero, zero, base two)
13	1,101	(one, one, zero, one, base two)
14	1,110	(one, one, one, zero, base two)
15	1,111	(one, one, one, one, base two)
16	10,000	(one, zero, zero, zero, zero, base two)

Answers for Part E

33. Two hundred seventy-five thousand, three hundred fifty-six

34. One million, two hundred ninety-one thousand, seven hundred seventy-seven

35. Nine trillion, nine hundred sixty billion

36. Twenty-eight quadrillion

Lesson 2

B. Give the value of each underlined digit. The value of a zero is 0 regardless of its place.

11. 2<u>3</u>,217,811 — 3 million
12. <u>1</u>67,400,000 — 100 million
13. <u>7</u>,431,133,000 — 7 billion
14. 9<u>8</u>,616,300,000 — 8 billion
15. 8,6<u>0</u>7,200,000,000 — zero
16. <u>2</u>3,100,000,000,000 — 20 trillion
17. 9<u>9</u>,900,000,000,000,000 — 9 quadrillion
18. <u>2</u>11,400,000,000,000,000 — 200 quadrillion

C. Round each number to the place indicated.

19. 15,410 (thousands) — 15,000
20. 74,593 (ten thousands) — 70,000
21. 13,654,899 (millions) — 14,000,000
22. 6,280,000,000 (billions) — 6,000,000,000
23. 277,879,102 (ten millions) — 280,000,000
24. 314,499,121 (millions) — 314,000,000
25. 4,510,000,000 (billions) — 5,000,000,000
26. 1,457,100,000 (hundred millions) — 1,500,000,000

D. Solve these reading problems.

27. The Oak Dale church cost $145,320 to build. Round the cost to the nearest thousand dollars. — $145,000

28. The Washington Monument stands in the center of the Mall in Washington, D.C. It was completed in 1884 at a cost of $1,187,710. Round the cost to the nearest hundred thousand dollars. — $1,200,000

29. One year in the 1990s, the government budget for the United States was $1,460,900,000,000. Write the digit in the hundred millions' place. — 9

30. Recently, the population of China was estimated to be 1,238,319,000. Write the place value of the digit 8. — millions

31. In the same year, the population of India was estimated to be 931,044,000. Write this number, using words. — Nine hundred thirty-one million, forty-four thousand

32. Light can travel one million miles in less than 5.4 seconds, one billion miles in about 1.5 hours, one trillion miles in a bit more than 62 days, and one quadrillion miles in about 170 years. Write one quadrillion, using digits. — 1,000,000,000,000,000

REVIEW EXERCISES

E. Write these numbers, using words. *(Lesson 1)* (See facing page.)

33. 275,356
34. 1,291,777
35. 9,960,000,000,000
36. 28,000,000,000,000,000

F. Write these numbers, using digits. *(Lesson 1)*

37. One hundred twenty-five thousand, sixty-seven — 125,067
38. Two million, seventeen thousand, one hundred twenty-one — 2,017,121
39. Twenty-five trillion, twenty-five — 25,000,000,000,025
40. Six quadrillion, two hundred trillion — 6,200,000,000,000,000

18   Chapter 1   Basic Mathematical Operations

3. Understanding Roman Numerals

Arabic numerals are the most common symbols used to represent numbers, but they were by no means the first numerals invented. The Egyptians and other people also devised numbering systems.

One early system was Roman numerals, which the Romans used when Christ lived on earth. This was the main system used by Europeans until the 1500s. Today we use Roman numerals on outlines and sometimes for dates.

In the Roman numeral system, seven letters are symbols that represent numbers. Roman numerals do not depend on place value. The value of a symbol is the same regardless of where it is in a numeral. These are the seven letters and their values.

$$I = 1 \quad V = 5 \quad X = 10 \quad L = 50$$
$$C = 100 \quad D = 500 \quad M = 1{,}000$$

Reading Roman Numerals

Observe the following rules when reading Roman numerals.

1. **When numerals are repeated, add the values. Never repeat V, L, or D.**
   III = 1 + 1 + 1 = 3   MM = M + M = 2,000

2. **When a smaller numeral follows a larger numeral, add the values.**
   VIII = 5 + 1 + 1 + 1 = 8
   MDCCC = 1,000 + 500 + 100 + 100 + 100 = 1,800

3. **When a smaller numeral precedes a larger numeral, subtract the smaller from the larger.** Never subtract V, L, D, or M, and never use more than one smaller numeral before a larger one.
   XC = 100 − 10 = 90   MCM = 1,000 + (1,000 − 100) = 1,900

4. **It is neither proper nor necessary to use the same numeral more than three times consecutively.** Instead, use subtraction as shown in rule 3 to write the next number.

5. **To show a subtracted value, use only the next smaller or second smaller numeral before a larger numeral.** Write 999 as CM + XC + IX (CMXCIX), not as IM. The following combinations are correct.
   I in IV and IX   X in XL and XC   C in CD and CM

6. **Placing a bar over a numeral or set of numerals multiplies its value by 1,000.** This bar is called a **vinculum** (vĭng´ kyə·ləm). Do not place the numeral I under the vinculum unless that I is needed as part of a larger number of thousands. Write 3,000 as MMM, not $\overline{\text{III}}$. Write 4,000 as $\overline{\text{IV}}$. Do not use the numeral M after a vinculum, except in the combination CM. Write 7,000 as $\overline{\text{VII}}$, not $\overline{\text{V}}$MM.

LESSON 3

Objectives
(New items are marked with a star.)

- To review writing Roman numerals up to millions.
- To teach that a *vinculum drawn over a Roman numeral increases its value 1,000 times. (Only the term is new.)

Review

1. *Give the place value and the value of each underlined digit.* (Lesson 2)

    a. 4<u>4</u>,799   (thousands; 4,000)

    b. 5<u>1</u>1,116,713
                 (ten millions; 10,000,000)

    c. 1<u>0</u>1,900,000,000   (ten billions; 0)

    d. <u>3</u>4,000,000,000,000,000 (ten quadrillions; 30,000,000,000,000,000)

2. *Round each number as indicated.* (Lesson 2)

    a. 41,117 to the nearest thousand
                                    (41,000)

    b. 34,520,400 to the nearest ten million   (30,000,000)

    c. 36,717,800,000 to the nearest billion   (37,000,000,000)

    d. 414,444,994 to the nearest hundred million   (400,000,000)

3. *Write these numbers, using words.* (Lesson 1)

    a. 40,090,010
    (forty million, ninety thousand, ten)

    b. 301,000,200,000   (three hundred one billion, two hundred thousand)

    c. 2,090,000,000,000,000
    (two quadrillion, ninety trillion)

4. *Write these numbers, using digits.* (Lesson 1)

    a. thirty-nine million, five hundred
                                    (39,000,500)

    b. six quadrillion, seven trillion, twenty-five million
                    (6,007,000,025,000,000)

Introduction

Ask the students to write the year Columbus discovered America in the way that the people in Columbus's day most likely did. Remind them that although Columbus was Italian and sailed for the Spanish, mathematical notation does not vary much by language.

When Columbus sailed to America, Europe was still largely using Roman numerals, the same as people of the Roman Empire did in Jesus' time. Roman numerals were in widespread use in Europe until the late 1500s, about one hundred years after Columbus sailed to America. In 1492, most people would have written the year as MCDXCII.

Roman numerals are still used occasionally, especially when writing outlines, chapter numbers, and dates.

Note: At the end of the class period, hand out the sheets *Lesson 3 Speed Test Preparation* (Addition). The purpose of these sheets is to help the students prepare for the 1-minute addition speed test in the next lesson.

Teaching Guide

1. **Observe the following rules when reading Roman numerals.** Apply the rules to various numbers.

    (1) When numerals are repeated, add the values. Never repeat V, L, or D.
       a. CC = 100 + 100 = 200
       b. XXX = 10 + 10 + 10 = 30

    (2) When a smaller numeral follows a larger numeral, add the values.
       a. XV = 10 + 5 = 15

T–19   Chapter 1   Basic Mathematical Operations

b. DCX = 500 + 100 + 10 = 610
c. MMMDCXXXIII = 3,000 + 500 + 100 30 + 3 = 3,633

(3) When a smaller numeral precedes a larger numeral, subtract the smaller from the larger. Never subtract V, L, D, or M, and never use more than one smaller numeral before a larger one.
  a. IX = 10 − 1 = 9
  b. XL = 50 − 10 = 40
  c. CD = 500 − 100 = 400
  d. CM = 1,000 − 100 = 900
  e. CMXLIX = (1,000 − 100) + (50 − 10) + (10 − 1) = 949
  f. MMMCMXLIV = 3,000 + (1,000 − 100) + (50 − 10) + (5 − 1) = 3,944

(4) It is neither proper nor necessary to use the same numeral more than three times consecutively. Instead, use subtraction as shown in rule 3 to write the next number. *Exception:* On clock faces with Roman numerals, 4 is commonly shown as IIII.

(5) To show subtracted values, use only the next smaller or second smaller numeral before a larger numeral. Write 999 as CM + XC + IX, not as IM. The following combinations are correct.
  I in IV and IX
  X in XL and XC
  C in CD and CM

(6) Placing a vinculum (bar) over a numeral or set of numerals multiplies its value by 1,000. Do not place the numeral I under the vinculum unless that I is needed as part of a larger number of thousands. 3,000 = MMM, not $\overline{\text{III}}$. 4,000 = $\overline{\text{IV}}$. Do not use the numeral M after a vinculum, except in the combination CM. 6,000 = $\overline{\text{VI}}$, not $\overline{\text{V}}$M.

  a. $\overline{\text{VIII}}$ = 8,000
  b. $\overline{\text{XXV}}$CDL = 25,450
  c. $\overline{\text{DCXI}}$CMIX = 611,909

2. **To write an Arabic numeral as a Roman numeral, start at the left; and one digit at a time, write the Roman equivalent for each digit with respect to its place value.** Be certain the students understand this concept. Students sometimes have difficulty trying to use the Roman numerals I and V to write place-value numbers.

  a. 245 = 200 + 40 + 5
     = CC + XL + V
     = CCXLV
  b. 777 = 700 + 70 + 7
     = DCC + LXX + VII
     = DCCLXXVII
  c. 949 = 900 + 40 + 9
     = CM + XL + IX
     = CMXLIX
  d. 3,289 = 3,000 + 200 + 80 + 9
     = MMM + CC + LXXX + IX
     = MMMCCLXXXIX
  e. 25,259
     = 20,000 + 5,000 + 200 + 50 + 9
     = $\overline{\text{XX}}$ + $\overline{\text{V}}$ + CC + L + IX
     = $\overline{\text{XXV}}$CCLIX

3. **To write a Roman numeral as an Arabic numeral, identify the value of each letter in the Roman numeral.** Be especially alert for any subtraction. Add all the values to write the Arabic numeral.

  a. CCCXLIV = CCC + XL + IV
     = 300 + 40 + 4 = 344
  b. CDLXXXIX = CD + LXXX + IX
     = 400 + 80 + 9 = 489
  c. MMCMIII = MM + CM + III
     = 2,000 + 900 + 3 = 2,903
  d. $\overline{\text{DX}}$CMIX = $\overline{\text{DX}}$ + CM + IX
     = 510,000 + 900 + 9 = 510,909

Changing Arabic Numerals to Roman Numerals

To write an Arabic numeral as a Roman numeral, start at the left; and one digit at a time, write the Roman equivalent for each digit with respect to its place value.

> **Example A:** Write 1,995 as a Roman numeral.
> 1,995 = 1,000 + 900 + 90 + 5
> = M + CM + XC + V
> = MCMXCV

> **Example B:** Write 120,400 as a Roman numeral.
> 120,400 = 100,000 + 20,000 + 400
> = $\overline{C}$ + $\overline{XX}$ + CD
> = $\overline{CXX}$CD

Changing Roman Numerals to Arabic Numerals

To write a Roman numeral as an Arabic numeral, identify the value represented by each letter in the Roman numeral. Be especially alert for any subtraction. Add all the values to write the Arabic numeral.

> **Example C:** Write MCDXCII as an Arabic numeral.
> MCDXCII = M + CD + XC + I + I
> = 1,000 + 400 + 90 + 1 + 1
> = 1,492

> **Example D:** Write $\overline{CX}$D as an Arabic numeral.
> $\overline{CX}$D = $\overline{C}$ + $\overline{X}$ + D
> = 100,000 + 10,000 + 500
> = 110,500

CLASS PRACTICE

Write these Roman numerals as Arabic numerals.

a. XVIII   18   b. CXCIX   199   c. CMIV   904   d. DXLVIII   548

e. MCMXCIX   1,999   f. MMMIV   3,004   g. $\overline{VIII}$X   8,010   h. $\overline{MMMCCC}$DCCV   3,300,705

Write these Arabic numerals as Roman numerals.

i. 48   XLVIII   j. 97   XCVII   k. 166   CLXVI   l. 499   CDXCIX

m. 777   DCCLXXVII   n. 2,222   MMCCXXII   o. 55,555 $\overline{LVDLV}$ p. 3,444,444 $\overline{MMMCDXLIV}$CDXLIV

20 Chapter 1 Basic Mathematical Operations

WRITTEN EXERCISES

A. Write these Roman numerals as Arabic numerals.

1. VIII   8   2. XXVII   27   3. MI   1,001   4. MMMV   3,005
5. $\overline{V}$ 5,000   6. $\overline{C}$ 100,000   7. $\overline{XIX}$CCC   19,300   8. $\overline{MMMCCCXXX}$ 3,300,030

B. Write these Arabic numerals as Roman numerals.

9. 22   XXII   10. 38   XXXVIII   11. 1,354 MCCCLIV   12. 1,997 MCMXCVII
13. 6,500 $\overline{VI}$D   14. 49,700 $\overline{XLIX}$DCC   15. 1,200,000 $\overline{MCC}$ 16. 2,150,000 $\overline{MMCL}$

C. Write the number of the rule that is broken by each of these Roman numerals.

17. DD   rule 1   18. IC   rule 5   19. $\overline{I}$ rule 6   20. CCCVC   rule 3 (or 5)

D. Solve these reading problems.

21. The prophetess Anna, who thanked the Lord when she saw the baby Jesus, was a widow 84 years old (Luke 2:37). Write her age as a Roman numeral.   LXXXIV

22. It took XLVI years to build the temple that stood in Jesus' day (John 2:20). Use an Arabic numeral to express the time spent in building the temple.   46 (years)

23. Plymouth, Massachusetts, was settled in 1620. Write this date as a Roman numeral.   MDCXX

24. The first Mennonite settlers in Lancaster County arrived in the year 1710. Write this date as a Roman numeral.   MDCCX

25. In 1997, the population of the United States was estimated to be two hundred sixty-seven million. Write this number as an Arabic numeral.   267,000,000

26. In 1998, the population of Pakistan was estimated to be 152,766,000. Write this number, using words.   One hundred fifty-two million, seven hundred sixty-six thousand

REVIEW EXERCISES (See facing page.)

E. Write the place value and the value of each underlined digit. (Lesson 2)

27. <u>2</u>88,117   28. 1<u>2</u>,000,500,000,000

F. Round each number to the place indicated. (Lesson 2)

29. 3,476 (thousands)   30. 214,512,999 (millions)
31. 485,192,144 (ten millions)   32. 17,311,161,000 (billions)

G. Write these numbers, using words. (Lesson 1)

33. 235,800   34. 16,050,000,000

H. Write these numbers, using digits. (Lesson 1)

35. Thirteen million, fifty thousand, sixty-five
36. Twenty-nine billion, six hundred thousand

An Ounce of Prevention

Be certain the students understand the concept that Roman numerals are not a place-value system. Students sometimes have difficulty trying to use the Roman numerals I and V to write place-value numbers.

Further Study

Although Roman numerals do not have ten digits, they are used for a ten-base numbering system. Computing with Roman numerals illustrates the clumsiness of writing numbers without the ten-digit and the place-value concepts. These concepts have greatly simplified both notation and calculation.

Computation shows the weakness of Roman numerals and the advantages of Arabic numerals. Try a few simple calculations with Roman numerals. Notice how the place-value arrangement greatly simplifies procedures such as carrying and borrowing.

$$\begin{array}{r} XXIX \\ + XIV \\ \hline XXX\ XIII = XLIII \end{array}$$

$$\begin{array}{r} XXXIII = XX\ \ XIII \\ - XIX = -X\ \ \ \ IX \\ \hline X\ \ \ \ IV = XIV \end{array}$$

$$\begin{array}{r} XXV \\ \times VII \\ \hline XXV \\ XXV \\ CXXV \\ \hline CLXXV \end{array}$$

Medieval Europeans used Roman numerals mainly to record numbers. For anything beyond simple calculations, they used the abacus.

Answers for REVIEW EXERCISES

27. hundred thousands; 200,000
28. trillions; 2,000,000,000,000
29. 3,000
30. 215,000,000
31. 490,000,000
32. 17,000,000,000
33. Two hundred thirty-five thousand, eight hundred
34. Sixteen billion, fifty million
35. 13,050,065
36. 29,000,600,000

T–21   Chapter 1 Basic Mathematical Operations

LESSON 4

Objectives

(New items are marked with a star.)

- To review addition terms.
- To teach *the commutative and associative laws of addition and multiplication, and to note that they do not apply to subtraction and division.
- To review situations where the students have been applying the commutative and the associative laws in arithmetic.

Review

1. Give *Lesson 4 Speed Test* (Addition).

    Speed tests in Lessons 4, 7, 10, and 13 test how well the students know the basic arithmetic facts. There are 100 problems on each of these tests. Give students exactly 1 minute to do each one. Do not expect them to complete the tests, for each one has enough problems so that very few students are able to do them all. The reason for including so many is to see exactly how many problems the students can do in 1 minute—something that cannot be deter-mined if a student finishes the test in less than 1 minute. If feasible, give the students the test several times over a number of days.

2. *Write these Roman numerals as Arabic numerals.* (Lesson 3)

    a. DCCVII   (707)

    b. MMCCCLXXII   (2,372)

    c. $\overline{\text{CXXIX}}$CCCII   (129,302)

    d. $\overline{\text{MCM}}$DCCL   (1,900,750)

3. *Round each number as indicated.* (Lesson 2)

    a. 587 to the nearest hundred   (600)

    b. 422,312 to the nearest ten thousand   (420,000)

    c. 17,449,300 to the nearest million   (17,000,000)

    d. 136,919,413,211 to the nearest ten billion   (140,000,000,000)

4. Using the Commutative and Associative Laws

God created a very orderly world. The sun and the moon rise and set at predictable times. Seasons will come and go in their regular cycle as long as the earth endures (Genesis 8:22). Numbers also follow orderly, logical patterns called laws of mathematics. These laws apply to both simple and complex calculations.

This lesson deals with two closely related mathematical laws: the commutative law and the associative law. These laws relate only to addition and multiplication. You have depended on these laws for many years without realizing it. Studying them now will prepare you to use them in more advanced calculations.

Commutative Law

The **commutative law** applies to both addition and multiplication. *Commutative* has the idea of "moving." Numbers to be added or multiplied can be moved to different positions—their order can be changed—and the result is the same. This is why $3 + 2$ gives the same sum as $2 + 3$, and 3×2 yields the same product as 2×3.

Commutative Laws

Addition: The order in which addends are added does not affect the sum.
$7 + 8 = 15$ and $8 + 7 = 15$ $3 + 5 + 2 + 1 = 11$ and $2 + 3 + 5 + 1 = 11$

Multiplication: The order in which factors are multiplied does not affect the product.
$7 \times 8 = 56$ and $8 \times 7 = 56$ $3 \times 5 \times 2 \times 1 = 30$ and $2 \times 5 \times 3 \times 1 = 30$

The commutative law provides a method for checking addition and multiplication problems. To check a sum, add the numbers in reverse order. To check a product, exchange the multiplicand and the multiplier. This check is quite reliable because you are not likely to get the same wrong answer both times.

Associative Law

The **associative law** applies to both addition and multiplication. *Associative* has the idea of "grouping." Numbers to be added or multiplied can be grouped in different ways, and the result is the same. This is why $(2 + 3) + 7$ gives the same sum as $2 + (3 + 7)$, and $3(4 \times 5)$ yields the same product as $(3 \times 4)5$.

Associative Laws

Addition: The way in which addends are grouped does not affect the sum.
$(5 + 6) + 7 + 8 = 26$ $5 + (6 + 7) + 8 = 26$

Multiplication: The way in which factors are grouped does not affect the product.
$5(6 \times 7) = 210$ $(5 \times 6)7 = 210$

Note: $5(6 \times 7)$ means 5 times the product of 6 and 7, or 5×42.

22 Chapter 1 Basic Mathematical Operations

The associative law can be used to simplify mental arithmetic. This is done by regrouping numbers to equal 10 or multiples of 10, as shown below.

$$7 + 8 + 3 = (7 + 3) + 8 = 10 + 8 = 18$$
$$8 \times 9 \times 5 = (8 \times 5)9 = 40 \times 9 = 360$$

The commutative and associative laws do not apply to subtraction or division. The answers to 9 − 3 and 3 − 9 are not the same. Neither are the answers to 9 ÷ 3 and 3 ÷ 9.

CLASS PRACTICE

Solve these addition problems. Check your work by using the commutative law. Write the correct term beside each part of problems a and b.

a. 256 addend	b. 45,513 addend	c. $1,231.67	d. 156,793
+ 212 addend	+ 18,908 addend	+ 2,674.23	+ 245,876
468 sum	64,421 sum	$3,905.90	402,669

Solve these addition problems. Look for pairs of addends that equal 10.

e. 6 + (2 + 8) + (4 + 6)   26   f. 7 + 9 + (6 + 4) + 2 + (7 + 3)   38
g. 8 + (5 + 5) + 9 + (4 + 6) + 9   46   h. 3 + (9 + 1) + 6 + (7 + 3) + (6 + 4)   39

Multiply. Look for pairs of factors that equal 10 or a multiple of 10.

i. 4 × 7 × 5   140   j. 8 × 5 × 6   240   k. 5 × 9 × 2   90

Tell whether each problem illustrates the commutative or the associative law.

l. a + (b + c) + d = a + b + (c + d)   m. r + s = s + r
      associative law   commutative law

WRITTEN EXERCISES

A. *Copy and solve these problems. Check by using the commutative law. Write the correct term beside each part of problems 1 and 2.*

1. 478 addend	2. 78,828 addend	3. 65,584	4. $2,173.45
+ 894 addend	+ 49,979 addend	+ 77,912	+ 6,721.98
1,372 sum	128,807 sum	143,496	$8,895.43

5. 8,763,332	6. 7,832,857	7. 9	8. 7
+ 7,712,999	+ 4,631,758	× 8	× 6
16,476,331	12,464,615	72	42

B. *Copy these addition problems. Place parentheses around pairs of addends that equal 10, and then solve.*

9. 7 + (6 + 4) + (2 + 8)   27   10. 9 + 4 + (5 + 5) + 7   30
11. (8 + 2) + (9 + 1) + 5 + 3   28   12. 6 + (6 + 4) + 8 + (3 + 7)   34

Lesson 4 T–22

Introduction

Write the following problems on the board. Challenge the students to mentally solve the second problem in each pair. How did they know the answer without calculating? These problems illustrate the commutative and the associative laws.

$$125 + 246 = 371$$
$$246 + 125 = \_\_\_ \quad (371)$$
$$12 \times 28 = 336$$
$$28 \times 12 = \_\_\_ \quad (336)$$
$$3 + (4 + 6) = 13$$
$$(3 + 4) + 6 = \_\_\_ \quad (13)$$
$$5(5 \times 8) = 200$$
$$(5 \times 5)8 = \_\_\_ \quad (200)$$

Teaching Guide

1. Review the addition terms.
   a. Addend—one of the numbers being added.
   b. Sum—the answer to an addition problem.

2. **The commutative law applies to both addition and multiplication.**

   Addition: The order in which addends are added does not affect the sum.

   Multiplication: The order in which factors are multiplied does not affect the product.

   This law applies regardless of how large or small the addends or factors are, and regardless of whether they are whole numbers, mixed numbers, fractions, or literal numbers. Checking addition problems by adding in reverse order and checking multiplication problems by reversing the factors are applications of this law.

   a. $\quad$ 361 $\qquad$ 186
   $\quad\quad$ + 186 $\quad$ + 361
   $\quad\quad\;\;$ 547 $\qquad\;\;$ 547

   b. $\quad\;$ 12 $\qquad\;\;$ 11
   $\quad\quad \times 11 \qquad \times 12$
   $\quad\quad\;\;$ 132 $\qquad\;\;$ 132

   c. $4 + 7 = 7 + 4$

   d. $x + y = y + x$

   e. $4 \times 9 = 9 \times 4$

   f. $xy = yx$

3. **The associative law applies to both addition and multiplication.**

   Addition: The way in which addends are grouped does not affect the sum.

   Multiplication: The way in which factors are grouped does not affect the product.

T–23   Chapter 1 Basic Mathematical Operations

Regrouping addends or factors to give answers of ten or a multiple of ten is an application of this law.

a. $5 + (5 + 7) + (3 + 8)$
    $= (5 + 5) + (7 + 3) + 8 = 28$

b. $(2 + 6) + (4 + 2) + (8 + 9) + 1$
    $= 2 + (6 + 4) + (2 + 8) + (9 + 1) = 32$

c. $7(2 \times 5) = (7 \times 2)5 = 70$

d. $(9 \times 4)(5 \times 3) = 9(4 \times 5)3 = 540$

e. $(x + y) + z = x + (y + z)$

f. $(xy)z = x(yz)$

Note: An expression such as $7 \times (2 \times 5)$ does not follow standard algebraic form, because the multiplication symbol is not to be used beside a parenthesis. An exception is made in exercises 13–16.

Further Study

Students may think this lesson fits in about second grade. The drilling of math facts in earlier grades does introduce the commutative concept by example. This lesson applies names to simple rules that they have known for five years or more.

But there is a purpose for this study. In high school mathematics such as algebra, students will need to apply these rules in ways which they have not heretofore. For example, could they differentiate between the true and the false statements below?

a. $a - b = b - a$

b. $a + (b + c) = (a + b) + c$

c. $ab = ba$

d. $\dfrac{a}{b} = \dfrac{b}{a}$

e. $a + b = b + a$

f. $a(bc)d = (ab)(cd)$

Statements *a* and *d* are false. The others are always true and are illustrations of the laws discussed in this lesson. A solid understanding of these laws will help students avoid many pitfalls in working with literal numbers. A lack of understanding of these "simple" rules is evident when upper graders violate these and similar rules in solving algebra problems.

Solutions for Exercises 23 and 24

23. $29 + 14 + 11 + 6 + 55$

24. $2 \times 5 \times 3 \times 2$

C. **Copy these multiplication problems. Place parentheses around pairs of factors that equal 10 or a multiple of 10, and then solve.**

13. 7 ×(8 × 5)   280
14. 2 ×(12 × 5)× 3   360
15. 3 ×(5 × 4)× 2   120
16. 9 × 3 ×(5 × 4)   540

D. **Write whether these problems illustrate the commutative law (C) or the associative law (A). The letters in problems 21 and 22 represent unknown numbers.**

17. 15 × 4 = 4 × 15   C
18. 12 + 9 + 6 = 12 + (9 + 6)   A
19. (5 × 6)(5 × 6) = 5 × 6 × 5 × 6   A
20. 8(7 × 5)5 = 8 × 7(5 × 5)   A
21. $a + (b + c) = (a + b) + c$ A
22. $a + b = b + a$ C

E. **Solve these reading problems.**

23. Joshua 15 lists the number of cities in the various geographical areas of Judah's inheritance. Find the sum of the cities in the following areas: southward toward Edom—29, in the valley—14, in the mountains—11, in the wilderness—6, and in other areas—55.   115 cities

24. When the people in the Ulrichs' church helped to put up a pole building, Sister Ulrich served homemade ice cream at the noon and evening meals. She made 5 freezers full of ice cream each time, using 3 quarts of milk per freezer. A quart of milk weighs about 2 pounds. What was the total weight of the milk she used for the two meals together?   60 pounds

25. Does the commutative law apply to subtraction? Write a sentence containing an example to explain your answer.   No. (Sample sentence: The answers to 5 – 2 and 2 – 5 are not the same.)

26. Does the commutative law apply to division? Write a sentence containing an example to explain your answer.   No. (Sample sentence: The answers to 8 ÷ 2 and 2 ÷ 8 are not the same.)

27. Grandfather was born in MCMXXXIII. Find his age in the year MM.   67 years

28. The *Martyrs Mirror* records that the apostle Paul was martyred in the year LXIX. How many years was that before MCMXCV?   1,926 years

REVIEW EXERCISES

F. **Write these Roman numerals as Arabic numerals.** *(Lesson 3)*

29. CCCXLVIII   348
30. MCCLXXXVII   1,287
31. $\overline{\text{MDCID}}$LV   1,601,555
32. $\overline{\text{CDIX}}$LXX   409,070

G. **Round each number to the place indicated.** *(Lesson 2)*

33. 321,400 (thousands)   321,000
34. 7,869,323 (hundred thousands)   7,900,000
35. 5,895,313 (millions)   6,000,000
36. 27,190,000,000 (ten billions)   30,000,000,000

5. Adding Mentally

Being able to add numbers mentally is a valuable benefit throughout life. Sharpening this skill is well worth the effort because you will often find the need to calculate when you do not have pencil and paper on hand.

Horizontal Addition

Horizontal addition is used less frequently than vertical addition because the addends are not aligned according to place value. To add horizontally, first add the ones and then the tens and so on, until you have found the entire sum. **Write the answer one digit at a time as you add.** Practice adding horizontally without copying the problems.

Horizontal addition is used to calculate balances in checkbook ledgers and to prove multicolumn ledgers. It is not strictly mental addition because the answer is written as it is being solved.

Example A

$$248 + 319 = \underline{}$$

Think: $8 + 9 = 17$ Write the 7 and carry the 1. $\underline{7}$ (partial sum)
         $1 + 4 + 1 = 6$ Write 6. $\underline{67}$ (partial sum)
         $2 + 3 = 5$ Write 5. $\underline{567}$ (sum)

Mental Addition

Mental addition is different from horizontal addition in that the answer is written only after the final calculation. The following shortcuts are often helpful in mental addition.

1. Add small numbers as complete units. Look for sets of addends that equal 10 or a multiple of 10. (Example B)

2. Add large numbers from left to right. (Example C)

Example B

$$9 + 14 + 7 + 8 + 12 = \underline{}$$

Think: $(9 + 7) + 14 + (8 + 12)$
        $= 16 + 14 + 20 = 50$

$$6 + 18 + 6 + 9 + 4 = \underline{}$$

Think: $(6 + 18) + (6 + 4) + 9$
        $= 24 + 10 + 9 = 43$

LESSON 5

Objectives

- To review horizontal addition.
- To review mental addition, including 3-digit problems and simple 4-digit problems.

Review

1. Give *Lesson 4 Speed Test* (Addition) a second time, or use *Lesson 3 Preparation for Speed Test* as a speed test.

2. *Solve these problems, and check by using the commutative law. Give the correct term for each part of exercise* a. (Lesson 4)

   a. 32 (addend)   b.   745
      + 79 (addend)   + 481
      (111) (sum)   (1,226)

   c. 21   d.   12
      × 3   × 9
      (63)   (108)

3. *Write these Roman numbers as Arabic numerals.* (Lesson 3)

   a. XXXV (35)
   b. CCLIX (259)
   c. $\overline{\text{IX}}$CDII (9,402)

4. *Write these Arabic numerals as Roman numerals.* (Lesson 3)

   a. 327 (CCCXXVII)
   b. 2,016 (MMXVI)
   c. 11,791 ($\overline{\text{XI}}$DCCXCI)

5. *Name the period that is underlined.* (Lesson 1)

   a. 2,<u>490</u>,214,000,000 (billions)
   b. <u>35</u>,000,182,356,045 (trillions)

6. *Write these numbers, using words.* (Lesson 1)

   a. 1,842,075
      (One million, eight hundred forty-two thousand, seventy-five)

   b. 18,930,051,000,123
      (Eighteen trillion, nine hundred thirty billion, fifty-one million, one hundred twenty-three)

T–25   Chapter 1   Basic Mathematical Operations

Introduction

Write the following addition problem on the board, and challenge the students to find the sum mentally. Ask the first student who has the correct answer to explain how he obtained it.

$$\begin{array}{r} 279 \\ +356 \\ \hline (635) \end{array}$$

Teaching Guide

1. **To add horizontally, first add the ones and then the tens and so on, until you have found the entire sum.** Write the answer one digit at a time as you add.

   a. 229 + 127   (356)
   b. 745 + 673   (1,418)
   c. $141.56 + $89.98   ($231.54)
   d. 352,321 + 438,932   (791,253)

2. **To add mentally, use one of the following shortcuts.**

   (1) Add small numbers as complete units. Look for sets of addends that equal 10 or a multiple of 10.

   a. 8 + 3 + 9 + 3 + 4 + 5
       = 8 + 10 + 9 + 5 = 32
   b. 7 + 9 + 13 + 11 + 12 + 15
       = 20 + 20 + 12 + 15 = 67
   c. 4 + 5 + 5 + 7 + 6 + 4
       = 4 + 10 + 7 + 10 = 31
   d. 7 + 3 + 6 + 8 + 2 + 9
       = 10 + 6 + 10 + 9 = 35

   (2) Add large numbers from left to right.

   a. 46 + 37
       Think: 46 + 30 = 76, + 7 = 83
   b. 38 + 28
       Think: 38 + 20 = 58, + 8 = 66
   c. 86 + 57   (143)
   d. 88 + 73   (161)
   e. 78 + 57   (135)
   f. 123 + 58   (181)
   g. 146 + 68   (214)
   h. 136 + 137   (273)

   Note: Mental addition is different from horizontal addition in that the answer is not written until the calculating is finished. Watch to see that pupils are writing answers only after they have finished calculating.

Lesson 5 25

Example C

$68 + 49 = \_\_\_$
Think: $68 + 40 = 108,$
$+ 9 = 117$

$249 + 456 = \_\_\_$
Think: $249 + 400 = 649,$
$+ 50 = 699,$
$+ 6 = 705$

$1,360 + 2,170 = \_\_\_$
Think: $1,360 + 2,000 = 3,360,$
$+ 100 = 3,460,$
$+ 70 = 3,530$

CLASS PRACTICE

Solve by horizontal addition. Do not copy the problems.

a. 226 + 134 360
b. 782 + 549 1,331
c. 4,581 + 6,382 10,963
d. 3,126 + 8,912 12,038

Solve these problems by mental addition.

e. 45 + 32 77
f. 61 + 38 99
g. 59 + 77 136
h. 86 + 57 143
i. 121 + 56 177
j. 254 + 78 332
k. 326 + 276 602
l. 455 + 375 830
m. 574 + 372 946
n. 761 + 646 1,407
o. 1,350 + 375 1,725
p. 2,450 + 725 3,175

WRITTEN EXERCISES

A. *Solve by horizontal addition. Do not copy the problems.*

1. 364 + 191 555
2. 479 + 176 655
3. 865 + 963 1,828
4. 768 + 695 1,463
5. 1,157 + 893 2,050
6. 2,671 + 812 3,483
7. 4,182 + 9,787 13,969
8. 6,812 + 7,172 13,984

B. *Solve these problems by mental addition.*

9. 46 + 27 73
10. 74 + 18 92
11. 58 + 78 136
12. 85 + 69 154
13. 78 + 43 121
14. 39 + 81 120
15. 67 + 98 165
16. 76 + 85 161
17. 115 + 89 204
18. 178 + 76 254
19. 295 + 189 484
20. 483 + 226 709
21. 437 + 313 750
22. 271 + 654 925
23. 821 + 765 1,586
24. 657 + 573 1,230
25. 1,200 + 540 1,740
26. 2,500 + 650 3,150
27. 1,350 + 1,120 2,470
28. 1,550 + 1,200 2,750

26 Chapter 1 *Basic Mathematical Operations*

C. Solve these reading problems. Solve problems 29–32 mentally.

29. One day Hillside Orchard sold 95 bushels of peaches in the forenoon and 128 bushels in the afternoon. How many bushels of peaches did they sell that day? 223 bushels

30. Hillside Orchard sold 454 bushels of peaches that week. The week before, they had sold 398 bushels. How many bushels altogether did they sell in those two weeks? 852 bushels

31. When David brought the ark of the covenant into Jerusalem, he numbered the children of Levi by their families (1 Chronicles 15:4–10). The numbers were as follows: 120 sons of Kohath, 220 sons of Merari, 130 sons of Gershom, 200 sons of Elizaphan, 80 sons of Hebron, and 112 sons of Uzziel. Mentally calculate the total in the six families of Levites. 862 Levites

32. Psalms 120–134 are called psalms of degrees. The Israelites sang them as they journeyed to Jerusalem to observe the various feasts. From the chart below, mentally find the total number of verses in these fifteen psalms. 101 verses

Psalms	120	121	122	123	124	125	126	127	128	129	130	131	132	133	134
Verses	7	8	9	4	8	5	6	5	6	8	8	3	18	3	3

33. The total number of the men of war in Israel was $\overline{\text{DCIII}}\text{DL}$ (Numbers 1:46). Write this number as an Arabic numeral. 603,550

34. The number of Israelites that Moses led out of Egypt has been estimated at 2,500,000. Write this number as a Roman numeral. $\overline{\text{MM}}\text{D}$

REVIEW EXERCISES

D. Solve these problems, and check by using the commutative law. Write the correct term beside each part of problem 35. *(Lesson 4)*

35. 18 addend
    + 42 addend
     60 sum

36. 13
    × 11
    143

E. Write these Roman numerals as Arabic numerals. *(Lesson 3)*

37. LXII   62

38. $\overline{\text{XX}}$CCCXI   20,311

F. Write these Arabic numerals as Roman numerals. *(Lesson 3)*

39. 296   CCXCVI

40. 19,824 $\overline{\text{XIX}}$DCCCXXIV

G. Name the period that is underlined. *(Lesson 1)*

41. 3,831,<u>011</u>   units (ones)

42. 9,300,<u>524</u>,643,000   millions

H. Write these numbers, using words. *(Lesson 1)* (See facing page.)

43. 15,000,921

44. 20,072,000,322,480

Lesson 5 T–26

Mental math skills are taught best by practice. Spend plenty of time with *Class Practice*. To ensure that students are using the shortcuts, have them say each step as they think.

Answers for Part H

43. Fifteen million, nine hundred twenty-one

44. Twenty trillion, seventy-two billion, three hundred twenty-two thousand, four hundred eighty

T–27 Chapter 1 Basic Mathematical Operations

LESSON 6

Objectives

(New items are marked with a star.)

- To review subtraction terms.
- To review vertical subtraction up to *9-digit minus 9-digit numbers. (Previous experience: subtracting 8-digit minus 8-digit numbers.)
- To review horizontal subtraction up to 5-digit minus 5-digit numbers.

Review

1. Solve by mental addition. (Lesson 5)
    a. 64 + 23   (87)
    b. 95 + 33   (128)
    c. 112 + 77   (189)
    d. 381 + 604   (985)
    e. 821 + 708   (1,529)
    f. 1,730 + 593   (2,323)

2. Tell whether each problem illustrates the commutative or the associative law. (Lesson 4)
    a. 13 + (5 + 6) + 8 = 13 + 5 + (6 + 8)
       (associative law)
    b. 12 × 22 = 22 × 12
       (commutative law)

3. Give the value of each underlined digit. (Lesson 2)
    a. 1,58<u>4</u>,018,421   (4,000,000)
    b. 432,8<u>0</u>1,876   (0)

4. Round each number as indicated. (Lesson 2)
    a. 552,956 to the nearest ten thousand   (550,000)
    b. 9,499,241,500 to the nearest billion   (9,000,000,000)

6. Reviewing Subtraction

Subtraction is taking one number away from another. It is the opposite of addition. Addition shows the result when two or more addends are combined. Subtraction shows the difference between two numbers.

Examples A and B illustrate the need for borrowing. This step is necessary when the digit in the minuend is smaller than the digit at the same place in the subtrahend. Practice borrowing mentally in the exercises in this lesson.

Because addition and subtraction are inverse operations, subtraction problems can be checked by addition. When the difference is added to the subtrahend, the sum should equal the minuend.

Example A

```
         7 13
        7 8̸ 3   minuend
       - 564    subtrahend
         219    difference
```

Check

```
         219    difference
       + 564    subtrahend
         783    minuend
```

Example B

```
       7 16 3 12 16
       $8̸6̸4̸.3̸6̸
      - 281.67
       $582.69
```

Check

```
       $582.69
     + 281.67
       $864.36
```

Horizontal Subtraction

Horizontal subtraction is calculated exactly like vertical subtraction. It is used in many checkbook ledgers. **Write the answer one digit at a time as you calculate.**

Example C

```
              375 - 286 =    ___
     Think:   15 - 6 = 9       9
              16 - 8 = 8      89
              2 - 2 = 0       89
```

Example D

```
              756 - 287 =    ___
     Think:   16 - 7 = 9       9
              14 - 8 = 6      69
              6 - 2 = 4      469
```

28 Chapter 1 Basic Mathematical Operations

CLASS PRACTICE

Solve these subtraction problems. Check your work by adding the subtrahend to the difference. Write the correct term beside each part of problems a and b.

a. 843 minuend
   − 172 subtrahend
    671 difference

b. 5,112 minuend
   − 2,307 subtrahend
    2,805 difference

c. 26,067
   − 11,337
    14,730

d. $78,024.46
   − 34,192.52
    $43,831.94

e. 6,821,278
   − 4,009,876
    2,811,402

f. 720,491,367
   − 342,783,025
    377,708,342

Solve these problems by horizontal subtraction.

g. 635 − 142   493   h. 1,673 − 542   1,131

i. $832.25 − $512.72 $319.53   j. 58,462 − 34,737   23,725

WRITTEN EXERCISES

A. *Solve these subtraction problems. Check your work by adding the subtrahend to the difference. Write the correct term beside each part of problems 1 and 2.*

1. 4,523 minuend
   − 2,361 subtrahend
    2,162 difference

2. 17,189 minuend
   − 6,998 subtrahend
    10,191 difference

3. 261,671
   − 76,182
    185,489

4. 786,325
   − 479,197
    307,128

5. $297.56
   − 179.71
    $117.85

6. $5,214.87
   − 2,616.89
    $2,597.98

7. 717,583
   − 389,124
    328,459

8. 226,542,203
   − 76,212,168
    150,330,035

9. $24,671.73
   − 15,777.19
    $8,894.54

10. $77,182.34
   − 38,534.76
    $38,647.58

11. 248,709,873
   − 179,323,175
    69,386,698

12. 3,782,812
   − 1,821,123
    1,961,689

B. *Solve these problems by horizontal subtraction.*

13. 275 − 149   126   14. 657 − 298   359

15. 1,156 − 436   720   16. 3,532 − 1,785   1,747

17. 15,342 − 12,917   2,425   18. 35,632 − 15,788   19,844

19. $285.73 − $179.76 $105.97   20. $358.87 − $282.76 $76.11

21. $1,643.76 − $698.34 $945.42   22. $4,527.21 − $2,716.93 $1,810.28

Lesson 6   T–28

Introduction

Ask the students, "What operation do you use to calculate how much larger one quantity is than another?" (subtraction) Subtraction shows how far apart two numbers are. This quantity is known as the difference.

Note: At the end of the class period, hand out the sheets *Lesson 6 Speed Test Preparation* (Subtraction).

Teaching Guide

1. Review the subtraction terms.
   a. Minuend—the number from which something is being subtracted.
   b. Subtrahend—the number that is being subtracted from the minuend.
   c. Difference—the answer to a subtraction problem.

2. Review vertical subtraction. Solve several problems on the board, reviewing the principles of subtraction, especially borrowing. Check by adding the difference to the subtrahend. The resulting sum should equal the minuend.

   a.   3,451   1,695
        – 1,756   + 1,756
        (1,695)   3,451

   b.   27,003   12,327
       – 14,676   + 14,676
       (12,327)   27,003

   c. $657.02 $169.59
       – 487.43   + 487.43
       ($169.59) $657.02

   d.   4,600,212   2,814,791
      – 1,785,421   + 1,785,421
      (2,814,791)   4,600,212

3. **Horizontal subtraction is done in the same way as vertical subtraction.** Write the answer one digit at a time, beginning with the ones' place. Horizontal subtraction is a bit more difficult because the digits of the minuend and the subtrahend are not in columns.

   a. 563 – 342   (221)
   b. 641 – 276   (365)
   c. 3,421 – 1,766   (1,655)
   d. 5,315 – 2,743   (2,572)
   e. 6,102 – 3,645   (2,457)

T–29 Chapter 1 Basic Mathematical Operations

Solutions for Part C

23. 603,550 – 601,730
24. 2,002,512 – 1,585,577
25. 71,275 – 57,764
26. 320,275 + 43,925 – 56,332
27. 120,000 – 90,331
28. 35,000 – 12,000

C. Solve these reading problems.
(Labels are optional with answers giving population figures.)

23. Twice in the Book of Numbers, God instructed Moses to number the men of Israel who were twenty years old and older. They numbered 603,550 the first time and 601,730 the second time (Numbers 2:32; 26:51). What was the difference?   1,820 men

24. In 1950 the population of Philadelphia was 2,002,512. The 1990 census showed a population of 1,585,577. What was the drop in population during this forty-year period?   416,935

25. A Christian publisher printed 71,275 tracts and sold 57,764 of them in three months. How many of the tracts were left?   13,511 tracts

26. At the beginning of one month, the publisher had a stock of 320,275 tracts. Over the following three months, 43,925 more tracts were printed and 56,332 were sold. How many tracts were on hand at the end of the period?   307,868 tracts

27. Christians in Russia suffered severely during the time of the communist revolution. In 1918, early in the revolution, about 120,000 Mennonites lived in Russia. In 1926, after the revolution, there were only 90,331 Mennonites in Russia. What was the drop in the Mennonite population in Russia over this time?   29,669

28. At the end of World War II, 35,000 Mennonites tried to flee from Russia and find freedom in the West. Only 12,000 were successful; the remainder either died or were forcibly returned to Russia. How many did not succeed in reaching the West?   23,000 people

REVIEW EXERCISES

D. Solve by mental addition. *(Lesson 5)*
29. 32 + 52   84
30. 89 + 64   153
31. 734 + 917   1,651
32. 1,428 + 876 2,304

E. Write *commutative law* or *associative law* to tell what each problem illustrates. *(Lesson 4)*
33. 30 + 14 = 14 + 30   commutative law
34. (2 × 7)6 × 6 = 2(7 × 6)6   associative law

F. Write the value of each underlined digit. *(Lesson 2)*
35. 1,3_8_6,018,000   80,000,000 (80 million)
36. 2,8_9_3,181,900,000   90,000,000,000 (90 billion)

G. Round each number to the place indicated. *(Lesson 2)*
37. 2,560,446 (millions)   3,000,000
38. 183,751,000 (ten millions) 180,000,000

7. Subtracting Mentally

Mental subtraction is much like mental addition. To subtract mentally, begin with the digit that has the highest place value in the subtrahend. Subtract that digit from the minuend. Then subtract the digit with the next highest place value. Repeat the process until the subtraction problem is solved. **Write the answer only after the final calculation.**

Example A	**Example B**
275 − 138 = ___	2,240 − 1,680 = ___
Think: 275 − 100 = 175,	Think: 2,240 − 1,000 = 1,240,
− 30 = 145,	− 600 = 640,
− 8 = 137	− 80 = 560

CLASS PRACTICE

Solve these problems by mental subtraction.

a. 76 − 24   52
b. 64 − 37   27
c. 187 − 99   88
d. 823 − 461   362
e. 165 − 76   89
f. 215 − 119   96
g. 328 − 147   181
h. 246 − 207   39
i. 86 − 39   47
j. 178 − 61   117
k. 571 − 354   217
l. 394 − 109   285
m. 653 − 532   121
n. 419 − 285   134
o. 3,650 − 1,825   1,825
p. 5,875 − 3,950   1,925

WRITTEN EXERCISES

A. *Solve these problems by mental subtraction. Write the answer only after the final calculation.*

1. 48 − 27   21
2. 87 − 46   41
3. 91 − 35   56
4. 52 − 43   9
5. 93 − 78   15
6. 75 − 38   37
7. 71 − 46   25
8. 83 − 58   25
9. 271 − 187   84
10. 365 − 182   183
11. 427 − 283   144
12. 645 − 387   258
13. 368 − 283   85
14. 477 − 179   298
15. 309 − 266   43
16. 507 − 249   258
17. 1,350 − 275   1,075
18. 1,525 − 650   875
19. 4,550 − 1,274   3,276
20. 6,750 − 2,475   4,275

B. *Solve these reading problems.*

21. "Or what king, going to make war against another king, sitteth not down first, and consulteth whether he be able with ten thousand to meet him that cometh against him with twenty thousand?" (Luke 14:31). Mentally find the difference between the numbers of soldiers in the two armies.   10,000 soldiers

LESSON 7

Objective

(New items are marked with a star.)

- To teach mental subtraction up to *4 digits minus 4 digits. (Previous experience: subtracting 3 digits from 3 digits mentally.)

Review

1. Give *Lesson 7 Speed Test* (Subtraction).
2. *Solve these subtraction problems.* (Lesson 6)

   a. $885.32
       − 316.68
       ($568.64)

   b. 3,452,986
       − 1,534,593
       (1,918,393)

   c. 532 − 291 (241)

   d. 12,638 − 9,266 (3,372)

   e. 43,169 − 28,263 (14,906)

3. *Solve by mental addition.* (Lesson 5)

   a. 39 + 67 (106)

   b. 1,846 + 7,404 (9,250)

4. *Write these numbers as Arabic numerals.* (Lesson 3)

   a. LVII (57)

   b. $\overline{\text{V}}$CMXIV (5,914)

5. *Write these numbers as Roman numerals.* (Lesson 3)

   a. 172 (CLXXII)

   b. 12,024 ($\overline{\text{XII}}$XXIV)

Solutions for Part B

21. 20,000 − 10,000

T–31   Chapter 1   Basic Mathematical Operations

Introduction

Review mental addition from Lesson 5. A review of the left-to-right method will serve as an introduction to the same process in mental subtraction.

- a. 32 + 47   (79)
- b. 73 + 18   (91)
- c. 45 + 81   (126)
- d. 121 + 32   (153)
- e. 230 + 79   (309)
- f. 382 + 38   (420)
- g. 1,200 + 850   (2,050)
- h. 1,540 + 1,215   (2,755)

Teaching Guide

To subtract mentally, begin with the digit that has the highest place value in the subtrahend. Subtract that digit from the minuend. Next subtract the value in the next largest place. Repeat the process until the subtraction problem is solved.

- a. 83 − 27
  Think: 83 − 20 = 63, − 7 = 56
- b. 93 − 44
  Think: 93 − 40 = 53, − 4 = 49
- c. 124 − 38   (86)
- d. 1,250 − 575   (675)
- e. 3,675 − 1,850   (1,825)
- f. 7,650 − 1,875   (5,775)

Note: Watch to see that pupils are writing answers only after they have finished calculating.

Mental math skills are taught best by practice. Spend plenty of time with *Class Practice*. Have the students say each step as they subtract mentally.

---

22. 200 − 70

23. 2,722 − 1,210

24. 2,548 − 1,379

22. At the third hour of the night, a company of 200 spearmen and 70 horsemen took the apostle Paul out of Jerusalem (Acts 23:23). Mentally find the difference between the numbers of soldiers in the two groups. 130 soldiers

23. If trees are spaced 4 feet apart, it takes 2,722 trees to plant an acre. At a 6-foot spacing, it takes 1,210 trees per acre. Find the difference in the number of trees per acre. 1,512 trees

24. For one job, a printer used 1,379 pounds of paper from a pallet having 2,548 pounds of paper. How many pounds of paper were left on the pallet? 1,169 pounds

The printing press that printed most Rod and Staff textbooks in the early 2000s.

25. Find Psalm CXLV:X and copy it. "All thy works shall praise thee, O LORD; and thy saints shall bless thee" (Psalm 145:10).

26. In Erla's Bible, the Old Testament contains MCIX pages. Write this number as an Arabic numeral. 1,109

REVIEW EXERCISES

C. Solve these subtraction problems. *(Lesson 6)*

27. $937.15
    − 402.77
    $534.38

28. 4,265,719
    − 1,836,724
    2,428,995

29. 582 − 346   236

30. 21,429 − 8,515   12,914

D. Solve by mental addition. *(Lesson 5)*

31. 335 + 827   1,162

32. 645 + 283   928

E. Write these Roman numerals as Arabic numerals. *(Lesson 3)*

33. XLIV   44

34. CCCXXIX   329

F. Write these Arabic numerals as Roman numerals. *(Lesson 3)*

35. 21,654   XXIDCLIV

36. 35,295   XXXVCCXCV

8. Reviewing Multiplication

Multiplication is a quick way to add equal addends. The multiplication problem 4 × 38 yields the same answer as an addition problem with 4 addends that are all 38.

$$\begin{array}{r} 38 \\ \times\ 4 \\ \hline 152 \end{array} \qquad 4 \times 38 = 152 \qquad \begin{array}{r} 38 \\ 38 \\ 38 \\ +\ 38 \\ \hline 152 \end{array}$$

Multiplication is indicated in three ways. The first is by the symbol ×, the second is by a raised dot (·), and the third is by writing the items to be multiplied directly beside each other. The expression 5a means "5 times a," and ab means "a times b." If the two items are numbers, one number is enclosed in parentheses for the sake of clarity. For example, 3(5) means "3 times 5," but 35 means "thirty-five."

$$3 \times 5 = 3 \cdot 5 = 3(5) = 15$$

To multiply when the multiplier has more than one digit, use the following steps.

1. Multiply the multiplicand by the digit in the ones' place in the multiplier.
2. Multiply the multiplicand by the digit in the tens' place in the multiplier. Place the first digit of this partial product under the digit in the tens' place of the first partial product.
3. If the multiplier has more digits, multiply by each one, always starting the partial product one place to the left of the previous partial product.
4. Add the partial products.

Example A

```
    275    multiplicand ┐
  × 138    multiplier   ┴─ factors
   2200    partial product
    825    partial product
    275    partial product
  37,950   product
```

Estimate for Example A

```
    300
  × 100
  30,000
```

LESSON 8

Objectives

- To review multiplication terms.
- To review the three ways of indicating multiplication: by using the signs × and ·, and by writing the factors directly beside each other (using parentheses to separate numbers).
- To review multiplying up to 4-digit multiplicands and 4-digit multipliers.
- To review estimating products by rounding both the multiplicand and the multiplier to the largest place in the numbers and multiplying mentally.
- To review checking the product of a multiplication problem by transposing the factors and by casting out nines.

Review

1. *Solve by mental subtraction.* (Lesson 7)
   a. 46 − 25 (21)
   b. 158 − 83 (75)
   c. 428 − 147 (281)
   d. 743 − 577 (166)
   e. 1,225 − 450 (775)
   f. 4,950 − 2,364 (2,586)

2. *Solve these subtraction problems. Use addition to check your answers.* (Lesson 6)

   a.   321   b.   3,827
       − 77   − 1,288
       (244)   (2,539)

   c. 3,725 − 1,392 (2,333)
   d. $941.58 − 736.84 ($204.74)

3. *Name the mathematical law illustrated in each problem. If no rule is illustrated, answer* none. (Lesson 4)
   a. 4 + 5 = 5 + 4 (commutative law)
   b. 9 − 5 = 2 × 2 (none)
   c. 17 × 5 × 3 = 17(5 × 3) (associative law)
   d. 7 + x = x + 7 (commutative law)

T–33   Chapter 1   Basic Mathematical Operations

Introduction

Write the following addition and multiplication problems on the board.

$$5 + 5 + 5 + 5 + 5 = 25$$
$$5 \times 5 \times 5 \times 5 \times 5 = 3{,}125$$

Notice how much larger the product of the multiplication problem is than the sum of the addition problem.

Teaching Guide

1. Review the multiplication terms.

    a. Multiplicand—the number being multiplied.

    b. Multiplier—the number of times the multiplicand is taken.

    c. Factor—multiplicand or multiplier.

    d. Product—the answer to a multiplication problem.

2. **Multiplication is indicated in three ways.** The first is by the symbol ×, the second is by a raised dot (·), and the third is by writing the items to be multiplied directly beside each other. Parentheses are used to separate numbers when they are placed directly beside each other.

    $$4 \times 5 = 4 \cdot 5 = 4(5)$$

3. **To multiply when the multiplier has more than one digit, use the following steps.**

    (1) Multiply the multiplicand by the digit in the ones' place in the multiplier.

    (2) Multiply the multiplicand by the digit in the tens' place in the multiplier. Place the first digit of this partial product under the digit in the tens' place of the first partial product.

    (3) If the multiplier has more digits, multiply by each one, always starting the partial product one place to the left of the previous partial product.

    (4) Add the partial products.

    Give practice with vertical multiplication, including problems with zeroes in the factors.

    a.   503   b.   674
        × 148   × 308
       (74,444)   (207,592)

Estimating Products

The product of a multiplication problem can easily be estimated by following the steps below. The estimate for Example A shows how these steps are applied.

1. Round the multiplicand and the multiplier to the largest place so that there is only one digit other than zero.

2. Multiply these rounded numbers mentally.

Estimation is useful in two ways. It gives an approximate answer when an exact answer is not needed, and it shows whether a calculated answer is sensible.

Checking Multiplication

There are two main ways to check multiplication problems. The more accurate way is to apply the commutative law. Transpose the factors and multiply, as shown in Example B. This check is quite reliable because you are not likely to get the same wrong answer both times.

Example B		**Example C**		
327	192	32̸7 →		3
× 192	× 327	× 1 9̸2 →		× 3
654	1344	654		9 → 0
2943	384	2943		
327	576	327		
62,784	62,784	6 2̸,7̸ 84 → 18 → 9 → 0		

Casting out nines is a quick way to check a product. Following are the steps to casting out nines as illustrated in Example C.

1. Cross out all the 9s in the multiplicand. Cross out any digits in the multiplicand whose sum equals 9.

2. Repeat step 1 for the multiplier and the product.

3. Add the digits of the multiplicand that were not crossed out. If the sum is greater than 9, add the digits of that sum. Repeat this step until the sum is a one-digit number. This is called the excess number. If the excess number is 9, change it to 0.

4. Repeat step 3 for the multiplier and the product.

5. Multiply the excess number for the multiplicand by the excess number for the multiplier. If the product of this multiplication is more than 9, find its excess number as in step 3. This excess number must be equal to the excess number of the product of the multiplication problem.

34 Chapter 1 Basic Mathematical Operations

Casting out nines will never show that a right answer is wrong. However, sometimes it could indicate that a wrong answer is right. Following are some errors that casting out nines will not reveal.

1. *If zeroes are added or dropped.* A zero in a number has no effect on its excess number. The numbers 44, 404, 440, and 4,004 all have the same excess number.

2. *If the product is wrong by a multiple of nine.* Since 9 equals 0 in casting out nines, this has the same effect as adding or dropping zeroes.

3. *If digits are transposed.* For example, 987 and 789 have the same excess numbers.

CLASS PRACTICE

Multiply. Check by using the commutative law.

| a. | 45 × 82 = 3,690 | b. | 89 × 57 = 5,073 | c. | 215 × 367 = 78,905 | d. | 825 × 639 = 527,175 |

Multiply. Check by casting out nines.

| e. | 245 × 188 = 46,060 | f. | 934 × 736 = 687,424 | g. | 2,165 × 827 = 1,790,455 | h. | 1,132 × 2,245 = 2,541,340 |

Estimate the products of these multiplication problems.

| i. | 4,632 × 6,192 = 30,000,000 | j. | 824 × 853 = 720,000 | k. | 5,184 × 9,811 = 50,000,000 | l. | 4,516 × 3,489 = 15,000,000 |

WRITTEN EXERCISES

A. *Multiply. Check by using the commutative law.*

| 1. | 78 × 67 = 5,226 | 2. | 93 × 56 = 5,208 | 3. | 215 × 163 = 35,045 | 4. | 274 × 314 = 86,036 |

B. *Multiply. Check by casting out nines.*

| 5. | 177 × 153 = 27,081 | 6. | 235 × 175 = 41,125 | 7. | 475 × 324 = 153,900 | 8. | 717 × 364 = 260,988 |

| 9. | 642 × 351 = 225,342 | 10. | 527 × 340 = 179,180 | 11. | 3,547 × 2,412 = 8,555,364 | 12. | 7,352 × 4,515 = 33,194,280 |

C. *Estimate the products of these multiplication problems.*

| 13. | 3,425 × 2,711 = 9,000,000 | 14. | 5,764 × 3,115 = 18,000,000 | 15. | 7,684 × 5,712 = 48,000,000 | 16. | 9,812 × 7,251 = 70,000,000 |

(Exact answers) (9,285,175)   (17,954,860)   (43,891,008)   (71,146,812)

c. 4,325   d. 8,273
   × 2,645   × 7,526
  (11,439,625) (62,262,598)

4. **To estimate the product of a multiplication problem, use the following steps.**

   (1) Round the multiplicand and the multiplier to the largest place so that there is only one digit other than zero.

   (2) Multiply these rounded numbers mentally.

   Practice estimating products in class. Be sure the pupils find estimates mentally.

   a. 6,345   6,000
      × 4,271   × 4,000
     (27,099,495) (24,000,000)

   b. 2,738   3,000
      × 7,256   × 7,000
     (19,866,928) (21,000,000)

5. **Multiplication can be checked by applying the commutative law.** Transpose the factors and multiply again.

   a.   805   263
       × 263   × 805
      (211,715)   (211,715)

   b.   907   146
       × 146   × 907
      (132,422)   (132,422)

6. **Multiplication can also be checked by casting out nines.** To check by this method, use the following steps.

   (1) Cross out all the 9s in the multiplicand. Cross out any digits in the multiplicand whose sum equals 9.

   (2) Repeat step 1 for the multiplier and the product.

   (3) Add the digits of the multiplicand that were not crossed out. If the sum is greater than 9, add the digits of that sum. Repeat this step until the sum is a one-digit number. This is called the excess number. If the excess number is 9, change it to 0.

   (4) Repeat step 3 for the multiplier and the product.

   (5) Multiply the excess number for the multiplicand by the excess number for the multiplier. If the product of this multiplication is more than 9, find its excess number as in step 3. This excess number must be equal to the excess number of the product of the multiplication problem.

   a.   3̶,652   →   7
        × 2,9̶5̶4   →   × 2
       10,788,008 → 23 → **5** ← 14

   b.   7,853   →   5
        × 6̶,7̶3̶2̶   →   × 0
       52,866,3̶96 → 27 → 9 → **0** ← 0

Casting out nines is not a foolproof check. It can wrongly indicate that an answer is correct in several circumstances.

(1) *If zeroes are added or dropped.* This is because a zero in a number has no effect on its excess number. The numbers 44, 404, 440, and 4,004 all have the same excess number.

(2) *If the product is wrong by a multiple of nine.* Since 9 equals 0 in casting out nines, this has the same effect as adding or dropping zeroes.

(3) *If digits are transposed.* For example, 987 and 789 have the same excess numbers.

T–35   Chapter 1   Basic Mathematical Operations

An Ounce of Prevention

This lesson contains five major teaching points, all of which are review from seventh grade. Do not spend too much time on any one point. Emphasize the calculation process and casting out nines.

Further Study

You have probably wondered, "How does casting out nines work?" Casting out nines is a quick way to determine what the last digit of the same number would be in a base nine system. Certain numbers in ones' place in the factors always produce certain numbers in ones' place of the product. Multiplying the excess numbers of the factors and casting out nines again verifies the ones' digit of the product. If it agrees with the excess number found for the product, the calculation is probably correct.

The examples below show the multiplication of the same numbers, expressed in the base ten and the base nine numeration systems. Notice the last digits of the factors and the product in the base nine calculation. They are the excess numbers of the base ten problem.

$$
\begin{array}{rcccr}
4{,}2\cancel{9}8_{\text{ten}} & \rightarrow & 5 & = & 5{,}80\,5_{\text{nine}} \\
\times\,408_{\text{ten}} & \rightarrow & \times\,3 & = & \times\,503_{\text{nine}} \\
\hline
34384 & & 15 \rightarrow 6 & & 18616 \\
0 & & & & 0 \\
\underline{17192} & & & & \underline{32427} \\
\cancel{1}{,}753{,}\cancel{5}\cancel{8}4_{\text{ten}} & \rightarrow & 15 \rightarrow 6 & = & 3{,}262{,}416_{\text{nine}}
\end{array}
$$

Solutions for Part D

17. 12 × 21

18. 12 × 1,403 × 1,112

19. Annex 4 zeroes to 10,000

20. 90,000 × 2,000

21. 1884 – 1848

22. $95,637.85 – $92,000.00

D. Solve these reading problems.

17. When the children of Israel dedicated the altar of the tabernacle, twelve princes, one from each of the twelve tribes, offered sacrifices upon the altar (Numbers 7:10–88). Among other things that each prince offered were 21 animals. How many animals were sacrificed at the dedication? 252 animals

18. The community of Milford contains 1,403 homes. If each home uses an average of 1,112 kilowatt-hours of electricity per month, how much electricity will the community of Milford use in one year? 18,721,632 kilowatt-hours

19. The number of the angels in heaven is described as ten thousand times ten thousand and thousands of thousands (Revelation 5:11). Multiply ten thousand times ten thousand mentally. 100,000,000

20. The Washington Monument, an obelisk built with Maryland marble, weighs 90,854 tons. Estimate the weight of the monument in pounds by rounding and then multiplying by 2,000. 180,000,000 pounds

21. The building of the Washington Monument was begun in 1848 but was not completed until 1884. How many years did it take to build the monument? 36 years

22. When the Masts began building their fertilizer warehouse, they estimated that the cost would be $92,000. The actual cost of the building was $95,637.85. By what amount did the cost overrun the estimate? $3,637.85

REVIEW EXERCISES

E. Solve by mental subtraction. (Lesson 7)

23. 55 − 32   23
24. 463 − 93   370
25. 1,625 − 1,175   450
26. 5,675 − 3,288   2,387

F. Solve these subtraction problems. Use addition to check your answers. (Lesson 6)

27. 272 − 94 = 178
28. 648 − 274 = 374
29. 6,735 − 4,178 = 2,557
30. 82,473 − 39,495 = 42,978
31. 5,821 − 1,744   4,077
32. 524,618 − 253,628   270,990

G. Name the mathematical law illustrated in each exercise. If no law is illustrated, write none. (Lesson 4)

33. 8 + 7 + 5 = 8 + (7 + 5)   associative law
34. 19 + 65 = 65 + 19   commutative law
35. 27 · 5 = 5 · 27   commutative law
36. 15 + 12 = 38 − 11   none

9. Multiplying Mentally

Like addition and subtraction, multiplication can often be done mentally. By regular practice and by applying a few mathematical rules, you can become skilled in mental multiplication.

Horizontal Multiplication

With a one-digit multiplier, multiply horizontally just as you would multiply vertically. **Write the answer one digit at a time as you calculate without copying the problem.** Horizontal multiplication is a step between written and mental multiplication.

> **Example**   Solve 8 × 356 horizontally.
> Think: 8 × 6 = 48   Write 8; carry 4.   _____8_ (partial product)
>   8 × 5 + 4 = 44   Write 4; carry 4.   ____48_ (partial product)
>   8 × 3 + 4 = 28   Write 28.   2,848 (product)

Mental Multiplication

Mental multiplication is different from horizontal multiplication in that the answer is not written until the calculations are finished. Here are a few helps for mental multiplication.

Help 1: To multiply by 10 or a power of 10, annex as many zeroes as there are in the multiplier.
 	
    10 × 343   = 3,430
    100 × 343   = 34,300
    1,000 × 343   = 343,000
    10,000 × 343 = 3,430,000
	Help 2: To multiply by a multiple of 10, multiply the nonzero digits, and annex as many zeroes as there are in the factors.
	
	    400 × 71 = ___
	Think: 4 × 71 = 284
	    Annex 2 zeroes.
	    400 × 71 = 28,400

Help 3: To multiply by a single digit, calculate from left to right.
 	
    6 × 345 = ___
Think: 6 × 300 = 1,800   1,800
    6 × 40 = 240   + 240
                              2,040
    6 × 5 = 30   + 30
                              2,070
	Help 4: To multiply a series of factors, first multiply the factors that yield multiples of 10. Then multiply the remaining factors.
	
	    9 × 15 × 4 = ___
	Think: 9(15 × 4)
	        = 9 × 60 = 540

LESSON 9

Objectives

- To review horizontal multiplication.
- To review the following mental multiplication skills.
  a. Multiplying by powers of 10.
  b. Multiplying *up to 4-digit multiplicands by 1-digit multipliers.
  c. Multiplying 2-digit multiplicands by multiples of 10.
  d. Looking for sets of factors that equal 10.

Review

1. *Solve these problems, and check by casting out nines.* (Lesson 8)

   a. 653
      × 821
      (536,113)

   b. 324
      × 903
      (292,572)

   c. 3,158
      × 3,865
      (12,205,670)

   d. 8,382
      × 5,381
      (45,103,542)

2. *Solve these problems mentally.* (Lessons 5, 7)

   a. 38 + 45 (83)
   b. 712 + 84 (796)
   c. 525 + 392 (917)
   d. 87 − 55 (32)
   e. 374 − 106 (268)
   f. 901 − 656 (245)

3. *Write these numbers, using words.* (Lesson 1)

   a. 219,000,112
      (two hundred nineteen million, one hundred twelve)
   b. 77,158,000,000,000,000
      (seventy-seven quadrillion, one hundred fifty-eight trillion)

T–37 Chapter 1 Basic Mathematical Operations

Introduction

Challenge the students to solve the problem 18 × 9 × 5 without using pencils. Did they find the answer to be 810? Have the students tell what method they used to find the answer. Multiplying 18 × 5 = 90 and then 90 × 9 = 810 is much simpler than 18 × 9 = 162 and 162 × 5 = 810.

Note: At the end of the class period, hand out the sheets *Lesson 9 Speed Test Preparation* (Multiplication).

Teaching Guide

1. **Horizontal multiplication is done in the same way as vertical multiplication.** Write the answer one digit at a time, beginning with the ones' place. Do not copy the problem.

    a. 7 × 4,342   (30,394)

    b. 8 × 6,812   (54,496)

    c. 7 × 17,784 (124,488)

    Note: Mental multiplication is different from horizontal multiplication in that the answer is not written until the calculating is finished. Watch to see that pupils are writing the answer only after they have finished calculating.

    Mental math is taught best by practice. Spend plenty of time with *Class Practice.* Have the students say each step as they multiply mentally.

2. **To multiply by 10 or a power of 10, annex as many zeroes as there are zeroes in the multiplier.** This should be quite familiar to the students by now.

    a. 100 × 341   (34,100)

    b. 1,000 × 537   (537,000)

    c. 10,000 × 718 (7,180,000)

3. **To multiply by a multiple of 10, multiply the nonzero digits, and annex as many zeroes as there are in the factors.**

    a. 20 × 27   (540)

    b. 300 × 13   (3,900)

    c. 60 × 12   (720)

    d. 800 × 32   (25,600)

CLASS PRACTICE

Solve by horizontal multiplication without copying the problems.

a. 8 × 753   6,024   b. 3 × 528   1,584

Solve mentally by annexing zeroes.

c. 100 × 1,649   164,900   d. 10,000 × 3,756   37,560,000

Solve by mental multiplication.

e. 20 × 18   360   f. 20 × 52   1,040   g. 300 × 24   7,200   h. 70 × 67   4,690
i. 50 × 14   700   j. 30 × 46   1,380   k. 40 × 32   1,280   l. 700 × 21   14,700

Multiply mentally, using the left-to-right method.

m. 6 × 78   468   n. 3 × 58   174   o. 7 × 44   308   p. 4 × 62   248
q. 9 × 176   1,584   r. 8 × 321   2,568   s. 7 × 1,853   12,971   t. 4 × 2,135   8,540

Solve mentally. First multiply the factors that yield multiples of 10.

u. (5 × 12) × 7   420   v. 9 × (6 × 5)   270   w. 11 × (15 × 4)   660   x. (15 × 2) × 8   240

WRITTEN EXERCISES

A. *Solve by horizontal multiplication without copying the problems.*

1. 7 × 342   2,394   2. 6 × 735   4,410   3. 4 × 926   3,704
4. 4 × 21,513   86,052   5. 3 × 193,261   579,783   6. 8 × 274,459   2,195,672

In parts B–E, do not write the answer until your calculations are complete.

B. *Solve mentally by annexing zeroes.*

7. 10 × 365   3,650   8. 1,000 × 2,311   2,311,000   9. 100 × 7,681   768,100
10. 10,000 × 3,515   35,150,000   11. 100 × 14,151   1,415,100   12. 10,000 × 87,653   876,530,000

C. *Solve by mental multiplication.*

13. 20 × 12   240   14. 40 × 14   560   15. 60 × 15   900
16. 30 × 43   1,290   17. 60 × 63   3,780   18. 80 × 71   5,680

D. *Multiply mentally, using the left-to-right method.*

19. 8 × 28   224   20. 4 × 128   512   21. 7 × 378   2,646
22. 8 × 427   3,416   23. 5 × 2,315   11,575   24. 6 × 3,526   21,156

E. *Solve mentally. First multiply the factors that yield multiples of 10.*

25. 7 × 5 × 6   210   26. 15 × 9 × 6   810   27. 13 × 8 × 5   520
28. 16 × 7 × 5   560   29. 15 × 4 × 5   300   30. 18 × 9 × 5   810

38   Chapter 1   Basic Mathematical Operations

F. Solve these reading problems mentally.

31. A legion in the Roman army consisted of about 6,000 soldiers. As the mob came to arrest Jesus in the Garden of Gethsemane, He told Peter that He could call for more than 12 legions of angels. At 6,000 per legion, how many angels would have been in 12 legions?   72,000 angels

32. Solomon dealt in horses and chariots from Egypt (1 Kings 10:26, 29). A chariot from Egypt cost 600 shekels. What was the value of Solomon's 1,400 chariots?   840,000 shekels

33. Solomon also had 12,000 horsemen (1 Kings 10:26). At 150 shekels per horse, what was the value of 6,000 horses?   900,000 shekels

34. On each day in November, a beef farmer fed his cattle one bale of hay per head. He had 36 beef cattle. How many bales did he feed his cattle in November?   1,080 bales

35. One day at a weekend Bible conference, the cooks served 478 people at noon and 386 in the evening. How many people were served in all?   864 people

36. In problem 35, how many more people were served at noon than in the evening?   92 people

REVIEW EXERCISES

G. Solve these problems, and check by casting out nines. (Lesson 8)

37.   841
    × 966
    812,406

38.   534
    × 732
    390,888

H. Solve these problems mentally. (Lessons 5, 7)

39. 77 + 43   120
40. 489 + 426   915
41. 93 − 45   48
42. 936 − 497   439

I. Write these numbers, using words. (Lesson 1)

43. 8,000,463,000
    Eight billion, four hundred sixty-three thousand

44. 28,000,655,000,000
    Twenty-eight trillion, six hundred fifty-five million

4. **To multiply by a single digit, calculate from left to right.** This is the same procedure as for adding and subtracting mentally.

    a. 7 × 36 (252)

    b. 9 × 76 (684)

    c. 6 × 241 (1,446)

    d. 8 × 363 (2,904)

5. **To multiply a series of factors, first multiply the factors that yield multiples of 10. Then multiply the remaining factors.** This is applying the commutative law.

    a. 7 × 8 × 5 (280)

    b. 9 × 6 × 5 (270)

    c. 14 × 7 × 5 (490)

    d. 15 × 7 × 6 (630)

An Ounce of Prevention

1. Students need plenty of practice with mental math. This lesson will not be taught in one day, but rather by frequent practice. Review mental math skills frequently.

2. Sometimes several mental shortcuts are possible. If the students suggest what seems to be a "shorter" shortcut than the one presented in the lesson, evaluate it for accuracy. Then help the students determine which method is better in this circumstance.

Solutions for Part F

31. 12 × 6,000

32. 1,400 × 600

33. 6,000 × 150

34. 30 × 36

35. 478 + 386

36. 478 − 386

T–39 Chapter 1 Basic Mathematical Operations

LESSON 10

Objectives

- To teach *the distributive law, and to apply it to multiplication problems with 2-digit multipliers. (The concept was introduced in Lesson 9 in working with 1-digit multipliers, but is now introduced formally and is used to solve more difficult problems.)
- To review the double-and-divide method including the use of this method to multiply by 5 and by 50.
- To review the shortcut for multiplying by 25.

Review

1. Give *Lesson 10 Speed Test* (Multiplication).
2. *Solve by mental multiplication.* (Lesson 9)
   a. 482 × 7 (3,374)
   b. 32,156 × 4 (128,624)
   c. 826 × 1,000 (826,000)
   d. 483 × 80 (38,640)
3. *Do these multiplications, and check by casting out nines.* (Lesson 8)

   a.   271   b.   7,324
       × 801   × 1,712
     (217,071)   (12,538,688)

4. *Do these subtractions, and check by addition.* (Lesson 6)

   a.   8,254   b.   53,704
       − 3,861   − 28,384
      (4,393)   (25,320)

5. *Round these numbers as indicated.* (Lesson 2)

   a. 23,852 to the nearest thousand (24,000)
   b. 90,398,764,000 to the nearest ten million (90,400,000,000)

10. Using the Distributive Law

In Lesson 9, you used the left-to-right method to multiply a number by a single digit. Example A reviews this method, showing how a complex problem can be broken down into three simpler problems: $5 \times 278 = (5 \times 200) + (5 \times 70) + (5 \times 8) = 1{,}390$.

Example A

$5 \times 278 =$ ___		
Think: $5 \times 200 = 1{,}000$	$1{,}000$	$5 \times 278 = 5(200 + 70 + 8)$
$5 \times 70 = 350$	$+ \ 350$	
	$1{,}350$	
$5 \times 8 = 40$	$+ \ \ \ 40$	
	$1{,}390$	

Note that the problem above can also be stated as $5 \times 278 = 5(200 + 70 + 8)$. This applies the mathematical principle known as the **distributive law**. The law has this name because a multiplier is distributed over a group of addends.

Distributive Law
Multiplying a combination of addends gives the same result as multiplying the addends separately and adding the products.
$$3(4 + 5) = (3 \times 4) + (3 \times 5) = 27$$

Applying the Distributive Law to a Group of Addends
To multiply a group of addends by a certain number, use either of the following methods.

1. Find the sum of the addends and then multiply.

    $5(6 + 3)$ $\qquad\qquad$ $(8 + 9)2$
    $= 5 \times 9 = 45$ $\qquad$ $= (17)2 = 34$

2. Multiply each addend separately and then add the products.

    $5(6 + 3)$ $\qquad\qquad$ $(8 + 9)2$
    $= (5 \times 6) + (5 \times 3)$ $\qquad$ $= (8 \times 2) + (9 \times 2)$
    $= 30 + 15 = 45$ $\qquad$ $= 16 + 18 = 34$

Applying the Distributive Law to a Sum
To multiply a larger number, you can break it down into smaller numbers to be multiplied separately. This use of the distributive law allows the simplifying of complex multiplication problems. Example A above is one illustration, and Example B on the following page is another. Note that the distributive law may be applied to either factor.

Chapter 1 Basic Mathematical Operations

Example B

$22 \times 13 = \underline{}$

Think: 22×13
$= (20 + 2)13$
$= (20 \times 13) + (2 \times 13)$
$= 260 + 26 = 286$

$22 \times 13 = \underline{}$

Think: 22×13
$= 22(10 + 3)$
$= (22 \times 10) + (22 \times 3)$
$= 220 + 66 = 286$

Double-and-Divide Method

The **double-and-divide** method can also simplify multiplication problems. With this method, one factor is multiplied by 2 and the other factor is divided by 2. Thus the problem 20×16 would be changed to $10 \times 32 = 320$.

Example C

$18 \times 17 = \underline{}$

Think: $18 \times 17 = 9 \times 34$
$= (9 \times 30) + (9 \times 4)$
$= 270 + 36 = 306$

Example D

$50 \times 28 = \underline{}$

Think: $100 \times 14 = 1{,}400$

As Example D shows, this method is especially useful for problems that have 5 or 50 as one of the factors. Doubling 5 gives 10, and doubling 50 gives 100.

A variation of the double-and-divide method can be used for problems in which one factor is 25. Multiply 25×4 and divide the other factor by 4. Thus 84×25 becomes $21 \times 100 = 2{,}100$.

Example E

$25 \times 44 = \underline{}$

Think: $100 \times 11 = 1{,}100$

Example F

$120 \times 25 = \underline{}$

Think: $30 \times 100 = 3{,}000$

CLASS PRACTICE

Use the distributive law to find the missing numbers.

a. $5 \times 17 = (5 \times 10) + (5 \times \underline{7})$

b. $21 \times 76 = (21 \times \underline{70}) + (21 \times 6)$

c. $7 \times 18 = (7 \times \underline{10}) + (7 \times \underline{8})$

d. $18 \times 23 = (18 \times 20) + (\underline{18} \times \underline{3})$

e. $150 \cdot 22 = (100 \cdot 22) + (50 \cdot \underline{22})$

f. $51 \cdot 187 = (51 \cdot 100) + (\underline{51} \cdot 87)$

Break one of the factors into two addends. Then solve the problems mentally.

g. 21×14 294
h. 22×17 374
i. 32×12 384
j. 22×21 462

Find the products mentally by applying the distributive law.

k. 22×12 264
l. 23×22 506
m. 34×21 714
n. 12×14 168

Lesson 10 T–40

Introduction

Write the problem 7(5 + 6) on the board. Ask the students, "Should I add 5 + 6 first and then multiply? Or should I multiply 7 × 5 and 7 × 6, and add the products together?"

If the students point out that order of operations requires doing the operation within parentheses first, give recognition to that principle. Then ask, "In this case, does it make any difference?"

Both methods will yield the same result. The distributive law codifies this principle. Two addends being multiplied by the same number can either be first added and then multiplied, or they can be multiplied separately and then the products added.

Teaching Guide

1. **To multiply a group of addends by a factor, use either of the following methods.**

   (1) Find the sum of the addends and then multiply.

   a. 9(5 + 6)
   = 9 × 11 = 99

   b. (7 + 2)8
   = 9 × 8 = 72

   (2) Multiply each addend separately and then add the products.

   a. 9(5 + 6)
   = (9 × 5) + (9 × 6) = 99

   b. (7 + 2)8
   = (7 × 8) + (2 × 8) = 72

   c. 5 × 17 = (5 × 10) + (5 × 7) = 85

   d. 9 × 16 = (9 × 10) + (9 × 6) = 144

   e. 6 × 29 = (6 × 20) + (6 × 9) = 174

   f. 25 × 44
   = (25 × 40) + (25 × 4) = 1,100

   g. 6(10 + 5) = 6 × 15
   = (6 × 10) + (6 × 5) = 90

   h. 12(10 + 9) = 12 × 19
   = (12 × 10) + (12 × 9) = 228

2. **The double-and-divide method can also simplify multiplication problems.** With this method, one factor is multiplied by 2 and the other factor is divided by 2. This is especially useful when one of the factors is 5, 50, or a number which ends in 5.

   a. 36 × 5 = 18 × 10 = 180

   b. 5 × 98 = 10 × 49 = 490

   c. 50 × 68 = 100 × 34 = 3,400

   d. 78 × 50 = 39 × 100 = 3,900

   e. 18 × 15 = 9 × 30 = 270

T–41 Chapter 1 Basic Mathematical Operations

3. **A variation of the double-and-divide method can be used for problems in which one factor is 25.** Multiply 25 by 4, and divide the other factor by 4.

   a. $16 \times 25 = 4 \times 100 = 400$
   b. $25 \times 48 = 100 \times 12 = 1{,}200$
   c. $25 \times 160 = 100 \times 40 = 4{,}000$
   d. $240 \times 25 = 60 \times 100 = 6{,}000$

Full Answers for Part B

(Two possibilities are given for each rewritten problem.)

7. $(10 \times 12) + (3 \times 12) = 156$ *or* $(13 \times 10) + (13 \times 2) = 156$
8. $(10 \times 11) + (5 \times 11) = 165$ *or* $(15 \times 10) + (15 \times 1) = 165$
9. $(10 \times 21) + (8 \times 21) = 378$ *or* $(18 \times 20) + (18 \times 1) = 378$
10. $(10 \times 22) + (9 \times 22) = 418$ *or* $(19 \times 20) + (19 \times 2) = 418$
11. $(10 \times 31) + (2 \times 31) = 372$ *or* $(12 \times 30) + (12 \times 1) = 372$
12. $(10 \times 32) + (3 \times 32) = 416$ *or* $(13 \times 30) + (13 \times 2) = 416$
13. $(10 \times 33) + (5 \times 33) = 495$ *or* $(15 \times 30) + (15 \times 3) = 495$
14. $(10 \times 21) + (4 \times 21) = 294$ *or* $(14 \times 20) + (14 \times 1) = 294$

Mental Solutions for Part E

31. 30×600; Annex 3 zeroes to 3×6

Use the double-and-divide method or the shortcut for multiplying by 25.

o. 32 × 25   800   p. 18 × 12   216   q. 96 × 50   4,800   r. 14 × 15   210

WRITTEN EXERCISES

A. **Use the distributive law to find the missing numbers.**

1. 4 × 18 = (4 × 10) + (4 × _8_)
2. 26 × 48 = (26 × 40) + (_26_ × 8)
3. 5 × 79 = (5 × _70_) + (5 × _9_)
4. 7 × 151 = (7 × 100) + (7 × _51_)
5. 15 · n = (10 · n) + (_5_ · n)
6. 43 · n = (40 · n) + (_3_ · _n_)

B. **Rewrite each multiplication problem by breaking one of the factors into two addends. Then use your rewritten problem to solve the problems mentally as in Example B.** (See facing page.)

7. 13 × 12   156   8. 15 × 11   165   9. 18 × 21   378   10. 19 × 22   418
11. 12 × 31   372   12. 13 × 32   416   13. 15 × 33   495   14. 14 × 21   294

C. **Find the products mentally by applying the distributive law.**

15. 14 × 12   168   16. 16 × 12   192   17. 26 × 11   286   18. 47 × 11   517
19. 41 × 21   861   20. 43 × 22   946   21. 23 × 13   299   22. 24 × 21   504

D. **Solve by using the double-and-divide method or the shortcut for multiplying by 25.**

23. 26 × 22   572   24. 32 × 22   704   25. 26 × 50   1,300   26. 50 × 48   2,400
27. 72 × 50   3,600   28. 88 × 50   4,400   29. 15 × 18   270   30. 35 × 18   630

E. **Solve these reading problems mentally.**

31. After feeding the five thousand, Jesus sent His disciples across the Sea of Galilee in a boat while He remained behind. Jesus came walking to them after they had rowed about 25 or 30 furlongs. The Greek furlong was equal to about 606 feet. Using the rounded measure of 600 feet for a furlong, find how many feet are in 30 furlongs.

    18,000 feet

42   Chapter 1   Basic Mathematical Operations

32. For a 3-week period one spring, Calvin spent about 45 minutes each day pruning the trees, shrubs, and vines. In these 18 work days, how many minutes in all did he spend at pruning?   810 minutes

33. Father drives about 50 miles per week in commuting to work. How many miles is that in 52 weeks?   2,600 miles

34. The Millers have a 25-mile round trip to church. One year they made the round trip 84 times. How many miles in all did they drive on these trips?   2,100 miles

35. The Shattocks have a 16-mile round trip to Milroy. They made the trip 6 times in one week and 4 times in the next week. How many miles did they travel on these trips?   160 miles

36. Rebekah's family pronounced a blessing on her before she left home to be Isaac's wife. In their blessing, they said, "Be thou the mother of thousands of millions" (Genesis 24:60). How many people would a thousand million be?   1,000,000,000 people
                                                                                                        (1 billion)

REVIEW EXERCISES

F. Solve by mental multiplication. (Lesson 9)

37. 5 × 734   3,670   38. 8 × 515   4,120

G. Do these multiplications, and check by casting out nines. (Lesson 8)

39.   452   40.   226
     × 346   × 139
    156,392   31,414

H. Do these subtractions, and check by addition. (Lesson 6)

41.   43,612   42.   466,017
    − 25,616   − 310,872
      17,996   155,145

I. Round each number to the place indicated. (Lesson 2)

43. 432,734,802 (ten millions)   44. 19,647,900,350 (billions)
                     430,000,000   20,000,000,000

Lesson 10 T–42

32. 18 × 45 = 9 × 90
33. 52 × 50 = 26 × 100
34. 21 × 100
35. 10 × 16
36. Annex 3 zeroes to 1,000,000

LESSON 11

Objectives

- To review division terms.
- To review short division.
- To review long division up to 7-digit dividends and 4-digit divisors.
- To review checking division problems by multiplication and by casting out nines.

Review

1. *Use the distributive law to find the missing numbers.* (Lesson 10)
   a. $9 \times 29 = (9 \times 20) + (9 \times \_\_)$ [9]
   b. $32 \times 54 = (32 \times \_\_) + (32 \times \_\_)$ [50, 4]
   c. $43 \times 22 = (43 \times 20) + (\_\_ \times \_\_)$ [43, 2]
   d. $39 \times 47 = (39 \times \_\_) + 39 \times \_\_$ [40, 7]

2. *Solve these problems mentally.* (Lessons 7, 9)
   a. $91 - 48$ (43)
   b. 6×418 (2,508)
   c. $275 - 169$ (106)
   d. $635 - 386$ (249)
   e. $5 \times 12,584$ (62,920)
   f. $10,000 \times 930$ (9,300,000)

3. *Write these Roman numerals as Arabic numerals.* (Lesson 3)
   a. CDLIX (459)
   b. $\overline{\text{IX}}$CCXVII (9,217)

11. Reviewing Division

Division is the inverse operation of multiplication. Just as multiplication is a quick method of repeated addition, so division is a quick method of repeated subtraction. When we divide 776 by 25, we find how many times 25 can be subtracted from 776.

Division is indicated in the three ways shown below. We read all three of these as "776 divided by 25."

$$25\overline{)776} \qquad \frac{776}{25} \qquad 776 \div 25$$

Long Division

Long division is used to solve problems that are too complex to solve mentally. Proper estimation (Example A) greatly reduces the amount of trial and error needed to find quotient figures. Example B shows the steps in solving a long-division problem.

Example A
Estimate the first quotient figure, and write it in the correct place.

$$\begin{array}{r} 2 \\ 157\overline{)3,786} \end{array}$$

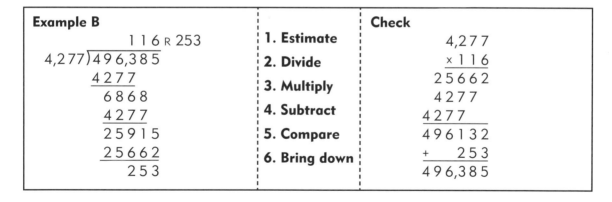

Long-division problems can be made shorter if both the dividend and the divisor end with zeroes. Cross out the same number of zeroes in both the dividend and the divisor. **If there is a remainder, add as many zeroes to the remainder as you crossed out in the dividend.** See Example C.

44   Chapter 1   Basic Mathematical Operations

Example C	**Check by multiplication**
$$\begin{array}{r} 1{,}851 \text{ R } 4{,}600 \\ 5{,}4\cancel{00})\overline{10{,}000{,}0\cancel{00}} \\ 54 \\ \overline{460} \\ 432 \\ \overline{280} \\ 270 \\ \overline{100} \\ 54 \\ \overline{46} \end{array}$$	$$\begin{array}{r} 1{,}851 \\ \times\, 5{,}400 \\ \hline 740400 \\ 9255 \\ \hline 9{,}995{,}400 \\ +4{,}600 \\ \hline 10{,}000{,}000 \end{array}$$

Short Division

Short division is used to solve problems with a 1-digit divisor. All the division steps are done mentally, and each quotient figure is written as it is calculated (Example D).

When dividing money, remember to place the decimal point in the quotient directly above the decimal point in the dividend, and to use the signs for dollars ($) and cents (¢). The check confirms the correct answer.

Example D	**Check (by casting out nines)**
$(7)(1)(3)$ $\$84.25 \text{ R } 3¢$ $7)\overline{\$589.78}(1)$	$7 \times 1 = 7, + 3 = 10 \rightarrow 1$ The excess number 1 matches the excess number of the dividend.

Checking Division

Division problems can be checked by multiplying the divisor times the quotient and adding the remainder. The result should equal the dividend (Examples B and C).

Division can also be checked by casting out nines (Example D). Cast out nines to find the excess numbers for the divisor, the dividend, the quotient, and the remainder. (They are shown in parentheses above.) Multiply the excess number of the quotient by that of the divisor, add the excess number of the remainder, and cast out nines from the result. The resulting number should match the excess number of the dividend.

Casting out nines will not reveal a wrong answer if you dropped or added zeroes or if you transposed two digits in the quotient.

CLASS PRACTICE

Copy these long-division problems. Estimate the first quotient figure, and place it properly above the dividend.

a. 1
   $376)\overline{674{,}921}$

b. 5
   $321)\overline{1{,}673{,}927}$

c. 8
   $4{,}331)\overline{3{,}891{,}045}$

Lesson 11 T–44

Introduction

Write the following problem on the board.

$$\begin{array}{r} 756 \\ \times\ ? \\ \hline 12{,}852 \end{array}$$

Ask the students how they would find the multiplier. (Divide the product by the multiplicand.) Division is the inverse operation of multiplication.

Now write the number 700 on the board and ask, "How many times could 100 be subtracted from 700?" When the students give the answer 7, ask, "How many of you subtracted 100 from 700 seven times to find the answer? How many divided 700 by 100 to find the answer?" Division tells how many times one number can be subtracted from another, or how many times one number is contained in another.

Teaching Guide

1. Review the division terms.
   a. Dividend—the number being divided.
   b. Divisor—the number by which the dividend is divided. It may be the multiplier or the multiplicand in the inverse operation.
   c. Quotient—the answer to a division problem.
   d. Remainder—a number that is left when the division does not work out evenly.

2. **Long division is used to solve problems that are too complex to solve mentally.** Proper estimation greatly reduces the amount of trial and error needed to find quotient figures.

   Practice estimating and placing the first quotient figure. Problem *a* below will come out correctly. Especially discuss problems *b* and *c*. In problem *b* the estimate is too high and needs to be lowered. In problem *c* the estimate is too low and needs to be increased. Estimation is needed to find each digit in the quotient, not only the first one.

   a.
   $$\begin{array}{r} 3 \\ 288\overline{)93{,}518} \\ \underline{86\ 4} \\ 7\ 1 \end{array}$$

   b.
   $$\begin{array}{r} 7\ (\text{wrong}) \\ 4{,}290\overline{)2{,}871{,}799} \\ \underline{3\ 003\ 0} \end{array} \quad \begin{array}{r} 6 \\ 4{,}290\overline{)2{,}871{,}799} \\ \underline{2\ 574\ 0} \\ 297\ 7 \end{array}$$

   c.
   $$\begin{array}{r} 2\ (\text{wrong}) \\ 6{,}511\overline{)19{,}590{,}211} \\ \underline{13\ 022} \\ 6\ 568 \end{array} \quad \begin{array}{r} 3 \\ 6{,}511\overline{)19{,}590{,}211} \\ \underline{19\ 533} \\ 57 \end{array}$$

T–45 Chapter 1 Basic Mathematical Operations

Use Example B to review the steps and calculations of long division. Then solve a few long-division problems to refresh the students' memories. The problem below includes all the calculations along with the check and the excess numbers for casting out nines.

```
        (0)         (0)      (0)
         738 R 270    738       0
d. 729)538,272  (0)  × 729     × 0
       510 3         6 642       0
        27 97        14 76     + 0
        21 87        516 6       0
         6 102       538,002
         5 832     +    270
           270       538,272
```

3. **Long division problems can be made shorter if both the dividend and the divisor end with a zero or with several zeroes.** Cross out the same number of zeroes in both the dividend and the divisor. If there is a remainder, you must add as many zeroes to the remainder as you crossed out in the dividend.

```
            15 0 R 3,000
a.  4,500)678,000
         45
         228
         225
          30
```

```
            43 0 R 1,000
b.  9,300)4,000,000
         3 72
         280
         279
          10
```

4. **Short division is used to solve problems with a 1-digit divisor.** All the division steps are done mentally, and each quotient figure is written as it is calculated.

Practice short division in class. Use the same steps as in long division, but write only the quotient as it is calculated.

```
        (2,235 R 4)
a.  6)13,414
```

```
        (3,973 R 1)
b.  7)27,812
```

```
        (10,973 R 8)
c.  9)98,765
```

Solutions for Part D

21. 30,442 ÷ 365
22. 45,360 ÷ 56

Copy and solve these long-division problems. Check your work by casting out nines. Show your check numbers. (Check numbers are in parentheses.)

d. (0) (4) 94 R 43 (7)
45)4,273

e. (7) (2) 1,307 R 215 (8)
736)962,167

f. (8) (5) 2,021 R 84 (3)
1,862)3,763,186

Copy and solve by short division.

g. (7) 763 R 1
5)3,816

h. (4) 271,142
3)813,426

i. $2,856.33
2)$5,712.66

WRITTEN EXERCISES

A. *Copy these long-division problems. Estimate the first quotient figure, and place it properly above the dividend.*

1. 5
911)464,511

2. 6
624)388,899

3. 5
488)2,515,122

4. 6
701)4,311,212

B. *Copy and solve these long-division problems. Check your work by casting out nines. Show your check numbers.* (Check numbers are in parentheses.)

5. (1) (8) 161 R 13 (4)
28)4,521   (3)

6. (1) (5) 140 R 12 (3)
46)6,452   (8)

7. (6) (8) 71 R 80 (8)
231)16,481   (2)

8. (0) (5) 1,787 R 236 (2)
315)563,141   (2)

9. (8) (4) 2,875 R 10 (1)
215)618,135   (6)

10. (4) (2) 1,523 R 371 (2)
841)1,281,214   (1)

11. (6) (7) 178 R 1,329 (6)
3,516)627,177   (3)

12. (8) (3) 228 R 1,534 (4)
4,166)951,382   (1)

C. *Copy and solve by short division.*

13. 388 R 1
7)2,717

14. 2,143 R 7
8)17,151

15. 7,253 R 6
9)65,283

16. 8,195 R 2
6)49,172

17. $18.95 R 7¢
8)$151.67

18. $45.78 R 1¢
5)$228.91

19. 37,865 R 3
4)151,463

20. 59,604
3)178,812

D. *Solve these reading problems.*

21. The Bible contains 30,442 verses. If you plan to read through the Bible in one year (365 days), how many verses will you need to read each day? (Round the answer up to the next whole verse.)   84 verses

22. A grain bin contains 45,360 pounds of shelled corn. At 56 pounds per bushel of shelled corn, how many bushels of corn are in the bin?   810 bushels

46   Chapter 1   Basic Mathematical Operations

23. When the Israelites left Egypt, there were 603,550 men of war in the congregation (Exodus 38:26). But when they rebelled at the border of Canaan, they had to return to the wilderness until all those men but Caleb and Joshua died. Since 603,548 men died in 38 years, what was the average number that died each year? (Round to the nearest whole number.)   15,883 men

24. The 603,550 men of war were from twelve of the thirteen tribes of Israel. (Two tribes were from Joseph.) The Levites were not counted because they did not go to battle, but rather kept the tabernacle. What was the average number of men of war in each of the twelve tribes? (Round to the nearest whole number.)   50,296 men

25. When the Israelites were about to enter Canaan 38 years later, God again told Moses to number the men of war (Numbers 26:51). The total number then was 601,730. Write this number as a Roman numeral.   DCIDCCXXX

26. In the days of King David, a census was taken against the commandment of the Lord. The number of men of war counted was 1,300,000. Write this number as a Roman numeral.   MCCC

REVIEW EXERCISES

E. Use the distributive law to find the missing numbers. *(Lesson 10)*

27. 19 × 12 = (19 × 10) + (19 × __2__)   28. 27 × 89 = (27 × 80 + (27 × __9__)

F. Solve these problems mentally. *(Lessons 7, 9)*

29. 86 − 32   54   30. 526 × 4   2,104
31. 10,000 × 537   5,370,000   32. 34 × 40   1,360

G. Write these Roman numerals as Arabic numerals. *(Lesson 3)*

33. XCII   92   34. MMCDLIX   2,459

An Ounce of Prevention

1. Students sometimes have problems remembering what to do with the excess numbers to check a division problem. Instead of repeating the division process, use the excess numbers for the inverse operation. Either way of checking is done by multiplying, but casting out nines lets you work with one-digit numbers. Multiply the excess number of the quotient by that of the divisor, and add to this product the excess number of the remainder. The result must equal the excess number of the dividend.

2. Emphasize that when zeroes are dropped in the dividend, they must be annexed to the remainder.

3. By eighth grade, the main reason for mistakes in long division is not a lack of understanding the procedure but a lack of proper care and accuracy. Emphasize casting out nines and reworking the problems if the check indicates that an answer is wrong.

4. Be careful about the amount of written work you assign. It is easy to overassign in lessons on long division.

---

23. $603{,}548 \div 38$
24. $603{,}550 \div 12$

T–47 Chapter 1 Basic Mathematical Operations

LESSON 12

Objectives

- To review using the divide-and-divide shortcut to solve division problems mentally, including *use of this method several times in the same division problem.
- To review the rules of divisibility.

Review

1. *Solve these division problems.* (Lesson 11)

   a. $65\overline{)34{,}972}$ (538 R 2)

   b. $382\overline{)267{,}988}$ (701 R 206)

   c. $576\overline{)2{,}910{,}261}$ (5,052 R 309)

2. *Use the distributive law to find the missing numbers.* (Lesson 10)

   a. $12 \times 36 = (12 \times 30) + (12 \times \_\_)$ [6]

   b. $81 \times 25 = (81 \times \_\_) + (81 \times \_\_)$
   [20, 5]

   c. $16 \times 54 = (16 \times 50) + (\_\_ \times 4)$ [16]

   d. $37 \times 78 = (37 \times \_\_) + (37 \times \_\_)$
   [70, 8]

3. *Solve these multiplication problems.* (Lesson 8)

   a. 823
       × 78
       (64,194)

   b. 5,903
       × 1,462
       (8,630,186)

4. *Find the sums, and check by using the commutative law.* (Lesson 4)

   a. 683
       + 482
       (1,165)

   b. $490.44
       + 629.16
       ($1,119.60)

12. Dividing Mentally, Part 1

Mental division, like all mental math, requires practice and perseverance. With frequent exercise, however, mental division will become very useful in everyday life.

Divide-and-Divide Method

In Lesson 10 you used the double-and-divide shortcut to solve multiplication problems mentally. Division has a shortcut with a similar name, the **divide-and-divide** method.

You saw in Lesson 11 that division problems can be expressed as fractions. The divide-and-divide shortcut is the same as reducing fractions to lowest terms or crossing out the same number of zeroes in the dividend and the divisor. *Divide-and-divide* means to divide both the dividend and the divisor by the same number.

Example A

$120 \div 24 = \underline{}$

Think: Divide both numbers by 12.

$10 \div 2 = 5$

Example B

$140 \div 35 = \underline{}$

Think: Divide both numbers by 7.

$20 \div 5 = 4$

Sometimes you can use the divide-and-divide method a second time to simplify the same problem even more.

Example C

$288 \div 32 = \underline{}$

Think: Divide both numbers by 2.

$144 \div 16 = \underline{}$

Think: Divide both numbers by 2 again.

$72 \div 8 = 9$

Rules of Divisibility

Knowing the rules of divisibility is helpful for using the divide-and-divide shortcut. Learn to recognize when the divisor and dividend have common factors.

48   Chapter 1   Basic Mathematical Operations

Divisor	Rule	Example
2	A number is divisible by 2 if it is an even number.	426 is divisible by 2 because it is an even number.
3	A number is divisible by 3 if the sum of its digits is divisible by 3.	258 is divisible by 3 because the sum of its digits (15) is divisible by 3.
4	A number is divisible by 4 if the last two digits are zeroes or are divisible by 4. Learn to recognize these multiples of 4: 52, 56, 60, 64, 68, 72, 76, 92, 96.	788 is divisible by 4 because the last two digits, 88, are divisible by 4.
5	A number is divisible by 5 if it ends with 0 or 5.	675 is divisible by 5 because it ends with 5.
6	A number is divisible by 6 if it is an even number and the sum of its digits is divisible by 3.	276 is divisible by 6 because it is even and the sum of its digits (15) is divisible by 3.
9	A number is divisible by 9 if the sum of its digits is divisible by 9.	288 is divisible by 9 because the sum of its digits (18) is divisible by 9.

CLASS PRACTICE

Say *yes* or *no* to tell whether each division will work out evenly.

a. 136 ÷ 4   yes   b. 485 ÷ 5   yes   c. 723 ÷ 6   no   d. 621 ÷ 9   yes
e. 750 ÷ 4   no   f. 3,438 ÷ 9   yes   g. 14,336 ÷ 4   yes   h. 47,996 ÷ 9   no

Solve mentally by using the divide-and-divide shortcut.

i. 198 ÷ 18   11   j. 288 ÷ 36   8   k. 256 ÷ 32   8   l. 324 ÷ 36   9
m. 112 ÷ 28   4   n. 168 ÷ 24   7   o. 432 ÷ 48   9   p. 224 ÷ 32   7

WRITTEN EXERCISES

A. Write *yes* or *no* to tell whether each division will work out evenly.

1. 208 ÷ 4   yes   2. 166 ÷ 3   no   3. 225 ÷ 5   yes   4. 495 ÷ 9   yes
5. 6,995 ÷ 2   no   6. 282 ÷ 6   yes   7. 510 ÷ 3   yes   8. 726 ÷ 4   no
9. 16,172 ÷ 6   no   10. 4,116 ÷ 6   yes   11. 3,172 ÷ 3   no   12. 27,200 ÷ 4   yes
13. 39,366 ÷ 9   yes   14. 99,995 ÷ 5   yes   15. 99,633 ÷ 3   yes   16. 21,040 ÷ 4   yes

Lesson 12   T–48

Introduction

Copy the problem 2,048 ÷ 256 on the board. Can the students find the quotient mentally?

Review the divide-and-divide method. Keep dividing both the dividend and the divisor until they can solve the problem mentally.

2,048 ÷ 256	(divide by 2)
1,024 ÷ 128	(divide by 2)
512 ÷ 64	(divide by 2)
256 ÷ 32	(divide by 2)
128 ÷ 16	(divide by 2)
64 ÷ 8	= 8

Note: At the end of the class period, hand out the sheets *Lesson 12 Speed Test Preparation* (Division).

Teaching Guide

1. **Divide-and-divide means to divide both the dividend and the divisor by the same number, the same as when reducing fractions.**

   a. 198 ÷ 22 = 99 ÷ 11 = 9
   b. 192 ÷ 24 = 96 ÷ 12 = 8
   c. 242 ÷ 22 = 121 ÷ 11 = 11

2. **Sometimes the divide-and-divide method can be used a second time to simplify the same problem even more.** Occasionally the numbers can be divided a third time, as shown in problem *b*.

   a. 648 ÷ 54 = 324 ÷ 27 = 108 ÷ 9 = 12
   b. 432 ÷ 72 = 216 ÷ 36 = 108 ÷ 18 = 56 ÷ 9 = 6

3. **The divide-and-divide shortcut becomes more useful if you remember the rules of divisibility and use them to find a common factor of the divisor and the dividend.** Discuss the chart in the pupil's lesson.

   Say yes or no to tell whether each division will work out evenly.

   a. 126 ÷ 3   (yes)
   b. 224 ÷ 4   (yes)
   c. 748 ÷ 9   (no)
   d. 1,385 ÷ 5   (yes)

Note: Watch to see that pupils are writing answers only after they have finished calculating.

Mental math skills are taught best by practice. Spend plenty of time with *Class Practice*. Have the students say each step as they divide mentally.

T–49 Chapter 1 Basic Mathematical Operations

Solutions for Part C

25. 288 ÷ 18 = 144 ÷ 9 = 48 ÷ 3
26. 90 ÷ 6 = 30 ÷ 2
27. 108 ÷ 18 = 54 ÷ 9
28. 144 ÷ 8 = 72 ÷ 4
29. 666 ÷ 12
30. 5,000 ÷ 12

B. Solve mentally by using the divide-and-divide shortcut.

17. 144 ÷ 24   6
18. 162 ÷ 18   9
19. 128 ÷ 16   8
20. 216 ÷ 18   12
21. 144 ÷ 18   8
22. 432 ÷ 36   12
23. 186 ÷ 62   3
24. 504 ÷ 24   21

C. Solve these reading problems. Try to solve problems 25–28 mentally.

25. One evening at Bible school there were 288 students in 18 classes. What was the average number of students in each class?   16 students

26. West Brook Mennonite School has 90 students in 6 classrooms. What is the average number of students per classroom?   15 students

27. One summer, Aaron worked part-time on a carpenter crew. One week he earned $108 for 18 hours of work. How much was Aaron's hourly pay? $6.00

28. One spring, the Weavers planted 144 acres of corn in 8 days. On the average, how many acres did they plant each day?   18 acres

29. The weight of the gold that came to King Solomon in one year was 666 talents (1 Kings 10:14). On the average, how many talents of gold came to Solomon each month? $55\frac{1}{2}$ talents

30. Assume that all 12 of the disciples helped to distribute food when Jesus fed the 5,000 men plus women and children. How many men would each disciple have served, to the nearest whole number?   417 men

REVIEW EXERCISES

D. Solve these division problems. *(Lesson 11)*

31. 852)194,463   228 R 207
32. 189)3,617,385   19,139 R 114

E. Use the distributive law to find the missing numbers. *(Lesson 10)*

33. 18 × 56 = (18 × 50) + (18 × 6)
34. 88 × 66 = (88 × 60) + (88 × 6)

F. Solve these multiplication problems. *(Lesson 8)*

35.   714
    × 85
   60,690

36.   904
    × 712
   643,648

G. Find the sums, and check by using the commutative law. *(Lesson 4)*

37.   285
    + 445
      730

38. $738.56
    + 714.25
    $1,452.81

50 Chapter 1 Basic Mathematical Operations

13. Dividing Mentally, Part 2

Double-and-Double Method

In Lesson 12 you used the divide-and-divide shortcut, which is the same as reducing fractions to lowest terms. The double-and-double method is the same as expanding fractions to higher terms. To use this shortcut, double both the dividend and the divisor.

The double-and-double method is especially useful for problems with the divisors 5, 50, and 500. The resulting divisors are 10, 100, and 1,000, which make division very simple.

Example A 90 ÷ 5 = ___ Think: 180 ÷ 10 = 18	**Example B** 71,000 ÷ 500 = ___ Think: 142,000 ÷ 1,000 = 142

The double-and-double shortcut is also beneficial when doubling the divisor results in a multiple of 10 or 100, as in the following examples.

Example C 180 ÷ 15 = ___ Think: 360 ÷ 30 = 12	**Example D** 10,500 ÷ 350 = ___ Think: 21,000 ÷ 700 = 30

Quadruple-and-Quadruple Method

To use the quadruple-and-quadruple shortcut, multiply both the dividend and the divisor by 4. This is especially useful if the divisor is 25, 250, or 2,500.

Example E 450 ÷ 25 = ___ Think: 1,800 ÷ 100 = 18	**Example F** 3,500 ÷ 250 = ___ Think: 14,000 ÷ 1,000 = 14

CLASS PRACTICE

Use the double-and-double shortcut to solve these problems mentally.

a. 400 ÷ 50   8   b. 215 ÷ 5   43   c. 405 ÷ 45   9   d. 12,500 ÷ 500   25
e. 475 ÷ 5   95   f. 4,800 ÷ 50   96   g. 6,000 ÷ 500   12   h. 26,000 ÷ 500   52

Use the quadruple-and-quadruple shortcut to solve these problems mentally.

i. 350 ÷ 25   14   j. 1,250 ÷ 250   5   k. 225 ÷ 25   9   l. 3,250 ÷ 250   13
m. 800 ÷ 25   32   n. 2,200 ÷ 25   88   o. 9,000 ÷ 250   36   p. 70,000 ÷ 2,500   28

WRITTEN EXERCISES

A. *Use the double-and-double shortcut to solve these problems mentally.*

1. 135 ÷ 5   27   2. 850 ÷ 5   170   3. 950 ÷ 50   19   4. 12,000 ÷ 50   240
5. 11,000 ÷ 500   22   6. 14,500 ÷ 500   29   7. 120 ÷ 15   8   8. 405 ÷ 45   9

LESSON 13

Objectives

- To review the double-and-double shortcut in division, to extend it to more difficult mental division problems, and to *use the same method for dividing by 500.

- To review the shortcut for dividing by 25, and to *use the same method for dividing by 250.

Review

1. Give *Lesson 13 Speed Test* (Division).

2. *Solve these problems mentally or horizontally.* (Lessons 5, 9, 12)

    a. 124 + 63   (187)
    b. 286 + 953   (1,239)
    c. 3 × 3,816   (11,448)
    d. 70 × 45   (3,150)
    e. 270 ÷ 18   (15)
    f. 126 ÷ 42   (3)
    g. 264 ÷ 22   (12)
    h. 328 ÷ 82   (4)

3. *Do these divisions, and check by casting out nines.* (Lesson 11)

    a. 34)52,187   (1,534 R 31)
    b. 316)87,135   (275 R 235)

Introduction

Write the problems 80 ÷ 5 = 16 and 120 ÷ 5 = 24 on the board. Can the students see any relationship between the dividend and the quotient in the first problem that is also true for the second problem? (The quotient can be found by dropping the zero from the dividend and doubling the remaining digit or digits.)

This is an application of the double-and-double division shortcut. In the first problem, 2 × 80 = 160 and 2 × 5 = 10, and 160 ÷ 10 = 16. Likewise, 2 × 120 = 240, 2 × 5 = 10, and 240 ÷ 10 = 24.

Teaching Guide

1. **To use the double-and-double method, double both the dividend and the divisor.** This method applies the same principle as expanding fractions. This method is often a shortcut if the divisor ends with a 5.

    a. 180 ÷ 15 = 360 ÷ 30 = 12
    b. 315 ÷ 35 = 630 ÷ 70 = 9
    c. 225 ÷ 45 = 450 ÷ 90 = 5
    d. 330 ÷ 55 = 660 ÷ 110 = 6

    Dropping a zero from the dividend and divisor is actually a divide-and-divide step that can be done with very little thought.

2. **The double-and-double method is especially useful for problems with the divisors 5, 50, and 500.** The resulting divisors are 10, 100, and 1,000, which make division very simple.

    a. 95 ÷ 5 = 190 ÷ 10 = 19
    b. 750 ÷ 50 = 1,500 ÷ 100 = 15

T–51 *Chapter 1 Basic Mathematical Operations*

The double-and-double shortcut is also beneficial when doubling the divisor results in a multiple of 10 or 100, as in the following examples.

a. 315 ÷ 35 = 630 ÷ 70 = 9

b. 3,600 ÷ 450 = 7,200 ÷ 900 = 8

3. **To use the quadruple-and-quadruple method, which applies the same principle, multiply both the dividend and the divisor by 4.** This shortcut is especially useful if the divisor is 25, 250, or 2,500.

a. 600 ÷ 25 = 2,400 ÷ 100 = 24

b. 2,100 ÷ 25 = 8,400 ÷ 100 = 84

An Ounce of Prevention

The vertical format of the exercises in Part C is not the usual arrangement for mental exercises. It is used here to avoid confusion with order-of-operation rules. The steps are intended to be done in the order they appear.

Mental Solutions for Part D

27. 1,800 ÷ 25 = 7,200 ÷ 100

28. 130 ÷ 5 = 260 ÷ 10

29. 260,000,000 ÷ 50 = 520,000,000 ÷ 100

30. 240 ÷ 15 = 480 ÷ 30

31. 50 × 44 = 100 × 22

32. 25 × 120 = 100 × 30

9. 360 ÷ 45   8   10. 280 ÷ 35   8   11. 240 ÷ 15   16   12. 210 ÷ 35   6

B. Use the quadruple-and-quadruple shortcut to divide each number by 25.

13. 300   12   14. 450   18   15. 625   25   16. 1,000   40
17. 1,200 48   18. 1,500 60   19. 2,000 80   20. 3,000   120

C. Find the answers mentally.

21.	6	22.	15	23.	28	24.	70	25.	12	26.	48
	× 5		− 7		− 14		÷ 10		× 9		− 14
	+ 6		× 10		÷ 7		− 4		÷ 2		÷ 2
	÷ 4		÷ 8		× 5		× 8		− 5		− 2
	× 7		+ 8		+ 20		÷ 6		÷ 7		÷ 3
	− 7		− 9		÷ 6		× 15		+ 7		× 12
	÷ 8		× 12		− 5		− 25		× 3		÷ 10
	× 4		− 12		× 6		÷ 7		− 6		÷ 2
	− 17		÷ 6		+ 7		× 5		− 6		× 3
	11		+ 8		+ 11		− 7		+ 15		− 3
			− 5		× 2		× 2		÷ 3		÷ 2
			19		÷ 12		× 0		× 4		× 4
					3		0		60		12

D. Solve these reading problems. See if you can do them all mentally.

27. The Westville congregation distributes 1,800 copies of the *Star of Hope* monthly. In October, 25 people helped in the distribution work. What was the average number of copies that each person distributed?   72 copies

28. The Millville Christian School has an enrollment of 130 students in 5 classrooms. What is the average number of students in each room?   26 students

29. The United States contains 50 states and had a population of about 260,000,000 in the 1990s. What was the average population per state?   5,200,000

30. A large congregation has 240 students in 15 Sunday school classes. What is the average number in each class?   16 students

31. Father bought 50 bales of hay that weighed an average of 44 pounds each. What was the total weight of the hay?   2,200 pounds

32. Father bought 25 calves that weighed an average of 120 pounds each. What was the total weight of the calves?   3,000 pounds

REVIEW EXERCISES

E. Solve these problems mentally or horizontally. *(Lessons 5, 9, 12)*

33. 220 ÷ 44   5   34. 112 ÷ 16   7   35. 4 × 1,943   7,772
36. 98 × 50   4,900   37. 295 + 766   1,061   38. 1,950 + 1,525   3,475

F. Do these divisions, and check by casting out nines. *(Lesson 11)*

39. 55)78,028   1,418 R 38   40. 44)25,793   586 R 9

52   *Chapter 1   Basic Mathematical Operations*

14. Reading Problems: Choosing the Correct Operation

Reading problems are a test of how well you can apply your mathematical skills to practical life. A set of reading problems may seem intimidating; but when broken down into small steps, they become much easier to solve. The challenge is to analyze a reading problem and decide what steps are needed to find the answer.

Here are three important steps in solving a reading problem.

1. Determine the actual question that the problem is asking.
2. Find the facts needed to answer the question.
3. Determine the correct operation or operations needed to answer the question.

The following lists show some of the terms typically used for the four operations in reading problems.

Addition	Multiplication	Subtraction	Division
in all	in all	how much larger	how many times larger
together	together	how much smaller	how many did each receive
altogether	altogether	find the difference	find the average
total	total		
the sum of	find the product		

In this lesson you will use number sentences to show the operations needed to solve your reading problems. Writing a number sentence is a logical approach to finding a solution. Choose a letter to represent the unknown amount that you need to find. Then write a number sentence showing how you will find that amount.

> Mark is 3 times as old as Samuel. Samuel is 6 years old and weighs 50 pounds. How old is Mark?

Step 1: Ask: **What is the actual question being asked?** The question is "How old is Mark?"

Step 2: Ask: **What facts are needed to answer the question?** The facts are that Mark is 3 times as old as Samuel, and that Samuel is 6 years old.

Step 3: Ask: **What operation or operations are needed to answer the question?** Since Mark is 3 times as old as Samuel, multiplication is needed to solve the problem.

Step 4: Write a number sentence and solve it to find the answer.

LESSON 14

Objectives

- To give practice with choosing the correct mathematical operations to solve reading problems.
- To give practice with writing number sentences to solve reading problems.

Review

1. *Solve by mental division.* (Lessons 12, 13)
   a. $255 \div 15$ (17)
   b. $850 \div 25$ (34)
   c. $216 \div 24$ (9)
   d. $775 \div 25$ (31)
   e. $660 \div 55$ (12)
   f. $540 \div 45$ (12)

2. *Use the distributive law to find the missing numbers.* (Lesson 10)
   a. $13 \times 63 = (13 \times 60) + (\_\_ \times 3)$ [13]
   b. $53 \times 14 = (53 \times \_\_) + (53 \times \_\_)$ [10, 4]

3. *Solve these subtraction problems.* (Lesson 6)

   a.   49,042   b. $342.00
       − 23,684   − 196.12
       (25,358)   ($145.88)

T–53   Chapter 1   Basic Mathematical Operations

Introduction

Read the following questions to the students. Ask them what operation they would use to answer each one. Ask them to explain why.

a. How much larger is item 1 than item 2?
   (Subtract. The question asks how much larger.)

b. How many times larger is item 1 than item 2?
   (Divide. The question asks how many times larger.)

c. What is the total weight of item 1 and item 2?
   (Add. The question asks for the total of 2 items.)

d. What is the total weight of a dozen units of item 1?
   (Multiply. The question asks for the total of 12 equal units.)

Teaching Guide

1. **The following four steps are important in solving a reading problem.**

   (1) Ask: **What is the actual question being asked?**

   (2) Ask: **What facts are needed to answer the question?**

   (3) Ask: **What operation or operations are needed to answer the question?**

   (4) Write a number sentence and solve it to find the answer. To do this, choose a letter to represent the unknown amount in the reading problem. Then write a number sentence showing how you will find that amount.

   Solve the reading problems below by following these four steps. Use the same approach to solve as many *Class Practice* problems as are needed to familiarize the students with the process.

   a. Michael picked up 12 rocks and placed them in a wheelbarrow. If the total weight of the rocks was 225 pounds, what was the average weight of each rock?
      (1) What is the actual question being asked?
         (What was the average weight of each rock?)
      (2) What facts are needed to answer the question?
         (12 rocks and 225 pounds)
      (3) What operation or operations are needed to answer the question?
         (Division is needed because 225 pounds is being divided among 12 rocks.)

Lesson 14 53

> Let m equal Mark's age. Since Mark is 3 times as old as Samuel, who is 6 years old, we can write the following number sentence.
>
> $m = 3 \times 6$
>
> $m = 18$ Mark is 18 years old.

A reading problem may contain facts that are not needed to find the answer. In the example above, Samuel's weight (50 pounds) is an unnecessary fact. Read the problem carefully, take notice of the question being asked, and use only the facts needed to answer it.

CLASS PRACTICE

Name the operation that would be used to answer each question.

a. How many times larger is the star Betelgeuse than the sun?   division

b. How many years longer was Solomon's reign over Israel than David's?   subtraction

c. At the stated price for each item, how much will 7 cost?   multiplication

d. What will be the total cost of the Bible, the concordance, and the Bible dictionary?   addition

Write a number sentence for each reading problem, and find the answer.

e. One Sunday the Spring Brook congregation distributed 225 copies of the *Star of Hope* door to door. If the average household has 3 members, how many persons (p) lived in the houses where the *Star* was placed? $p = 3 \times 225 = 675$ persons

f. Nine persons distributed the 225 copies of the *Star*. What was the average number of copies (c) that each person distributed? $c = 225 \div 9 = 25$ copies

g. The meetinghouse for the Spring Brook Mennonite Church was built mainly with volunteer labor. The total cost of materials was about $125,000. If the floor space is 3,100 square feet, what was the average cost (c) per square foot? (Round your answer to the next higher cent.) $c = \$125,000 \div 3,100 = \$40.33$

h. The Chunnel, a 50-kilometer (31-mile) tunnel under the English Channel, was opened for use in May 1995. The cost (c) of the tunnel was more than $300,000,000 per kilometer. At $300,000,000 per kilometer, what was the cost of the tunnel? $c = 50 \times \$300,000,000 = \$15,000,000,000$

WRITTEN EXERCISES

A. Name the operation that would be used to answer each question.

1. At 60 pounds per bushel, what was the total weight of the soybeans?   multiplication

2. How many times more corn did the Browns harvest than soybeans?   division

3. What was the combined enrollment in the two schools?   addition

4. How many items are in 10 cases if each case has a dozen items?   multiplication

5. How many more dozen ears of corn did Mark pick today than last week?   subtraction

6. What did the groceries in the shopping cart cost altogether? addition
7. How many quarts did Mother can per family member? division
8. How many more bushels did the Brubakers sell yesterday than today? subtraction

B. Choose the correct number sentence for each problem, and solve.

9. What is the total amount of apples (a) picked if David picked 7 bushels and Susan picked 8 bushels? 15 bushels
   - (a.) $a = 8 + 7$
   - b. $a = 8 - 7$
   - c. $a = 8 \times 7$
   - d. $a = 8 \div 7$

10. What is the difference (d) between the number of boys and girls in the classroom if there are 18 boys and 15 girls? 3 students
    - a. $d = 18 + 15$
    - (b.) $d = 18 - 15$
    - c. $d = 18 \times 15$
    - d. $d = 18 \div 15$

11. How many cans (c) of corn are in 14 cases that contain 25 cans each? 350 cans
    - a. $c = 14 + 25$
    - b. $c = 25 - 14$
    - (c.) $c = 14 \times 25$
    - d. $c = 25 \div 14$

12. What is the average weight (w) of each rabbit if 20 rabbits weigh a total of 160 pounds? 8 pounds
    - a. $w = 20 - 160$
    - b. $w = 160 + 20$
    - c. $w = 160 \times 20$
    - (d.) $w = 160 \div 20$

13. How many acres (a) of corn did Nelson harvest when he harvested 1,800 bushels at an average of 150 bushels per acre? 12 acres
    - a. $a = 1,800 + 150$
    - b. $a = 1,800 - 150$
    - c. $a = 1,800 \times 150$
    - (d.) $a = 1,800 \div 150$

14. How many pounds (p) of cherries were picked in all if 6 persons picked 40 pounds each? 240 pounds
    - (a.) $p = 6 \times 40$
    - b. $p = 6 \div 40$
    - c. $p = 40 - 6$
    - d. $p = 6 + 40$

C. Write number sentences for these reading problems, and solve. In exercises 17–26, choose your own letter for each number sentence.
(Allow variation of letters.)

15. A machine performs 9,350 foot-pounds of work each second. What is the horsepower (h) produced if one horsepower is 550 foot-pounds of work per second?
    $h = 9,350 \div 550 = 17$ horsepower

16. The city of St. Augustine, Florida, was established in 1565. Find the age (a) of this city in the year 2000.
    $a = 2000 - 1565 = 435$ years

17. In a census taken while David was king, it was found that the 5 sons of Bela, who was Benjamin's son, had a posterity numbering 22,034 (1 Chronicles 7:7). What was the average number of descendants from each of Bela's sons, to the nearest whole number?
    $d = 22,034 \div 5 = 4,407$ descendants

(4) Write a number sentence and solve it to find the answer. (Let n equal the average weight of each rock. $n = 225 \div 12 = 18\frac{3}{4}$ lb.)

b. The estimated cost to put up a church building was $115,000. Due to unanticipated expenses, the final cost was $125,000. What was the amount of the building cost overrun?

(1) What is the actual question being asked?
(What was amount of the building cost overrun?)

(2) What facts are needed to answer the question?
($115,000 and $125,000)

(3) What operation or operations are needed to answer the question?
(Subtraction is needed because the answer will be the difference between the estimate of $115,000 and the final cost of $125,000.)

(4) Write a number sentence and solve it to find the answer. (Let c equal the difference in cost. $c = \$125,000 - \$115,000 = \$10,000$)

2. **A reading problem may contain facts that are not needed to find the answer.** Read the problem carefully, take notice of the question being asked, and use only the facts needed to answer the question.

18. Uncle Dennis purchased some second-hand furniture at an auction. He paid $120 for a bed, $67 for a dresser, and $162.50 for a table. He had $450.23 in his checking account at the time. What was his total bill? $b = \$120 + \$67 + \$162.50 = \$349.50$

19. Aunt Marie canned 192 quarts of applesauce and 120 quarts of cherries. At 3 pounds of apples per quart of applesauce, how many pounds of apples did she use? $p = 192 \times 3 = 576$ pounds

20. If a van travels 16 miles on each gallon of gasoline, how many gallons will it use to travel 700 miles? (Round your answer to the next higher gallon.) $g = 700 \div 16 = 44$ gallons

21. Noah's ark had a length of 300 cubits, a width of 50 cubits, and a height of 30 cubits (Genesis 6:15). If the ark was a rectangular solid, its volume was equal to the product of the length, the width, and the height. What was the volume of the ark if it was a rectangular solid? (Label your answer "cubic cubits.") $v = 300 \times 50 \times 30 = 450{,}000$ cubic cubits

22. For the month of October, Miller Electric had a beginning checkbook balance of $4,289.33. The income deposited that month amounted to $7,454.25, and the total expenses paid were $9,684.49. What was the ending balance for October? $b = \$4{,}289.33 + \$7{,}454.25 - \$9{,}684.49 = \$2{,}059.09$

23. The Upper Dale Mennonite Church building was constructed at a cost of $130,000. With an area of 3,168 square feet, find the cost per square foot (to the nearest cent). $c = \$130{,}000 \div 3{,}168 = \$41.04$

24. David built a 150-foot fence along one edge of his property at a cost of $5.75 per foot. What was the total cost of the fence? $c = 150 \times \$5.75 = \$862.50$

25. One historian estimates that 3,000 to 5,000 Mennonites immigrated to America between 1707 and 1754. If about 4,000 Mennonites moved to America in that 47-year period, what was the average number of Mennonite immigrants each year? (Drop any remainder in your answer.) $i = 4{,}000 \div 47 = 85$ immigrants

26. The same historian estimates that 3,000 Amish people immigrated to America in the years 1815 through 1880. What was the average number of Amish immigrants per year in that 65-year period? (Drop any remainder in your answer.) $i = 3{,}000 \div 65 = 46$ immigrants

REVIEW EXERCISES

D. Solve by mental division. *(Lessons 12, 13)*

27. 168 ÷ 24   7   28. 308 ÷ 28   11   29. 360 ÷ 15   24   30. 400 ÷ 25   16

E. Use the distributive law to find the missing numbers. *(Lesson 10)*

31. 24 × 23 = (24 × 20) + (24 × __3__)   32. 67 × 120 = (67 × 100) + (67 × __20__)

F. Solve these subtraction problems. *(Lesson 6)*

33.   4,603
  − 1,862
    2,741

34. $23,085.55
 − 16,812.35
   $6,273.20

56   Chapter 1   Basic Mathematical Operations

15. Chapter 1 Review

A. Copy and solve. Write the correct term for each blank. *(Lessons 4, 6, 8, 11)*

Addition: Check your work by using the commutative law.

1. 563   addend   2.   6,834   addend
   +935   addend   +7,523   addend
   1,498 sum   14,357   sum

Subtraction: Check your work by addition.

3. 7,523 minuend   4. 23,161   minuend
   −4,346 subtrahend   −16,345   subtrahend
   3,177   difference   6,816   difference

Multiplication: Check your work by casting out nines.

5.   864   multiplicand 6.   637   multiplicand
    × 87   multiplier   × 83   multiplier
   75,168 product   52,871 product

Division: Check your work by casting out nines.

   quotient 127 R 19 remainder   quotient 88 R 44 remainder
7. divisor 45)5,734 dividend   8. divisor 86)7,612 dividend

B. Solve by horizontal computation or short division. *(Lessons 5, 6, 9, 11)*

9. 5,585 + 385   5,970   10. 6,356 + 868   7,224
11. 9,346 − 4,252   5,094   12. 34,266 − 16,755   17,511
13. 3 × 32,263   96,789   14. 6 × 34,372   206,232

       $35.32 R 4¢ $69.62 R 1¢
15. 7)$247.28   16. 4)$278.49

C. Solve mentally. Follow the directions for each set. *(Lesson 4)*

Look for pairs of addends that equal 10.
17. 4 + 3 + 7 + 6 + 2 + 8 + 1 + 9   40   18. 5 + 6 + 4 + 8 + 4 + 6 + 5 + 5   43

Look for sets of factors that equal multiples of 10.
19. 3 × 5 × 7 × 4   420   20. 7 × 3 × 8 × 5   840

D. Do these additions and subtractions mentally. *(Lessons 5, 7)*

21. 456 + 871   1,327   22. 1,830 + 1,960   3,790   23. 1,457 + 1,284   2,741
24. 485 − 239   246   25. 1,275 − 325   950   26. 1,650 − 775   875

LESSON 15

Objective

- To review the material taught in Chapter 1 (Lessons 1–14).

Teaching Guide

1. Lesson 15 reviews the material taught in Lessons 1–14. Be sure to discuss any parts with which your students had special difficulty. The problems in each lesson are arranged so that each odd-numbered problem is of the same type and difficulty as the next even-numbered problem. Thus 1 and 2 are a pair, 3 and 4 are a pair, and so on. This is especially useful in review lessons because you can assign either the even-numbered or the odd-numbered problems, and use the others for class practice. If you do assign all the exercises, page through the lessons in Chapter 1 and select problems for class review.

    This odd-even pattern applies for all the review lessons in this text.

2. **Math problems lend themselves well to board work.** Review lessons are especially good for board work because you usually do not need to demonstrate how to do the problems on the board yourself.

3. **Be sure to review the new concepts introduced in this chapter.** The main ones are shown below, along with the exercises in this lesson that review those concepts.

Lesson number and new concept	Exercises in Lesson 15
1—Numbers to the quadrillions' place.	43, 44
4—Commutative and associative laws.	53, 54
10—Distributive law.	55, 56
12—Using the divide-and-divide method several times in the same problem.	38

T-57   Chapter 1   Basic Mathematical Operations

Answers for Exercises 43–46

43. Sixty-seven quadrillion, eight hundred trillion, nine billion, eight hundred million

44. 68,000,000,002,000,000

45. 60,000,000,000,000 (60 trillion)

46. 80,000,000,000,000 (80 trillion)

E. Multipy mentally. *(Lesson 9)*

27. 1,000 × 2,761   2,761,000
28. 10,000 × 84,346   843,460,000
29. 20 × 14   280
30. 60 × 12   720

F. Multiply mentally, using the method indicated. *(Lessons 9, 10)*

Use the left-to-right method.
31. 6 × 256   1,536
32. 4 × 1,232   4,928

Use the distributive law.
33. 15 × 13   195
34. 26 × 22   572

Use the double-and-divide shortcut.
35. 28 × 16   448
36. 35 × 22   770

G. Divide mentally, using the method indicated. *(Lessons 12, 13)*

Use the divide-and-divide method.
37. 216 ÷ 24   9
38. 256 ÷ 16   16

Use the double-and-double method.
39. 20,000 ÷ 500   40
40. 255 ÷ 15   17

Use the quadruple-and-quadruple method.
41. 700 ÷ 25   28
42. 3,000 ÷ 25   120

H. Follow the directions for these numerals. *(Lessons 1–3)* (See facing page)

43. Use words to write the number 67,800,009,800,000,000.
44. Use digits to write this number: Sixty-eight quadrillion, two million.
45. Write the value of the 6 in 162,192,000,000,000.
46. Write the value of the 8 in 100,080,000,111,000,000.
47. Round 2,776,414 to the nearest million.   3,000,000
48. Round 481,634,151 to the nearest ten million.   480,000,000
49. Write $\overline{\text{MCM}}$DCLXXXIII as an Arabic numeral.   1,900,683
50. Write $\overline{\text{MDCMXXXIII}}$ as an Arabic numeral.   1,500,933
51. Write 2,267 as a Roman numeral.   MMCCLXVII
52. Write 125,000 as a Roman numeral. $\overline{\text{CXXV}}$

I. Write the missing word or number. *(Lessons 4, 10)*

53. The equation 7(6 × 2) = (7 × 6)2 illustrates the ___ law.   associative
54. The equation 5 + 4 = 4 + 5 illustrates the ___ law.   commutative
55. (10 × 5) + (8 × 5) = _18_ × 5
56. (20 × 9) + (_5_ × 9) = 25 × 9

58   Chapter 1   Basic Mathematical Operations

J. Do these exercises. *(Lessons 8, 12)*

57. Is 43,515,121 divisible by 9?   no
58. Is 2,002,512 divisible by 4?   yes
59. Estimate the product for exercise 5. Show your work. $90 \times 900 = 81,000$
60. Estimate the product for exercise 6. Show your work. $80 \times 600 = 48,000$

K. Write a number sentence for each statement. You do not need to find the solution. *(Lesson 14)*

61. The number of pecks (p) of apples picked is equal to 12 bushels times 4 pecks in a bushel. $p = 12 \times 4$
62. The number of miles (m) driven is equal to 7 hours times an average rate of 54 miles per hour. $m = 7 \times 54$

L. Write a number sentence for each problem, and solve. *(Lesson 14)*

63. In a recent year, the United States Postal Service delivered nearly 580,000,000 pieces of mail daily. If this mail was delivered by 690,000 employees, what was the average number of pieces (p) delivered by each employee? (Drop any remainder.)
    $p = 580,000,000 \div 690,000 = 840$ pieces
64. How many pieces (p) of mail were delivered in a year if 580,000,000 pieces were delivered each day for 295 days? $p = 580,000,000 \times 295 = 171,100,000,000$ pieces
65. In the same year, the postal service delivered 92,100,000,000 pieces of first-class mail and 10,300,000,000 pieces of second-class mail. How many more pieces (p) of first-class mail were handled than pieces of second-class mail?
    $p = 92,100,000,000 - 10,300,000,000 = 81,800,000,000$ pieces
66. The postal service also reported that it had more than 39,000 post offices, stations, and branches. What is the average number of pieces (p) of first-class mail handled in each post office if 92,100,000,000 pieces went through 39,000 post offices? (Answer to the nearest whole number.) $p = 92,100,000,000 \div 39,000 = 2,361,538$ pieces
67. In 1860, the Pony Express carried mail from Saint Joseph, Missouri, to Sacramento, California, at a rate of $10 per ounce. At that rate, what would be the postage (p) on a 40-ounce textbook? (In the 1990s, 30 ounces of first-class mail could be mailed anywhere in the United States for $4.00, and textbooks could be mailed for $2.28.)
    $p = 40 \times \$10 = \$400$
68. In the 1990s, a 32-ounce package could be mailed anywhere in the United States for $3.00. What was the cost (c) per ounce of mailing a 32-ounce package? (Round your answer to the next higher cent.) $c = \$3.00 \div 32 = 10¢$

---

16. Chapter 1 Test

LESSON 16

Objective

- To test the students' mastery of the concepts in Chapter 1.

Teaching Guide

1. Correct Lesson 15.
2. Review any areas of special difficulty.
3. Administer the test.

Following are a few pointers for testing:

1. Only the test, scratch paper, pencils, and an eraser should be on each student's desk.
2. Steps should be taken to minimize the temptation for dishonesty and the likelihood of accidentally seeing other students' answers. Following are some suggestions.
   a. Desk tops should be level. If the desks are very close to each other, have the students keep their work directly in front of them on the desk.
   b. Students should not look around more than necessary during test time.
   c. No communication should be allowed.
   d. As a rule, students should remain seated during the whole test period. It is a good idea to sharpen a few extra pencils and have them on hand.
   e. Students should hand in their tests before going on to any other work.
3. Encourage the students to do their work carefully and to go back over it if they have time. Do not allow them to hand in their tests too soon. On the other hand, some students are so meticulous that they can hardly finish their tests. If you have this problem, set a time when you will collect all the tests. Once 90 percent of the tests are completed, the rest of them should be finished in the next five or ten minutes. Of course, there are exceptions for slower students.
4. A test is different from homework. Students should realize that they must rely on their own knowledge as they work. The teacher should not help them except to make sure they understand all instructions.

Evaluating the Results

1. If you check the tests in class, have the students check each other's work. Spot-check the corrected tests.
2. Tests are valuable tools in determining what the students have grasped. Are there any places where the class is uniformly weak? If so, reinforcement is needed.
3. One effective way to discover the general performance of the class is to find the class median. This is done by arranging all the scores in order from highest to lowest and finding the middle score. If there are an even number of students, find the average of the two middle scores.

Value of the Test Score

Test scores should carry considerable weight in determining report card grades. It is suggested that the test grade average have a value at least equal to that of the homework grade average. That is, the report card grade would be the homework grade average plus the test grade average, divided by 2.

Disposition of the Corrected Tests

1. As a rule, students should have the privilege to see their tests. Review any weak points, and answer any questions about why an answer is wrong.

2. The teacher may use his discretion about whether the student should be allowed to keep his test permanently. Some teachers prefer to collect them again so that the students' younger siblings will have no chance of seeing the tests in later years.

"Who hath measured the waters in the hollow of his hand, and meted out heaven with the span?" Sometimes this verse is used to describe God's greatness, in that He can hold the oceans in His palm, and He can span the heavens with the reach of His hand. Indeed, He is a mighty God!

The context also implies man's smallness. Who could begin to measure the oceans with a teaspoon? Who could measure the heavens with a ruler? Man's measuring systems pale in the presence of the infinity of God.

Chapter 2
English and Metric Measures

We use measures to express quantities and distances. For example, by giving the length of a certain table as 6 feet, we indicate its size in such a way that other people know just how long it is. Without units of measure such as the foot, it would be almost impossible to indicate exact sizes and amounts.

God established some measures of time when He said, "Let there be lights in the firmament of the heaven to divide the day from the night; and let them be for signs, and for seasons, and for days, and years" (Genesis 1:14). Soon people developed systems of measure to use in building, selling, and other activities of life. At first they used nonstandard units such as the cubit, which is the length of the forearm. But since these were not exact, people eventually developed standard units of measure. These are units such as the foot and the yard, which have a fixed size.

Two main systems of measure are used in the world today. The United States and a few other nations use the English system, which developed in Europe during the Middle Ages. Most other nations use the International System of Measures, often called the metric system. This chapter discusses both of these systems.

Who hath . . .
meted out heaven with the span?   *Isaiah 40:12*

17. English Units of Linear Measure and Weight

Units of Linear Measure

Linear measure is the measure of distance. It is often used in construction, sewing, and transportation. Common English units of linear measure are shown in the table below.

English Units of Linear Measure

(Know these by memory.)
1 foot (ft.)	=	12 inches (in.)	(Less common units)		
1 yard (yd.)	=	3 feet	1 fathom	=	6 feet
1 yard	=	36 inches	1 rod	=	$16\tfrac{1}{2}$ feet
1 mile (mi.)	=	5,280 feet	1 league	=	3 miles
1 mile	=	1,760 yards			

Units of Weight

In finding weight, we measure the force of gravity on an object in terms of standard units. Many products such as groceries and feed are sold according to weight.

The ounce, the pound, and the ton are the three English units of weight in common use today. (See the table.) The long ton is used to weigh minerals such as iron ore and coal when they are mined. The short ton is so commonly used in the United States that it is simply referred to as the ton.

English Units of Weight

(Know these by memory.)
1 pound (lb.)	=	16 ounces (oz.)	(Less common unit)		
1 ton (short ton)	=	2,000 pounds	1 long ton	=	2,240 pounds

The rules for finding equivalent measures are the same for all measures. To find the equivalent amount in a larger unit of measure, divide by the number of smaller units in the larger unit. To find the equivalent amount in a smaller unit of measure, multiply by the number of smaller units in the larger unit.

> **To change from larger units to smaller units, multiply.**
> **To change from smaller units to larger units, divide.**

LESSON 17

Objectives

- To review the concept of standard units of measure.
- To review these English units of linear measure.
  - inch
  - foot
  - yard
  - mile
- To review English units of weight (ounce, pound, ton), and to teach that *1 long ton = 2,240 pounds. (On the chapter test, it is assumed that students have memorized all the common English equivalents presented in this chapter.)

Review

1. *Solve by mental division.* (Lesson 13)
   a. 245 ÷ 5 (49)
   b. 540 ÷ 45 (12)
   c. 32,000 ÷ 50 (640)
   d. 850 ÷ 25 (34)

2. *Solve these multiplication problems mentally.* (Lesson 9)
   a. 143 × 5 (715)
   b. 930 × 7 (6,510)
   c. 716 × 30 (21,480)
   d. 840 × 80 (67,200)

3. *Write these numbers, using words.* (Lesson 1)
   a. 35,000,088,620
      (thirty-five billion, eighty-eight thousand, six hundred twenty)
   b. 73,000,540,000,000,000
      (seventy-three quadrillion, five hundred forty billion)

4. *Write these numbers, using digits.* (Lesson 1)
   a. four hundred fifty-one billion, two hundred nineteen thousand
      (451,000,219,000)
   b. sixty-six quadrillion, eight hundred three trillion
      (66,803,000,000,000,000)

Introduction

Do the students know why the unit called the foot has that name? They should remember from former years that the foot received its name from the length of a man's foot.

What difficulties would arise in purchasing land if it was measured by the length of men's feet? Perhaps you and one of the students could each measure in "feet" the length of the room. Is it the same? Buying a plot of land measuring 300 by 500 feet of this type could cause serious disagreements. To avoid such problems, the foot was determined to be the length of the king's foot.

Although this would be more satisfactory, what was the potential problem? When a new king came to the throne, very likely a new "foot" was instituted. Therefore, all previous land deeds would either need to be recalculated using the new king's foot or else have noted exactly how long the previous king's foot was.

Because of the confusion which resulted from having a changeable foot, a standard foot was eventually established, which was maintained from generation to generation and from century to century. Because the foot and all other recognized measurements are now standard, it is possible to know what is meant by a foot or a pound.

T–61   Chapter 2   *English and Metric Measures*

Note the following points in the introduction to the chapter.

Units of standard size are used to measure quantities. By comparing a chalkboard with a yardstick, we can determine that it is four yards long, a measure we can understand. (Measure your chalkboard.) If time permits, you could try to visualize what the students' lives would be like without measures.

Standard units of measure have a predetermined size. Present linear measures are based on the meter, which in the 1790s was defined as one ten-millionth of the distance from the equator to the North Pole. More recently it was defined as the distance light travels in a vacuum in 1/299,792,458 of a second.

Teaching Guide

1. **Linear measure is the measure of distance.** Review the common English units of linear measure and their abbreviations as shown in the lesson. Students are to know by memory the relationships between the common units.

2. **In finding weight, we measure the force of gravity on an object in terms of standard units.** Three English units of weight are in general use today: the ounce, the pound, and the ton. The long ton is used to weigh minerals, such as iron ore and coal, when they are mined. The short ton is so commonly used in the United States that it is simply referred to as the ton. Students are to know by memory the relationships between the common units.

3. **To find equivalent measures, follow these two rules.**
   To change from larger units to smaller units, multiply.
   To change from smaller units to larger units, divide.

   Find these equivalents.

   a. 132 in. = ___ ft.
   (Think: inches to feet; smaller unit to larger unit; divide.
   132 ÷ 12 = 11 ft.)

   b. 350 yd. = ___ ft.
   (Think: yards to feet; larger unit to smaller unit; multiply.
   3 × 350 = 1,050 ft.)

   c. 52,800 ft. = ___ mi. (10)

   d. 112 oz. = ___ lb. (7)

   e. 28,000 lb. = ___ tons (14)

   f. 65 ft. = ___ in. (780)

   g. 25 lb. = ___ oz. (400)

   h. 45 tons = ___ lb. (90,000)

   i. 34 in. = ___ ft. ($2\frac{10}{12} = 2\frac{5}{6}$)

   j. 8 lb. 5 oz. ___ lb. ($8\frac{5}{16}$)

Lesson 17

Example A	Example B
9 ft. = ___ in. 9 × 12 = 108 in.	9,000 lb. = ___ tons 9,000 ÷ 2,000 = $4\frac{1,000}{2,000}$ = $4\frac{1}{2}$ tons
Example C	Example D
15 ft. 9 in. = ___ in. 15 × 12 = 180, + 9 = 189 in.	8 lb. 12 oz. = ___ lb. 8 lb. 12 oz. = $8\frac{12}{16}$ lb. = $8\frac{3}{4}$ lb.

CLASS PRACTICE

Find these equivalents.

a. 4 ft. = __48__ in.
b. 6 lb. = __96__ oz.
c. 24 tons = ____ lb.   48,000
d. 168 in. = __14__ ft.
e. 2 lb. 9 oz. = __41__ oz.
f. 27 yd. 1 ft. = __82__ ft.
g. 3 mi. 180 ft. = ____ ft.   16,020
h. 6 tons 1,200 lb. = ____ lb.   13,200

WRITTEN EXERCISES

A. *Write the abbreviations for these units by memory.*

1. inch   in.
2. foot   ft.
3. yard   yd.
4. mile   mi.

B. *Write these equivalents by memory.*

5. 1 ft. = __12__ in.
6. 1 yd. = __36__ in.
7. 1 mi. = ____ ft.   5,280
8. 1 yd. = __3__ ft.
9. 1 in. = $\frac{1}{12}$ ft.
10. 1 ft. = $\frac{1}{3}$ yd.
11. 1 lb. = __16__ oz.
12. 1 ton = ____ lb.   2,000
13. 1 oz. = $\frac{1}{16}$ lb.
14. 1 lb. = $\frac{1}{2000}$ ton

C. *Find these equivalents.*

15. 5 yd. = __15__ ft.
16. 75 ft. = __25__ yd.
17. 4 mi. = ____ yd.   7,040
18. 14 mi. = ____ yd.   24,640
19. 31,680 ft. = __6__ mi.
20. 324 in. = __27__ ft.
21. 15 lb. = __240__ oz.
22. 384 oz. = __24__ lb.
23. 66,000 lb. = __33__ tons
24. 8 lb. = __128__ oz.
25. 3,500 yd. = ____ ft.   10,500
26. 5 ft. 6 in. = __66__ in.
27. 4 tons 1,800 lb. = $4\frac{9}{10}$ tons
28. 9 lb. 7 oz. = $9\frac{7}{16}$ lb.
29. 14 tons 700 lb. = $14\frac{7}{20}$ tons
30. 25 yd. 2 ft. 9 in. = __933__ in.

62   Chapter 2   English and Metric Measures

D. Solve these reading problems.

31. Daniel 4 describes how the Lord punished Nebuchadnezzar for his great pride in his capital city, Babylon. Archaeologists have discovered that the walls Nebuchadnezzar built around Babylon were 45.5 yards thick. How many feet thick were the walls?
    136.5 feet

32. Archaeologists also discovered the ruins of Nebuchadnezzar's palace. The great banquet hall was 168 feet long. How many yards is that?
    56 yards

33. When the Conestoga Drive Mennonite Church was moved to a higher location to avoid flood damage, the engineer estimated the weight of the building to be 50 tons. How many pounds is that?
    100,000 pounds

34. The Reedville congregation is planning to build a new meetinghouse. The architect is to design an auditorium having 2 sections of benches with 16 benches per section. Each of the 32 benches is to be 18 feet long. How many people can the benches seat if each person has 24 inches of seating space? (Try to solve this problem mentally.)
    288 people

35. The meetinghouse will be 50 feet wide. How many rows of 48-inch-wide plywood will it take to cover the width of the church floor? (Count a part row as a whole row.)
    13 rows

36. A parking lot has room for 4 rows of cars. Three of the rows each have 24 parking spaces plus 3 spaces marked for handicapped drivers. The fourth row has 22 parking spaces. What is the total number of parking spaces?
    103 spaces

REVIEW EXERCISES

E. Solve by mental division. *(Lesson 13)*

37. 470,000 ÷ 500   940
38. 3,250 ÷ 25   130

F. Solve these multiplication problems mentally. *(Lesson 9)*

39. 463 × 50   23,150
40. 1,335 × 60   80,100

G. Write these numbers using words. *(Lesson 1)*

41. 31,532,000,000   Thirty-one billion, five hundred thirty-two million
42. 14,280,560,000,000,000   Fourteen quadrillion, two hundred eighty trillion, five hundred sixty billion

H. Write these numbers, using digits. *(Lesson 1)*

43. Twenty-one trillion, eight hundred forty-two thousand   21,000,000,842,000
44. One quadrillion, fifteen billion, nine hundred million   1,000,015,900,000,000

Further Study

The English system of measures includes a considerable number of archaic units. We are familiar with several of these because they are still used in some old land deeds (such as the rod or perch). Others were in general use during the seventeenth century and are mentioned in the King James Bible.

Following is a list of archaic English linear measures. Notice the applications for each. Also notice that because 1 nautical mile equals $\frac{1}{60}$ of a degree, it equals 1 minute of the circumference of the earth at the equator.

Other English Units of Linear Measure

General

1 rod, perch, or pole	= $16\frac{1}{2}$ feet
1 furlong*	= 40 rods ($\frac{1}{8}$ mile)
1 league	= 3 miles

Surveyor's Units

1 link	= 7.92 inches
1 chain	= 66 feet

Engineer's Units

1 link	= 1 foot
1 chain	= 100 feet

Nautical Units

1 span	= 9 inches
1 fathom	= 8 spans (6 feet)
1 cable's length	= 120 fathoms (720 feet)
1 nautical mile	= 1.1516 statute miles

(also called geographic mile or international mile)

1 nautical league	= 3.45 miles (3 nautical miles)
1 degree	= 69.169 miles (60 nautical miles)

*This is an archaic measure of the English system, equivalent to about 660 feet. The word *furlong* was used in the King James translation of the Bible for the Greek measure *stadion,* which is equivalent to about 606 feet, and closer to $\frac{1}{9}$ mile. This math series teaches the 606-foot furlong as a Bible measure.

English measures also include several systems of weight: avoirdupois weight, troy weight, and apothecaries' weight. Gold is measured by troy weight. The ounce and pound in the troy system are not the same as those in the avoirdupois system. Apothecaries' weight differs from troy weight only in that 20 grains = 1 scruple and 3 scruples = 1 dram.

Avoirdupois Weight

1 dram	= 27.24275 grains
1 ounce	= 16 drams
1 pound	= 16 ounces
1 hundredweight	= 100 pounds
1 short ton	= 2,000 pounds

Troy Weight

1 carat	= 3.086 grains
1 pennyweight	= 24 grains
1 troy ounce	= 24 pennyweight
1 troy pound	= 12 troy ounces

In addition, the English system includes the following units of weight.

1 stone	= 14 pounds
1 hundredweight	= 112 pounds
1 long ton	= 2,240 pounds

Solutions for Part D

31. 45.5×3

32. $168 \div 3$

33. $50 \times 2{,}000$

34. $32 \times 18 \div 2 = 32 \times 9$

35. 50 ft. $\div$ 4 ft. = $12\frac{1}{2}$

36. $3(24 + 3) + 22$

T-63 Chapter 2 English and Metric Measures

LESSON 18

Objective

- To review the English units of capacity listed below. (The equivalents in the lesson are to be memorized.)

  liquid measure
      teaspoon
      tablespoon
      fluid ounce
      cup
      pint
      quart
      gallon

  dry measure
      pint
      quart
      peck
      bushel

Review

1. *Find these equivalents.* (Lesson 17)
   a. 3,520 yd. = ___ mi. (2)
   b. 20 lb. = ___ oz. (320)
   c. 3 mi. 500 ft. = ___ ft. (16,340)
   d. 5 yd. 2 ft. = ___ yd. ($5\frac{2}{3}$)

2. *Round each number as indicated.* (Lesson 2)
   a. 35,601 to the nearest ten thousand (40,000)
   b. 12,314,623 to the nearest million (12,000,000)

3. *Give the place value and the value of each underlined digit.* (Lesson 2)
   a. 2<u>2</u>1,684,834,000
      (ten billions; 20,000,000,000)
   b. <u>4</u>,032,583,000,000,000
      (quadrillions; 4,000,000,000,000,000)

4. *Tell what mathematical operation is needed to solve each problem.* (Lesson 14)
   a. The Deer Run Mennonite School is 35 feet wide. The length is 20 feet more than twice the width. How long is the building?
      (multiplication, addition)
   b. The school property is a rectangle that is twice as long as it is wide. The property is 450 feet long. How wide is it? (division)
   c. The parking area is 40 feet longer than it is wide. If the length is 75 feet, how wide is it? (subtraction)

18. English Units of Capacity

Many things are sold by capacity, the amount of space they fill. The units of liquid and dry measure are the ones most frequently used in the English system of measure for capacity.

Units of Liquid Measure

The common units of liquid measure are the pint, the quart, and the gallon. They are used to measure liquids such as water, milk, and gasoline.

The pint, quart, and gallon are also known as household measures. Smaller household measures include the cup, the tablespoon, and the teaspoon. Household measures are often used for both liquid and dry items such as milk, sugar, and flour. The fluid ounce is also in common use, but it is actually an apothecaries' unit of capacity.

English Units of Liquid Measure
(Know these by memory.)
1 tablespoon (tbsp.)	=	3 teaspoons (tsp.)
1 fluid ounce (fl. oz.)	=	2 tablespoons
1 cup	=	8 fluid ounces
1 pint (pt.)	=	2 cups
1 quart (qt.)	=	2 pints
1 gallon (gal.)	=	4 quarts

Units of Dry Measure

Dry measure is used for solids such as fresh fruits, vegetables, and grains. The pint, the quart, the peck, and the bushel are units of dry measure.

The pint and the quart are units of both liquid and dry measure. However, they do not have the same volume. The dry pint and the dry quart are slightly larger than the liquid pint and the liquid quart.

English Units of Dry Measure
(Know these by memory.)
1 quart	=	2 pints
1 peck (pk.)	=	8 quarts
1 bushel (bu.)	=	4 pecks

The rules for changing units of weight and linear measure also apply to units of capacity.

> **To change from larger units to smaller units, multiply.**
> **To change from smaller units to larger units, divide.**

64   Chapter 2   English and Metric Measures

Example A	**Example B**
7 gal. = ___ qt. 7 × 4 = 28 qt.	44 pk. = ___ bu. 44 ÷ 4 = 11 bu.
Example C	**Example D**
7 pk. 3 qt. = ___ qt. 7 × 8 = 56, + 3 = 59 qt.	8 bu. 2 pk. = ___ bu. 8 bu. 2 pk. = $8\frac{2}{4}$ bu. = $8\frac{1}{2}$ bu.

To find the equivalent for a measure other than the next larger or smaller unit, calculate the equivalent one step at a time.

Example E	**Example F**
8 gal. = ___ pt. 8 × 4 = 32 qt., × 2 = 64 pt.	48 qt. = ___ bu. 48 ÷ 8 = 6 pk., ÷ 4 = $1\frac{1}{2}$ bu.

CLASS PRACTICE

Give these equivalents by memory.

a. 1 cup = __8__ fl. oz.   b. 1 pk. = __8__ qt.   c. 1 tbsp. = __3__ tsp.
d. 1 bu. = __4__ pk.   e. 1 fl. oz. = __2__ tbsp.   f. 1 gal. = __4__ qt.

Find these equivalents.

g. 24 qt. = __48__ pt.   h. 11 tbsp. = __33__ tsp.
i. 7 pt. = __14__ cups   j. 8 bu. = __32__ pk.
k. 6 cups 3 fl. oz. = __51__ fl. oz. l. 10 gal. 1 qt. = __41__ qt.
m. 58 cups = __29__ pt.   n. 72 qt. = __9__ pk.

WRITTEN EXERCISES

A. *Write the abbreviations for these units by memory.*

1. teaspoon   tsp.   2. tablespoon   tbsp.   3. fluid ounce   fl. oz.
4. pint   pt.   5. quart   qt.   6. gallon   gal.
7. peck   pk.   8. bushel   bu.

B. *Write these equivalents by memory.*

9. 1 tbsp. = __3__ tsp.   10. 1 fl. oz. = __2__ tbsp.   11. 1 cup = __8__ fl. oz.
12. 1 pt. = __2__ cups   13. 1 qt. = __2__ pt.   14. 1 gal. = __4__ qt.
15. 1 pk. = __8__ qt.   16. 1 bu. = __4__ pk.   17. 1 pk. = __$\frac{1}{4}$__ bu.
18. 1 qt. = __$\frac{1}{4}$__ gal.   19. 1 fl. oz. = __$\frac{1}{8}$__ cup   20. 1 cup = __$\frac{1}{2}$__ pt.

Lesson 18 T-64

Introduction

By what units of measure are the following items usually sold?

sewing thread and yarn (yard)

medicine (fluid ounce)

fabric (yard—the linear yard, not the square yard)

coal (ton)

milk (gallon, liquid quart, hundred weight)

salad dressing (liquid quart)

peaches (bushel)

strawberries (dry quart)

Point out that the pint and the quart can be either liquid or dry measures. Milk and salad dressing are measured in liquid quarts. Strawberries and blueberries are measured in dry quarts. These dry units are larger than the liquid units. For example, if a quart box used to measure strawberries were sealed and filled with milk, it would hold more than a liquid quart of milk.

Teaching Guide

1. **Many things are sold by capacity, the amount of space they fill.** The units of liquid and dry measure are the ones most frequently used in the English system of measure for capacity.

2. **The common units of liquid measure are the pint, the quart, and the gallon.** They are used to measure liquids such as water, milk, and gasoline. The pint, quart, and gallon are also known as household measures—a group of units that include the cup, the tablespoon, and the teaspoon.

   Review the units in the lesson. You might wish to mention that there is a small liquid measure called the gill, which is equal to $\frac{1}{4}$ pint or $\frac{1}{2}$ cup. See *Further Study* for more types of measures.

3. **The common units of dry measure are the pint, the quart, the peck, and the bushel.** These are often used to measure fresh fruits and vegetables. Be sure students understand that the pint and the quart are both liquid and dry measures. However, they do not have the same volume. The dry pint and the dry quart are slightly larger than the liquid pint and the liquid quart.

4. **To find equivalent measures, remember these two rules.**

   To change from larger units to smaller units, multiply.

   To change from smaller units to larger units, divide.

   Find these equivalents.

   a. 24 qt. = ___ gal. (6)

   b. 56 qt. = ___ pk. (7)

   c. 3 cups = ___ fl. oz. (24)

   d. 40 pk. = ___ bu. (10)

   e. 4 fl. oz. = ___ tbsp. (8)

   f. 16 bu. = ___ pk. (64)

   g. 3 cups 4 oz. = ___ cups ($3\frac{1}{2}$)

   h. 7 qt. 1 pt. = ___ qt. ($7\frac{1}{2}$)

   i. 15 gal. = ___ cups (240)

   j. 4 bu. = ___ pt. (256)

T–65   Chapter 2   *English and Metric Measures*

Further Study

There are a number of English systems of measure for capacity. Several groups are listed below.

Liquid Capacity

1 pint	= 4 gills
1 quart	= 2 pints
1 gallon	= 4 quarts
1 barrel (liquid)	= 31.5 gallons
1 barrel (petroleum)	= 42 gallons

Dry Capacity

1 quart	= 2 pints
1 peck	= 8 quarts
1 bushel	= 4 pecks
1 barrel (dry; about 3.28 bu.)	= 7,056 cu. in.

Household Capacity

1 tablespoon	= 3 teaspoons
1 cup	= 16 tablespoons
1 pint	= 2 cups
1 quart	= 2 pints
1 gallon	= 4 quarts

Apothecaries' Fluid Measure

1 fluid dram	= 60 minims
1 fluid ounce	= 8 fluid drams
1 pint	= 16 fluid ounces
1 gallon	= 8 pints

Shipping Capacity

1 barrel bulk	= 5 cubic feet
1 displacement ton	= 7 barrels bulk = 35 cubic feet
1 shipping ton	= 8 barrels bulk = 40 cubic feet
1 register ton	= 100 cubic feet

Note that the tons in this group are units of capacity rather than weight. The displacement ton is the unit by which the size of ships is measured.

Mental Solutions for Part D

35. 10 qt. = $2\frac{1}{2}$ gal.; $2\frac{1}{2} \times 60 = 120 + 30$

36. $3 \times 8\frac{1}{3} = 25$ lb. per min.; $25 \times 60 = 100 \times 15$

37. 750×16 cups × 2 servings; $750 \times 2 = 1{,}500$; $1{,}500 \times 16 = 3{,}000 \times 8$

38. $40{,}000 \times \frac{3}{4} = 30{,}000$ cups; $30{,}000 \div 16 = 15{,}000 \div 8 \ldots$

39. $180 \div 9 = 20$ hogs fed from 1 acre; $240 \div 20$

40. $90 \times 80 = 7200$ bu. needed; $7{,}200 \div 160 = 360 \div 8 \ldots$

Lesson 18

C. Find these equivalents.

21. 5 qt. = __10__ pt.
22. 32 fl. oz = __4__ cups
23. 7 bu. = __28__ pk.
24. 15 tsp. = __5__ tbsp.
25. 5 gal. 2 qt. = __22__ qt.
26. 12 bu. 3 pk. = __51__ pk.
27. 7 qt. 1 pt. = __15__ pt.
28. 1 pt. 1 cup = __3__ cups
29. 6 pk. 3 qt. = __$6\frac{3}{8}$__ pk.
30. 7 gal. 3 qt. = __$7\frac{3}{4}$__ gal.
31. 5 cups 4 tbsp. = __$5\frac{1}{4}$__ cups
32. 7 bu. 2 pk. = __$7\frac{1}{2}$__ bu.
33. 128 cups = __8__ gal.
34. 96 qt. = __3__ bu.

D. Solve these reading problems. See how many you can solve mentally.

35. Water flows from a kitchen faucet at a rate of 10 quarts per minute. At that rate, how many gallons would flow from it in an hour?   150 gallons

36. In the same house, water flows from the bathtub faucet at a rate of 3 gallons per minute. Water weighs $8\frac{1}{3}$ pounds per gallon. What is the weight of the water that would flow from the faucet in 1 hour? (The answer is simple to find mentally if you calculate in a certain sequence.)   1,500 pounds

37. The Musselmans have a 750-gallon milk tank on their dairy farm. If the average amount of milk in a serving of cereal is $\frac{1}{2}$ cup, how many servings of cereal can be made with 750 gallons of milk?   24,000 servings

38. A chicken drinks an average of $\frac{3}{4}$ cup of water per day. At that rate, how many gallons per day are needed to supply 40,000 chickens with water?   1,875 gallons

39. A hog may eat 9 bushels of corn before it is ready for market. If a field of corn produces 180 bushels per acre, how many acres of corn are needed to feed 240 hogs?   12 acres

40. Beef cattle may eat 90 bushels of corn each before they are ready for market. If a field of corn yields 160 bushels of corn per acre, how many acres of corn are needed to feed 80 head of beef cattle?   45 acres

REVIEW EXERCISES

E. Find these equivalents. (Lesson 17)
41. 14 ft. = __168__ in.
42. 10,560 ft. = __2__ mi.

F. Round each number to the place indicated. (Lesson 2)
43. 2,812 (thousands)   3,000
44. 533,011 (ten thousands)   530,000

G. Write the place value and the value of each underlined digit. (Lesson 2)
45. <u>3</u>,465,870,000,000   trillions; 3,000,000,000,000
46. <u>8</u>,241,070,000,000,000   quadrillions; 8,000,000,000,000,000

19. English Units of Area

Area is the amount of surface within a given boundary. Linear measure has one dimension: length. Area has two dimensions: length and width.

The square at the right measures 1 inch long and 1 inch wide, or 1 square inch. The term *square* denotes both length and width. One square foot is the area of a square 1 foot long and 1 foot wide.

A larger unit of area that is often used in building is named the square. A square is equal to 100 square feet. Roofing and siding materials are measured in squares.

The acre is an even larger unit that is used to measure land area. It is equal to 43,560 square feet. A square with sides of $208\frac{3}{4}$ feet contains almost exactly one acre. Since the square and the acre are units of square measure, they are not preceded by the word *square*.

Area includes both length and width; therefore, the relationships between units of area are different from the respective linear measures. Remember this difference when changing measures of area from one unit to another.

English Units of Area
(Know these by memory.)

1 square foot (sq. ft.)	=	144 square inches (sq. in.)
1 square yard (sq. yd.)	=	9 square feet
1 square	=	100 square feet
1 acre (a.)	=	43,560 square feet
1 square mile (sq. mi.)	=	640 acres

Remember the rule for changing measures from one unit to another.

To change from larger units to smaller units, multiply.

To change from smaller units to larger units, divide.

Example A

32 sq. yd. = ___ sq. ft.
32 × 9 = 288 sq. ft.

Example B

3,520 a. = ___ sq. mi.
3,520 ÷ 640 = $5\frac{320}{640}$ = $5\frac{1}{2}$ sq. mi.

Example C

$4\frac{1}{3}$ sq. yd. = ___ sq. ft.
$4\frac{1}{3}$ × 9 = (4 × 9) + ($\frac{1}{3}$ × 9)
= 39 sq. ft.

Example D

5 sq. mi. 100 a. = ___ a.
5 × 640 = 3,200,
+ 100 = 3,300 a.

LESSON 19

Objectives

- To review these English units of area.
    - square inch
    - square foot
    - square yard
    - acre
    - square mile
- To teach the relationships listed below. (These equivalents are to be memorized.)
    - 1 square foot = 144 square inches
    - 1 square yard = 9 square feet
    - 1 acre = 43,560 square feet
    - 1 square mile = 640 acres
    - *1 square = 100 square feet

Review

1. *Find these equivalents.* (Lessons 17, 18)

    a. 8 pk. = ___ qt. (64)

    b. 11 cups = ___ fl. oz. (88)

    c. 8 tbsp. 2 tsp. = ___ tsp. (26)

    d. 44 oz. = ___ lb. ($2\frac{3}{4}$)

    e. 9 ft. 5 in. = ___ in. (113)

    f. 4 mi. = ___ yd. (7,040)

2. *Solve by long division, and check by casting out nines.* (Lesson 11)

    a. 54)6,324 (117 R 6)

    b. 378)274,903 (727 R 97)

3. *Write these Arabic numerals as Roman numerals.* (Lesson 3)

    a. 132   (CXXXII)

    b. 54,500   ($\overline{\text{LIV}}$D)

4. *Write these Roman numerals as Arabic numerals.* (Lesson 3)

    a. MCDI   (1,401)

    b. $\overline{\text{MMCXIC}}$ (2,111,100)

T–67 Chapter 2 *English and Metric Measures*

Introduction

When someone wants to sell a field, why would he not advertise it as a "1,000-foot field"?

The description gives the measurement of only one side. The other side of the field could measure 100 feet or 1,300 feet. He needs to use a unit that describes the entire field.

An item such as a field, which includes both length and width, must be described in terms of its area. Fields such as the one mentioned above are measured in acres. If the 1,000-foot by 100-foot field is rectangular, it has an area of about $2\frac{1}{3}$ acres. If the 1,000-foot by 1,300-foot field is rectangular, it has an area of about 30 acres.

Have the students name several other items that are measured with units of area. Several examples are flooring, roofing, and ceiling tile.

Solutions for Part D

25. $72 \times 144 \div 9$
26. $182 \div 9$
27. $\frac{1}{2} \times 43{,}560$

Teaching Guide

1. **Area is the amount of surface within a given boundary.** Common English units of area are the square inch, the square foot, the square yard, and the square mile. The acre is commonly used to measure real estate. The units and relationships shown in the lesson are to be memorized.

2. **Because area includes both length and width, the relationships between units of area are different from the respective linear measures.** Note that the relationship of a smaller unit of area to a larger unit is the square of the relationship of the respective linear measures.
   1 sq. ft. = 144 sq. in.
   (12 × the 12 inches in a foot)
   1 sq. yd. = 9 sq. ft.
   (3 × the 3 feet in a yard)
   1 sq. yd. = 1,296 sq. in.
   (36 × the 36 inches in a yard)

3. **To find equivalent measures, remember these two rules.**
   To change from larger units to smaller units, multiply.
   To change from smaller units to larger units, divide.

   Find these equivalents.

   a. 15 sq. ft. = ___ sq. in. (2,160)
   b. 11 sq. mi. = ___ a. (7,040)
   c. 32 sq. yd. = ___ sq. ft. (288)
   d. 1,872 sq. in. = ___ sq. ft. (13)
   e. 810 sq. ft. = ___ sq. yd. (90)
   f. 304,920 sq. ft. = ___ a. (7)
   g. 20 sq. ft. = ___ sq. yd. ($2\frac{2}{9}$)
   h. 720 a. = ___ sq. mi. ($1\frac{80}{640} = 1\frac{1}{8}$)
   i. 3 sq. yd. 1 sq. ft. = ___ sq. ft. (28)
   j. 5 sq. yd. 4 sq. ft. = ___ sq. yd. ($5\frac{4}{9}$)

CLASS PRACTICE

Find these equivalents.

a. 8 sq. yd. = __72__ sq. ft.
b. 3 a. = ____ sq. ft.   130,680
c. 11 sq. mi. = 7,040 a.
d. 1,872 sq. in. = __13__ sq. ft.
e. 12 sq. ft. 97 sq. in. = 1,825 sq. in.
f. 4 a. 4,580 sq. ft. = ____ sq. ft. 178,820
g. 30 sq. yd. 7 sq. ft. = __277__ sq. ft.
h. 3 sq. mi. 440 a. = 2,360 a.

WRITTEN EXERCISES

A. *Write the abbreviations for these units by memory.*

1. square inch   sq. in.
2. square foot   sq. ft.
3. acre   a.
4. square mile   sq. mi.

B. *Write these equivalents by memory.*

5. 1 sq. ft. = __144__ sq. in.
6. 1 sq. yd. = __9__ sq. ft.
7. 1 acre = ____ sq. ft.   43,560
8. 1 sq. mi. = __640__ a.

C. *Find these equivalents.*

9. 6 sq. ft. = __864__ sq. in.
10. 9 sq. yd. = __81__ sq. ft.
11. 25 sq. mi. = ____ a.   16,000
12. 108 sq. yd. = __972__ sq. ft.
13. 2,880 sq. in. = __20__ sq. ft.
14. 217,800 sq. ft. = __5__ a.
15. 75 sq. yd. = __675__ sq. ft.
16. 2,304 sq. in. = __16__ sq. ft.
17. 10,880 a. = __17__ sq. mi.
18. 15 sq. ft. = 2,160 sq. in.
19. 2 sq. ft. = $\frac{2}{9}$ sq. yd.
20. 500 a. = $\frac{25}{32}$ sq. mi.
21. 2 sq. yd. = 2,592 sq. in.
22. 1 sq. mi. = ____ sq. ft.   27,878,400
23. 4 sq. yd. 6 sq. ft. = __42__ sq. ft.
24. 4 sq. mi. 65 a. = 2,625 a.

D. *Solve these reading problems.*

25. Sandra is helping Mother sew a quilt made of small patches. After the seams are sewn, each patch measures 9 square inches. The quilt has an area of 72 square feet. How many patches will it take to sew the quilt?   1,152 patches

26. Sandra's bedroom floor has an area of 182 square feet. How many square yards of carpet will be needed to cover the floor? $20\frac{2}{9}$ square yards

27. In 1 Samuel 14:1–14, we read of Jonathan and his armor bearer leading the battle against a Philistine garrison. They slew about 20 men within a half acre of ground. If that acre was equal to the English acre, within how many square feet of ground did the battle take place?   21,780 square feet

68   Chapter 2   English and Metric Measures

28. The table of shewbread in the tabernacle was 2 cubits long and 1 cubit wide (Exodus 25:23), or about 4.5 square feet. This is equal to how many square inches?   **648 square inches**

29. Samuel calculated that each tire on the family van had a surface area of 30 square inches on the road. When the Stauffer family is in the van, the total weight of the vehicle and its passengers is 5,400 pounds. What is the average weight on each square inch of the tires?   **45 pounds**

30. Each time the tires make one revolution, the van travels 70 inches. When the van travels 60 miles per hour, it is moving 88 feet per second. At that speed, how many times does a tire revolve each second? (Round to the nearest whole number.)   **15 times**

REVIEW EXERCISES

E. Find these equivalents. *(Lessons 17, 18)*

31. 3 mi. = ____ ft.   **15,840**

32. 15 yd. = __45__ ft.

33. 6,160 yd. = __$3\frac{1}{2}$__ mi.

34. 9 ft. 8 in = __116__ in.

35. 15 gal. = __60__ qt.

36. 36 tsp. = __12__ tbsp.

F. Solve by long division, and check by casting out nines. *(Lesson 11)*

37. 48)3,906   **81 R 18**

38. 712)633,872   **890 R 192**

G. Write these Arabic numbers as Roman numerals. *(Lesson 3)*

39. 209   **CCIX**

40. 816   **DCCCXVI**

H. Write these Roman numbers as Arabic numerals. *(Lesson 3)*

41. $\overline{\text{CLXXII}}\text{CM}$ **172,900**

42. $\overline{\text{MDCCCXI}}\text{LII}$ **1,811,052**

Lesson 19 T–68

Further Study

Following are some English units of area not included in the lesson.

Surveyor's Square Measure

1 square link	= 62.73 square inches
1 square rod	= 30.25 square yards
1 square chain	= 484 square yards
1 acre	= 160 square rods
1 acre	= 10 square chains
1 section	= 640 acres (1 square mile)
1 township	= 36 sections

---

28. 4.5×144

29. $5{,}400 \div (4 \times 30)$

30. $88 \times 12 \div 70$

LESSON 20

Objectives

- To review these units of time.
  - second
  - minute
  - hour
  - day
  - week
  - month
  - year
  - decade
  - century
  - millennium
- To review the relationships between different units of time.

Review

1. *Find these equivalents.* (Lessons 18, 19)
   a. 720 sq. in. = ___ sq. ft. (5)
   b. 99 sq. ft. = ___ sq. yd. (11)
   c. 6,400 a. = ___ sq. mi. (10)
   d. 2 a. 20,000 sq. ft. = ___ sq. ft. (107,120)
   e. 9 bu. = ___ pk. (36)
   f. 7 gal. 1 qt. = ___ qt. (29)

2. *Solve mentally by the divide-and-divide method.* (Lesson 12)
   a. 128 ÷ 32 (4)
   b. 442 ÷ 34 (13)
   c. 270 ÷ 18 (15)
   d. 308 ÷ 28 (11)

3. *Tell whether each problem illustrates the commutative or the associative law.* (Lesson 4)
   a. 3 × 18 = 18 × 3 (commutative)
   b. (2 + 4) + 8 = 2 + (4 + 8) (associative)

20. Units of Time

God is eternal; He never had a beginning, and He will never have an end. He has no need to measure time as man does.

When God created the earth, He created units of time based on the movements of heavenly bodies. Time will continue until God brings it to a close at the end of the age (Revelation 10:5, 6).

1 day = 1 rotation of the earth on its axis
1 month = 1 revolution of the moon around the earth (approximately)
1 year = 1 revolution of earth around the sun

Units of Time
(Know these by memory.)

1 minute (min.) = 60 seconds (sec.)		1 year	= 12 months
1 hour (hr.) = 60 minutes		1 standard year	= 365 days
1 day = 24 hours		1 leap year	= 366 days
1 week (wk.) = 7 days		1 decade	= 10 years
1 month (mo.) = about 30 days		1 century	= 100 years
1 year (yr.) = about 52 weeks		1 millennium	= 1,000 years

To change from larger units to smaller units, multiply.
To change from smaller units to larger units, divide.

Example A

15 mo. = ___ yr.
$15 \div 12 = 1\frac{3}{12} = 1\frac{1}{4}$ yr.

Example B

3 wk. = ___ hr.
$3 \times 7 = 21$ days, $\times 24 = 504$ hr.

Time Zones

Clock settings are based on the assumption that the sun is at its highest point in the sky at 12:00 noon. However, if every clock were set to 12:00 noon when the sun is directly overhead, there would be many degrees of variation between clocks.

To eliminate this confusion, the earth was divided into 24 time zones in 1884. All areas within one time zone have the same clock time. In this way, large areas have the same time and the sun is still nearly overhead at 12:00 noon.

Each time zone covers about 15 degrees of the earth's circumference. However, the boundary lines of the time zones zigzag in many places in order to follow political and geographical boundaries.

North America touches eight time zones. Newfoundland Standard Time is only a half hour ahead of Atlantic Standard Time.

70   Chapter 2   English and Metric Measures

To change time from one time zone to the next, subtract one hour when moving from east to west and add one hour when moving from west to east. The abbreviation A.M. means ante meridiem (before noon) and is used for times between midnight and noon. The abbreviation P.M. means post meridiem (after noon) and is used for times between noon and midnight. Use the terms noon and midnight only for 12:00 noon and 12:00 midnight.

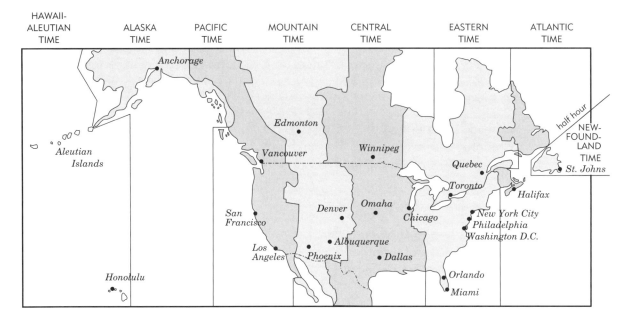

Time Zone	City	Time
Newfoundland	St. Johns, Newfoundland	1:30 P.M.
Atlantic (AST)	Halifax, Nova Scotia	1:00 P.M.
Eastern (EST)	New York City, New York	12:00 Noon
Central (CST)	Chicago, Illinois	11:00 A.M.
Mountain (MST)	Denver, Colorado	10:00 A.M.
Pacific (PST)	Los Angeles, California	9:00 A.M.
Alaska	Anchorage, Alaska	8:00 A.M.
Hawaii-Aleutian	Honolulu, Hawaii	7:00 A.M.

> **To change from one time zone to the next:**
> **Subtract one hour when moving from east to west.**
> **Add one hour when moving from west to east.**

Example C
2:00 A.M. in Halifax, Nova Scotia; ___ in Anchorage, Alaska
Anchorage, Alaska, is 5 time zones west of Halifax.
Subtract 5 hours from 2:00.
5 hours earlier than 2:00 A.M. = 9:00 P.M. (on the previous evening)

Lesson 20   T–70

Introduction

Have the students name the unit of time that is suggested by each of the following natural cycles.

    from one full moon to another (month)
    from one summer to the next (year)
    from one sunrise to the next (day)

Except for the week, which was established by God at Creation, the basic units of time relate to the movements of the celestial bodies. The other units, such as the hour and the minute, name divisions of a day. Longer units, such as the decade and the century, name multiples of years. Thus the units of time are based on the movements of heavenly bodies, just as God stated in Genesis 1:14: "And God said, Let there be lights in the firmament of the heaven to divide the day from the night; and let them be for signs, and for seasons, and for days, and years."

Teaching Guide

1. **The main units of time are based on the movements of heavenly bodies.**

   The day is based on the rotation of the earth on its axis.

   The month is based on the revolution of the moon around the earth.

   The year is based on the revolution of the earth around the sun.

2. **Units of time include years, months, weeks, days, hours, minutes, and seconds.** Students should know these units and their relationships by memory.

3. **To find equivalent measures, remember these two rules.**

   To change from larger units to smaller units, multiply.

   To change from smaller units to larger units, divide.

   Give practice with finding equivalent measures of time, and especially with changing compound measures (*c–f* below) and changing to a unit other than the next larger or smaller unit (*g* and *h* below).

   a. 20 min. = ___ sec. (1,200)

   b. 250 yr. = ___ decades (25)

   c. 5 wk. 6 days = ___ days (41)

   d. 7 yr. 3 mo. = ___ mo. (87)

   e. 23 yr. = ___ decades ($2\frac{3}{10}$)

   f. 6 hr. 30 min. = ___ hr. ($6\frac{1}{2}$)

   g. 3 hr. = ___ sec. (10,800)

   h. 15 decades = ___ mo. (1,800)

4. **North America touches eight time zones. To change time from one zone to the next, subtract one hour when moving from east to west and add one hour when moving from west to east.** East-to-west on a map can be likened to the right-to-left of subtraction on a

T–71   Chapter 2   *English and Metric Measures*

number line; likewise, west-to-east corresponds to the left-to-right of addition. Give special attention to cases like *e* and *f* below, which involve changes between A.M. and P.M.

a. 1:00 P.M. in Vancouver, British Columbia; ___ in Toronto, Ontario (4:00 P.M.)

b. 9:00 A.M. in Washington, D.C.; ___ in Honolulu, Hawaii   (4:00 A.M.)

c. 7:25 P.M. in Denver, Colorado; ___ in New York City, New York (9:25 P.M.)

d. 12:35 A.M. in Anchorage, Alaska; ___ in Halifax, Nova Scotia (5:35 A.M.)

e. 9:00 P.M. in Honolulu, Hawaii; ___ in St. Johns, Newfoundland (3:30 A.M.)

f. 2:25 A.M. in St. Johns, Newfoundland; ___ in Winnipeg, Manitoba   (11:55 P.M.)

Further Study

1. God established the basic unit for measuring time when He called the evening and the morning the first day. Since then the sun has been used to measure time. The units we use are sexagesimal units; that is, they are based on sixty instead of ten. It takes sixty seconds to make one minute and sixty minutes to make one hour. The length of the day is determined by the time it takes for the earth to make one complete rotation on its axis—24 hours. Because the year is not exactly 365 days long, one day is added every fourth year. But this is still not exact, so the leap year is dropped on the century years (1700, 1800, 1900) except for the ones divisible by 400. Therefore, 1600 and 2000 are leap years.

2. Three of the less common units of time are shown below.

    1 fortnight = 2 weeks
    (fourteen nights)

    1 millisecond = 0.001 second

    1 microsecond = 0.000001 second
    (one-millionth)

Example D

11:00 A.M. in Los Angeles, California; ___ in New York City.
New York City is 3 time zones east of Los Angeles, California.
Add 3 hours to 11:00.
3 hours later than 11:00 A.M. = 2:00 P.M.

CLASS PRACTICE

Find these equivalents.

a. 3 yr. = <u>1,095</u> days
b. 288 hr. = <u>12</u> days
c. 8 decades 8 yr. = <u>88</u> yr.
d. 9 centuries 25 yr. = <u>$9\frac{1}{4}$</u> centuries
e. 6 hr. = ____ sec.   21,600
f. 2 wk. = <u>336</u> hr.

Change these times as indicated.

g. 5:00 P.M. in Honolulu, Hawaii; ___ in Denver, Colorado   8:00 P.M.
h. 7:00 A.M. in Honolulu, Hawaii; ___ in St. Johns, Newfoundland   1:30 P.M.
i. 11:00 P.M. in Toronto, Ontario; ___ in Anchorage, Alaska   7:00 P.M.
j. 9:00 P.M. in Miami, Florida; ___ in Toronto, Ontario   9:00 P.M.

WRITTEN EXERCISES

A. *Write the abbreviations for these units by memory.*

1. hour   hr.
2. month   mo.
3. week   wk.
4. year   yr.
5. minute   min.
6. second   sec.

B. *Write these equivalents by memory.*

7. 1 min. = <u>60</u> sec.
8. 1 day = <u>24</u> hr.
9. 1 wk. = <u>7</u> days
10. 1 yr. = <u>365</u> days
11. 1 decade = <u>10</u> yr.
12. 1 millennium = <u>1,000</u> yr.

C. *Find these equivalents.*

13. 12 hr. = <u>720</u> min.
14. 7 wk. = <u>49</u> days
15. 7 yr. = <u>84</u> mo.
16. 168 hr. = <u>7</u> days
17. 540 sec. = <u>9</u> min.
18. 700 yr. = <u>7</u> centuries
19. 168 mo. = <u>14</u> yr.
20. 600 hr. = <u>25</u> days
21. 8 centuries = <u>800</u> yr.
22. 5 millennia = <u>5,000</u> yr.
23. 10 hr. 50 min. = <u>$10\frac{5}{6}$</u> hr.
24. 20 min. 15 sec. = <u>$20\frac{1}{4}$</u> min.
25. 4 millennia = <u>40</u> centuries
26. 7 centuries = <u>70</u> decades

72   Chapter 2   English and Metric Measures

D. Change these times as indicated. Be sure to use correct labels when crossing noon or midnight.

27. 4:00 P.M. in San Francisco; ___ in Halifax, Nova Scotia   8:00 P.M.
28. 2:30 A.M. in Omaha, Nebraska; ___ in Miami, Florida   3:30 A.M.
29. 7:35 P.M. in St. Johns, Newfoundland; ___ in Chicago, Ill.   5:05 P.M.
30. 11:00 P.M. in Philadelphia, Pennsylvania; ___ in Honolulu, Hawaii 6:00 P.M.
31. 11:30 P.M. in the Aleutian Islands; ___ in Orlando, Florida   4:30 A.M.
32. 9:45 A.M. in Winnipeg, Manitoba; ___ in St. Johns, Newfoundland 12:15 P.M.

E. Solve these reading problems.

33. About six millennia have passed since God created the heavens and the earth (Genesis 1). How many decades is that?   600 decades

34. A generation is considered to be 30 years. Using this assumption, how many generations have passed in the six millennia since God created man?   200 generations

35. Traveling at 45 miles per hour is traveling $\frac{3}{4}$ mile per minute. How many feet is that per second?   66 feet

36. Noah waited 40 days after the ark rested on Mt. Ararat before sending out a raven (Genesis 8). Apparently 7 days later, he sent out a dove that soon returned; and when he released her again after another 7 days, she returned with an olive leaf. After waiting 7 more days, he released the dove again. It did not return this time, so Noah knew the waters had abated. How many weeks passed from the grounding of the ark until Noah's conclusion that the waters were gone? $8\frac{5}{7}$ weeks

37. Malvern Dairy sells a gallon jug of milk for $2.25 and a 16-ounce bottle of milk for 49¢. How much more expensive is it to purchase a gallon of milk in 16-ounce bottles than to buy a gallon jug of milk? $1.67

38. Mother bought a gallon of vinegar. When she bakes, she often measures vinegar by the tablespoon. How many tablespoons are in a gallon?   256 tbsp.

REVIEW EXERCISES

F. Find these equivalents. *(Lessons 18, 19)*

39. 52 pk. = __13__ bu.   40. 56 pt. = __28__ qt.
41. 3,200 a. = __5__ sq. mi.   42. 3 a. = ____ sq. ft.   130,680

G. Solve mentally by the divide-and-divide method. *(Lesson 12)*

43. 462 ÷ 22   21   44. 224 ÷ 16   14

H. Write *commutative law* **or** *associative law* **to tell what each problem illustrates.** *(Lesson 4)*

45. 9 + 23 = 23 + 9   46. (3 × 6) × 9 = 3 × (6 × 9)
    commutative law   associative law

Lesson 20   T–72

Solutions for Part E

33. 6 × 1,000 ÷ 10

34. 6 × 1,000 ÷ 30

35. $\frac{3}{4}$ × 5,280 ÷ 60

36. (40 + 7 + 7 + 7) ÷ 7

37. (8 × 0.49) − $2.25

38. 2 (tablespoons) × 8 (fluid ounces) × 16 (cups)

T-73   Chapter 2   *English and Metric Measures*

LESSON 21

Objectives

- To review the common metric prefixes, and to introduce *some metric prefixes whose meanings indicate more than 1,000 (*kilo-*) and less than $\frac{1}{1,000}$ (*milli-*). (Students are not expected to memorize these seldom-used prefixes.)
- To review the basic metric unit of linear measure (the meter) and its related units.
- To review the method for changing from one unit of measure to another.
- To review the English/metric equivalents shown below. (Students are expected to know the equivalents taught in this chapter for the test. The equivalents are not exact, but they are accurate enough for most uses.)

    1 foot = 0.3 meter
    1 meter = 3.28 feet.
    1 inch = 2.54 centimeters
    1 centimeter = 0.39 inch

Review

1. Give Lesson 21 Quiz (English Measures).
2. *Change these times as indicated.* (Lesson 20)
    a. 12:25 A.M. in New York City; ___ in Los Angeles, California (9:25 P.M.)
    b. 3:00 P.M. in Honolulu, Hawaii; ___ in Halifax, Nova Scotia (9:00 P.M.)
3. *Find these equivalents.* (Lessons 17, 19)
    a. 14 lb. = ___ oz. (224)
    b. 2 ft. 9 in. = ___ in. (33)
    c. 9 sq. yd. = ___ sq. ft. (81)
    d. 864 sq. in. = ___ sq. ft. (6)
4. *Solve mentally by the double-and-double method.* (Lesson 13)
    a. 935 ÷ 5 (187)
    b. 43,500 ÷ 500 (87)
    c. 245 ÷ 35 (7)
    d. 225 ÷ 15 (15)
5. *Solve by mental addition.* (Lesson 5)
    a. 45 + 72 (117)
    b. 32 + 57 (89)
    c. 142 + 453 (595)
    d. 1,600 + 455 (2,055)

21. Metric Units of Linear Measure

The English and metric systems of measures are the two main systems used today. The official name of the metric system is the International System of Units.

The metric system is easy to learn because its units have a uniform scale of relationships based on the decimal system (system of tens). The prefixes listed below indicate the relationship of a basic unit to larger and smaller units. The boldface prefixes are the ones most commonly used.

Prefixes Used With Metric Measures

exa- (E) = 1,000,000,000,000,000,000
peta- (P) = 1,000,000,000,000,000
tera- (T) = 1,000,000,000,000
giga- (G) = 1,000,000,000
mega- (M) = 1,000,000
kilo- (k) = 1,000
hecto- (h) = 100
deka- (dk) = 10
1
deci- (d) = 0.1
centi- (c) = 0.01
milli- (m) = 0.001
micro- (μ) = 0.000 001
nano- (n) = 0.000 000 001
pico- (p) = 0.000 000 000 001
femto- (f) = 0.000 000 000 000 001
atto- (a) = 0.000 000 000 000 000 001

Know the boldface prefixes, abbreviations, and values by memory.

The abbreviation for *micro-* is the twelfth letter of the Greek alphabet. Its name is *mu* and its sound is like the English *m*.

Basic Units of Metric Measure

meter . . . . length
gram . . . . weight
liter . . . . capacity
square meter . . area
cubic meter . . . volume

Metric Units of Linear Measure

UNIT	ABBREVIATION	VALUE
kilometer	km	**1,000 m**
hectometer	hm	100 m
dekameter	dkm	10 m
meter	m	**1 m**
decimeter	dm	0.1 m
centimeter	cm	**0.01 m**
millimeter	mm	**0.001 m**

The table at the left shows the main units of linear measure, with boldface for the ones most frequently used.

Because the metric system is a decimal system, converting from one unit to another is very simple. All multiplication and division is by a power of 10 (such as 10, 100, or 1,000). Therefore, all conversions can be made simply by moving the decimal point as shown in the following steps.

74 Chapter 2 English and Metric Measures

1. Decide which operation is needed. To change a larger unit to a smaller unit, multiply. To change a smaller unit to a larger unit, divide.

2. Decide how many units larger or smaller the change involves. The decimal point will be moved that number of places.

3. Move the decimal point right to multiply and left to divide. You may need to annex zeroes to the right when multiplying (Example A) or to the left when dividing (Example B).

Example A	**Example B**
2.4 m = ___ cm	850 m = ___ km
2 units smaller; multiply by 100	3 units larger; divide by 1,000
100 × 2.4 = 240 mm	850 ÷ 1,000 = 0.85 km

In North America, where both metric and English measures are used, it is helpful to know the relationships between the common units. Therefore, each metric lesson in this chapter has some English-metric equivalents for you to memorize.

English/Metric Linear Equivalents
1 meter = 3.28 feet
1 centimeter = 0.39 inch
1 foot = 0.3 meter
1 inch = 2.54 centimeters

To change meters to feet, multiply the number of meters by 3.28. To change centimeters to inches, multiply the number of centimeters by 0.39.

To change feet to meters, multiply the number of feet by 0.3. To change inches to centimeters, multiply the number of inches by 2.54.

Example C	**Example D**
14 m = ___ ft.	16 cm = ___ in.
14 × 3.28 = 45.92 ft.	16 × 0.39 = 6.24 in.
Example E	**Example F**
37 ft. = ___ m	25 in. = ___ cm
37 × 0.3 = 11.1 m	25 × 2.54 = 63.5 cm

CLASS PRACTICE

Find these equivalents.

a. 18 m = <u>0.018</u> km

b. 30 dkm = <u>3,000</u> dm

c. 25 cm = <u>250</u> mm

d. 34 mm = <u>0.034</u> m

e. 76 hm = <u>7,600</u> m

f. 55,000 cm = <u>0.55</u> km

Introduction

Review the numbers used to express relationships between English units learned in this text.

2—cups in a pint; pints in a quart
3—feet in a yard
4—cups in a quart; quarts in a gallon; pecks in a bushel
8—quarts in a peck; fluid ounces in a cup
9—square feet in a square yard
12—inches in a foot
16—ounces in a pound; tablespoons in a cup
36—inches in a yard
144—square inches in a square foot
1,760—yards in a mile
5,280—feet in a mile

About 200 years ago, scientists in France developed a system of measures based on the decimal system. All units were related by powers of ten (10, 100, 1,000, and so on). This system is known as the metric system.

Teaching Guide

1. **The metric system is the most widely used system of measures in the world today.** Its official name is the International System of Measures, but this text uses the more common term *metric system*.

2. **In the metric system, all units of measure are related by powers of ten.** Prefixes are used to show the relationship of each derived unit to the base unit. Review the prefixes in the lesson.

Note: The abbreviation for *deka-* appears to be changing from *dk* to *da*. You may prefer to teach the *da* form, such as *dam* for "dekameter," *dag* for "dekagram," and *dal* for "dekaliter."

3. **The main metric units of linear measure are the kilometer, the hectometer, the dekameter, the meter, the decimeter, the centimeter, and the millimeter.** Of these units, the kilometer, the meter, the centimeter, and the millimeter are most frequently used.

    Use a metric ruler to measure various objects in the room.

4. **Because the metric system is a decimal system, all conversions can be made by simply moving the decimal point.** Discuss the steps in the lesson; then practice making conversions by using the problems below and *Class Practice* problems a–f.

    a. 560 mm = 0.56 m
       (Move decimal three places to left.)

    b. 785 cm = 7.85 m
       (Move decimal two places to left.)

    c. 1.55 km = 1,550 m
       (Move decimal three places to right.)

    d. 12,300 mm = 0.0123 km
       (Move decimal six places to left.)

    e. You could give this problem for

T–75   Chapter 2 English and Metric Measures

extra credit. The length of the meter was established as $\frac{1}{10,000,000}$ the distance from the North Pole to the equator. If that measurement had been exact, what would be the circumference of the earth around the poles (polar circumference)?

Answer: If the distance between the North Pole and the equator were 10,000,000 meters, the total distance around the earth would be 40,000,000 meters (4 × 10,000,000) or 40,000 kilometers. In reality, the polar circumference of the earth is 40,008.6 kilometers. The scientists calculating the polar circumference erred by 8.6 kilometers—a miscalculation of only 0.0215 percent.

5. **In North America, it is helpful to know the relationships between common English and metric units.** Pupils are to memorize several relationships in each lesson.

6. **To make conversions between meters and feet, use the following rules.**

    To change meters to feet, multiply the number of meters by 3.28.

    To change feet to meters, multiply the number of feet by 0.3.

    a. 13 m = ___ ft.   (42.64)
    b. 200 ft. = ___ m   (60)
    c. 43 m = ___ ft.   (141.04)
    d. 244 ft. = ___ m   (73.2)

    To change centimeters to inches, multiply the number of centimeters by 0.39.

    To change inches to centimeters, multiply the number of inches by 2.54.

    e. 8 cm = ___ in.   (3.12)
    f. 5 in. = ___ cm   (12.7)
    g. 61 cm = ___ in.   (23.79)
    h. 32 in. = ___ cm   (81.28)

Solutions for Part G

33. 25 × 0.71

34. 400 ÷ 60

Find these English/metric equivalents.

g. 17 m = <u>55.76</u> ft.
h. 200 m = <u>656</u> ft.
i. 50 cm = <u>19.5</u> in.
j. 67 cm = <u>26.13</u> in.
k. 430 ft. = <u>129</u> m
l. 72 ft. = <u>21.6</u> m
m. 34 in. = <u>86.36</u> cm
n. 58 in. = ____ cm   147.32

WRITTEN EXERCISES

A. *Write the abbreviations for these units by memory.*
1. centimeter   cm
2. millimeter   mm
3. kilometer   km
4. dekameter   dkm
5. hectometer   hm
6. decimeter   dm

B. *Write the prefix that represents each value.*
7. 1,000   kilo-
8. 100   hecto-
9. 10   deka-
10. 0.1   deci-
11. 0.01   centi-
12. 0.001   milli-

C. *Write these equivalents by memory. Memorize these equivalents well.*
13. 1 ft. = <u>0.3</u> m
14. 1 m = <u>3.28</u> ft.

D. *Write the next larger unit in sequence.*
15. meter   dekameter
16. hectometer   kilometer

E. *Find these equivalents.*
17. 1 m = <u>100</u> cm
18. 25 m = ____ mm   25,000
19. 300 cm = <u>3</u> m
20. 8 dkm = <u>800</u> dm
21. 5 km = <u>500</u> dkm
22. 65 cm = <u>0.65</u> m
23. 420 mm = <u>42</u> cm
24. 7 km = ____ cm   700,000

F. *Find these English/metric equivalents.*
25. 18 m = <u>59.04</u> ft.
26. 300 m = <u>984</u> ft.
27. 60 cm = <u>23.4</u> in.
28. 82 cm = <u>31.98</u> in.
29. 4 ft. = <u>1.2</u> m
30. 350 ft. = <u>105</u> m
31. 12 in. = <u>30.48</u> cm
32. 76 in. = ____ cm   193.04

G. *Solve these reading problems.*

33. Earthworms gradually cover the surface of the soil with earth from below. In one area, they brought up an average depth of 7.1 millimeters of soil annually. How many centimeters of soil would these tiny "plowmen" bring up in 25 years?   17.75 cm

34. Some moles can dig burrows at the rate of 4 meters per hour. Find how many centimeters that is in one minute, to the nearest tenth.   6.7 cm

76 Chapter 2 *English and Metric Measures*

35. Traveling at 40 miles per hour, a driver should signal at least 300 feet before turning. How many meters should he signal before turning?   90 m

36. The northern bamboo weed grows very rapidly. If a stalk grows 2.1 meters in 30 days, how many centimeters is that per day?   7 cm

37. Sister Anne and one of her daughters mend clothing together for about 50 minutes each week. How many hours are spent together, mending in the 52 weeks of a year? (Express any remainder as a fraction.) $43\frac{1}{3}$ hours

38. The Conley family spends 20 minutes per day in their family worship. In 365 days, how many hours is that? (Express any remainder as a fraction.) $121\frac{2}{3}$ hours

Every day will I bless thee; and I will praise thy name for ever and ever.
Psalm 145:2

REVIEW EXERCISES

H. Change these times as indicated. (Lesson 20)

39. 6:15 P.M. Halifax, Nova Scotia; ___ in Anchorage, Alaska   1:15 P.M.

40. 2:25 A.M. in Chicago, Illinois; ___ in Miami, Florida   3:25 A.M.

I. Find these equivalents. (Lessons 19, 20)

41. 2 sq. mi. = 1,280 a.   **42.** 846 sq. ft. = 94 sq. yd.

43. 528 min. = $8\frac{4}{5}$ hr.   **44.** 9 wk. 4 days = 67 days

J. Solve mentally by the double-and-double method. (Lesson 13)

45. 540 ÷ 45   12   **46.** 240 ÷ 15   16

47. 170 ÷ 5   34   **48.** 360 ÷ 45   8

K. Solve by mental addition. (Lesson 5)

49. 205 + 861   1,066   **50.** 2,850 + 450   3,300

An Ounce of Prevention

Logically, if you take a metric measure and convert it to the English equivalent, and then convert that English measure into its metric equivalent, the result should be the same as the original metric measure. That does not hold true with these equivalents, because they are rounded for ease of memory.

---

35. 300×0.3

36. $210 \div 30$

37. $52 \times 50 \div 60$

38. $365 \times 20 \div 60$

Further Study

The French scientists also tried to establish a decimal-based time system, but it was a total failure. To establish a 10-day week was to change the 7-day cycle established by God at the Creation. People had difficulty adjusting to such a week, so it was dropped (after ten years!). Changing months and years to a decimal system was virtually impossible because such a system does not match the celestial movements established by God.

The metric system is appealing because of its logic and its simplicity of calculation. However, it does have the drawback of being more theoretical than practical. The English system was designed for convenience; hence the units are of practical size. But the metric system is based on powers of ten rather than on need; therefore, many units (such as the dekameter, decimeter, dekagram, and decigram) are seldom used. In a sense, the English system of measures was developed to fit man; but the metric system was developed to fit a theory, and then man tried to adapt to it.

The metric system is used in most nations of the world today, with the United States remaining one of the few exceptions. So entrenched is the English system that most people in the United States would rather continue dealing with its "illogical" relationships than change to a less familiar system. (The ease of conversions by computer has probably contributed to this.) Therefore, conversion of the United States to the metric system has not been accomplished.

T–77   Chapter 2   *English and Metric Measures*

LESSON 22

Objectives

- To review the basic metric unit of weight (the gram) and its related units, including the metric ton.
- To review changing from one metric unit of weight to another.
- To review the English/metric equivalents shown below. (These equivalents are to be memorized.)
    1 pound = 0.45 kilograms
    1 kilogram = 2.2 pounds.

Review

1. *Find these equivalents.* (Lessons 18, 20, 21)
    a. 15 bu. = ___ pk.   (60)
    b. 14 qt. 1 pt. = ___ pt.   (29)
    c. 9 centuries = ___ yr.   (900)
    d. 6 days = ___ hr.   (144)
    e. 45 m = ___ dkm   (4.5)
    f. 275 cm = ___ dkm   (0.275)

2. *Change these times as indicated.* (Lesson 20)
    a. 4:00 P.M. Eastern Standard Time; ___ Pacific Standard Time
        (1:00 P.M.)
    b. 11:00 P.M. Alaska Standard Time; ___ Newfoundland Standard Time
        (4:30 A.M.)

3. *Solve by horizontal subtraction.* (Lesson 6)
    a. 22,913 – 15,753   (7,160)
    b. $3,078.52 – 871.49   ($2,207.03)

22. Metric Units of Weight

The basic metric unit of weight is the gram. One gram is the weight of one cubic centimeter of water. It is a very small unit, equal to less than $\frac{1}{28}$ of an ounce.

The boldface units are the most frequently used. The milligram is a very small measure, commonly used to designate amounts in medical weight. Many food packages are labeled with weight in grams. The kilogram, 1,000 grams, equals approximately 2.2 pounds. It probably has the most frequent application in everyday life. The metric ton, 1,000 kilograms, is used to weigh very heavy objects, as is the English ton.

Metric Units of Weight

UNIT	ABBREVIATION	VALUE
metric ton	**MT**	**1,000 kg**
kilogram	**kg**	**1,000 g**
hectogram	hg	100 g
dekagram	dkg	10 g
gram	**g**	**1 g**
decigram	dg	0.1 g
centigram	cg	0.01 g
milligram	**mg**	**0.001 g**

The steps for finding metric weight equivalents are the same as for linear equivalents as shown in Lesson 21.

1. Decide which operation is needed. To change a larger unit to a smaller unit, multiply. To change a smaller unit to a larger unit, divide.

2. Decide how many units larger or smaller the change involves. The decimal point will be moved that number of places.

3. Move the decimal point right to multiply and left to divide. You may need to annex zeroes to the right when multiplying (Example A) or to the left when dividing (Example B).

To change kilograms to metric tons or metric tons to kilograms, multiply or divide by 1,000. Do this by moving the decimal point three places right or left.

Example A	**Example B**
24.5 kg = ___ g	95 kg = ___ MT
3 units smaller; multiply by 1,000	Divide by 1,000.
1,000 × 24.5 = 24,500 g	95 ÷ 1,000 = 0.095 MT

78 Chapter 2 *English and Metric Measures*

English/Metric Weight Equivalents

1 kilogram = 2.2 pounds   1 pound = 0.45 kilogram

To change kilograms to pounds, multiply the number of kilograms by 2.2. To change pounds to kilograms, multiply the number of pounds by 0.45. Be sure to memorize these equivalents.

Example C
45 kg = ___ lb.
45 × 2.2 = 99 lb.

Example D
140 lb. = ___ kg
140 × 0.45 = 63 kg

CLASS PRACTICE

Find these equivalents.

a. 29 kg = ___ g   29,000
b. 2.18 g = 2,180 mg
c. 9 MT = 9,000 kg
d. 2,087 g = 2.087 kg
e. 3 kg = ___ mg   3,000,000
f. 2.08 MT = 2,080 kg
g. 390 g = 0.39 kg
h. 14 kg = 0.014 MT
i. 29 kg = 63.8 lb.
j. 84 lb. = 37.8 kg
k. 306 lb. = 137.7 kg
l. 180 kg = 396 lb.

WRITTEN EXERCISES

A. Write the abbreviations for these units by memory.

1. milligram   mg
2. kilogram   kg
3. hectogram   hg
4. decigram   dg
5. centigram   cg
6. metric ton   MT

B. Write these equivalents by memory. Memorize these equivalents well.

7. 1 kg = 2.2 lb.
8. 1 lb. = 0.45 kg

C. Find these equivalents.

9. 15 kg = ___ g   15,000
10. 15 g = ___ mg   15,000
11. 6,000 g = 6 kg
12. 8,000 mg = 8 g
13. 7.5 kg = 7,500 g
14. 9.3 g = 9,300 mg
15. 85 MT = ___ kg   85,000
16. 4.3 kg = 4,300 g
17. 255 mg = 0.255 g
18. 760 kg = 0.76 MT
19. 7 MT = 7,000 kg
20. 3.1 kg = ___ mg   3,100,000

D. Find these English/metric equivalents.

21. 15 kg = 33 lb.
22. 95 kg = 209 lb.
23. 120 lb. = 54 kg
24. 1,200 kg = 2,640 lb.

Lesson 22 T–78

Introduction

Why is *millipede* a name meaning 1,000 legs, yet a *millimeter* is not 1,000 meters but $\frac{1}{1,000}$ meter? The prefixes *milli-* and *kilo-* both mean "thousand." But when the prefixes were chosen, *kilo-* was arbitrarily established to mean 1,000 and *milli-* to mean $\frac{1}{1,000}$. So the word *kilogram* should make us think "thousand grams," and *milligram* should make us think "thousandth of a gram."

Teaching Guide

1. **The main metric units of weight are the kilogram, the hectogram, the dekagram, the gram, the decigram, the centigram, and the milligram.** Of these units, the kilogram, the gram, and the milligram are most frequently used while the others have almost no practical value.

2. **All conversions between metric units of weight can be made by simply moving the decimal point.** The procedure is the same as for making conversions between metric units of linear measure. Use *Class Practice* and the exercises below to drill this concept.

   a. 795 g = ___ kg (0.795)
   b. 7.8 g = ___ mg (7,800)
   c. 1,990 mg = ___ kg (0.00199)
   d. 13 kg = ___ g (13,000)
   e. 2.3 MT = ___ kg (2,300)
   f. 791 kg = ___ MT (0.791)

   Make special mention of the relationship between kilograms and metric tons. That relationship is 1,000 rather than the usual 10 when going from one unit to the next larger unit.

   g. 15 MT = ___ kg (15,000)
   h. 2.3 MT = ___ kg (2,300)
   i. 9,870 kg = ___ MT (9.87)
   j. 980 kg = ___ MT (0.98)

3. **To make conversions between kilograms and pounds, use the following rules.** Pupils are to memorize the numbers for making these conversions.

   To change kilograms to pounds, multiply the number of kilograms by 2.2.

   To change pounds to kilograms, multiply the number of pounds by 0.45.

   Give practice with finding English and metric equivalents.

   a. 72 lb. = ___ kg (32.4)
   b. 55 kg = ___ lb. (121)

T-79 Chapter 2 *English and Metric Measures*

Further Study

Lesson 21 states that the metric system has a scientific basis. The original standard for the meter was a bar of platinum constructed in 1793. In 1875, the International Bureau of Weights and Measures made a new standard from a bar of platinum alloy. More recently the meter has been defined as the distance light travels in a vacuum in $\frac{1}{299,792,458}$ of a second.

Metric units of volume, capacity, and weight are interrelated in their scientific definitions, which are based on pure water at its maximum density (4°C or 39.2° F). The following table shows these relationships.

Distilled Water at 4°C at Sea Level

Volume		Capacity		Weight
1 cm^3	=	1 ml	=	1 g
1 dm^3	=	1 l	=	1 kg
1 m^3	=	1 kl	=	1 MT

The gram is one of the most basic units of the metric system. It is used to measure mass, force, and weight. The relationship of the three is that, at sea level, 1 gram of mass is affected by 1 gram of gravitational force, which results in 1 gram of weight.

Solutions for Part E

25. 800×0.142
26. $75,000 \div 20$
27. 35×0.567
28. $20 \times 10 \times 2.26$
29. $\frac{5}{48} \times 24$
30. $1,550 \times 1,000$

E. Solve these reading problems.

25. A panda born in a zoo weighed 142 grams. Its mother was 800 times as heavy. Find the mother panda's weight in kilograms. **113.6 kg**

26. Marlin's father delivers coal in the winter. One week he delivered a total of 75 metric tons to 20 houses. On the average, how many kilograms of coal was that per family? **3,750 kg**

27. Rosebank Grocery has 35 boxes of oats cereal in stock. Each box contains 567 grams of cereal. What is the weight in kilograms of the total stock of oats cereal? **19.845 kg**

28. One month before the canning season began, Rosebank Grocery placed an order for 20 bales of sugar. Each bale contained 10 bags, and each bag held 2,260 grams of sugar. How many kilograms of sugar did the store order? **452 kg**

29. Hezekiah, king of Judah, became seriously ill. In response to his prayer for recovery, God moved the sundial backward ten degrees (2 Kings 20:11). If ten degrees was equal to $\frac{5}{48}$ of a day, how many hours backward did the sundial move? **$2\frac{1}{2}$ hours**

30. The Martins took a ferry to visit friends on a large island. Brother Wilmer and his son Merlin calculated by the ferry's displacement that the ship weighed 1,550 metric tons. How many kilograms is that? **1,550,000 kg**

REVIEW EXERCISES

F. Find these equivalents. *(Lessons 20, 21)*

31. 2 millennia = **2,000** yr.
32. 360 hr. = **15** days
33. 980 mm = **0.98** m
34. 70 dm = **7,000** mm

G. Change these times as indicated. *(Lesson 20)*

35. 5:00 P.M. Alaska Standard Time; ___ Atlantic Standard Time **10:00 P.M.**
36. 2:00 A.M. in Washington, D.C.; ___ in Honolulu, Hawaii **9:00 P.M.**

H. Solve these problems mentally. *(Lesson 13)*

37.	38.	39.	40.
4	16	11	5
× 7	− 9	− 5	− 2
÷ 2	+ 12	× 9	× 13
− 6	× 2	÷ 2	− 6
× 6	+ 4	+ 5	÷ 11
÷ 16	÷ 7	÷ 4	× 0
+ 3	÷ 3	− 4	+ 6
6	**2**	**4**	**6**

I. Solve by horizontal subtraction. *(Lesson 6)*

41. 74,842 − 36,662   **38,180**
42. $1,499.32 − 583.78   **$915.54**

23. Metric Units of Capacity

Units of capacity are based on volume, the amount an object can contain. The basic metric unit of capacity is the liter. One liter is the volume of one kilogram of water at its greatest density (4 degrees Celsius or about 39 degrees Fahrenheit). The liter is about 5 percent larger than the quart.

The liter is always the same size, whether used for liquid or dry measure.

The milliliter, the liter, and the kiloliter are the most common units of metric capacity. The centiliter, the deciliter, the dekaliter, and the hectoliter are seldom used.

Metric Units of Capacity

UNIT	ABBREVIATION	VALUE
kiloliter	kl	**1,000 l**
hectoliter	hl	100 l
dekaliter	dkl	10 l
liter	l	**1 l**
deciliter	dl	0.1 l
centiliter	cl	0.01 l
milliliter	ml	**0.001 l**

The steps for finding metric capacity equivalents should be familiar to you. They are the same as those used in Lessons 21 and 22.

1. Decide which operation is needed. To change a larger unit to a smaller unit, multiply. To change a smaller unit to a larger unit, divide.

2. Decide how many units larger or smaller the change involves. The decimal point will be moved that number of places.

3. Move the decimal point right to multiply and left to divide. You may need to annex zeroes to the right when multiplying (Example A) or to the left when dividing (Example B).

Example A	**Example B**
9.6 kl = ___ l	1,140 ml = ___ l
3 units smaller; multiply by 1,000	3 units larger; divide by 1,000
1,000 × 9.6 = 9,600 l	1,140 ÷ 1,000 = 1.14 l

LESSON 23

Objectives

- To review the basic metric unit of capacity (the liter) and its related units.
- To review changing from one metric unit of capacity to another.
- To teach the English/metric equivalents shown below. (These equivalents are to be memorized.)
  - 1 liquid quart = 0.95 liter
  - 1 liter = 1.06 liquid quarts
  - *1 gallon = 3.8 liters
  - *1 liter = 0.26 gallon

Review

1. Give Lesson 23 Quiz (English and Metric Measures).
2. *Find these equivalents.* (Lessons 19, 21, 22)
   a. 225 sq. ft. = ___ sq. yd. (25)
   b. 2 sq. ft. 95 sq. in. = ___ sq. in. (383)
   c. 34 km = ___ dkm (3,400)
   d. 83 dm = ___ hm (0.083)
   e. 18 mg = ___ g (0.018)
   f. 54 MT = ___ kg (54,000)
3. *Solve by mental subtraction.* (Lesson 7)
   a. 87 − 63 (24)
   b. 106 − 88 (18)
   c. 843 − 792 (51)
   d. 4,250 − 375 (3,875)

T–81 Chapter 2 *English and Metric Measures*

Introduction

Review the term *unit of capacity*. Such a unit measures volume, the amount of space something occupies. Have students volunteer the common English units of capacity (gallon, quart, pint, cup). The cubic yard, used to measure concrete, is also a unit of capacity.

Have the students ever seen anything that was measured in metric units of capacity? Perhaps the most common example in the United States is soft drinks sold in 2-liter bottles.

Do they have experience with purchasing gasoline by the liter? How many liters would it take to fill the fuel tank in a family car—20, 40, 60, 80, or 100? Most tanks hold 60 to 80 liters. The liter is just a bit larger than the quart; hence, filling a gasoline tank by the liter is much like filling it by the quart.

Note: If your school is in a country that uses the metric system, you will probably want to reverse the train of thought. Have your students ever been in a country such as the United States where gasoline is sold by the gallon? Did they think the pump meter was quite slow?

Teaching Guide

1. **The liter is the basic unit of capacity in the metric system.** Review the liter, its related units, and their abbreviations.

2. **All conversions between metric units of capacity can be made by simply moving the decimal point.** The procedure is the same as for making other conversions between metric units. Use *Class Practice* and the exercises below to drill this concept.

    a. 135 ml = ___ *l*   (0.135)
    b. 47 *l* = ___ kl   (0.047)
    c. 95 kl = ___ *l*   (95,000)
    d. 3.4 *l* = ___ ml   (3,400)

3. **To make conversions between liters, quarts, and gallons, use the following rules.** Pupils are to memorize the numbers for making these conversions.

    To change liters to quarts, multiply the number of liters by 1.06.

    To change quarts to liters, multiply the number of quarts by 0.95.

    To change liters to gallons, multiply the number of liters by 0.26.

    To change gallons to liters, multiply the number of gallons by 3.8.

    a. 12 qt. = ___ *l*   (11.4)
    b. 150 qt. = ___ *l*   (142.5)
    c. 75 *l* = ___ qt.   (79.5)
    d. 775 *l* = ___ qt.   (821.5)
    e. 95 *l* = ___ gal.   (24.7)
    f. 900 *l* = ___ gal.   (234)
    g. 125 gal. = ___ *l*   (475)
    h. 750 gal. = ___ *l*   (2,850)

English/Metric Capacity Equivalents

1 liter = 1.06 quarts   1 quart = 0.95 liter

To change liters to quarts, multiply the number of liters by 1.06. To change quarts to liters, multiply the number of quarts by 0.95. Be sure to memorize these equivalents.

Example C	**Example D**
55 l = ___ qt.	27 gal. = ___ l
55 × 1.06 = 58.3 qt.	27 × 3.8 = 102.6 l

The relationships between liters and gallons are shown below. To change liters to gallons, multiply the number of liters by 0.26. To change gallons to liters, multiply the number of gallons by 3.8.

   1 liter = 0.26 gallon   1 gallon = 3.8 liters

CLASS PRACTICE

Find these English and metric equivalents.

a. 20 l = <u>21.2</u> qt.
b. 46 qt. = <u>43.7</u> l
c. 211 qt. = ___ l 200.45
d. 350 l = <u>371</u> qt.
e. 38 gal. = <u>144.4 l</u>
f. 150 l = <u>39</u> gal.
g. 24 kl = <u>6,240</u> gal.
h. 65 l = <u>16.9</u> gal.

WRITTEN EXERCISES

A. *Write the abbreviations for these units by memory.*

1. milliliter   ml
2. kiloliter   kl
3. hectoliter   hl
4. deciliter   dl
5. centiliter   cl
6. dekaliter   dkl

B. *Write these equivalents by memory. Memorize these equivalents well.*

7. 1 l = <u>1.06</u> qt.
8. 1 qt. = <u>0.95</u> l
9. 1 l = <u>0.26</u> gal.
10. 1 gal. = <u>3.8</u> l

C. *Find these equivalents.*

11. 15 l = ___ ml   15,000
12. 35 kl = ___ l 35,000
13. 4.1 kl = <u>4,100</u> l
14. 7.7 l = <u>7,700</u> ml
15. 0.9 l = <u>900</u> ml
16. 85 ml = <u>0.085</u> l

82 Chapter 2 *English and Metric Measures*

D. Find these English/metric equivalents. See how many you can do without referring to the tables of equivalents.

17. 25 qt. = <u>23.75</u> *l*
18. 18 qt. = <u>17.1</u> *l*
19. 100 *l* = <u>106</u> qt.
20. 76 qt. = <u>72.2</u> *l*
21. 71 *l* = <u>75.26</u> qt.
22. 275 *l* = <u>291.5</u> qt.
23. 7 gal. = <u>26.6</u> *l*
24. 18 gal. = <u>68.4</u> *l*
25. 400 *l* = <u>104</u> gal.
26. 700 *l* = <u>182</u> gal.

E. Solve these reading problems.

27. A large tank contains 1,074 kiloliters of gasoline. If it takes an average of 75 liters to fill the tank of a car, how many cars can be fueled from the large tank?   14,320 cars

28. Huge supertankers can hold more than 475,000 kiloliters of petroleum, the source of oil and gasoline. How many times larger is 475,000 kiloliters than an 80-liter automobile fuel tank?   5,937,500 times

29. A tablespoon holds about 15 milliliters. How many tablespoons are in a liter? $66\frac{2}{3}$ tbsp

30. How many 175-milliliter servings of milk are in a 2-liter container? (Drop any remainder.)   11 servings

31. A truck delivered a load of dog food weighing 3 metric tons. How long would that amount feed a dog that eats 750 grams per day?   4,000 days

32. The Ulrich children walk 1 kilometer to school in 20 minutes. On the average, how many meters per minute do they walk?   50 m

REVIEW EXERCISES

F. Find these equivalents. *(Lessons 19, 21, 22)*

33. 2,160 sq. in. = <u>15</u> sq. ft.
34. 18 sq. yd. 8 sq. ft. = <u>170</u> sq. ft.
35. 4.8 km = ___ cm   480,000
36. 56.1 cm = <u>561</u> mm
37. 250 mg = <u>0.25</u> g
38. 40 MT = ___ kg   40,000

G. Solve by mental subtraction. *(Lesson 7)*

39. 110 − 94   16
40. 862 − 677   185
41. 419 − 131   288
42. 647 − 239   408

Lesson 23 T–82

Further Study

1. One liter is equal to one cubic decimeter or 1,000 cubic centimeters. Because of this relationship, one cubic centimeter is equal to one milliliter.

2. In 1960, the metric system became the base of a broadened system of weights and measures named the *International System of Units*, commonly referred to with the initials SI for *Systeme International*. Some of the metric units, including the liter, the metric ton, and the use of degrees for measuring angles, were abandoned by the broader system. But because these and several other units are so widely used, they continue to be common in the United States as "metric" units.

Solutions for Part E

27. $1{,}074{,}000 \div 75$

28. $475{,}000{,}000 \div 80$

29. $1{,}000 \div 15$

30. $2{,}000 \div 175$

31. $3{,}000{,}000 \div 750$

32. $1{,}000 \div 20$

T-83   Chapter 2   English and Metric Measures

LESSON 24

Objectives

- To review the basic metric unit of area (the square meter), its related units, and the use of the exponent 2 to mean "square."
- To review changing from one metric unit of area to another.
- To review the English/metric equivalents shown below. (These equivalents are to be memorized.)
  1 acre = 0.4 hectare
  1 hectare = 2.5 acres

Review

1. *Find these equivalents.* (Lessons 20, 22, 23)
   a. 336 days = ___ wk.   (48)
   b. 12 centuries 56 yr. = ___ yr.   (1,256)
   c. 78 kg = ___ lb.   (171.6)
   d. 110 lb. = ___ kg   (49.5)
   e. 26 kl = ___ l (26,000)
   f. 70 l = ___ qt.   (74.2)
   g. 52 gal. = ___ l (197.6)
   h. 440 l = ___ gal.   (114.4 with liter/gallon equivalent; 116.6 if done in two steps using the memorized liter/quart equivalent)

2. *Change these times as indicated.* (Lesson 20)
   a. 8:30 P.M. in Denver, Colorado; ___ in St. Johns, Newfoundland   (12:00 midnight)
   b. 2:35 A.M. Eastern Standard Time; ___ Alaska Standard Time   (10:35 P.M.)

3. *Multiply, and check by casting out nines.* (Lesson 8)

   a.   845
      × 603
      (509,535)

   b.   2,056
      × 827
      (1,700,312)

24. Metric Units of Area

Area is the surface within given boundaries. It has two dimensions: length and width.

For most of the metric square units, the abbreviations are the same as those for the respective linear units, with the exponent 2 added. The abbreviation for meter is m, and the abbreviation for square meter is m^2 instead of *sq. m*. (The exponent 2 means "square" and indicates the two dimensions, length and width.) The most common metric units and their abbreviations are shown at the top of page 84.

Metric measure of area also has a unit that is not named with *square*. The *hectare* is a square measure used for large areas, as the acre is in English measure. Its name is based on a little-used metric unit called the are (pronounced like *air* or *are*). The are is 100 square meters, and the hectare (100 ares) is 1 square hectometer or 10,000 square meters.

Metric units in the preceding lessons increased by powers of ten. Because area has two dimensions, a *square* decimeter is 10 × 10 times as much area as a *square* centimeter. For a linear unit whose measure is 100 times as long as another, the corresponding square measure is 100 × 100 (10,000) times as great. Any power of ten that shows the relationship between linear units is squared to show the relationship of the respective units of square measure.

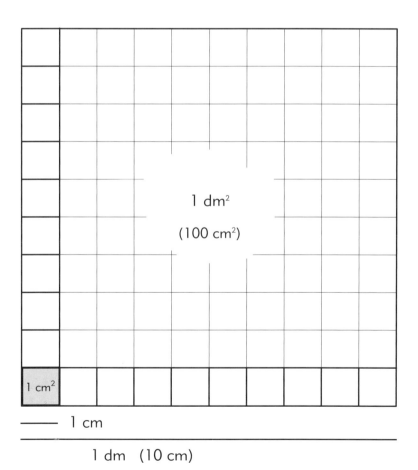

Metric Units of Area

UNIT	ABBREVIATION	VALUE
square kilometer	km²	100 ha or 1,000,000 m²
hectare	ha	10,000 m²
square meter	m²	1 m²
square centimeter	cm²	0.0001 m²
square millimeter	mm²	0.000001 m²

When changing square measures, give special attention to the number of places to move the decimal point. In measures of length, weight, and capacity, each step in the table of units is ten times greater or less than the one next to it. When measures are squared, one step is 100 times greater or less than the next. Not every step is listed in the table above, so some changes will multiply or divide by powers of 100. Each power of 100 moves two places in the decimal system, so all changes will be an even number of places.

The steps for finding metric equivalents also apply to area. These steps are similar to those used in previous lessons.

1. Decide which operation is needed. To change a larger unit to a smaller unit, multiply. (Move the decimal point to the right.) To change a smaller unit to a larger unit, divide. (Move the decimal point to the left.)

2. Move the decimal point the correct number of places. A simple way to find the correct number of places is to compare the values of the units involved. How many places does the digit 1 move in the stated value? Move your decimal point that many places. (Powers of 100 always move an even number of places.) You may need to annex zeroes to the right when multiplying (Example A) or to the left when dividing (Example B).

Example A
25 km² = ___ ha
Multiply by 100.
25 × 100 = 2,500 ha

Example B
5,600 cm² = ___ m²
Move 4 decimal places to divide by 10,000.
5,600 ÷ 10,000 = 0.56 m²

English/Metric Area Equivalents

1 hectare = 2.5 acres   1 acre = 0.4 hectare

To change hectares to acres, multiply the number of hectares by 2.5. To change acres to hectares, multiply the number of acres by 0.4. Be sure to memorize these equivalents.

Example C
74 ha = ___ a.
74 × 2.5 = 185 a.

Example D
150 a. = ___ ha
150 × 0.4 = 60 ha

Introduction

Review English units of area by asking the students, "What unit of measure would you use if you were discussing the size of the following things?"

a carpet (square yard)
a farm (acre)
a nation (square mile)
a lawn (square foot or square yard)
the plastic used to cover a book (square inch)

Now call attention to the list of metric units of area. What metric units would be used to measure the same areas?

a carpet (square meter)
a farm (hectare)
a nation (square kilometer)
a lawn (square meter)
the plastic used to cover a book
(square centimeter)

Teaching Guide

1. **Most metric units of area are squares formed by using linear units.** The most common metric units of area are the square centimeter, the square meter, and the square kilometer.

2. **The hectare, like the English acre, is a square measure; therefore, the term *square* does not precede *hectare*.** A hectare is actually a square hectometer. It measures plots of land just as the acre does. The term *hectare* is derived from a little-used metric unit called the are, which is equal to a square dekameter.

3. **For most of the metric square units, the abbreviations are the same as those for the respective linear units except that the exponent 2 is added.** An exception is the hectare, which is already a square unit. Point out the metric abbreviations shown in the lesson.

    In some math textbooks, the exponent 2 is used in both English and metric abbreviations for units of square measure, resulting in expressions such as "ft.^{2}" and "mi.^{2}" However, this does not appear to be a widespread practice.

4. **All conversions between metric units of area can be made by simply moving the decimal point.** The procedure is the same as for making other conversions between metric units. Use *Class Practice* and the exercises below to drill this concept.

    a. 220 ha = ___ km^2 (2.2)
    b. 45,000 m^2 = ___ ha (4.5)
    c. 750 cm^2 = ___ m^2 (0.075)
    d. 4,500 m^2 = ___ km^2 (0.0045)
    e. 3.2 km^2 = ___ ha (320)
    f. 7.7 ha = ___ m^2 (77,000)

T–85 Chapter 2 English and Metric Measures

5. To make conversions between hectares and acres, use the following rules. Pupils are to memorize the numbers for making these conversions.

To change hectares to acres, multiply the number of hectares by 2.5.

To change acres to hectares, multiply the number of acres by 0.4.

a. 7 a. = ___ ha (2.8)

b. 640 a. = ___ ha (256)

c. 15 ha = ___ a. (37.5)

d. 100 ha = ___ a. (250)

The following diagram shows the relationships of ares, hectares, and square kilometers.

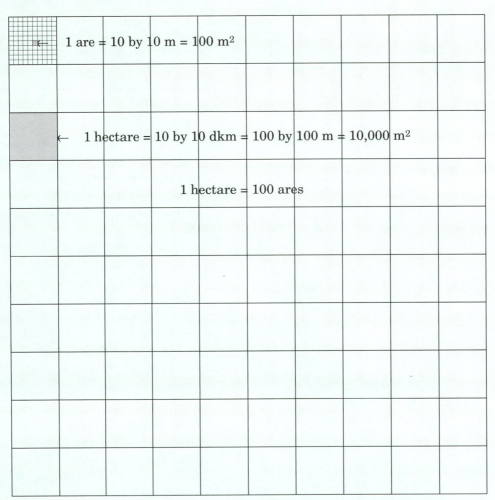

1 are = 10 by 10 m = 100 m²

1 hectare = 10 by 10 dkm = 100 by 100 m = 10,000 m²

1 hectare = 100 ares

1 square kilometer = 10 by 10 hm = 1,000 by 1,000 m = 1,000,000 m²
1 square kilometer = 100 hectares

Solutions for Part E

23. 15,000 ÷ 10,000

24. 44 × 10,000

25. 30,000 ÷ 180

CLASS PRACTICE

Find these equivalents.

a. 12 m² = ____ cm²   120,000
b. 2.34 km² = 234 ha
c. 76,000 m² = 0.076 km²
d. 28 ha = ____ m²   280,000
e. 81 ha = 202.5 a.
f. 210 a. = 84 ha
g. 4.82 ha = 12.05 a.
h. 887.5 a. = 355 ha

WRITTEN EXERCISES

A. *Write the abbreviations for these units by memory.*

1. square millimeter   mm²
2. square centimeter   cm²
3. square meter   m²
4. hectare   ha

B. *Write these equivalents by memory.*

5. 1 ha = 2.5 a.
6. 1 a. = 0.4 ha

C. *Find these equivalents.*

7. 15 ha = ____ m²   150,000
8. 3 km² = 300 ha
9. 90,000 cm² = 9 m²
10. 60,000 m² = 6 ha
11. 4.2 ha = ____ m²   42,000
12. 5.7 km² = 570 ha
13. 2,500 cm² = 0.25 m²
14. 95 ha = 0.95 km²
15. 10,500 cm² = 1.05 m²
16. 55 km² = 5,500 ha

D. *Find these English/metric equivalents. See how many you can do without looking at the tables.*

17. 25 ha = 62.5 a.
18. 172 ha = 430 a.
19. 90 a. = 36 ha
20. 500 a. = 200 ha
21. 98 ha = 245 a.
22. 1 km² = 250 a.

E. *Solve these reading problems.*

23. Brother Lehman calculated the school property to be 15,000 square meters. How many hectares is that?   1.5 ha

24. The smallest independent country, Vatican City, is within the city of Rome, Italy. Vatican City has an area of 44 hectares. How many square meters are in 44 hectares?   440,000 m²

25. The second smallest independent country, Monaco, borders France and the Mediterranean Sea. Monaco has an area of 1.8 square kilometers and a population of 30,000. What is the average population of Monaco per hectare? (Round to the nearest whole number.)   167

86   Chapter 2   *English and Metric Measures*

26. The storm windows in the Wilson residence have an area of 2,800 square centimeters each. How many storm windows can they cut from a piece of Plexiglas measuring 3 square meters? (Drop any remainder.)   10 windows

27. The Lamberts are tearing down an unused chimney that rests on a wall of the house. The chimney has 48 courses of bricks. Each course has 6 bricks, and each brick weighs 1,575 grams. What is the total weight of the bricks in kilograms?   453.6 kg

28. The Lamberts estimate that the chimney was built 175 years ago. How many months ago was that?   2,100 months

REVIEW EXERCISES

F. Find these equivalents. *(Lessons 20, 22, 23)*

29. 5 decades 7 yr. = __57__ yr.
30. 744 hr. = __31__ days
31. 2.8 kg = __2,800__ g
32. 4.65 kg = __10.23__ lb.
33. 420 l = __0.42__ kl
34. 34 gal. = __129.2__ l

G. Change these times as indicated. *(Lesson 20)*

35. 3:00 P.M. in St. Johns, Newfoundland; ___ in Albuquerque, New Mexico   11:30 A.M.
36. 1:30 A.M. in Los Angeles, California; ___ in Toronto, Ontario   4:30 A.M.

H. Multiply. Check by casting out nines. *(Lesson 8)*

37. 934	38. 854	39. 1,240	40. 1,834
× 762	× 649	× 1,531	× 1,356
711,708	554,246	1,898,440	2,486,904

26. 30,000 ÷ 2,800
27. 48 × 6 × 1.575
28. 175 × 12

LESSON 25

Objectives

- To review adding and subtracting compound English measures.
- To review adding and subtracting compound metric measures by first changing them to like units.

Review

1. *Find these equivalents.* (Lessons 17, 19, 21, 23, 24)
   a. 27 ft. 3 in. = ___ in. (327)
   b. 9 sq. mi. = ___ a. (5,760)
   c. 130 cm = ___ hm (0.013)
   d. 10 km = ___ m (10,000)
   e. 28 gal. = ___ l (106.4)
   f. 42 l = ___ gal. (10.92 if calculated with liter/gallon equivalent; 11.13 if done in two steps with memorized liter/quart equivalent)
   g. 11,000 m² = ___ ha (1.1)
   h. 1.87 km² = ___ ha (187)
   i. 4.8 ha = ___ a. (12)
   j. 620 a. = ___ ha (248)

2. *Solve by mental multiplication.* (Lesson 9)
   a. 6 × 172 (1,032)
   b. 8 × 4,805 (38,440)

25. Adding and Subtracting Compound Measures

In the English measuring system, people often use compound measures. They say that a pail holds 2 gallons 2 quarts, that a baby weighs 7 pounds 9 ounces, or that a person is 5 feet 3 inches tall. If one board is 11 feet 5 inches long and another is 9 feet 9 inches long, addition is required to find the total length of the two boards.

Metric measures are usually not expressed in the compound form because the decimal relationship of units makes conversion very easy. For example, instead of saying that you are 1 meter 55 centimeters tall, you would say either 1.55 meters or 155 centimeters. However, different metric units must sometimes be added or subtracted.

To add compound English measures, follow these steps.
1. Add each unit separately.
2. Simplify the answer by changing smaller units to larger units when possible. See Example A.

To subtract compound English measures, follow these steps.
1. If borrowing is required, borrow 1 from the next larger unit and change it to the correct number of smaller units. Add that amount to the number of smaller units already in the minuend.
2. Subtract each unit separately.

In Example B, 1 minute is changed to 60 seconds and added to the 15 seconds already there. The result (75) is used as the minuend.

Example A	**Example B**
4 lb. 12 oz.	5 75 (60 + 15)
+ 4 lb. 8 oz.	6̶ min. 1̶5̶ sec.
8 lb. 20 oz. = 9 lb. 4 oz.	− 3 min. 25 sec.
	2 min. 50 sec.

To add or subtract metric measures, use the following steps.
1. Change all measures to the unit desired for the answer.
2. Add or subtract as indicated. Carry or borrow just as you do when you add or subtract other decimals.

Example C	**Example D**
17 m + 225 cm = ___ m	1.25 kg − 925 g = ___ g
225 cm = 2.25 m	1.25 kg = 1,250 g
17 m + 2.25 m = 19.25 m	1,250 g − 925 g = 325 g

88   Chapter 2   *English and Metric Measures*

CLASS PRACTICE

Solve these problems.

a. 8 lb. 9 oz.
   + 3 lb. 4 oz.
   11 lb. 13 oz.

b. 4 cups 4 fl. oz.
   + 5 cups 6 fl. oz.
   10 cups 2 fl. oz.

c. 10 ft. 9 in.
   − 6 ft. 6 in.
   4 ft. 3 in.

d. 12 gal. 1 qt.
   − 8 gal. 3 qt.
   3 gal. 2 qt.

e. 14 m + 45 cm = 1,445 cm

f. 8.8 MT + 1,400 kg = 10.2 MT

g. 4 l − 500 ml = 3,500 ml

h. 28 km − 14 hm = 26.6 km

WRITTEN EXERCISES

A. *Solve these problems involving English measures.*

1. 3 ft. 8 in.
    + 4 ft. 7 in.
    8 ft. 3 in.

2. 6 yd. 2 ft.
    + 4 yd. 2 ft.
    11 yd. 1 ft.

3. 4 lb. 13 oz.
    + 9 lb. 14 oz.
    14 lb. 11 oz.

4. 5 bu. 3 pk.
    + 4 bu. 2 pk.
    10 bu. 1 pk.

5. 7 gal. 3 qt.
    + 5 gal. 2 qt.
    13 gal. 1 qt.

6. 5 pk. 5 qt.
    + 3 pk. 4 qt.
    9 pk. 1 qt.

7. 7 ft. 2 in.
    − 3 ft. 6 in.
    3 ft. 8 in.

8. 8 yd. 1 ft.
    − 3 yd. 2 ft.
    4 yd. 2 ft.

9. 8 lb. 4 oz.
    − 3 lb. 9 oz.
    4 lb. 11 oz.

10. 6 bu. 1 pk.
    − 4 bu. 3 pk.
    1 bu. 2 pk.

11. 8 gal. 1 qt.
    − 2 gal. 3 qt.
    5 gal 2 qt.

12. 7 pk. 1 qt.
    − 4 pk. 3 qt.
    2 pk. 6 qt.

B. *Solve these problems involving metric measures.*

13. 20 cm + 40 mm = 240 mm

14. 2.2 kg + 475 g = 2,675 g

15. 1,141 kg + 4.5 MT = 5.641 MT

16. 290 l + 3.7 kl = 3.99 kl

17. 2.7 MT − 990 kg = 1.71 MT

18. 2.9 kl − 1,160 l = 1.74 kl

19. 5.5 km − 775 m = 4,725 m

20. 7.7 l − 75 ml = 7.625 l

C. *Solve these reading problems.*

21. Last week Father worked the following hours: Monday, 8 hours 15 minutes; Tuesday, 9 hours 45 minutes; Wednesday, 8 hours 30 minutes; Thursday, 10 hours 15 minutes; and Friday, 8 hours 30 minutes. What was the total number of hours worked?
    $45\frac{1}{4}$ hours

22. Lucy is 9 years 2 months old. Her brother Galen's age is 4 years 5 months. How much younger is Galen than Lucy?
    4 years 9 months

23. The Weavers harvested 144 metric tons of shelled corn from a 15-hectare field and 17,800 kilograms from a 2-hectare field. What was the total yield in metric tons?
    161.8 metric tons

Lesson 25   T–88

Introduction

Suppose you have two packs of meat whose weights are 1 pound 12 ounces and 2 pounds 14 ounces. Why would it not be proper to say that their total weight is 3 pounds 26 ounces? (The number of ounces, 26, is greater than 16, the number of ounces in a pound.)

When adding compound measures, it is often necessary to simplify the sum.

Teaching Guide

1. **To add compound English measures, use the following steps.**

    (1) Add each unit separately.

    (2) Simplify the answer by changing smaller units to larger units when possible.

    a.   5 lb.   9 oz.
       + 3 lb. 11 oz.
       (8 lb. 20 oz. = 9 lb. 4 oz.)

    b.   7 yr.   9 mo.
       + 8 yr. 11 mo.
       (15 yr. 20 mo. = 16 yr. 8 mo.)

    c.   8 bu.   3 pk.
       + 7 bu.   2 pk.
       (15 bu.   5 pk. = 16 bu. 1 pk.)

2. **To subtract compound English measures, use the following steps.**

    (1) If borrowing is required, borrow one from the next larger unit and change it to the correct number of smaller units. Add that amount to the number of smaller units already in the minuend.

    (2) Subtract each unit separately.

    a. $\quad$ 5 $\quad$ 13 (12 + 1)
       $\quad$ $\cancel{6}$ ft. $\cancel{1}$ in.
       $-$ 2 ft.   5 in.
       (3 ft.   8 in.)

    b. $\quad$ 2 $\quad$ 5 (4 + 1)
       $\quad$ $\cancel{3}$ gal. $\cancel{1}$ qt.
       $-$ 1 gal.   3 qt.
       (1 gal.   2 qt.)

    c. $\quad$ 4 $\quad$ 9 (7 + 2)
       $\quad$ $\cancel{5}$ wk. $\cancel{2}$ days
       $-$ 1 wk.   5 days
       (3 wk.   4 days)

3. **To add or subtract metric measures, use the following steps.**

    (1) Change all units to the unit desired for the answer.

Solutions for Part C

21. $8\frac{1}{4} + 9\frac{3}{4} + 8\frac{1}{2} + 10\frac{1}{4} + 8\frac{1}{2}$

22. 8 yr. 14 mo. − 4 yr. 5 mo.

23. 144 + 17.8

T–89 Chapter 2 English and Metric Measures

(2) Add or subtract as indicated. Carry or borrow just as you do when you add or subtract other decimals.

a. 7 m + 124 cm = ___ cm (824)
b. 12 MT + 3,543 kg = ___ kg (15,543)
c. 2 kl − 567 l = ___ kl (1.433)
d. 6 km − 2,650 m = ___ m (3,350)

24. $(1\tfrac{1}{3} + 2\tfrac{1}{3} + 1\tfrac{2}{3} + 2\tfrac{2}{3}) \times \$1.95$

25. $14{,}181 \div 640$

26. $247{,}501 \div 640$

24. Mother is buying four pieces of remnant cloth. The pieces have these lengths: 1 yard 1 foot; 2 yards 1 foot; 1 yard 2 feet; and 2 yards 2 feet. At $1.95 per yard, what is the cost of the material? $15.60

25. Hong Kong is one of the most heavily populated regions in the world. It has an average population density of 14,181 people per square mile. What is the average population per acre? (Round your answer to the nearest whole number.)   22

26. The city of Victoria, Hong Kong, is one of the most densely populated places in the world, with an average population density of 247,501 people per square mile. What is the average population of Victoria per acre? (Round your answer to the nearest whole number.)   387

REVIEW EXERCISES

D. Find these equivalents. *(Lessons 17, 19, 21, 23, 24)*

27. 9,600 a. = __15__ sq. mi.   28. 82 sq. ft. = ____ sq. in.   11,808
29. 2,640 mm = 0.264 dkm   30. 5.2 hm = 5,200 dm
31. 238 *l* = 61.88 gal.   32. 3.5 kl = 3,710 qt.
33. 60 ha = 150 a.   34. 735 a. = 294 ha

E. Solve by mental multiplication. *(Lesson 9)*

35. 1,000 × 345.2   345,200   36. 7 × 3,810   26,670

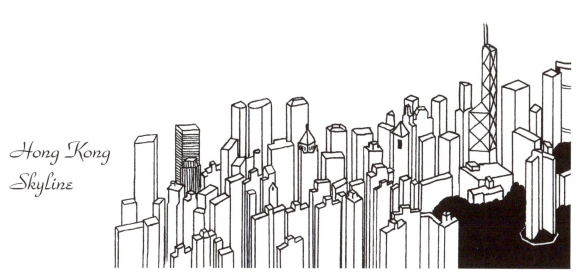

Hong Kong Skyline

The earth is the LORD's and the fulness thereof; the world, and they that dwell therein.   Psalm 24:1

90   Chapter 2   *English and Metric Measures*

26. Multiplying and Dividing Compound Measures

Multiplying Compound English Measures

To multiply compound English measures, follow the same steps as when adding them.
1. Multiply each unit separately.
2. Simplify the answer by changing as many smaller units as possible to larger units.

Example A
```
 4 tons    750 lb.
        x  5
20 tons  3,750 lb.  =  21 tons  1,750 lb.
```

Example B
```
 5 ft.    7 in.
      x  9
45 ft.   63 in.  =  50 ft.  3 in.
```

Dividing Compound English Measures

To divide compound English measures, use these steps:
1. Divide the amount of the larger unit.
2. If there is a remainder, change it to the smaller unit.
3. Add the remainder to the smaller unit.
4. Divide the smaller unit.
5. If there is a remainder, express it as a fraction in lowest terms. In Example D, the remainder of 6 would result in the fraction $\frac{6}{9}$, which reduces to $\frac{2}{3}$.

Example C
```
        2 ft.  5 in.
6)14 ft.  6 in.
  12
   2 ft.  6 in.
       = 30 in.
         30
```

Example D
```
         10 pk.  4⅔ qt.
9)95 pk.  2 qt.
  90
   5 pk.  2 qt.
       =  42 qt.
          36
           6
```

Multiplying and Dividing Metric Measures

To multiply or divide metric measures, use these steps:
1. Change all units to the unit desired for the answer.
2. Multiply or divide as indicated. Calculate the same as in any problem involving decimals.

LESSON 26

Objectives

- To review multiplying and dividing compound English measures.
- To review multiplying and dividing compound metric measures by first changing them to like units.

Review

1. *Add or subtract these compound measures.* (Lesson 25)

    a. 12 bu.   3 pk.
        − 3 bu.   3 pk.
        (9 bu.)

    b.   5 lb. 14 oz.
        + 6 lb.   2 oz.
        (12 lb.)

    c.   3 ft.   3 in.
        − 1 ft. 11 in.
        (1 ft.   4 in.)

    d. 54 g + 89 mg = ___ mg (54,089)

    e. 34.5 km − 21 dkm = ___ km
       (34.29)

2. *Find these equivalents.* (Lessons 18, 22, 24)

    a. 15 fl. oz. = ___ tbsp. (30)
    b. 5 pk. 3 qt. = ___ qt. (43)
    c. 36 kg = ___ lb. (79.2)
    d. 840 lb. = ___ kg (378)
    e. 68.2 ha = ___ a. (170.5)
    f. 500 a. = ___ ha (200)

3. *Calculate mentally by applying the distributive law.* (Lesson 10)

    a. 22 × 26 (572)
    b. 14 × 32 (448)
    c. 12 × 88 (1,056)
    d. 13 × 62 (806)

T–91 Chapter 2 *English and Metric Measures*

Introduction

Write the following multiplication problems on the board. Solve the first and second problems in the manner illustrated, which yields the correct answers. Then solve the third problem by using the same method. Ask the students, "Is the answer to problem c correct? What is wrong?"

a.   29
    × 3
    ―――
    87

b.   2 m 9 dm
       × 3
    ―――――――
    8 m 7 dm

c.   2 ft. 9 in.
         × 3
    ――――――――――
    8 ft. 7 in.

The third answer is wrong because there are not 10 inches in a foot; the normal multiplication process cannot be used in this problem. (The correct answer is 8 ft. 3 in.) Carrying as we normally do in multiplication assumes a decimal system; that is, that each place to the left of another has ten times its value. The English measures have values other than ten from one unit to another. Therefore, we multiply the different units as separate multiplication problems.

See *Further Study* for a demonstration of how the problem above could be solved by calculating with base twelve numbers.

Teaching Guide

1. **To multiply compound English measures, use the following steps.** (They are much the same as the steps for addition.)

   (1) Multiply each unit separately.

   (2) Simplify the answer by changing as many smaller units as possible to larger units.

   a.   5 ft. 6 in.
             × 7
       ――――――――――
       (35 ft. 42 in. = 38 ft. 6 in.)

   b.   3 lb. 5 oz.
             × 6
       ――――――――――
       (18 lb. 30 oz. = 19 lb. 14 oz.)

   c.   6 qt. 3 cups
             × 5
       ―――――――――――
       (30 qt. 15 cups = 33 qt. 3 cups)

2. **To divide compound English measures, use the following steps.**

   (1) Divide the amount of the larger unit.

   (2) If there is a remainder, change it to the smaller unit.

   (3) Add the remainder to the smaller unit.

   (4) Divide the smaller unit.

   (5) If there is a remainder, express it as a fraction in lowest terms.

   a.   (1 ft. 10 in.)
       8)14 ft. 8 in.

   b.   (3 gal. 3$\frac{4}{5}$ qt.)
       5)19 gal. 3 qt.

   c.   (2 lb. 10$\frac{1}{3}$ oz.)
       6)15 lb. 14 oz.

3. **To multiply or divide metric measures, use the following steps.**

   (1) Change all units to the unit desired for the answer.

Lesson 26

Example E	**Example F**
4 × 7 m 17 cm = ___ cm	78 kg 300 g ÷ 9 = ___ kg
7 m 17 cm = 717 cm	78 kg 300 g = 78.3 kg
4 × 717 = 2,868 cm	78.3 ÷ 9 = 8.7 kg

CLASS PRACTICE

Solve these problems.

a. 4 ft. 8 in.
    × 3
    ―――
    14 ft.

b. 3 days 12 hr.
    × 7
    ―――――
    24 days 12 hr.

c. 6 gal. 2 qt.
    × 11
    ―――――
    71 gal. 2 qt.

d. 2 hr. 36 min.
    4)10 hr. 24 min.

e. 2 lb. 4 oz.
    7)15 lb. 12 oz.

f. 1 bu. 1⅔ pk.
    9)12 bu. 3 pk.

g. 5 × 10 m 14 cm = <u>50.7</u> m

h. 8 × 25 MT 824 kg = ____ MT 206.592

i. 8 × 32 kl 90 *l* = ____ *l* 256,720

j. 15 hm 50 m ÷ 5 = <u>310</u> m

k. 95 *l* 360 ml ÷ 8 = <u>11.92</u> *l*

l. 66 g 920 mg ÷ 7 = <u>9,560</u> mg

WRITTEN EXERCISES

A. *Solve these problems involving English measures.*

1. 7 ft. 4 in.
    × 6
    ―――
    44 ft.

2. 6 lb. 15 oz.
    × 7
    ―――――
    48 lb. 9 oz.

3. 4 pk. 6 qt.
    × 5
    ―――
    23 pk. 6 qt.

4. 4 bu. 2 pk.
    × 6
    ―――
    27 bu.

5. 2 ft. 9 in.
    7)19 ft. 3 in.

6. 3 yd. 2 ft.
    5)18 yd. 1 ft.

7. 4 bu. 2⅓ pk.
    6)27 bu. 2 pk.

8. 5 gal. 1¾ qt.
    8)43 gal. 2 qt.

B. *Solve these problems involving metric measures.*

9. 8 × 45 cm 6 mm = <u>3,648</u> mm

10. 7 × 6 kg 575 g = ____ g 46,025

11. 9 × 7 MT 260 kg = <u>65.34</u> MT

12. 6 × 8 kl 360 *l* = <u>50.16</u> kl

13. 5 × 7 km 720 m = <u>38.6</u> km

14. 8 × 2 *l* 175 ml = ____ ml 17,400

15. 75 cm 6 mm ÷ 9 = <u>84</u> mm

16. 3 kg 600 g ÷ 5 = <u>720</u> g

17. 4 MT 600 kg ÷ 25 = <u>184</u> kg

18. 16 kl 260 *l* ÷ 30 = <u>542</u> *l*

19. 453 m 90 cm ÷ 60 = <u>7.565</u> m

20. 4 *l* 662 ml ÷ 3 = <u>1.554</u> *l*

92   Chapter 2   English and Metric Measures

C. Solve these reading problems.

21. The Petres have an encyclopedia set with 29 volumes. If the average weight of each volume is 1 pound 14 ounces, what is the weight of the whole set?   54 pounds 6 ounces

22. Mother canned 42 quarts of applesauce. The weight of the sauce in each jar was 3 pounds 1 ounce. What was the weight of the 42 quarts of applesauce?   128 lb. 10 oz.

23. Susan was 1 foot 8 inches tall when she was born. Twelve years later, her height is 5 feet 2 inches. What was her average growth per year? $3\frac{1}{2}$ inches

24. Uncle James planted a tree that was 2 feet 4 inches tall. Thirty-two years later, the tree was 63 feet tall. What was the average growth of the tree per year?
   1 foot $10\frac{3}{4}$ inches (or $22\frac{3}{4}$ in.)

25. The basement walls of the Petres' house are capped with pieces of cut granite. The entire front wall has three pieces with these lengths: 14 feet 6 inches, 4 feet 4 inches, and 12 feet 11 inches. What is the total length of the three pieces?   31 feet 9 inches

26. The seventh and eighth grade classroom is 28 feet 3 inches long and 21 feet 7 inches wide. How much greater is the length than the width?   6 feet 8 inches

REVIEW EXERCISES

D. Add or subtract these compound measures. *(Lesson 25)*

27. 25 yd. 2 ft.
    − 18 yd. 1 ft.
    ───────────
      7 yd. 1 ft.

28.   8 lb. 5 oz.
    − 5 lb. 12 oz.
    ───────────
      2 lb. 9 oz.

29. 86 kl + 301 l = ____ l 86,301

30. 22 kg − 870 g = 21.13 kg

E. Find these equivalents. *(Lessons 18, 22, 24)*

31. 56 fl. oz. = __7__ cups

32. 15 qt. 1 pt. = __31__ pt.

33. 88 kg = 193.6 lb.

34. 420 lb. = 189 kg

35. 84 ha = 210 a.

36. 665 a. = 266 ha

F. Calculate mentally by applying the distributive law. *(Lesson 10)*

37. 15 × 34   510

38. 22 × 61   1,342

(2) Multiply or divide as indicated. Calculate the same as in any problem involving decimals.

a. 6 × 5 m 16 cm = ___ cm   (3,096)
b. 8 × 21 kl 150 l = ___ kl   (169.2)
c. 25 m 16 cm ÷ 4 = ___ m   (6.29)
d. 35 kg 250 g ÷ 5 = ___ g   (7,050)

Further Study

In reference to the illustration in the introduction, the standard multiplication process could be used to solve the problem if numbers in a base twelve system were used. The students actually use the same principle when changing units in this lesson. Therefore, this lesson is a practical application of working with number systems other than base ten—a concept that is taught in Chapter 12.

$$\begin{array}{r} 2 \\ 2\text{ ft.}\quad 9\text{ in.} \\ \underline{\times\, 3} \\ 8\text{ ft.}\quad 3\text{ in.} \end{array} \quad (3 \times 9 = 23_{\text{twelve}})$$

Solutions for Part C

21. 29 × 1 lb. 14 oz.
22. 42 × 3 lb. 1 oz.
23. (62 − 20) ÷ 12 or (4 ft. 14 in. − 1 ft. 8 in.) − 12
24. (62 ft. 12 in. − 2 ft. 4 in.) ÷ 32 or (756 − 28) ÷ 32
25. 14 ft. 6 in. + 4 ft. 4 in. + 12 ft. 11 in.
26. 27 ft. 15 in. − 21 ft. 7 in.

T-93   Chapter 2   English and Metric Measures

LESSON 27

Objectives

- To review making conversions between metric and English measures.
- To emphasize the English/metric equivalents that are to be memorized in this chapter.

Review

1. Give Lesson 27 Quiz (Metric Measures).

2. *Multiply or divide these compound measures.* (Lesson 26)

   a. 25 ft. 5 in.
       $\underline{\times 6}$
       (152 ft. 6 in.)

   b. (2 hr. 49 min.)
       8)$\overline{22\text{ hr. }32\text{ min.}}$

   c. 55 km 4 dkm ÷ 32 = ___ dkm (172)

   d. 41 × 91 MT 27 kg = ___ MT
       (3,732.107)

3. *Add or subtract these compound measures.* (Lesson 25)

   a. 14 gal.   2 qt.
       $\underline{+\ 8\text{ gal.}\quad 3\text{ qt.}}$
       (23 gal.   1 qt.)

   b. 15 pk.   3 qt.
       $\underline{-12\text{ pk.}\quad 7\text{ qt.}}$
       (2 pk.   4 qt.)

   c. 35.9 kl + 242 l = ___ l (36,142)

   d. 8 hm − 64 m = ___ m (736)

4. *Find these equivalents.* (Lessons 19, 23)

   a. 7,920 sq. in. = ___ sq. ft. (55)

   b. 34 sq. yd. 2 sq. ft. = ___ sq. ft.
       (308)

   c. 75 gal. = ___ l (285)

   d. 1,500 l = ___ gal.
       (390 with liter/gallon;
       397.5 with liter/quart ÷ 4)

5. *Solve by long division, and check by casting out nines.* (Lesson 11)

   a. (1,186 R 47)
       56)$\overline{66,463}$

   b. (1,280 R 320)
       502)$\overline{642,880}$

27. Converting Between English and Metric Measures

Because metric and English units are both in wide use, it is often necessary to find equivalent metric or English measures. Lessons 21–24 contain relationships for you to memorize. This lesson includes those and several other relationships for making conversions between English and metric measures. Be sure to memorize the ones in boldface in the tables below. They have all been covered in previous lessons, except the ton/kg equivalent.

By using the equivalency tables below, you can make conversions between English and metric measures in two simple steps.

1. Choose the correct relationship. To change a metric unit to an English unit, use the "Metric to English" table. To change an English unit to a metric unit, use the "English to Metric" table.

2. Multiply the number of units by the relationship between the two units.

Metric to English	English to Metric
Length	*Length*
1 cm = 0.39 in.	**1 in. = 2.54 cm**
1 m = 39.4 in.	**1 ft. = 0.3 m**
1 m = 3.28 ft.	1 yd. = 0.91 m
1 km = 0.62 mi.	1 mi. = 1.61 km
Weight	*Weight*
1 g = 0.035 oz.	1 oz. = 28.3 g
1 kg = 2.2 lb.	**1 lb. = 0.45 kg**
1 MT = 2,205 lb.	**1 ton = 907 kg**
1 MT = 1.1 tons	1 ton = 0.91 MT
Capacity	*Capacity*
1 l = 1.06 qt. (liquid)	1 gal. = 3.8 l
1 l = 0.26 gal.	**1 qt. = 0.95 l**
1 ml = 0.07 tbsp.	1 tbsp. = 15 ml
Area	*Area*
1 m² = 1.2 sq. yd.	1 sq. yd. = 0.84 m²
1 ha = 2.5 a.	**1 a. = 0.4 ha**
1 km² = 0.39 sq. mi.	1 sq. mi. = 2.59 km²

Chapter 2 English and Metric Measures

Example A
7 mi. = ___ km
1 mi. = 1.61 km
7 × 1.61 = 11.27 km

Example B
8 MT = ___ tons
1 MT = 1.1 tons
8 × 1.1 = 8.8 tons

CLASS PRACTICE

Find these equivalents.

a. 4 ft. = 1.2 m
b. 14 m = 45.92 ft.
c. 27 mi. = 43.47 km
d. 92 ft. = 27.6 m
e. 35 gal. = 133 l
f. 22 ha = 55 a.
g. 81 l = 85.86 qt.
h. 90 g = 3.15 oz.
i. 26 in. = 66.04 cm
j. 30 sq. yd. = 25.2 m^2

WRITTEN EXERCISES

A. Find these equivalents.

1. 17 cm = 6.63 in.
2. 14 ft. = 4.2 m
3. 11 in. = 27.94 cm
4. 40 mi. = 64.4 km
5. 240 g = 8.4 oz.
6. 12 oz. = 339.6 g
7. 76 lb. = 34.2 kg
8. 77 tons = 70.07 MT
9. 65 l = 16.9 gal.
10. 90 l = 23.4 gal.
11. 3 qt. = 2.85 l
12. 2,000 l = 520 gal.
13. 45 km^2 = 17.55 sq. mi.
14. 300 sq. yd. = 252 m^2
15. 12 sq. mi. = 31.08 km^2
16. 650 m^2 = 780 sq. yd.
17. 25 MT = 27.5 tons
18. 240 gal. = 912 l
19. 72 ha = 180 a.
20. 17 yd. = 15.47 m

B. Solve these reading problems.

21. As the Haldemans traveled in Canada, they observed the metric road signs used there. One sign they saw said, "Toronto—150 kilometers." How many miles were they from Toronto?   93 miles

22. Another sign said, "Road Construction—300 meters." How many feet were they from the road construction area?   984 feet

23. After he bought gasoline, Father said he had put 65 liters into the tank. How many gallons was that?   16.9 gallons

24. At one bridge, a sign read, "Maximum weight—13,000 kilograms." How many English tons was that?   14.3 tons

Introduction

Give an ungraded quiz to see if the students remember the common metric/English equivalents.

    a. 12 ft. = ___ m (3.6)
    b. 100 l = ___ qt. (106)
    c. 10 m = ___ ft. (32.8)
    d. 5 a. = ___ ha (2)
    e. 10 kg = ___ lb. (22)

When the students are either finished or stumped, have them open their textbooks and find the correct equivalents. English/metric tables are convenient tools for making these conversions.

Teaching Guide

By the use of equivalency tables, we can make conversions between English and metric measures in two simple steps.

(1) Choose the correct relationship. To change a metric unit to an English unit, use the "Metric to English" table. To change an English unit to a metric unit, use the "English to Metric" table.

(2) Multiply the number of units by the relationship between the two units.

Change these metric measures to English measures.

    a. 12 m = ___ in. (472.8)
    b. 275 g = ___ oz. (9.625)
    c. 79 MT = ___ tons (86.9)
    d. 95 l = ___ gal. (24.7)
    e. 80 ml = ___ tbsp. (5.6)
    f. 30 m^2 = ___ sq. yd. (36)

Change these English measures to metric measures.

    a. 76 in. = ___ cm (193.04)
    b. 3 tons = ___ kg (2,721)
    c. 17 tons = ___ MT (15.47)
    d. 29 gal. = ___ l (110.2)
    e. 12 sq. yd. = ___ m^2 (10.08)
    f. 53 a. = ___ ha (21.2)

Solutions for Part B

21. 150 × 0.62
22. 300 × 3.28
23. 65 × 0.26
24. 13 × 1.1

T–95 *Chapter 2 English and Metric Measures*

25. 4,900 × 0.45
26. 42 × 2.5

25. "Our van weighs 4,900 pounds," Father informed the children. How many kilograms is that? 2,205 kg

26. The Haldemans stayed with a family that lived on a 42-hectare farm. How many acres were on the farm? 105 acres

REVIEW EXERCISES

C. Multiply or divide these compound measures. *(Lesson 26)*

27. 12 hr. 10 min.
     × 4
     48 hr. 40 min.

28. 21 lb. 10 oz.
     × 9
     194 lb. 10 oz.

29. 8 kl 370 l ÷ 9 = 930 l

30. 4 × 32 kg 180 g = ___ kg   128.72

D. Add or subtract these compound measures. *(Lesson 25)*

31. 34 qt. 1 pt.
   + 16 qt. 3 pt.
   52 qt.

32. 6 yr. 290 days
   − 2 yr. 310 days
   3 yr. 345 days

33. 1.78 dkm + 22 mm = ____ mm 17,822   **34.** 21 kl − 365 l = ____ kl   20.635

E. Find these equivalents. *(Lessons 19, 23)*

35. 34 kl = ____ l 34,000

36. 84 qt. = 79.8 l

37. 8,320 a. = 13 sq. mi.

38. 5 a. = ____ sq. ft.   217,800

F. Solve by long division, and check by casting out nines. *(Lesson 11)*

39. 27)16,409   607 R 20

40. 714)478,025   669 R 359

A just weight and balance are the LORD's.
Proverbs 16:11

28. Converting Between Fahrenheit and Celsius Temperatures

Two temperature scales are in widespread use. One is the Celsius scale, which is named after its Swedish inventor Anders Celsius. On this scale, water freezes at 0 degrees and boils at 100 degrees. The Celsius scale is most common in nations using the metric system.

The other temperature scale was devised by Gabriel Fahrenheit in the early 1700s. On this scale, the freezing point of water is 32 degrees and its boiling point is 212 degrees. The Fahrenheit scale is commonly used in the United States.

Because we encounter temperature readings in both scales, it is useful to know how to change temperatures from one scale to the other.

To change a Celsius temperature to Fahrenheit, use the following formula. (Memorize it.)

$$F = \tfrac{9}{5}C + 32$$

This formula is applied by using the following steps.

1. Multiply $\tfrac{9}{5}$ times the degrees Celsius. Round the answer to the nearest whole number.

2. Add 32 to the product found in step 1.

3. Label the result of step 2 as degrees Fahrenheit (°F).

Example A	**Example B**
4°C = ___ °F	75°C = ___ °F
$F = \tfrac{9}{5}C + 32$	$F = \tfrac{9}{5}C + 32$
$\tfrac{9}{5} \times 4 = 7\tfrac{1}{5}$ (round to 7)	$\tfrac{9}{5} \times 75 = 135$
7 + 32 = 39°F	135 + 32 = 167°F

To change from Fahrenheit to Celsius, use the following formula. (Memorize it.)

$$C = \tfrac{5}{9}(F - 32)$$

This formula is applied by using the following steps:

1. Subtract 32 from the Fahrenheit temperature.

2. Multiply $\tfrac{5}{9}$ times the difference found in step 1. Round the answer to the nearest whole number.

3. Label the result of step 2 as degrees Celsius (°C).

LESSON 28

Objectives

- To review Fahrenheit and Celsius temperatures, including equivalents between the two scales. (This lesson does not include any exercise that requires calculation with negative numbers. That will be taught in Chapter 10.)
- To *memorize the formulas for converting between Fahrenheit and Celsius temperatures.

Review

1. *Find these equivalents.* (Lesson 27)
    a. 25 in. = ___ cm (63.5)
    b. 12 tons = ___ MT (10.92)
    c. 350 l = ___ gal. (91)
    d. 1,500 a. = ___ ha (600)

2. *Multiply or divide these compound measures.* (Lesson 26)
    a. 6 yr. 12 wk.
        $\times 9$
        (56 yr. 4 wk.)
    b. 7)33 hr. 36 min. (4 hr. 48 min.)
    c. 8 × 26 MT 360 kg = ___ MT (210.88)
    d. 4 l 374 ml ÷ 6 = ___ ml (729)

3. *Find these equivalents.* (Lessons 20, 24)
    a. 1,095 days = ___ yr. (3)
    b. 9 millennia 7 centuries = ___ centuries (97)
    c. 1,200 cm^2 = ___ m^2 (0.12)
    d. 76 ha = ___ m^2 (760,000)

4. *Change these times as indicated.* (Lesson 20)
    a. 11:30 A.M. in Edmonton, Alberta; ___ in Miami, Florida (1:30 P.M.)
    b. 11:50 A.M. in Washington D.C.; ___ in Anchorage, Alaska (7:50 A.M.)

5. *Solve mentally by the divide-and-divide method.* (Lesson 12)
    a. 416 ÷ 16 (26)
    b. 308 ÷ 14 (22)
    c. 294 ÷ 42 = (7)
    d. 672 ÷ 32 (21)

T-97   Chapter 2   *English and Metric Measures*

Introduction

Bring a Celsius–Fahrenheit thermometer to class. Have the students give the freezing point and boiling point of water on both scales. (Freezing point: 32°F, 0°C; boiling point: 212°F, 100°C)

How can there be 180 degrees between the freezing and boiling points on the Fahrenheit scale, but only 100 degrees between those points on the Celsius scale? One degree Fahrenheit equals $\frac{5}{9}$ degree Celsius. Conversely, $\frac{9}{5}$ degrees Fahrenheit equals 1 degree Celsius. On the thermometer, point out that the marks on the Celsius scale are farther apart than the marks on the Fahrenheit scale. One Celsius degree is a greater range of temperature than one Fahrenheit degree.

Teaching Guide

1. **Two temperature scales in widespread use are the Celsius scale and the Fahrenheit scale.** On the Celsius scale, which is commonly used in conjunction with the metric system, water freezes at 0 degrees and boils at 100 degrees. On the Fahrenheit scale, water freezes at 32 degrees and boils at 212 degrees.

2. **To change a Celsius temperature to Fahrenheit, use the following formula.** Students are to memorize this formula.

    $F = \frac{9}{5}C + 32$

    a. 43°C = ___°F   (109)
    b. 12°C = ___°F   (54)
    c. 78°C = ___°F   (172)
    d. 17°C = ___°F   (63)
    e. 175°C = ___°F   (347)
    f. 900°C = ___°F   (1,652)

3. **To change a Fahrenheit temperature to Celsius, use the following formula.** Students should also memorize this formula.

    $C = \frac{5}{9}(F - 32)$

    a. 85°F = ___°C   (29)
    b. 47°F = ___°C   (8)
    c. 98°F = ___°C   (37)
    d. 150°F = ___°C   (66)
    e. 1,200°F = ___°C   (649)
    f. 3,000°F = ___°C   (1,649)

Solutions for Part C

17. $\frac{9}{5} \times 2 + 32$

18. $\frac{5}{9}(75 - 32)$

Lesson 28 97

Example C	**Example D**
41°F = ___ °C	300°F = ___ °C
C = $\frac{5}{9}$(F − 32)	C = $\frac{5}{9}$(F − 32)
41 − 32 = 9	300 − 32 = 268
$\frac{5}{9}$ × 9 = 5°C	$\frac{5}{9}$ × 268 = 148$\frac{8}{9}$
	Round to 149°C.

CLASS PRACTICE

For each Celsius temperature, find the Fahrenheit equivalent.

a. 10°C = __50__ °F   b. 45°C = __113__ °F
c. 65°C = __149__ °F   d. 14°C = __57__ °F
e. 62°C = __144__ °F   f. 95°C = __203__ °F

For each Fahrenheit temperature, find the Celsius equivalent.

g. 77°F = __25__ °C   h. 140°F = __60__ °C
i. 194°F = __90__ °C   j. 44°F = __7__ °C
k. 100°F = __38__ °C   l. 120°F = __49__ °C

WRITTEN EXERCISES

A. *For each Celsius temperature, find the Fahrenheit equivalent.*

1. 5°C = __41__ °F   2. 35°C = __95__ °F
3. 65°C = __149__ °F   4. 110°C = __230__ °F
5. 21°C = __70__ °F   6. 19°C = __66__ °F
7. 450°C = __842__ °F   8. 3,800°C = __6,872__ °F

B. *For each Fahrenheit temperature, find the Celsius equivalent.*

9. 95°F = __35__ °C   10. 68°F = __20__ °C
11. 32°F = __0__ °C   12. 167°F = __75__ °C
13. 74°F = __23__ °C   14. 121°F = __49__ °C
15. 2,000°F = __1,093__ °C   16. 10,000°F = __5,538__ °C

C. *Solve these reading problems.*

17. Pedro Calderon lives in the mountains of Guatemala, where it becomes quite chilly. One Sunday morning it was 2 degrees Celsius. What was the Fahrenheit temperature to the nearest degree?   36°F

18. That Sunday noon as the Calderon family walked home from worship services, Father read on a thermometer that the temperature was 75 degrees Fahrenheit. What was the Celsius temperature to the nearest degree?   24°C

98   Chapter 2   *English and Metric Measures*

19. Because of the high altitude where the Calderon family lives, the boiling point of water is 195 degrees Fahrenheit. To the nearest degree, what is that temperature on the Celsius scale?   91°C

20. Along the Pacific coast where Pedro's cousin lives, the temperature often reaches 32 degrees Celsius. To the nearest degree, what temperature is that on the Fahrenheit scale?   90°F

21. One of the higher mountains near the Calderons' home has an elevation of 10,500 feet. What is its elevation in meters?   3,150 meters

22. Pedro's family farms a plot of land with an area of 3 hectares. How many acres of land is that?   7.5 acres

REVIEW EXERCISES

D. Find these equivalents. *(Lesson 27)*

23. 9 tbsp. = <u>135</u> ml   24. 110 g = <u>3.85</u> oz.

E. Multiply or divide these compound measures. *(Lesson 26)*

25. 8 × 32 hm 82 m = ____ m   26,256   26. 15 kl 162 l ÷ 42 = <u>0.361</u> kl

F. Find these equivalents. *(Lessons 20, 24)*

27. 4,500 m² = <u>0.45</u> ha   28. 54.5 ha = <u>0.545</u> km²

29. 10 yr. 2 mo. = <u>122</u> mo.   30. 1,320 hr. = ____ min.   79,200

G. Change these times as indicated. *(Lesson 20)*

31. 5:30 P.M. in Toronto, Ontario; ___ in Honolulu, Hawaii   12:30 P.M.

32. 2:30 A.M. in Vancouver, British Columbia; ___ in Philadelphia, Pennsylvania   5:30 A.M.

H. Solve mentally by the divide-and-divide method. *(Lesson 12)*

33. 216 ÷ 18   12   34. 360 ÷ 24   15

Further Study

1. The Fahrenheit scale was devised by Gabriel Daniel Fahrenheit (1686–1736), the inventor of the mercury thermometer. For zero on his scale, Fahrenheit used the coldest point he could achieve—the temperature obtained when salt is used to melt ice. He established 96 degrees as the temperature "when the thermometer is held in the mouth or under the armpit." (Why he chose 96 degrees is unknown. Normal body temperature is now considered as 98.6°F.) On the Fahrenheit scale, the freezing point of water was found to be 32 degrees.

    This scale has a certain amount of logic even though it is not based on the freezing and the boiling points of water. It has been suggested that in the area of Germany where Fahrenheit lived, the winter temperature was seldom colder than 0°F, and the summer temperature was seldom higher than 100°F. This makes the Fahrenheit scale seem particularly suitable for those living in the middle latitudes, since 0–100 degrees is the normal range of outdoor temperatures.

2. Two other temperature scales have been devised for scientific purposes. They are based on absolute zero, the coldest possible temperature.

    The Kelvin scale places 0 degrees at absolute zero and uses Celsius units. This puts the freezing point of water at 273°K. The Kelvin scale was adopted by the International System of Units.

    The Rankine scale begins at absolute zero and uses Fahrenheit units. This puts the freezing point of water at 492°R.

---

19. $\frac{5}{9}(195 - 32)$

20. $\frac{9}{5} \times 32 + 32$

21. $10{,}500 \times 0.3$

22. 3×2.5

T-99 Chapter 2 English and Metric Measures

LESSON 29

Objective

- To review the relationships between distance, rate, and time, and the formulas that express these relationships.

Review

1. *Find these temperature equivalents to the nearest degree. (Lesson 28)*

   a. 55°C = ___°F (131)

   b. 32°C = ___°F (90)

   c. 33°F = ___°C (1)

   d. 86°F = ___°C (30)

   e. 12°C = ___°F (54)

   f. 190°F = ___°C (88)

2. *Find these equivalents. (Lessons 21, 27)*

   a. 280 mm = ___ dm (2.8)

   b. 11 km = ___ dkm (1,100)

   c. 250 gal. = ___ l (950)

   d. 442 m = ___ ft. (1,449.76)

3. *Add or subtract these compound measures. (Lesson 25)*

   a. 23 hr. 30 min.
       + 12 hr. 55 min.
       (36 hr. 25 min.)

   b. 14 lb.
       − 3 lb. 10 oz.
       (10 lb. 6 oz.)

   c. 12 km − 44 dkm = ___ km (11.56)

   d. 41 kg + 160 g = ___ g (41,160)

4. *Say yes or no to tell whether each division will work out evenly. (Lesson 12)*

   a. 417 ÷ 6 (no)

   b. 815 ÷ 5 (yes)

   c. 144 ÷ 9 (yes)

29. Distance, Rate, and Time

Travel always involves three facts: the distance traveled, the rate of speed, and the amount of time. In the English system of measure, the rate is stated as a linear unit and a unit of time combined by the word *per*, such as "miles per hour." In the metric system, a slash (/) usually replaces *per*, as in "70 km/h." However, both abbreviations are read in the same way; *70 km/h* is read as "70 kilometers per hour."

Abbreviations for Rates of Speed

m.p.h.	=	miles per hour
m.p.m.	=	miles per minute
f.p.s.	=	feet per second
km/h	=	kilometers per hour
km/m	=	kilometers per minute
m/s	=	meters per second

Multiplication and division are inverse operations. The formulas for distance, rate, and time illustrate this fact. The rate of speed times the travel time equals the distance. The rate equals the distance divided by the time, and the time equals the distance divided by the rate.

These three facts are stated as formulas below. Memorize these formulas.

$$\text{distance} = \text{rate} \times \text{time} \qquad \text{rate} = \frac{\text{distance}}{\text{time}} \qquad \text{time} = \frac{\text{distance}}{\text{rate}}$$

$$d = rt \qquad\qquad r = \frac{d}{t} \qquad\qquad t = \frac{d}{r}$$

Example A
$d = \underline{}$ mi.
$r = 52$ m.p.h.
$t = 45$ hr.

Use formula $d = rt$
$d = 45 \times 52$
$d = 2{,}340$ mi.

Example B
$d = 2{,}340$ mi.
$r = \underline{}$ m.p.h.
$t = 45$ hr.

Use formula $r = \dfrac{d}{t}$
$r = \dfrac{2{,}340}{45}$
$r = 52$ m.p.h.

Example C
$d = 2{,}340$ mi.
$r = 52$ m.p.h.
$t = \underline{}$ hr.

Use formula $t = \dfrac{d}{r}$
$t = \dfrac{2{,}340}{52}$
$t = 45$ hr.

100   Chapter 2   English and Metric Measures

CLASS PRACTICE

Find the missing parts. Write any remainder as a fraction.

	Distance	Rate	Time
a.	210 mi.	__30__ m.p.h.	7 hr.
b.	15 mi.	$2\frac{1}{2}$ m.p.h.	__6__ hr.
c.	__378__ mi.	54 m.p.h.	7 hr.
d.	__1,152__ km	96 km/h	12 hr.
e.	468 km	65 km/h	$7\frac{1}{5}$ hr.
f.	17 mi.	__34__ m.p.h.	$\frac{1}{2}$ hr.
g.	648 ft.	72 f.p.s.	__9__ sec.
h.	__180__ m	30 m/s	6 sec.
i.	342 ft.	$28\frac{1}{2}$ f.p.s.	12 sec.

Homing pigeons can fly up to 97 km/h.

WRITTEN EXERCISES

A. Write and memorize these formulas.

1. The formula for finding distance.   distance = rate × time $(d = rt)$
2. The formula for finding rate.   rate = $\frac{\text{distance}}{\text{time}}$ $(r = \frac{d}{t})$
3. The formula for finding time.   time = $\frac{\text{distance}}{\text{rate}}$ $(t = \frac{d}{r})$

B. Find the missing parts. Write any remainder as a fraction.

	Distance	Rate	Time
4.	__300__ mi.	50 m.p.h.	6 hr.
5.	__384__ mi.	48 m.p.h.	8 hr.
6.	__232__ mi.	58 m.p.h.	4 hr.
7.	__624__ mi.	78 m.p.h.	8 hr.
8.	90 mi.	__45__ m.p.h.	2 hr.
9.	162 mi.	__54__ m.p.h.	3 hr.
10.	450 mi.	__150__ m.p.h.	3 hr.
11.	1,420 mi.	__355__ m.p.h.	4 hr.
12.	140 km	35 km/h	__4__ hr.
13.	400 km	80 km/h	__5__ hr.
14.	280 km	80 km/h	$3\frac{1}{2}$ hr.
15.	690 km	92 km/h	$7\frac{1}{2}$ hr.
16.	__10__ mi.	40 m.p.h.	$\frac{1}{4}$ hr.
17.	525 ft.	__75__ f.p.s.	7 sec.
18.	__900__ m	30 m/s	30 sec.

Gavest thou... wings and feathers unto the ostrich?

What time she lifteth up herself on high, she scorneth the horse and his rider.

Job 39:13, 18

Lesson 29   T-100

Introduction

Ask students to suggest typical speeds for the following items.

   fast train   (110 miles per hour)

   running man   (18 miles per hour)

   light   (186,000 miles per second)

   turtle   (2 feet per minute)

   duck hawk   (175 miles per hour)

A rate of speed is a relationship between what two units of measure? (It is always a relationship between a linear unit and a unit of time.)

Teaching Guide

1. **Travel always involves three facts: the distance traveled, the rate of speed, and the amount of time.** In the English system, the rate is stated as a linear unit and a unit of time combined by the word *per*, such as "miles per hour." In the metric system, a slash (/) usually replaces *per*, as in "70 km/h." However, both abbreviations are read in the same way; *70 km/h* is read as "70 kilometers per hour."

2. **Working with distance, rate, and time is one application of the fact that multiplication and division are inverse operations.** The rate of speed times the travel time equals the distance. The rate equals the distance divided by the time, and the time equals the distance divided by the rate. This is illustrated by the three formulas used in working with distance, rate, and time.

3. **Distance, rate, and time are calculated by the following formulas.**

   distance = rate × time

   $$\text{rate} = \frac{\text{distance}}{\text{time}}$$

   $$\text{time} = \frac{\text{distance}}{\text{rate}}$$

   Give practice with using the three formulas to calculate distance, rate, and time.

	Distance	Rate	Time
a.	190 mi.	___ m.p.h.	4 hr. ($47\frac{1}{2}$)
b.	765 ft.	85 f.p.s.	___ sec. (9)
c.	4 mi.	___ m.p.m.	6 min. ($\frac{2}{3}$)
d.	___ km	59 km/h	3 hr. (177)
e.	75 m	5 m/s	___ sec. (15)
f.	___ m	16 m/s	5 sec. (80)

T–101   Chapter 2   *English and Metric Measures*

Further Study

1. The different ways of designating speed are general practices rather than established rules. The one given as the general practice for English measures is occasionally used to express metric measures. Conversely, the one given as the general practice for metric measures is sometimes used in newer books for English measures. With the growing influence of metric units and symbols, the *per* designation seems to be decreasing in usage and the slash form increasing.

2. The term *velocity* is closely related to speed. However, there is a technical difference between the two. *Speed* means a rate of travel in any direction and not necessarily in a straight line. *Velocity* means a rate of travel in a stated straight direction. Thus the speed and velocity of an object are not always the same.

Solutions for Part C

19. $350 \div 8$
20. $\frac{1}{3} \times 115$
21. $5{,}280 \div 169$
22. $5{,}280 \div 220$
23. $\$55.00 \div 350$
24. 560×0.3

Lesson 29

C. Solve these reading problems.

19. Brother Sanford traveled 350 miles by train in 8 hours. What was the average rate of speed? $43\frac{3}{4}$ m.p.h.

20. A conductor told Brother Sanford that the train would reach its greatest speed between Providence and Newport, Rhode Island, when it would travel 115 miles per hour. At that rate, how far would the train travel in 20 minutes? $38\frac{1}{3}$ miles

21. The speed of 115 miles per hour is equal to almost 169 feet per second. At that rate, how many seconds would it take to travel 1 mile? (Round your answer to the nearest whole number.)   31 seconds

22. The conductor described some railroad cars designed to travel 150 miles per hour, which is 220 feet per second. How many seconds would it take to travel 1 mile at that rate?   24 seconds

23. The cost of the train ticket was $55.00. What was the cost per mile for the 350-mile trip? (Round to the nearest cent.)   16¢

24. The train that Brother Sanford traveled on had a total length of 560 feet. How many meters long was the train?   168 m

REVIEW EXERCISES

D. Find these equivalents. Round temperatures to the nearest degree. (Lessons 21, 27, 28)

25. 490 dkm = __4.9__ km
26. 93 dm = __9,300__ mm
27. 14 tons = __12.74__ MT
28. 38 m² = __45.6__ sq. yd.
29. 44°C = __111__ °F
30. 45°F = __7__ °C

E. Add or subtract these compound measures. (Lesson 25)

31.   25 ft. 5 in.
    + 8 ft. 8 in.
     34 ft. 1 in.

32.   10 lb. 4 oz.
    − 7 lb. 8 oz.
      2 lb. 12 oz.

33. 8 dkm − 234 cm = __7,766__ cm
34. 18 kl + 32 l = ____ kl   18.032

F. Write yes or no to tell whether each division problem will work out evenly. (Lesson 12)

35. 3,400 ÷ 4   yes
36. 465 ÷ 9   no

102   Chapter 2   *English and Metric Measures*

30. Understanding Bible Measures

Both English and metric units are relatively new. The metric system was developed about 200 years ago, and the English system developed gradually since about 1,000 years ago. Neither system of measures was in use when the Bible was being written.

The Old Testament gives many measures, especially in relation to the temple and temple worship. Many of these were not standard units; rather they were based on parts of the body such as the fingers (finger and handbreadth), the hand (span), and the forearm (cubit). As time went on, however, some standardization developed. In 2 Samuel 14:26, we read of a shekel "after the king's weight," indicating that King David had a standard shekel weight.

Israel used a system that today is called the Mesopotamian weight system. This still does not mean that measures were standard across all of Mesopotamia. Rather, although similar measures were used, each city or area had its own standard.

Below is a list of Bible measures, along with their approximate English and metric equivalents. Note that these equivalents are only approximate. Measures in Bible times could vary from time to time and from place to place. However, using these equivalents should give fairly close approximations of the quantities indicated in the Scriptures.

Hebrew Measures (Used mainly in the Old Testament)

Bible unit	Approximate relationship	English equivalent	Metric equivalent	Scripture reference
Length				
finger	$\frac{1}{4}$ handbreadth	$\frac{3}{4}$ in.	1.9 cm	Jer. 52:21
handbreadth	$\frac{1}{6}$ cubit	3 in.	7.6 cm	1 Kings 7:26
span	$\frac{1}{2}$ cubit	9 in.	23 cm	1 Sam. 17:4
cubit	2 spans	18 in. ($1\frac{1}{2}$ ft.)	46 cm	1 Sam. 17:4
reed	7 cubits	$10\frac{1}{2}$ ft.	3.25 m	Ezek. 40:5
Weight				
gerah	$\frac{1}{20}$ shekel	$\frac{1}{50}$ oz.	0.6 g	Ex. 30:13
shekel	20 gerahs	$\frac{2}{5}$ oz.	11.3 g	Josh. 7:21
pound	50 shekels	$1\frac{1}{4}$ lb.	566 g	Ezra 2:69
talent	3,000 shekels	75 lb.	34 kg	2 Sam. 12:30
Liquid Capacity				
log	$\frac{1}{12}$ hin	0.9 pt.	$0.33\ l$	Lev. 14:10–24
hin	$\frac{1}{6}$ bath	$5\frac{1}{3}$ qt.	$5\ l$	Ezek. 4:11
bath	6 hins	8 gal.	$30.3\ l$	Ezek. 45:11
homer	10 baths	80 gal.	$303\ l$	Ezek. 45:11–14

LESSON 30

Objectives

- To review Bible measures and their English and metric equivalents. (Students are not expected to memorize the equivalents listed in this lesson.)
- To review that many Bible measures were not standard units and therefore do not have exact English and metric equivalents.

Review

1. *Find the missing parts.* (Lesson 29)

	Distance	Rate	Time
a.	___ mi.	36 m.p.h.	$\frac{3}{4}$ hr. (27)
b.	780 km	___ km/h	12 hr. (65)
c.	240 mi.	48 m.p.h.	___ hr. (5)
d.	44 km	___ km/h	11 hr. (4)

2. *Find these temperature equivalents to the nearest degree.* (Lesson 28)

    a. 66°C = ___°F (151)
    b. 99°F = ___°C (37)

3. *Multiply or divide these compound measures.* (Lesson 26)

    a. 23 gal. 3 qt.
       $\underline{ \times 9 }$
       (213 gal. 3 qt.)

    b. $\overset{(4 \text{ lb. } 6 \text{ oz.})}{7\overline{)30 \text{ lb. } 10 \text{ oz.}}}$

    c. 6 × 21 m 14 cm = ___ m (126.84)
    d. 9 kl 560 l ÷ 8 = ___ l (1,195)

4. *Find these equivalents.* (Lesson 22)

    a. 290 g = ___ kg (0.29)
    b. 11 MT = ___ kg (11,000)

T-103 Chapter 2 English and Metric Measures

Introduction

Ruth 2:17 says that after Ruth gleaned in the field of Boaz the first day, she had one ephah of barley. Have the students guess how much grain that was in modern units. You might give the following possibilities.

   a. 4 quarts
   b. 1 peck
   c. 3 pecks
   d. 1 bushel

Answer *c* (3 pecks) is closest to the correct amount. One ephah is about $3\frac{1}{2}$ pecks, which would have been about 40 pounds of barley. Studying Bible measures is beneficial because it helps us to understand the quantities mentioned in the Bible.

Teaching Guide

1. **Many units of measure in the Bible were not standard as our units are today.** Especially in Old Testament times, each city often had its own system of measures. Certain units may have had the same names as those of a neighboring city, but they probably were not exactly the same. For this reason, we can give only approximate English and metric equivalents for most Bible measures.

2. **To make conversions between Bible units and modern units, use the tables in the lesson in the same way as the tables for making English/metric conversions.** Students are not expected to memorize the equivalents listed in this lesson.

   Use *Class Practice* and the problems below as exercises in class.

   a. When the children of Israel crossed the Jordan, they followed the ark of the covenant at a distance of 2,000 cubits. How many feet was that? How many meters?
   ($2,000 \times 1\frac{1}{2}$ ft. = 3,000 ft.)
   ($2,000 \times 46$ cm = 92,000 cm = 920 meters)

   b. In 1 Chronicles 29:7, the leaders of Israel gave 10,000 talents of silver for the building of the temple. If silver is valued at $560 per pound, what was the value of that silver in today's money?
   ($10,000 \times 75$ lb. = 750,000 lb.
   $750,000 \times \$560 = \$420,000,000$)

Solutions for CLASS PRACTICE

a. $3,000 \times 75 = 225,000$ pounds; $3,000 \times 34 = 102,000$ kilograms

b. $5 \times 1\frac{1}{2} = 7\frac{1}{2}$ feet; $5 \times 0.46 = 2.3$ meters

Dry Capacity

cab		$2\frac{3}{4}$ pt	1.5 l	2 Kings 6:25
omer	$\frac{1}{10}$ ephah	.5 pt.	2.8 l	Ex. 16:36
ephah	10 omers	$3\frac{1}{4}$ pk.	28.2 l	Ex. 16:36
homer	10 ephahs	.8 bu.	282 l	Ezek. 45:11–14

By New Testament times, Roman units of measure were the widespread standard. The New Testament does not refer to as many measures, because the temple and temple worship were fulfilled in Christ.

Roman Measures (Used in the New Testament)

Type of measure	Name of measure	English equivalent	Metric equivalent	Scripture reference
weight	pound	$\frac{3}{4}$ lb.	340 g	John 19:39
dry capacity	bushel	$\frac{1}{4}$ bu.	.9 l	Matt. 5:15
dry capacity	measure*	13 bu.	453 l	Luke 16:7
linear	furlong	606 ft.	185 m	Luke 24:13
linear	mile	4,848 ft.	1,478 m	Matt. 5:41

*Several different Greek units of capacity are translated "measure" in the New Testament.

To change a Bible measure to English or metric units, multiply the number of units in the Bible measure by the desired equivalent.

Example A

In preparing materials for the temple, David gathered 100,000 talents of iron (1 Chronicles 29:7). How many tons of iron was that? How many metric tons?

100,000 × 75 lb. = 7,500,000 lb.
7,500,000 ÷ 2,000 = 3,750 tons

100,000 × 34 kg = 3,400,000 kg
3,400,000 ÷ 1,000 = 3,400 MT

Example B

Og, king of Bashan, had a bedstead of iron that measured 9 cubits by 4 cubits. What were the dimensions of the bedstead in feet? in meters?

9 × $1\frac{1}{2}$ ft. = $13\frac{1}{2}$ ft. long
4 × $1\frac{1}{2}$ ft. = 6 ft. wide

9 × 46 cm = 414 cm = 4.14 m long
4 × 46 cm = 184 cm = 1.84 m wide

CLASS PRACTICE

Solve these reading problems.

a. David gave 3,000 talents of gold for the building of the temple (1 Chronicles 29:4). How many pounds of gold was that? How many kilograms?   225,000 lb.; 102,000 kg

b. In 1 Chronicles 11:23, Benaiah slew an Egyptian who was 5 cubits tall. What was his height in feet? in meters? $7\frac{1}{2}$ feet; 2.3 meters

104　*Chapter 2　English and Metric Measures*

WRITTEN EXERCISES

A. Solve these problems by changing Bible measures to English measures.

1. Judas agreed to betray Jesus for 30 pieces of silver, which were probably shekels (Matthew 26:14–16). What was the total weight of this silver in ounces?　　12 ounces

2. A homer is thought to have been the normal load that a donkey could carry. At 56 pounds per bushel, how many pounds of wheat are in a homer?　　448 pounds

3. By scheming and dishonesty, Gehazi obtained 2 talents of silver from Naaman. What was the weight in ounces of the silver that Gehazi received?　　2,400 ounces

4. In Leviticus 27:16, a homer of barley seed is valued at 50 shekels of silver. How many bushels are in 4 homers?　　32 bushels

5. Numbers 11:32 says that the children of Israel feasted greedily on the quail that the Lord sent, bringing God's wrath upon themselves. "He that gathered least gathered ten homers." How many bushels of quail was the least amount gathered?　　80 bushels

6. The molten sea in Solomon's temple held 2,000 baths of water (1 Kings 7:26). How many gallons of water did it contain?　　16,000 gallons

7. Solomon had 300 shields of beaten gold, each made with 3 Hebrew pounds of gold (1 Kings 10:17). How many English pounds of gold were in the 300 shields altogether?　　1,125 pounds

8. Isaiah declared that the Lord would punish the sin of Judah by sending a famine. He prophesied that a 10-acre vineyard would produce only 1 bath of grape juice (Isaiah 5:10). How many gallons of juice is that per acre?　　$\frac{4}{5}$ gallon

9. In the same prophecy, Isaiah said that a homer of seed would yield only an ephah of grain. An ephah is one-tenth of a homer. At that rate of production, if Father planted 500 bushels of wheat, how much wheat would he harvest?　　50 bushels

10. When the people returned to Judah after the Babylonian captivity, they willingly gave an offering to rebuild the temple. Among the gifts were 5,000 Hebrew pounds of silver (Ezra 2:69). What would this amount be in English pounds?　　6,250 pounds

B. Solve these problems by changing Bible measures to metric measures.

11. Joseph's brothers sold him for 20 pieces of silver (Genesis 37:28), which may have been shekels. If that is true, what was the weight of the silver in grams?　　226 grams

12. Solomon made 10 lavers of brass for the temple, and each laver held 40 baths (1 Kings 7:38). How many liters of water could the lavers hold altogether?　　12,120 liters

13. In Ezekiel 4:11, Ezekiel was told to drink only $\frac{1}{6}$ hin of water per day during a fast. What fraction of a liter of water was Ezekiel allowed to drink?　　$\frac{5}{6}$ liters

14. Ezekiel 41 gives the dimensions of a temple that Ezekiel saw in a vision. The length of the house was 100 cubits. What was its length in meters?　　46 meters

15. The thickness of the walls in part of this temple was 5 cubits (Ezekiel 41:12). How thick was this wall in meters?　　2.3 meters

200

Lesson 30 T-104

An Ounce of Prevention

This lesson is designed to work in localities where either English or metric measures prevail. You may want to assign only Part A or Part B, depending on the system that your class is more familiar with.

Further Study

In the Mesopotamian system, each type of measure had four different units: the common measure, the heavy measure (twice the common measure), the royal measure (5% more than the common measure), and the royal heavy measure (5% more than the heavy measure.) The list below shows the units that would have been derived from the shekel.

Unit	English equivalent
common shekel	0.4 oz.
heavy shekel	0.8 oz.
king's shekel	0.42 oz.
king's heavy shekel	0.84 oz.

Solutions for WRITTEN EXERCISES

A. Changing Bible Measures to English Measures.

1. $30 \times \frac{2}{5}$
2. 8×56
3. $2 \times 75 \times 16$
4. 4×8
5. 10×8
6. $2{,}000 \times 8$
7. $300 \times 3 \times 1\frac{1}{4}$
8. $8 \div 10$
9. $\frac{1}{10} \times 500$
10. $5{,}000 \times 1\frac{1}{4}$

B. Changing Bible Measures to Metric Measures.

11. 20×11.3
12. $10 \times 40 \times 30.3$
13. $\frac{1}{6} \times 5$
14. 100×0.46
15. 5×0.46

T–105   Chapter 2   English and Metric Measures

16. 4 × 500 × 3.25
17. 25,000 × 0.00325
18. (2 × 1,478) − 2,000
19. 100 × 453
20. 100 × 0.34

16. In Ezekiel 40–42, Ezekiel used a reed to measure the temple. Around the temple were four walls that were each 500 reeds long (Ezekiel 42:20). If this reed was 3.25 meters long, what was the length in meters of the entire wall around the temple? 6,500 meters

17. In Ezekiel 48:8, a section of land measured 25,000 reeds wide. At 3.25 meters per reed, how many kilometers are in 25,000 reeds? 81.25 kilometers

18. "And whosoever shall compel thee to go a mile, go with him twain" (Matthew 5:41). Two Roman miles would be how many meters longer than 2 kilometers? 956 meters

19. In Luke 16:7, a debtor owed 100 measures of wheat. How many liters of wheat was that? 45,300 liters

20. When Joseph of Arimathaea and Nicodemus buried Christ, Nicodemus brought "about an hundred pound weight" of spices to wrap in the burial clothes (John 19:39). How much weight in kilograms is 100 Roman pounds? 34 kilograms

REVIEW EXERCISES

C. Find the missing parts. Write any remainder as a fraction. *(Lesson 29)*

	Distance	Rate	Time
21.	84 km	_40_ km/h	$2\frac{1}{10}$ hr.
22.	_144_ mi.	36 m.p.h.	4 hr.
23.	816 ft.	54 f.p.s.	$15\frac{1}{9}$ sec.
24.	168 m	$18\frac{2}{3}$ m/s	9 sec.

D. Find these temperature equivalents to the nearest degree. *(Lesson 28)*

25. 72°F = _22_ °C   26. 68°C = _154_ °F

E. Multiply or divide these compound measures. *(Lesson 26)*

27.   9 ft. 10 in.
          × 9
         ―――――――
         88 ft. 6 in.

28. 8)79 lb. 8 oz.   = 9 lb. 15 oz.

29. 5 × 48 kg 160 g = ____ g   240,800

30. 8 kl 610 l ÷ 7 = _1.23_ kl

F. Find these equivalents. *(Lesson 22)*

31. 0.12 g = _120_ mg   32. 2,400 mg = ____ kg   0.0024

31. Reading Problems: Choosing the Necessary Facts

Reading problems are much like the math problems that we need to solve in everyday life. They usually do not include symbols indicating the operations needed to solve them. Rather, they come as a body of facts from which we must choose the necessary information to arrive at a correct conclusion. Therefore, reading problems are practical mathematics that we must approach with care in order to find the right answers.

The first step in solving a reading problem correctly is to read it carefully, being certain that you understand the question to be answered. Skimming has a place in reviewing a book before deciding to read it, but it has no place in solving a reading problem. **To solve a reading problem correctly, read it carefully and make sure you understand it.**

The next step in solving a reading problem is to choose the correct information. Facts relating to elevation are interesting but usually not necessary for solving a reading problem about temperatures. Likewise, the temperature in New York is interesting but not necessary when the question asks you to change a temperature in Arizona from Fahrenheit to Celsius. **To solve a reading problem correctly, carefully choose the facts needed to answer the question.**

Example
"There was a certain creditor which had two debtors: the one owed five hundred pence, and the other fifty" (Luke 7:41). One penny was a working man's daily wage. If the man owing the greater debt worked 22 days per month, how many month's wages did he owe his creditor?

Consider the following questions about the reading problem above.
1. **What is the actual question being asked?** The question is, "If the man owing the greater debt worked 22 days per month, how many month's wages did he owe his creditor?"
2. **What facts are needed to answer the question?** The facts needed are 500 pence, 1 penny per day, and 22 days per month. All the other facts, though interesting, are not necessary to answer the question.

Solution: 500 pence = 500 days' labor
$500 \div 22$ days per month = $22\frac{8}{11}$ months

CLASS PRACTICE

Choose the facts you will use to solve each problem. Then find the answer.

a. The Riverside congregation purchased 100 Spanish Bibles for a Guatemala mission. The supplier allowed a volume discount of $324, which brought the cost of the Bibles to $973. Shipping charges were $343. What was the regular price of each Bible?
Facts: 100 Bibles; $324; $973; Answer: $12.97

LESSON 31

Objective

- To give practice with recognizing the information needed to solve reading problems.

Review

1. Give Lesson 31 Quiz (Practice With Measures).

2. *Find these equivalents.* (Lesson 30)
   a. 40 ephahs = ___ homers (4)
   b. 20 talents = ___ pounds (1,500)
   c. 20 cubits = ___ meters (9.2)
   d. 25 Hebrew pounds = ___ English pounds ($31\frac{1}{4}$)

3. *By memory, give these formulas for distance, rate, and time.* (Lesson 29)

   a. For finding distance.
      (distance = rate × time)
   b. For finding rate. (rate = $\frac{distance}{time}$)
   c. For finding time.
      (time = $\frac{distance}{rate}$)

4. *Find these equivalents.* (Lessons 23, 27)
   a. 9.2 l = ___ ml (9,200)
   b. 28 l = ___ qt. (29.68)
   c. 45 yd. = ___ m (40.95)
   d. 72.5 a. = ___ ha (29)

Introduction

A plumber was called to fix a water leak in a house. After spending quite a bit of time accessing the pipes inside a wall, he discovered that the hoses on a washing machine upstairs were not tight enough. Those leaking hoses were causing the problems on the first floor.

Has something similar ever happened when you tried to solve a reading problem? Perhaps you got a solution that was not sensible, or you simply could not decide on a way to find the answer. When solving a reading problem, as with most other activities, it is crucial to consider the entire scope of the problem. Then you must choose the facts that are needed to solve the problem.

Teaching Guide

1. **Reading problems are much like the math problems that we need to solve in everyday life.** Such practical problems come as a body of facts from which we must choose the necessary items to arrive at a correct conclusion. Likewise, reading problems include information from which we must select the necessary facts in order to find the answers.

2. **To solve a reading problem correctly, read it carefully and make sure you understand it.** Skimming is useful for certain purposes, but not for solving a reading problem.

3. **To solve a reading problem correctly, carefully choose the facts needed to answer the question.** Not every number in a problem will be needed to find the solution. Use the lesson example, the

T–107 Chapter 2 English and Metric Measures

reading problems below, and *Class Practice* to illustrate.

a. One morning the temperature was 53°F, a northwest wind was blowing at 25 miles per hour, and there was a cloud ceiling at 3,000 feet. What was the speed of the wind in kilometers per hour?
   (1) *What is the actual question being asked?* The question is, "What is the speed of the wind in kilometers per hour?
   (2) *What facts are needed to answer the question?* The main fact is that the wind speed was 25 miles per hour. (One also needs to know that 1 mile = 1.61 kilometers.)
   Solution: 25 × 1.61 = 40.25 km/h

b. There are 4 teachers at the Westbridge Christian School. Grades 1–3 have 15 students, grades 4–6 have 12 students, and grades 7–10 have 11 students. The students come from 11 families. To the nearest whole number, what is the average number of students per grade?
   (1) *What is the actual question being asked?* The question is, "To the nearest whole number, what is the average number of students per grade?"
   (2) *What facts are needed to answer the question?* The facts needed are the number of grades (10) and the numbers of students in the different rooms (15, 12, 11).
   Solution: 15 + 12 + 11 = 38; 38 ÷ 10 = $3\frac{8}{10}$ = 4 students to the nearest whole number

Solutions for CLASS PRACTICE

a. ($973 + 324) ÷ 100

b. 100 × $11.75

c. 245 = 128

d. (242 − 55) × $0.87

Solutions for WRITTEN EXERCISES

1. 10,275 ÷ 2,000

2. 3 × ($6.98 ÷ 2)

3. 6,000 − (25 × 175)

4. $9\frac{1}{2}$ × $15

5. 500 − 320

6. 80 × $1\frac{1}{2}$ ÷ 2

7. 40,000 ÷ 800

b. At the same time, the congregation purchased 100 *Church Hymnals* for use in their own worship services. After a 25 percent discount, the new hymnals cost $11.75 each. What was the total cost of the hymnals? Facts: 100 hymnals; $11.75; Answer: $1,175

c. A fuel deliveryman came to a house and filled the tank with heating fuel that cost $0.87 per gallon. After he put in 128 gallons, the fuel tank contained 245 gallons. How many gallons had been in the tank before he came? Facts: 128 gal.; 245 gal.; Answer: 117 gal.

d. At the next stop, the deliveryman put fuel into a 275-gallon tank. The tank contained 55 gallons when he came and 242 gallons when he left. At $0.87 per gallon, what was the charge for the fuel he delivered? Facts: 55 gal.; 242 gal.; $0.87; Answer: $162.69

WRITTEN EXERCISES

A. Write the facts you will use to solve each problem. Then find the answer.

1. The Old Testament tabernacle was set on 105 sockets. There were 100 sockets of silver and 5 sockets of brass. The 100 silver sockets weighed about 10,275 pounds. How many tons was this? Fact: 10,275 lb.; Answer: $5\frac{11}{80}$ tons

2. A 5-pound box of 10-penny nails that regularly sold for $6.98 was on sale for half price. What was the cost of 3 boxes of these nails? Facts: $6.98; half price; 3 boxes; Answer: $10.47

3. Marlin's father paid $132 for a 6,000-foot roll of drip tape to irrigate 40 rows of sweet corn. He used 175 feet for each of the 25 longest rows. How many feet were left on the roll? Facts: 6,000 ft.; 175 ft.; 25 rows; Answer: 1,625 ft.

4. Three men worked $9\frac{1}{2}$ hours each at painting the interior of a new house. Each man's pay was $15 per hour. How much did one painter earn? Facts: $9\frac{1}{2}$ hr.; $15; Answer: $142.50

5. Instead of the usual 480 tons of corn silage, the Groffs this fall harvested only 320 tons from their 50 acres of cropland. Since they need about 500 tons annually for their 75 steers, how many tons will they need to buy so that they will have enough feed until next fall?
Facts: 320 tons; 500 tons; Answer: 180 tons

6. When Mother cut off sweet corn for freezing, she got an average of 1 quart from every 10 ears. She put the corn into bags holding $1\frac{1}{2}$ pints each. If she put 80 bags into the freezer, how many quarts did she freeze?
Facts: $1\frac{1}{2}$ pt.; 80 bags; Answer: 60 qt.

7. Father bought 4 new tires for the family car. The tires came with a 40,000-mile warranty and were purchased in January 1999. Since the car is driven about 800 miles each month, how many months can the tires be expected to last?
Facts: 40,000 mi.; 800 mi.; Answer: 50 mo.

108 Chapter 2 *English and Metric Measures*

8. In a highway construction zone, a single lane 12 feet wide is bounded on both sides by concrete barriers. Traveling down the center of this lane is a truck with a flatbed trailer 8 feet wide and 40 feet long. If a load $9\frac{1}{2}$ feet wide is on the center of the trailer, how many inches are left to spare on each side?   Facts: 12 ft.; $9\frac{1}{2}$ ft.; Answer: 15 in.

9. During a heavy thunderstorm that produces 1 inch of rainfall, an estimated 1,000,000 raindrops fall on every 3 square feet of the ground. On an area of similar size, about how many drops fall during a storm bringing 2.2 inches of rain?   (See facing page.)

10. Sister Joy offered to make 2 gallons 2 quarts of fruit salad for the meal at the Bible conference. For every 2 quarts of salad, she will need 3 oranges and 2 apples. How many oranges will she need for the salad?

11. Boston, Massachusetts, received 16.5 inches of rain by September 15 one year. That was 11.8 inches less than the 28.3 inches normally received by that date. How many centimeters of rain had Boston received from January 1 to September 15 that year?

12. In the same period, Burlington, Vermont, received 18.1 inches of rain. That was 6.2 inches less than the 24.3 inches normally received. In centimeters, how much rain fell at Burlington by September 15?   Fact: 18.1 in.; Answer: 45.974 cm

13. By September 14, the normal precipitation is 31.2 inches at Providence, Rhode Island; 24.3 inches at Burlington, Vermont; and 28.9 inches at Portland, Maine. How much more rain falls at Providence than at Burlington in the period given? Answer to the nearest whole centimeter.   Facts: 31.2 in.; 24.3 in.; Answer: 18 cm

14. The distance between Burlington, Vermont, and Providence, Rhode Island, is 311 miles by main highways. The distance between Burlington and Portland, Maine, is 263 miles. How many kilometers is it from Burlington to Providence?
     Fact: 311 mi.; Answer: 500.71 km

REVIEW EXERCISES

B. Find the missing parts. Write any remainder as a fraction. *(Lesson 29)*

	Distance	Rate	Time
15.	14 mi.	$3\frac{1}{2}$ m.p.h.	4 hr.
16.	450 m	18 m/s	25 sec.
17.	462 km	84 km/h	$5\frac{1}{2}$ hr.
18.	578 ft.	$48\frac{1}{6}$ f.p.s.	12 sec.

C. Find these equivalents. *(Lessons 23, 27, 30)*

19. 2.43 kl = 2,430 l
20. 345 ml = 0.345 l
21. 745 l = 789.7 qt.
22. 85 qt. = 80.75 l
23. 350 mi. = 563.5 km
24. 35 sq. yd. = 29.4 m²
25. 25 gal. = 95 l
26. 72 g = 2.52 oz.
27. 4 homers = 32 bu.
28. 5 baths = 40 gal.

Lesson 31   T–108

8. $(12 - 9\tfrac{1}{2}) \div 2$
9. Facts: 1 in.; 1,000,000 raindrops; 2.2 in.
   Answer: 2,200,000 raindrops   ($2.2 \times 1{,}000{,}000$)
10. Facts: 2 gallons 2 quarts; 3 oranges for 2 quarts
    Answer: 15 oranges   (10 quarts $\div 2 \times 3$)
11. Fact: 16.5 in.
    Answer: 41.91 cm   (16.5×2.54)
12. 18.1×2.54
13. $(31.2 - 24.3) \times 2.54$
14. 311×1.61

T–109 *Chapter 2 English and Metric Measures*

LESSON 32

Objective

- To review the material taught in Chapter 2 (Lessons 17–31).

Teaching Guide

1. Lesson 32 reviews the material taught in Lessons 17–31. For pointers on using these review lessons, see *Teaching Guide* for Lesson 15.
2. Be sure to review all the formulas assigned for memorization.

    Celsius/Fahrenheit conversion

    $F = \frac{9}{5}C + 32$

    $C = \frac{5}{9}(F - 32)$

    Distance, rate, and time

    distance = rate × time

    $\text{rate} = \frac{\text{distance}}{\text{time}}$

    $\text{time} = \frac{\text{distance}}{\text{rate}}$

    Review all the relationships between English units listed in the text.

    Review all the relationships between metric units listed in the text.

    Review the following equivalents between English and metric units.

    1 meter = 3.28 feet
    1 foot = 0.3 meter
    1 kilogram = 2.2 pounds
    1 pound = 0.45 kilogram
    1 liter = 1.06 quarts
    1 quart = 0.95 liters
    1 hectare = 2.5 acres
    1 acre = 0.4 hectares

Lesson number and new concept	Exercises in Lesson 32
17—Long ton.	None
21—Metric prefixes greater than *-kilo* and less than *-milli*.	None
23—Gallon/liter and liter/gallon equivalents.	13, 14, 33, 34
28—Fahrenheit/Celsius and Celsius/Fahrenheit conversion formulas.	55–58

32. Chapter 2 Review

A. Write the abbreviations for these units by memory. *(Lessons 17–24)*

1. inch — in.
2. fluid ounce — fl. oz.
3. square inch — sq. in.
4. second — sec.
5. centimeter — cm
6. milligram — mg
7. decigram — dg
8. square millimeter — mm²

B. Write these equivalents by memory. *(Lessons 17–24)*

9. 1 fl. oz. = _2_ tbsp.
10. 1 cup = _8_ fl. oz.
11. 1 km = _1,000_ m
12. 1 hm = ___ mm   100,000
13. 1 gal. = _3.8_ l
14. 1 l = _0.26_ gal.
15. 1 ft. = _0.3_ m
16. 1 m = _3.28_ ft.
17. 1 cm = _0.39_ in.
18. 1 inch = _2.54_ cm
19. 1 kg = _2.2_ lb.
20. 1 lb. = _0.45_ kg
21. 1 ha = _2.5_ a.
22. 1 a. = _0.4_ ha

C. Find these equivalents. *(Lessons 17–24)*

23. 25,000 lb. = _12½_ tons
24. 9 lb. = _144_ oz.
25. 31 pk. = _7¾_ bu.
26. 40 fl. oz = _5_ cups
27. 8 bu. 1 pk. = _33_ pk.
28. 360 pt. = _5⅝_ bu.
29. 900 yr. = _9_ centuries
30. 3 millennia = _3,000_ yr.
31. 720 g = _0.72_ kg
32. 15 MT = ___ kg   15,000
33. 14 gal. = _53.2_ l
34. 90 l = _23.4_ gal.
35. 3,100 cm² = _0.31_ m²
36. 88 ha = _0.88_ km²

D. Make these English/metric conversions. You should be able to remember the equivalents without consulting the English/metric tables. *(Lesson 27)*

37. 45 m = _147.6_ ft.
38. 650 km = _403_ mi.
39. 90 ft. = _27_ m
40. 85 mi. = ___ km   136.85
41. 78 ha = _195_ a.
42. 61 a. = _24.4_ ha

E. Find the equivalents for these Bible measures. *(Lesson 30)*

43. 70 shekels = _28_ oz.
44. 20 talents = _680_ kg
45. 12 homers = _96_ bu.
46. 7 ephahs = _197.4_ l

Chapter 2 English and Metric Measures

F. Solve these problems involving compound measures. *(Lessons 25, 26)*

47. 6 ft. 9 in.
    + 7 ft. 9 in.
    ─────────
     14 ft. 6 in.

48. 8 pk. 2 qt.
    − 5 pk. 3 qt.
    ─────────
     2 pk. 7 qt.

49. 8 yd. 2 ft.
    × 7
    ─────────
     60 yd. 2 ft.

50. 9 lb. 13½ oz.
     8)78 lb. 12 oz.

51. 50 cm + 90 mm = __590__ mm

52. 3.8 kg − 987 g = __2,813__ g

53. 265 m 50 cm ÷ 50 = __5.31__ m

54. 14 *l* 790 ml ÷ 3 = __4.93__ *l*

G. Find these temperature equivalents to the nearest degree. *(Lesson 28)*

55. 58°F = __14__ °C

56. 102°F = __39__ °C

57. 34°C = __93__ °F

58. 115°C = __239__ °F

H. Find the missing facts. Write any remainders as fractions. *(Lesson 29)*

	Distance	Rate	Time
59.	__595__ mi.	85 m.p.h.	7 hr.
60.	__406__ mi.	58 m.p.h.	7 hr.
61.	396 mi.	__44__ m.p.h.	9 hr.
62.	570 km	60 km/h	__9½__ hr.
63.	920 ft.	__92__ f.p.s.	10 sec.
64.	__1,125__ m	25 m/s	45 sec.

I. Answer these questions. *(Lessons 20, 29)*

65. If it is 3:00 P.M. in Hailfax, Nova Scotia, what time is it in Honolulu, Hawaii? 9:00 A.M.

66. If it is 10 A.M. in Anchorage, Alaska, what time is it in St. Johns, Newfoundland? 3:30 P.M.

67. What is the formula for finding rate when distance and time are known?
    rate = distance ÷ time ($r = \frac{d}{t}$)

68. What is the formula for finding time when distance and rate are known?
    time = distance ÷ rate ($t = \frac{d}{r}$)

J. First write the facts you will use to solve each problem. Then find the answer. *(Lesson 31)*

69. Antelopes, a marvel of God's creation, are very nimble-footed animals. One of the fastest antelopes is the lesser kudu, which weighs about 230 pounds. It can jump 6 feet into the air and cover 30 feet in a single leap. How far in meters can a lesser kudu move with a single leap? Fact: 30 ft.; Answer: 9 m

70. The greater kudu is noted for its corkscrew horns, which grow as long as 5 feet. This antelope may reach a height of 60 inches at the shoulders and a weight of 600 pounds. What is the greater kudu's weight in kilograms? Fact: 600 lb.; Answer: 270 kg

Solutions for Part J

69. 30 × 0.3
70. 600 × 0.45

T–111 Chapter 2 English and Metric Measures

71. 97 × 0.62
72. 300 ÷ 25
73. 120 inches ÷ 40
74. (33 − 24) × 2.54

Lesson 32

71. Antelopes rank among the swiftest runners on land. They can reach speeds as high as 97 kilometers per hour or 27 meters per second. How many miles per hour can these antelopes run? (Answer to the nearest whole number.) Fact: 97 km/h; Answer: 60 m.p.h.

72. The tiny royal antelope stands about 25 centimeters (10 inches) tall, and it can leap about 3 meters in a single jump. How many times its height is this leap? Facts: 25 cm; 3 m; Answer: 12 times

73. The impala, a small antelope standing about 40 inches tall, can make leaps as much as 10 feet high and 30 feet long. These great leaps confuse and startle its main predators, lions and wild dogs, often allowing the impala to escape. The impala can jump how many times as high as its own height? Facts: 40 in.; 10 ft.; Answer: 3 times

74. The gentle dorcas gazelle, a favorite pet of wandering Arabs, is about 2 feet tall. The Grants gazelle stands about 33 inches high at the shoulder, and its horns may grow to a length of 30 inches or more. How many centimeters taller is a Grants gazelle than a dorcas gazelle? (Answer to the nearest whole centimeter.) Facts: 2 ft.; 33 in.; Answer: 23 cm

African Antelope (Impala)

All things bright and beautiful,
  All creatures great and small,
All things wise and wonderful,
  The Lord God made them all.
      —C. F. Alexander

33. Chapter 2 Test

LESSON 33

Objective

- To test the students' mastery of the concepts in Chapter 2.

Teaching Guide

1. Correct Lesson 32.
2. Review any areas of special difficulty.
3. Administer the test. For pointers on giving tests, see *Teaching Guide* for Lesson 16.

Factoring numbers breaks computation down to more convenient levels.

Fan and breeze separate parts of the grain stalk for useful purposes.

The Word of God parts the soul and spirit, joints and marrow, and thoughts and intents of the heart for His working.

Chapter 3
Factoring and Fractions

Fractions are used to express quantities that are less than 1. Common fractions have two parts, the numerator and the denominator. Understanding common fractions is the foundation for working with decimals and percents, which are other methods of expressing fractional parts.

It may seem that working with decimals is easier than working with common fractions because operations with decimals are the same as with whole numbers. However, using common fractions is sometimes faster, especially when calculating mentally. Gaining skill in working with common fractions and with decimal fractions lets you choose the better method of calculation for any problem.

Thou shalt fan them, and the wind shall carry them away, and the whirlwind shall scatter them:

and thou shalt rejoice in the LORD, and shalt glory in the Holy One of Israel.
Isaiah 41:16

34. Finding Prime Factors

The factors of a number are the whole numbers by which it is divisible. A number that is divisible only by itself and 1 is a **prime number**; examples are 3 and 5. A number that is divisible by numbers other than itself and 1 is a **composite number** (kəm·pŏz´ ĭt); examples are 8 and 12. There are many more composite numbers than prime numbers. The number 1 is considered neither prime nor composite.

> Know the first row of primes by memory.
>
> Prime numbers up to 100.
> **2, 3, 5, 7, 11, 13, 17, 19,**
> 23, 29, 31, 37, 41, 43, 47, 53, 59, 61, 67, 71, 73, 79, 83, 89, 97

Composite numbers help to simplify work with fractions. They can be divided to reduce fractions to lowest terms and to cancel in multiplication and division.

Prime Factors

Any composite number is the product of two or more prime numbers. The prime numbers that make up a composite number are called **prime factors**. Therefore, multiplying two or more prime factors yields a composite number.

Division by primes is used to divide a number into its prime factors. Follow these steps.
1. Divide the composite number by its smallest prime factor.
2. Repeat step 1 until the quotient is a prime number.
3. State the prime factors by writing the composite number, the equal sign, and then the factors in order from smallest to largest, with multiplication signs in between. If you divided correctly, the prime factors will be in that order. (See the examples below.)

Example A	**Example B**
Find the prime factors of 30.	Find the prime factors of 48.
2)30 3)15 5	2)48 2)24 2)12 2)6 3
30 = 2 × 3 × 5	48 = 2 × 2 × 2 × 2 × 3

Exponents

An exponent is a raised number that indicates how often the base (the number it follows) is a factor. For example, the multiplication 2 × 2 × 2 can be expressed as 2^3. The expression 3^4 indicates the multiplication 3 × 3 × 3 × 3 (Example C).

LESSON 34

Objectives

- To review the definitions of prime numbers and composite numbers.
- To review finding the prime factors of composite numbers by the factoring process.
- To memorize *the prime numbers from 2 to 19.
- To learn *the use of exponents to express prime factors when the same ones are repeated.

Review

1. *Find the English and metric equivalents of these Bible measures.* (Lesson 30)

   a. 2 baths (16 gal.; 60.6 l)

   b. 4 ephahs (13 pk.; 112.8 l)

2. *Solve these problems involving compound English and metric measures.* (Lesson 26)

   a. 12 yd. 1 ft.
      $\underline{\times 8}$
      (98 yd. 2 ft.)

   b. (3 gal. 3 qt.)
      $5\overline{)18\text{ gal. 3 qt.}}$

   c. 34 hm 44 m ÷ 21 = ___ m (164)

   d. 7 × 45.8 g 160 mg = ___ mg
      (321,720)

3. *Find these equivalents.* (Lesson 18)

   a. 14 pt. = ___ cups (28)

   b. 20 qt. = ___ pk. ($2\frac{1}{2}$)

4. *Give the value of each underlined digit.* (Lesson 2)

   a. 44<u>2</u>,366 (40,000)

   b. 19<u>0</u>,747,040,000 (0)

T–115   Chapter 3   Factoring and Fractions

Introduction

Ask, "Could you divide 20 pencils equally among any number of people other than 20?"

The answer is yes. You could give 2 people 10 pencils each, 4 people 5 pencils each, 5 people 4 pencils each, or 10 people 2 pencils each.

Now ask, "Could you divide 23 pencils equally among any group other than 23 people?" No, that would be impossible.

Twenty is a composite number because it is divisible by numbers other than itself and 1. Twenty-three is a prime number because it is not divisible by any number other than itself and 1.

Answers for CLASS PRACTICE g–l

g. $44 = 2 \times 2 \times 11$
h. $57 = 3 \times 19$
i. $64 = 2 \times 2 \times 2 \times 2 \times 2 \times 2$
j. $24 = 2 \times 2 \times 2 \times 3$
k. $56 = 2 \times 2 \times 2 \times 7$
l. $39 = 3 \times 13$

Answers for Part B

9. $38 = 2 \times 19$
10. $26 = 2 \times 13$
11. $92 = 2 \times 2 \times 23$
12. $33 = 3 \times 11$
13. $72 = 2 \times 2 \times 2 \times 3 \times 3$
14. $84 = 2 \times 2 \times 3 \times 7$
15. $76 = 2 \times 2 \times 19$
16. $45 = 3 \times 3 \times 5$

Teaching Guide

1. **A prime number is divisible only by itself and 1.** Have the students see how many prime numbers they can identify. Following are the prime numbers up to 100.

   2, 3, 5, 7, 11, 13, 17, 19, 23, 29, 31, 37, 41, 43, 47, 53, 59, 61, 67, 71, 73, 79, 83, 89, 97

2. **A composite number is divisible by numbers other than itself and 1.** Have the students name some composite numbers. (Note that 1 is neither prime nor composite.) Following are the composite numbers up to 40.

   4, 6, 8, 9, 10, 12, 14, 15, 16, 18, 20, 21, 22, 24, 25, 26, 27, 28, 30, 32, 33, 34, 35, 36, 38, 39, 40

3. **The prime factors of a composite number can be found through division by primes.** Discuss the steps in the lesson.

   a. Prime factors of 24

   $$\begin{array}{r} 2\overline{)24} \\ 2\overline{)12} \\ 2\overline{)6} \\ 3 \end{array}$$

   $24 = 2 \times 2 \times 2 \times 3$

   b. Prime factors of 168

   $$\begin{array}{r} 2\overline{)168} \\ 2\overline{)84} \\ 2\overline{)42} \\ 3\overline{)21} \\ 7 \end{array}$$

   $168 = 2 \times 2 \times 2 \times 3 \times 7$

4. **When any prime factor is repeated, an exponent can be used to express that factor.** Review the principle of exponents, and then practice using exponents to rewrite the

Lesson 34

When the prime factors of a number include a repeating factor, an exponent may be used to express that factor. Write the factor once, and write an exponent that indicates the number of times that factor is repeated.

Example C
Write $3 \times 3 \times 3 \times 3 = 81$ as an equation with an exponent.

$3^4 = 81$

Example D
Find the prime factors of 96. Express repeating factors with exponents.

$2\overline{)96}$
$2\overline{)48}$
$2\overline{)24}$
$2\overline{)12}$ $96 = 2 \times 2 \times 2 \times 2 \times 2 \times 3$
$2\overline{)6}$ $96 = 2^5 \times 3$
3

CLASS PRACTICE

Identify each number as prime or composite. Try to answer without looking at the list of prime numbers.

a. 6 — composite
b. 13 — prime
c. 27 — composite
d. 54 — composite
e. 67 — prime
f. 91 — composite

Find the prime factors of these composite numbers. (See facing page.)

g. 44   h. 57   i. 64   j. 24   k. 56   l. 39

Express these multiplications by using exponents to show repeating factors.

m. $2 \times 4 \times 4$ — 2×4^2
n. $3 \times 3 \times 7 \times 7 \times 7$ — $3^2 \times 7^3$
o. $2 \times 4 \times 5 \times 5$ — $2 \times 4 \times 5^2$
p. $2 \times 2 \times 10 \times 10$ — $2^2 \times 10^2$
q. $2 \times 2 \times 2 \times 8$ — $2^3 \times 8$
r. $3 \times 3 \times 3 \times 6$ — $3^3 \times 6$

WRITTEN EXERCISES

A. *Identify each number as prime (P) or composite (C). Try to answer without looking at the list of prime numbers.*

1. 7   P
2. 86 C
3. 41 P
4. 97 P
5. 71 P
6. 35 C
7. 83 P
8. 69 C

B. *Divide by primes to find the prime factors of these composite numbers. Do not use exponents in your answers.* (See facing page.)

9. 38
10. 26
11. 92
12. 33
13. 72
14. 84
15. 76
16. 45

C. *Write these multiplications with exponents to show repeating factors.*

17. $2 \times 2 \times 2 \times 3$ — $2^3 \times 3$
18. $2 \times 3 \times 5 \times 5 \times 11$ — $2 \times 3 \times 5^2 \times 11$
19. $5 \times 7 \times 13 \times 13$ — $5 \times 7 \times 13^2$
20. $7 \times 7 \times 9 \times 9 \times 11$ — $7^2 \times 9^2 \times 11$

116 Chapter 3 Factoring and Fractions

D. Divide by primes to find the prime factors of these composite numbers. Express any repeating factors with exponents.

21. 16 $16 = 2^4$ 22. 36 $36 = 2^2 \times 3^2$ 23. 64 $64 = 2^6$ 24. 125 $125 = 5^3$

25. 48 $48 = 2^4 \times 3$ 26. 81 $81 = 3^4$ 27. 80 $80 = 2^4 \times 5$ 28. 88 $88 = 2^3 \times 11$

E. Solve these reading problems.

29. Luke 3:23 records that Jesus was about 30 years old at the time of His baptism. Write the prime factors of 30. $30 = 2 \times 3 \times 5$

30. Uzziah, who was king for 52 years, reigned longer than any other king of Judah. Write the prime factors of 52, using exponents. $52 = 2^2 \times 13$

31. During a heavy snowfall, a certain area received 24 inches of snow. Some distance to the south, another area received snowfall equaling the largest prime factor of 24. How much snow was that?   3 inches

32. A tree may have about 100,000 leaves. One breezy autumn afternoon, the number of leaves Gilbert saw falling in 10 seconds was equal to the largest prime factor of 100,000. How many leaves did he see falling?   5 leaves

33. The Whitsells had a 15-hectare corn field that produced 144 metric tons of shelled corn. What was the average yield in kilograms per hectare?   9,600 kilograms

34. The first day Ruth gleaned in Boaz's field, she gathered about an ephah of barley (Ruth 2). She continued gleaning until the end of harvest. If she gathered an ephah each day for two weeks (12 days), how many bushels of grain did she glean? $9\frac{3}{4}$ bushels
(Computed with the ephah/peck equivalent; $9\frac{3}{5}$ if computed with the homer/bushel equivalent)

REVIEW EXERCISES

F. Find these equivalents. *(Lessons 18, 30)*

35. 5 bu. = __20__ pk.

36. 12 gal. 1 qt. = __49__ qt.

37. 10 ephahs = __282__ l

38. 24 bu. = __96__ Roman bushels

G. Solve these problems involving compound measures. *(Lesson 26)*

39. 8 ft. 7 in.
    × 6
    51 ft. 6 in.

40. $9\overline{)15 \text{ lb. } 6 \text{ oz.}}$ 1 lb. $11\frac{1}{3}$ oz.

H. Write the value of each underlined digit. *(Lesson 2)*

41. 1̲0,865,903   10,000,000 (10 million)

42. 7̲92,847,040,000   700,000,000,000 (700 billion)

CHALLENGE EXERCISES

43. The area of a rectangular room is 221 square feet. Its dimensions are two consecutive prime numbers, which are the prime factors of 221. What are the dimensions?   13 ft. by 17 ft.

44. The volume of a certain room is 1,001 cubic feet. Its dimensions are three consecutive prime numbers. What are the dimensions of the room?   7 ft. by 11 ft. by 13 ft.

prime factors of various numbers.

a. $25 = 5 \times 5 = 5^2$
b. $81 = 3 \times 3 \times 3 \times 3 = 3^4$
c. $24 = 2 \times 2 \times 2 \times 3 = 2^3 \times 3$
d. $56 = 2 \times 2 \times 2 \times 7 = 2^3 \times 7$
e. $52 = 2 \times 2 \times 13 = 2^2 \times 13$
f. $168 = 2 \times 2 \times 2 \times 3 \times 7 = 2^3 \times 3 \times 7$

Solutions for Exercises 33 and 34

33. $144{,}000 \div 15$
34. $12 \times 3\frac{1}{4} \div 4$

T–117   Chapter 3   Factoring and Fractions

LESSON 35

Objectives

- To review finding the greatest common factor of a set of numbers.
- To review the two methods for finding the lowest common multiple of a set of numbers.

Review

1. Give Lesson 35 Quiz (Distance, Rate, and Time).
2. *Find these equivalents.* (Lessons 19, 27)
   a. 18 sq. ft. = ___ sq. in.   (2,592)
   b. 16 sq. yd. 8 sq. ft. = ___ sq. yd.   ($16\frac{8}{9}$)
   c. 14 g = ___ oz.   (0.49)
   d. 57 sq. mi. = ___ km²   (147.63)
4. *Write these Roman numerals as Arabic numerals.* (Lesson 3)
   a. CDLIX   (459)
   b. DCXXXI   (631)
5. *Write these Arabic numerals as Roman numerals.* (Lesson 3)
   a. 2,044   (MM$\overline{\text{XLIV}}$)
   b. 990,231   ($\overline{\text{CMXCCCXXXI}}$)

Introduction

Find all the factors of 32.
   (1, 2, 4, 8, 16, 32)

Find all the factors of 48.
   (1, 2, 3, 4, 6, 8, 12, 16, 24, 48)

What is the largest factor that these composite numbers have in common? (16) Sixteen is the *greatest common factor* of 32 and 48. Knowing the greatest common factor is useful in reducing fractions. Dividing both terms of $\frac{32}{48}$ by 16 reduces the fraction to $\frac{2}{3}$.

Teaching Guide

1. **The greatest common factor (g.c.f.) is the largest whole number that is a factor of two different numbers.** The greatest common factor may be the smaller of the two numbers themselves. The greatest common factor of 4 and 6 is 2. Students should be able to find the greatest common factors of the following sets mentally.
   a. 8, 12   (4)
   b. 6, 15   (3)
   c. 8, 24   (8)

2. **To find the greatest common factor of two numbers, use the following steps.**
   (1) Look at the smaller number. If it is a factor of the larger number, it is the greatest common factor. If not, go to step 2.
      a. g.c.f. of 4, 12 = ___   (4)
      b. g.c.f. of 15, 45 = ___   (15)
   (2) Use division by primes to find the prime factors of each number. List them from smallest to largest.
   (3) Cross out all the factors that are not common, and one in each pair of common factors. When exponents are used, cross out the factor with the larger exponent. (No exponent after a number means the exponent is 1.)
   (4) Multiply all the remaining factors. The product is the greatest common factor. If the two numbers have only one common factor, that is the greatest common factor.
      c. g.c.f. of 6, 8 = ___
         6 is not a factor of 8.
         2)6   2)8 $6 = 2 \times 3$
          3   2)4 $8 = 2^3$
                     2   g.c.f. = (2)

35. Finding Greatest Common Factors and Lowest Common Multiples

Greatest Common Factors

A common factor is a number that is a factor in two different numbers. Both numbers are divisible by the common factor. The largest whole number that is a factor of two different numbers is the **greatest common factor**, abbreviated g.c.f.

Factors of 12: 2, 3, 4, 6
Factors of 16: 2, 4, 8
Common factors of 12 and 16: 2, 4
g.c.f. = 4

Because the greatest common factor is a factor of both numbers, it will never be larger than the smaller of the two numbers. The following steps tell how to find the greatest common factor of two numbers.

1. Look at the smaller of the two numbers. If it is a factor of the larger number, it is the greatest common factor. If not, go to step 2.

2. Use division by primes to find the prime factors of each number. List them from smallest to largest.

3. Cross out all the factors that are not common, and one in each pair of common factors. When exponents are used, cross out the factor with the larger exponent. (No exponent after a number means the exponent is 1.)

4. Multiply all the remaining factors. The product is the greatest common factor. If the two numbers have only one common factor, that is the greatest common factor.

Example A
Find the greatest common factor of 30 and 105.

2)30   3)105
3)15   5)35
   5   7

$30 = \cancel{2} \times 3 \times 5$
$105 = \cancel{3} \times \cancel{5} \times \cancel{7}$
g.c.f. = 3 × 5 = 15

Example B
Find the greatest common factor of the numerator and denominator in $\frac{36}{54}$.

2)36   2)54
2)18   3)27
3)9   3)9
   3   3

$36 = 2 \times 2 \times 3 \times 3$, or $36 = \cancel{2^2} \times 3^2$
$54 = 2 \times 3 \times 3 \times 3$, or $54 = 2 \times \cancel{3^3}$
g.c.f. = 2×3^2 = 18

118   Chapter 3   Factoring and Fractions

Lowest Common Multiples

The **lowest common multiple** (l.c.m.) is the smallest number that is a multiple of a set of numbers. It is possible for the largest number in the set to be the lowest common multiple.

Multiples of 8 = 8, 16, 24, 32, 40, 48, . . .
Multiples of 12 = 12, 24, 36, 48, 60, . . .
Common multiples of 8 and 12: 24, 48, . . . l.c.m. = 24

The lowest common multiple can be found by trial and error or by factoring (dividing by primes). It is usually faster to first use the trial-and-error method mentally. If that does not work, then use the factoring method. The steps below tell how to use the trial-and-error method.

1. Look at the larger of the two numbers. Is it a multiple of the smaller number? If not, go to step 2.
2. Multiply the larger number by 2. Is the answer a multiple of the smaller number? If not, go to step 3.
3. Multiply the larger number by 3. Is the answer a multiple of the smaller number? If not, use the factoring method.

Example C (Trial and error)
Find the lowest common multiple of 15 and 20.
(1) 20 is not a multiple of 15
(2) 2 × 20 = 40—not a multiple of 15
(3) 3 × 20 = 60—a multiple of 15

l.c.m. = 60

The following steps tell how to find the lowest common multiple of two numbers by the factoring method.

1. Find and list the prime factors of both numbers. Use exponents to show repeating factors.
2. Identify the common factors. If there are no common factors, multiply the two original numbers to find the lowest common multiple.
3. Cross out one factor in each pair of common factors. When exponents are used, cross out the factor with the smaller exponent. Remember, no exponent after a number means the exponent is 1.

Example D (Factoring)
Find the lowest common multiple of the denominators in $\frac{1}{36}$ and $\frac{1}{40}$.
40, 2 × 40 (80), and 3 × 40 (120) are not multiples of 36

$2 \overline{)36}$　　　　　$2 \overline{)40}$
$2 \overline{)18}$　　　　　$2 \overline{)20}$
$3 \overline{)9}$　　　　　　$2 \overline{)10}$
　3　　　　　　　　5

$36 = 2 \times 2 \times 3 \times 3$, or $36 = 2^2 \times 3^2$
$40 = 2 \times 2 \times 2 \times 5$, or $40 = 2^3 \times 5$
l.c.m. = $2^3 \times 3^2 \times 5 = 360$

4. Multiply all the remaining prime factors. The product is the lowest common multiple.

CLASS PRACTICE

Find the greatest common factor of each pair.

  a. 14, 35   7   b. 48, 72   24   c. 16, 96   16   d. 25, 95   5

d. g.c.f. of 20, 28 = ___
   20 is not a factor of 28.

   $$\begin{array}{cc} 2\underline{)20} & 2\underline{)28} \\ 2\underline{)10} & 2\underline{)14} \\ 5 & 7 \end{array}$$

   $20 = 2^2 \times 5$
   $28 = 2^2 \times 7$
   g.c.f. = (4)

3. **The lowest common multiple (l.c.m.) is the smallest number that is a multiple of a set of numbers.** It can be the largest number in the set. The lowest common multiple of 2 and 3 is 6, and of 4 and 6 is 12. This is useful in finding a common denominator when adding or subtracting fractions.

   The lowest common denominator can be found for more than just two numbers. Students should be able to find the following answers mentally.

   a. l.c.m. of 6, 9 = ___ (18)
   b. l.c.m. of 5, 8, 10 = ___ (40)
   c. l.c.m. of 12, 16 = ___ (48)

4. **One way to find the lowest common multiple is by trial and error.** It is usually faster to first use this method mentally. If that does not work, then use the factoring method. The steps below tell how to use the trial-and-error method.

   (1) Look at the larger of the two numbers. Is it a multiple of the smaller number? If not, go to step 2.

   (2) Multiply the larger number by 2. Is the answer a multiple of the smaller number? If not, go to step 3.

   (3) Multiply the larger number by 3. Is the answer a multiple of the smaller number? If not, use the factoring method.

   The following problems can be solved with these three trial-and-error steps.

   a. l.c.m. of 10, 20 = ___ (20)
   b. l.c.m. of 10, 15 = ___ (30)
   c. l.c.m. of 15, 25 = ___ (75)

5. **Another way to find the lowest common multiple is by factoring.** The following steps tell how to find the lowest common multiple of two numbers by the factoring method.

   (1) Find and list the prime factors of both numbers. Use exponents to show repeating factors.

   (2) Identify the common factors. If there are no common factors, multiply the two original numbers to find the lowest common multiple.

   (3) Cross out one factor in each pair of common factors. When exponents are used, cross out the factor with the smaller exponent. Remember, no exponent after a number means the exponent is 1.

   (4) Multiply all the remaining prime factors. The product is the lowest common multiple.

   Most of the following problems will be solved by the factoring method. See how many of your students detect that problem a is much more easily solved by the trial-and-error method (in four steps).

   a. l.c.m. of 40, 50 = ___

   $$\begin{array}{cc} 2\underline{)40} & 2\underline{)50} \\ 2\underline{)20} & 5\underline{)25} \\ 2\underline{)10} & 5 \\ 5 & \end{array}$$

   $40 = 2^3 \times 5$
   $50 = 2 \times 5^2$
   l.c.m. = $2^3 \times 5^2$
   l.c.m. = (200)

   b. l.c.m. of 13, 21 = ___

   $$\begin{array}{cc} 1\underline{)13} & 3\underline{)21} \\ 13 & 7 \end{array}$$

   $13 = 1 \times 13$
   $21 = 3 \times 7$
   l.c.m. = $3 \times 7 \times 13$
   l.c.m. = (273)

T-119   Chapter 3   Factoring and Fractions

c. l.c.m. of 45, 50 = ___

    3)45   2)50
    3)15   5)25
      5   5

    $45 = 3^2 \times 5$
    $50 = 2 \times 5^2$
    l.c.m. $= 2 \times 3^2 \times 5^2$
    l.c.m. = (450)

d. lowest common denominator for $\frac{1}{35}$ and $\frac{1}{60}$ = ___

    5)35   2)60
      7   2)30
             3)15
               5

    $35 = 5 \times 7$
    $60 = 2^2 \times 3 \times 5$
    l.c.m. $= 2^2 \times 3 \times 5 \times 7$
    l.c.m. = (420)

An Ounce of Prevention

1. Make certain the students do not confuse greatest common factors and lowest common multiples. Here are two simple rules to remember. The greatest common factor is equal to or a factor of the smaller number in the pair. The lowest common multiple is equal to or a multiple of the larger number in the pair.

2. It may also be confusing to remember which factors to cross off and what to multiply.

    a. The **g.c. factor** is a portion that is contained within both numbers. Eliminate any factors that are not common, and the larger exponents. Multiply only what is common to both numbers.

    b. The **l.c. multiple** must be large enough to cover both numbers as factors. Eliminate only what is duplicated, and the smaller exponent. Include in the multiplication all factors that are not common.

Further Study

1. Finding the lowest common multiple of a set of numbers is simple if you know the greatest common factor. Divide the smaller of the numbers by the greatest common factor, and multiply the result times the other number.

    a. l.c.m. of 35, 50 = ___
        g.c.f. = 5
        l.c.m. = 35 ÷ 5 × 50 = 350

    b. l.c.m. of 12, 14 = ___
        g.c.f. = 2
        l.c.m. = 12 ÷ 2 × 14 = 84

2. Finding the greatest common factor may be easier by using a trial-and-error method similar to that used for finding least common multiples, except that division is used instead of multiplication.

    (1) Look at the smaller number. Is it a factor of the larger number? If not, go to step 2.

    (2) Divide the smaller number by 2. Is the answer a factor of the larger number? If not, go to step 3.

    (3) Divide the smaller number by 3. Is the answer a factor of the larger number? If not, go to step 4.

    (4) Divide the smaller number by 4. Is the answer a factor of the larger number? If the greatest common factor is not yet found, use division by primes to find it.

    a. g.c.f. of 30, 50 = ___
        30 ÷ 2 = 15 (not g.c.f.)
        30 ÷ 3 = 10 (g.c.f.)

    b. g.c.f. of 27, 45 = ___
        27 ÷ 3 = 9 (g.c.f.)

    c. g.c.f. of 32, 40 = ___
        32 ÷ 2 = 16 (not g.c.f.)
        32 ÷ 4 = 8 (g.c.f.)

Find the lowest common multiple of each pair.

e. 15, 25   75
f. 16, 48   48
g. 9, 15   45
h. 20, 45   180
i. 14, 34   238
j. 48, 54   432
k. 19, 38   38
l. 18, 26   234

WRITTEN EXERCISES

A. Find the greatest common factor of the numbers in each fraction.

1. $\frac{10}{15}$ 5
2. $\frac{18}{30}$ 6
3. $\frac{42}{70}$ 14
4. $\frac{45}{75}$ 15
5. $\frac{48}{80}$ 16
6. $\frac{21}{63}$ 21
7. $\frac{35}{42}$ 7
8. $\frac{60}{84}$ 12

B. Find the lowest common multiple of the denominators in each pair.

Use the trial-and-error method.

9. $\frac{1}{12}, \frac{1}{18}$ 36
10. $\frac{1}{14}, \frac{1}{21}$ 42
11. $\frac{1}{18}, \frac{1}{30}$ 90
12. $\frac{1}{60}, \frac{1}{80}$ 240

Use the factoring method.

13. $\frac{1}{14}, \frac{1}{16}$ 112
14. $\frac{1}{15}, \frac{1}{18}$ 90
15. $\frac{1}{20}, \frac{1}{24}$ 120
16. $\frac{1}{28}, \frac{1}{35}$ 140

Use the trial-and-error method first and the factoring method if necessary.

17. $\frac{1}{23}, \frac{1}{46}$ 46
18. $\frac{1}{21}, \frac{1}{33}$ 231
19. $\frac{1}{25}, \frac{1}{30}$ 150
20. $\frac{1}{24}, \frac{1}{36}$ 72

C. Solve these reading problems.

21. Find the greatest common factor of the numerator and denominator $\frac{17}{51}$.   17
22. Find the greatest common factor of the numerator and denominator $\frac{18}{48}$.   6
23. At the Grassy Meadows Mennonite School, the total number of students in grades 7–9 is the lowest common multiple of the numbers of students in grades 8 and 9. Eighth grade has 5 students, and ninth grade has 3 students. How many students are in seventh grade?   7 students
24. Find the lowest common multiple of 14 and 18.   126
25. The tire on the Stauffers' van has a treaded area of 594 square inches. How many square feet is that? $4\frac{1}{8}$ square feet (594 ÷ 144)
26. Abraham Lincoln delivered his Gettysburg address in 1863. What year was $1\frac{1}{4}$ centuries after 1863?   1988 (1863 + 125)

REVIEW EXERCISES

D. Find these equivalents. *(Lessons 19, 27)*

27. 18 sq. ft. = __2__ sq. yd.
28. 5 sq. mi. = 3,200 a.
29. 23 tons = 20.93 MT
30. 12 m = 39.36 ft.

E. Write the Roman numerals as Arabic numerals and the Arabic numerals as Roman numerals. *(Lesson 3)*

31. MCMLXXXI   1,981
32. $\overline{\text{MMCLIV}}$CC   2,154,200
33. 1,734   MDCCXXXIV
34. 24,409 $\overline{\text{XXIV}}$CDIX

36. Using Common Fractions to Express Parts

Fractions are used to express quantities less than 1. A **common fraction**, often simply called a fraction, has two parts that are called the **terms** of the fraction. The **denominator** is the lower number; this term shows into how many parts a whole is divided. The **numerator** is the upper number; this term shows how many parts of the denominator are represented by the fraction. The numerator can be any number, and the denominator can be any number except zero.

Fractions are expressions that show division because their value is equal to the numerator divided by the denominator. For example, if 3 cookies were divided equally among 4 people, each person would receive $\frac{3}{4}$ of a cookie because 3 divided by 4 equals $\frac{3}{4}$.

The following definitions and examples show some expressions that are used in working with fractions.

Expressions   Examples

Proper fraction—A fraction with a numerator smaller than the denominator. The value is less than 1. $\quad\frac{3}{4}\quad\frac{2}{7}\quad\frac{1}{16}$

Improper fraction—A fraction with a numerator equal to or greater than the denominator. The value is equal to or greater than 1. $\quad\frac{4}{3}\quad\frac{8}{8}\quad\frac{12}{4}$

Whole number—A number that does not include a fraction. $\quad 3\quad 14\quad 278$

Mixed number—A whole number and a fraction together. $\quad 1\frac{1}{4}\quad 5\frac{3}{8}$

Unit fraction—A proper fraction whose numerator is 1. $\quad\frac{1}{3}\quad\frac{1}{12}$

Complex fraction—An expression that includes a fraction in the numerator, in the denominator, or in both terms. $\quad\frac{1\frac{1}{2}}{5}\quad\frac{\frac{1}{5}}{\frac{3}{4}}$

Expanding and Reducing Fractions

To **expand** a fraction, multiply both of its terms by the same number. This changes it to an equivalent fraction, which has the same value.

$$\frac{3 \times 3}{4 \times 3} = \frac{9}{12}$$

To **reduce** a fraction, divide both terms by the same number. This also changes the fraction to an equivalent fraction.

$$\frac{15 \div 3}{18 \div 3} = \frac{5}{6}$$

A fraction is **reduced to lowest terms** when the two terms have no common factors other than 1. To reduce a fraction to lowest terms, divide the numerator and denominator by their greatest common factor.

$$\frac{18 \div 9}{27 \div 9} = \frac{2}{3}$$

LESSON 36

Objectives

- To review the meaning and use of fractions.
- To review expanding and reducing fractions, with emphasis on reducing fractions to lowest terms.
- To review reducing fractions to lowest terms by dividing the numerator and the denominator by their greatest common factor.
- To review changing improper fractions to whole or mixed numbers.
- To introduce *complex fractions and unit fractions.

Review

1. *Find the greatest common factor of each pair.* (Lesson 35)
    a. 20, 32 (4)
    b. 23, 69 (23)
    c. 28, 98 (14)
    d. 66, 44 (22)

2. *Find the lowest common multiple of each pair.* (Lesson 35)
    a. 14, 16 (112)
    b. 42, 48 (336)
    c. 35, 15 (105)
    d. 45, 99 (495)

3. *Find these temperature equivalents to the nearest degree.* (Lesson 28)
    a. 31°C = ___°F (88)
    b. 87°C = ___°F (189)
    c. 95°F = ___°C (35)
    d. 54°F = ___°C (12)

4. *Change these times as indicated.* (Lesson 20)
    a. 12:30 A.M. in Vancouver, British Columbia; ___ in Washington, D.C. (3:30 A.M.)
    b. 12:00 midnight in Albuquerque, New Mexico; ___ in St. Johns, Newfoundland (3:30 A.M.)

5. *Find these equivalents.* (Lesson 20)
    a. 3 yr. 16 wk. = ___ wk. (about 172)
    b. 312 hr. = ___ days (13)

6. *Tell whether each problem illustrates the commutative or the associative law.* (Lesson 4)
    a. $a + (b + c) = (a + b) + c$
        (associative)
    b. $4 \times 7 \times 3 = 7 \times 4 \times 3$
        (commutative)

Introduction

If 3 apples are divided among 6 children, what part of an apple will each child receive?

Each child gets $\frac{3}{6}$ apple. The students will readily recognize that $\frac{3}{6}$ can be reduced to $\frac{1}{2}$. Each child will receive $\frac{1}{2}$ apple.

Fractions are used to express part of a whole.

Teaching Guide

1. **Fractions are used to express numbers that are not whole numbers.** Examples include $\frac{1}{2}$, $\frac{3}{4}$, and $\frac{2}{10}$.

2. **Common fractions have two parts, which are called the terms of the fraction.** The denominator is the lower number; this term shows into how many parts a whole is divided. The numerator is the upper number; this term shows how many parts of the denominator are represented by the fraction. The numerator can be any number, and the denominator can be any number except zero.

    a. $\frac{1}{4}$ indicates 1 part of a whole that is divided into 4 equal parts.

    b. $\frac{3}{8}$ indicates 3 parts of a whole that is divided into 8 equal parts.

    c. $\frac{0}{5}$ is a valid fraction because $0 \div 5 = 0$.

    d. $\frac{5}{0}$ ($5 \div 0$) is not a valid fraction, because division by zero is not allowed.

3. **Fractions are expressions that show division because their value is equal to the numerator divided by the denominator.**

    a. The fraction $\frac{3}{8}$ can represent 3 items divided into 8 parts.

    b. The fraction $\frac{4}{5}$ can represent 4 items divided into 5 parts.

4. **The following terms are used in working with fractions.**

    Proper fraction—A fraction with a numerator smaller than the denominator. The value is less than 1.

    Improper fraction—A fraction with a numerator equal to or greater than the denominator. The value is equal to or greater than 1.

Improper Fractions, Whole Numbers, and Mixed Numbers

An improper fraction should usually be changed to a whole or mixed number. To do this, divide the numerator by the denominator and express the remainder as a fraction.

$$\frac{5}{3} = 5 \div 3 = 1\frac{2}{3}$$

To change a whole number to an improper fraction, write a fraction using the whole number as the numerator and 1 as the denominator.

$$5 = \frac{5}{1}$$

To change a mixed number to an improper fraction, multiply the denominator of the fraction times the whole number, and add the numerator. The result is the numerator of the improper fraction. The denominator is the same as that of the original fraction.

Example A

$3\frac{1}{8} = \frac{?}{8}$

$8 \times 3 + 1 = 25$

$3\frac{1}{8} = \frac{25}{8}$

Comparing Fractions

To compare fractions, first express them as **like fractions**. These are fractions with a **common denominator** (the same denominator). The steps for comparing fractions are given below.

1. Find the lowest common multiple for the denominators of the fractions.
2. Express the fractions as like fractions, using the lowest common multiple found in step 1 for the denominators.
3. See which of the numerators is larger. Then write > ("is greater than") or < ("is less than") between the fractions. Note that the smaller end of this symbol (the point) is turned toward the smaller number.

Example B

Compare $\frac{5}{8}$ and $\frac{2}{3}$.

l.c.m. of 3 and 8 = 24

$\frac{5 \times 3 = 15}{8 \times 3 = 24}$ $\frac{2 \times 8 = 16}{3 \times 8 = 24}$

$\frac{15}{24} < \frac{16}{24}$, so $\frac{5}{8} < \frac{2}{3}$

CLASS PRACTICE

Expand each fraction by multiplying both the numerator and the denominator by the number in parentheses.

a. $\frac{2}{3}$ (12) $\frac{24}{36}$ b. $\frac{7}{8}$ (6) $\frac{42}{48}$ c. $\frac{3}{5}$ (9) $\frac{27}{45}$ d. $\frac{4}{9}$ (8) $\frac{32}{72}$

Reduce these fractions to lowest terms by dividing both the numerator and the denominator by the greatest common factor.

e. $\frac{25}{35}$ $\frac{5}{7}$ f. $\frac{9}{18}$ $\frac{1}{2}$ g. $\frac{16}{72}$ $\frac{2}{9}$ h. $\frac{54}{90}$ $\frac{3}{5}$

Express these improper fractions as whole or mixed numbers.

i. $\frac{15}{5}$ 3   j. $\frac{21}{4}$ $5\frac{1}{4}$ k. $\frac{23}{2}$ $11\frac{1}{2}$ l. $\frac{38}{9}$ $4\frac{2}{9}$

Express these whole or mixed numbers as improper fractions.

m. $4\frac{1}{2}$ $\frac{9}{2}$ n. 7 $\frac{7}{1}$ o. $2\frac{7}{8}$ $\frac{23}{8}$ p. $6\frac{2}{5}$ $\frac{32}{5}$

122 Chapter 3 Factoring and Fractions

Compare each set of fractions, and place < or > between them.

q. $\frac{1}{2}$ ___ $\frac{2}{9}$ $\frac{9}{18} > \frac{4}{18}$ r. $\frac{5}{8}$ ___ $\frac{5}{7}$ $\frac{35}{56} < \frac{40}{56}$

s. $\frac{11}{15}$ ___ $\frac{7}{9}$ $\frac{33}{45} < \frac{35}{45}$ t. $\frac{3}{20}$ ___ $\frac{1}{8}$ $\frac{6}{40} > \frac{5}{40}$

WRITTEN EXERCISES

A. Label each item as *whole number, proper fraction, improper fraction, mixed number,* **or** *complex fraction.*

1. $\frac{\frac{1}{4}}{\frac{1}{6}}$ complex fraction   2. 759   whole number

3. $\frac{7}{4}$ improper fraction   4. $15\frac{2}{7}$ mixed number

B. Expand each fraction by multiplying both the numerator and the denominator by the number in parentheses.

5. $\frac{3}{4}$ (8) $\frac{24}{32}$ 6. $\frac{5}{6}$ (5) $\frac{25}{30}$

7. $\frac{3}{8}$ (12) $\frac{36}{96}$ 8. $\frac{7}{9}$ (20) $\frac{140}{180}$

C. Reduce these fractions to lowest terms.

9. $\frac{9}{12}$ $\frac{3}{4}$ 10. $\frac{8}{16}$ $\frac{1}{2}$

11. $\frac{34}{85}$ $\frac{2}{5}$ 12. $\frac{21}{54}$ $\frac{7}{18}$

13. $\frac{26}{39}$ $\frac{2}{3}$ 14. $\frac{49}{63}$ $\frac{7}{9}$

15. $\frac{54}{72}$ $\frac{3}{4}$ 16. $\frac{27}{81}$ $\frac{1}{3}$

D. Express these improper fractions as whole or mixed numbers.

17. $\frac{7}{5}$ $1\frac{2}{5}$ 18. $\frac{12}{3}$ 4

19. $\frac{17}{2}$ $8\frac{1}{2}$ 20. $\frac{29}{3}$ $9\frac{2}{3}$

E. Express these whole or mixed numbers as improper fractions.

21. 15 $\frac{15}{1}$ 22. $2\frac{1}{5}$ $\frac{11}{5}$

23. $3\frac{3}{4}$ $\frac{15}{4}$ 24. $4\frac{2}{7}$ $\frac{30}{7}$

F. Compare each set of fractions, and write < or > between them.

25. $\frac{3}{4}$ ___ $\frac{5}{8}$ $\frac{6}{8} > \frac{5}{8}$ 26. $\frac{5}{6}$ ___ $\frac{7}{8}$ $\frac{20}{24} < \frac{21}{24}$

27. $\frac{7}{12}$ ___ $\frac{9}{16}$ $\frac{28}{48} > \frac{27}{48}$ 28. $\frac{17}{20}$ ___ $\frac{13}{15}$ $\frac{51}{60} < \frac{52}{60}$

Whole number—A number that does not include a fraction.

Mixed number—A whole number and a fraction together.

Unit fraction—A proper fraction whose numerator is 1.

Complex fraction—An expression that includes a fraction in the numerator, in the denominator, or in both terms.

5. **Fractions can be changed to equivalent fractions in two ways.**

   a. *By expanding them.* This is done by multiplying both the numerator and the denominator by the same number.

   b. *By reducing them.* This is done by dividing both the numerator and the denominator by the same number.

   A fraction is reduced to lowest terms when the two terms have no common factors other than 1. To reduce a fraction to lowest terms, divide the numerator and denominator by their greatest common factor.

   Have students expand the following fractions by the factors in brackets. Stress that the expanded fractions have the same values as the original fractions.

   a. $\frac{3}{4}$ [7] ($\frac{21}{28}$)
   b. $\frac{4}{7}$ [12] ($\frac{48}{84}$)
   c. $\frac{7}{12}$ [8] ($\frac{56}{96}$)
   d. $\frac{5}{9}$ [11] ($\frac{55}{99}$)

   Practice reducing fractions to lowest terms by dividing the numerator and denominator by their greatest common factor.

   e. $\frac{6}{9}$ ($\frac{2}{3}$; g.c.f. = 3)
   f. $\frac{12}{16}$ ($\frac{3}{4}$; g.c.f. = 4)
   g. $\frac{27}{45}$ ($\frac{3}{5}$; g.c.f. = 9)
   h. $\frac{21}{49}$ ($\frac{3}{7}$; g.c.f. = 7)

6. **The form of improper fractions, whole numbers, and mixed numbers can be changed in the following ways.**

   a. To change an improper fraction to a whole or mixed number, divide the numerator by the denominator and express the remainder as a fraction.

   b. To change a whole number to an improper fraction, write a fraction with the whole number as the numerator and 1 as the denominator.

   c. To change a mixed number to an improper fraction, multiply the denominator of the fraction times the whole number and add the numerator. The result is the numerator of the improper fraction. The denominator is the same as that of the original fraction.

   Give practice with changing improper fractions to whole or mixed numbers (*a–c* below) and changing whole or mixed numbers to improper fractions (*d–f* below).

   a. $\frac{7}{4}$ ($1\frac{3}{4}$)
   b. $\frac{11}{3}$ ($3\frac{2}{3}$)
   c. $\frac{19}{5}$ ($3\frac{4}{5}$)
   d. 16 ($\frac{16}{1}$)
   e. $4\frac{1}{6}$ ($\frac{25}{6}$)
   f. $8\frac{3}{7}$ ($\frac{59}{7}$)

7. **It is sometimes useful to compare fractions to see which one is larger.** The steps for comparing fractions are given below.

   (1) Find the lowest common multiple for the denominators of the fraction.

   (2) Express the fractions as like fractions, using the lowest common

T-123   Chapter 3   Factoring and Fractions

multiple found in step 1 for the denominators.

(3) See which of the numerators is larger. Then write > ("is greater than") or < ("is less than") between the fractions. Note that the smaller end of this symbol (the point) is turned toward the smaller number.

   a. $\frac{2}{3}$ — $\frac{5}{8}$ ($\frac{16}{24} > \frac{15}{24}$, so $\frac{2}{3} > \frac{5}{8}$)
   b. $\frac{7}{10}$ — $\frac{11}{15}$ ($\frac{21}{30} < \frac{22}{30}$, so $\frac{7}{10} < \frac{11}{15}$)
   c. $\frac{13}{16}$ — $\frac{17}{20}$ ($\frac{65}{80} < \frac{68}{80}$, so $\frac{13}{16} < \frac{17}{20}$)
   d. $\frac{9}{14}$ — $\frac{13}{21}$ ($\frac{27}{42} > \frac{26}{42}$, so $\frac{9}{14} > \frac{13}{21}$)

Further Study

1. One old method for expressing fractions was to separate the numerator and the denominator with a colon. Thus, the fraction $\frac{3}{4}$ was expressed as 3:4. The same form is still used in writing ratios.

2. Fractions were used in ancient Egypt, but the Egyptians worked only with unit fractions. For example, the fraction $\frac{3}{4}$ would have been expressed as $\frac{1}{2} + \frac{1}{4}$. The fraction $\frac{7}{8}$ would have been expressed as $\frac{1}{2} + \frac{1}{4} + \frac{1}{8}$.

Solutions for Exercises 33 and 34

33. $\frac{5}{9}(69 - 32)$
34. $\frac{9}{5} \times 8 + 32$

Lesson 36   123

G. Solve these reading problems.

29. In Numbers 28:12, God instructed the Israelites to bring flour mingled with oil for a meat offering when they offered a bullock or a ram. If they offered a ram, they were to bring $\frac{2}{10}$ deal of flour. Reduce $\frac{2}{10}$ to lowest terms. $\frac{1}{5}$

30. In Numbers 28:12, the Israelites were also instructed to bring $\frac{3}{10}$ deal of flour along with the offering of a bullock. Expand $\frac{3}{10}$ by multiplying both terms by 5. $\frac{15}{50}$

31. Two and one-half tribes of the twelve tribes of Israel asked to receive their inheritance of land east of the Jordan River. Write a complex fraction comparing the number of tribes receiving land east of the Jordan (the numerator) with all the tribes (the denominator). $\frac{2\frac{1}{2}}{12}$

32. One week David slept 60 of the 168 hours in a week. Write a fraction stating what part of the week David was awake. Also write the fraction in lowest terms. $\frac{108}{168} = \frac{9}{14}$

33. One morning in December, it was 69°F in Miami, Florida. To the nearest whole degree, what was the temperature on the Celsius scale?   21°C

34. That same morning it was 8°C in Portland, Oregon. To the nearest whole degree, what was the temperature on the Fahrenheit scale?   46°F

REVIEW EXERCISES

H. Find the greatest common factor of each pair. (Lesson 35)

35. 18, 24   6
36. 32, 96   32

I. Find the lowest common multiple of each pair. (Lesson 35)

37. 15, 21   105
38. 32, 40   160

J. Find these temperature equivalents to the nearest degree. (Lesson 28)

39. 16°C = __61__ °F
40. 110°F = __43__ °C

K. Change these times as indicated. (Lesson 20)

41. 6:45 P.M. in Dallas, Texas; ___ in St. Johns, Newfoundland   9:15 P.M.
42. 12:00 noon in Philadelphia, Pennsylvania; ___ in Albuquerque, New Mexico   10:00 A.M.

L. Find these equivalents. (Lesson 20)

43. 84 days = __12__ wk.
44. 1,440 min. = __24__ hr.

M. Write commutative law or associative law to tell what each problem illustrates. (Lesson 4)

45. $a \times b \times c = b \times a \times c$

    commutative law

46. $2 + (8 + 14) = (2 + 8) + 14$

    associative law

124 Chapter 3 *Factoring and Fractions*

37. Adding and Subtracting Common Fractions

The methods you practiced in the last several lessons are part of the normal process of adding and subtracting fractions. If you have **unlike fractions** (fractions with different denominators), find the lowest common multiple of the denominators and expand one or both of the fractions to equivalent fractions.

Adding Fractions
To add fractions or mixed numbers, follow these steps.
1. Change any unlike fractions to like fractions. Use the lowest common multiple of the denominators as the common denominator for the addends.
2. Add the numerators. Use the denominator of the addends as the denominator for the sum.
3. Add any whole numbers in the addends.
4. Change the answer to simplest form, with no improper fraction and with any proper fraction in lowest terms.

Example A	Example B
$\frac{5}{8}$ $+ \frac{7}{8}$ $\frac{12}{8} = 12 \div 8 = 1\frac{4}{8} = 1\frac{1}{2}$	$1\frac{2}{3} = 1\frac{10}{15}$ $+ 2\frac{4}{5} = 2\frac{12}{15}$ $3\frac{22}{15} = 3 + 1\frac{7}{15} = 4\frac{7}{15}$

Subtracting Fractions
To subtract fractions or mixed numbers, follow these steps.
1. Change any unlike fractions to like fractions.
2. If necessary, borrow 1 from the whole number. Change the 1 to an improper fraction and add it to the fraction in the minuend.
3. Subtract the numerators. Use the denominator of the minuend and subtrahend as the denominator for the difference.
4. Subtract any whole numbers.
5. Change the answer to simplest form.

Example C	Example D
$\frac{5}{6} = \frac{25}{30}$ $-\frac{1}{5} = \frac{6}{30}$ $\frac{19}{30}$	$3\frac{3}{8} = 3\frac{9}{24} = 2\frac{24}{24} + \frac{9}{24} = 2\frac{33}{24}$ $-1\frac{7}{12} = 1\frac{14}{24} \qquad\qquad = 1\frac{14}{24}$ $\qquad\qquad\qquad\qquad\qquad\qquad 1\frac{19}{24}$

LESSON 37

Objective

- To review addition and subtraction with fractions, whole numbers, and mixed numbers.

Review

1. Give Lesson 37 Quiz (Factoring Numbers).

2. *Expand each fraction by the number in brackets.* (Lesson 36)
   a. $\frac{5}{9}$ [4] ($\frac{20}{36}$)
   b. $\frac{6}{11}$ [5] ($\frac{30}{55}$)
   c. $\frac{2}{5}$ [8] ($\frac{16}{40}$)
   d. $\frac{7}{8}$ [6] ($\frac{42}{48}$)

3. *Reduce these fractions to lowest terms.* (Lesson 36)
   a. $\frac{14}{56}$ ($\frac{1}{4}$)
   b. $\frac{15}{25}$ ($\frac{3}{5}$)
   c. $\frac{28}{60}$ ($\frac{7}{15}$)
   d. $\frac{24}{30}$ ($\frac{4}{5}$)

4. *Find the greatest common factor of each pair.* (Lesson 35)
   a. 39, 52 (13)
   b. 42, 63 (21)

5. *Find the lowest common multiple of each pair.* (Lesson 35)
   a. 15, 24 (120)
   b. 18, 32 (288)

6. *Find the missing parts. Write any remainder as a fraction.* (Lesson 29)

Distance	Rate	Time
a. 124 mi.	___ m.p.h.	2 hr. (62)
b. ___ m	15.5 m/s	8 sec. (124)
c. 795 km	60 km/h	___ hr. ($13\frac{1}{4}$)
d. ___ km	86 km/h	$15\frac{1}{2}$ hr. (1,333)

7. *Find these equivalents.* (Lesson 21)
   a. 21 dm = ___ km (0.0021)
   b. 4.5 hm = ___ mm (450,000)
   c. 65 ft. = ___ m (19.5)
   d. 215 m = ___ ft. (705.2)

8. *Add mentally.* (Lesson 5)
   a. 43 + 64 (107)
   b. 974 + 54 (1,028)
   c. 865 + 468 (1,333)
   d. 3,750 + 675 (4,425)

Introduction

How much money would you have altogether if you had 5 American dollars and 5 British pounds? These numbers cannot be added together because the money they represent does not have the same value. To add these amounts, both kinds of money must be expressed either as dollars or as pounds.

Adding and subtracting unlike fractions is like adding dollars and pounds. Fractions must first be changed to like fractions before they can be added.

Teaching Guide

1. **To add fractions or mixed numbers, follow these steps.**

    (1) Change any unlike fractions to like fractions. Use the lowest common multiple of the denominators as the common denominator of the addends.

    (2) Add the numerators. Use the denominator of the addends as the denominator of the sum.

    (3) Add any whole numbers in the addends.

    (4) Change the answer to simplest form, with no improper fraction and with any proper fraction in lowest terms.

    a. $\begin{array}{r} \frac{7}{8} = \frac{7}{8} \\ + \frac{3}{4} = \frac{6}{8} \\ \hline \frac{13}{8} = 1\frac{5}{8} \end{array}$

    b. $\begin{array}{r} 2\frac{4}{5} = 2\frac{16}{20} \\ + 1\frac{3}{4} = 1\frac{15}{20} \\ \hline 3\frac{31}{20} = 4\frac{11}{20} \end{array}$

    c. $\begin{array}{r} 3\frac{7}{12} = 3\frac{7}{12} \\ + 4\frac{2}{3} = 4\frac{8}{12} \\ \hline 7\frac{15}{12} = 8\frac{3}{12} = 8\frac{1}{4} \end{array}$

2. **To subtract fractions or mixed numbers, follow these steps.**

    (1) Change any unlike fractions to like fractions.

    (2) If necessary, borrow 1 from the whole number. Change the 1 to an improper fraction and add it to the fraction in the minuend.

    (3) Subtract the numerators. Use the denominator of the minuend and subtrahend as the denominator for the difference.

    (4) Subtract any whole numbers.

Solutions for Part D

21. $4 + 3\frac{3}{4} + 2\frac{1}{2} + 2 + 1\frac{1}{2}$

22. $\frac{1}{2} + \frac{1}{3} + \frac{1}{4}$

Lesson 37

CLASS PRACTICE

Add these fractions. Express the answers in simplest form.

a. $\frac{1}{2}$
    $+\frac{1}{6}$
    $\overline{\frac{2}{3}}$

b. $\frac{4}{7}$
    $+\frac{3}{4}$
    $\overline{1\frac{9}{28}}$

c. $4\frac{3}{5}$
    $+6\frac{1}{2}$
    $\overline{11\frac{1}{10}}$

d. $6\frac{5}{9}$
    $+8\frac{5}{6}$
    $\overline{15\frac{7}{18}}$

Subtract these fractions. Express the answers in simplest form.

e. $\frac{3}{4}$
    $-\frac{3}{7}$
    $\overline{\frac{9}{28}}$

f. $\frac{7}{11}$
    $-\frac{1}{2}$
    $\overline{\frac{3}{22}}$

g. 8
    $-2\frac{1}{4}$
    $\overline{5\frac{3}{4}}$

h. $3\frac{1}{8}$
    $-1\frac{3}{4}$
    $\overline{1\frac{3}{8}}$

WRITTEN EXERCISES

A. Give the lowest common multiple that would serve as a common denominator for each pair of fractions.

1. $\frac{1}{4}, \frac{3}{8}$ 8
2. $\frac{1}{6}, \frac{5}{8}$ 24
3. $\frac{2}{3}, \frac{3}{5}$ 15
4. $\frac{1}{9}, \frac{1}{12}$ 36

B. Add these fractions. Express the answers in simplest form.

5. $\frac{5}{6}$
    $+\frac{1}{12}$
    $\overline{\frac{11}{12}}$

6. $\frac{3}{5}$
    $+\frac{1}{3}$
    $\overline{\frac{14}{15}}$

7. $\frac{5}{9}$
    $+\frac{5}{6}$
    $\overline{1\frac{7}{18}}$

8. $\frac{9}{10}$
    $+\frac{11}{15}$
    $\overline{1\frac{19}{30}}$

9. $\frac{11}{16}$
    $+\frac{11}{12}$
    $\overline{1\frac{29}{48}}$

10. $1\frac{3}{4}$
    $+2\frac{5}{6}$
    $\overline{4\frac{7}{12}}$

11. $6\frac{5}{7}$
    $+1\frac{1}{4}$
    $\overline{7\frac{27}{28}}$

12. $3\frac{5}{9}$
    $+2\frac{3}{5}$
    $\overline{6\frac{7}{45}}$

C. Subtract these fractions. Express the answers in simplest form.

13. $\frac{3}{4}$
    $-\frac{5}{8}$
    $\overline{\frac{1}{8}}$

14. $\frac{5}{6}$
    $-\frac{1}{4}$
    $\overline{\frac{7}{12}}$

15. $\frac{7}{8}$
    $-\frac{5}{6}$
    $\overline{\frac{1}{24}}$

16. $\frac{9}{14}$
    $-\frac{1}{2}$
    $\overline{\frac{1}{7}}$

17. $\frac{7}{9}$
    $-\frac{3}{5}$
    $\overline{\frac{8}{45}}$

18. $3\frac{3}{4}$
    $-2\frac{5}{6}$
    $\overline{\frac{11}{12}}$

19. 6
    $-1\frac{1}{4}$
    $\overline{4\frac{3}{4}}$

20. 4
    $-2\frac{7}{8}$
    $\overline{1\frac{1}{8}}$

D. Solve these reading problems.

21. The women of the Brighton congregation made five dresses for a family in Mexico. They used 4 yards of material for the mother, and $3\frac{3}{4}$, $2\frac{1}{2}$, 2, and $1\frac{1}{2}$ yards for the girls. How much material was needed for the five dresses? $13\frac{3}{4}$ yards

22. The Old Testament Law required that a drink offering be included with a burnt sacrifice. A drink offering of $\frac{1}{2}$ hin was to be offered with a bullock, $\frac{1}{3}$ hin with a ram, and $\frac{1}{4}$ hin with a lamb. How many hins would be used for these three sacrifices? $1\frac{1}{12}$ hins

126 Chapter 3 Factoring and Fractions

23. Sister Phebe found a partly used bag of fertilizer with "$34\frac{1}{2}$ lb." marked on it. She applied the fertilizer as follows: $5\frac{1}{4}$ pounds to the cabbage patch, $10\frac{1}{2}$ pounds to the sweet corn, $2\frac{1}{2}$ pounds to the cucumbers, and $\frac{3}{4}$ pound to the lettuce bed. How much fertilizer was left? $15\frac{1}{2}$ pounds

24. When the Sommers family had guests in their home for three days, Mother estimated that she would use 15 pounds of sausage. She served $3\frac{1}{4}$ pounds on the first day, $4\frac{1}{4}$ pounds on the second day, and $2\frac{3}{4}$ pounds on the third day. How much less than her estimate did Mother use? $4\frac{3}{4}$ pounds

25. One snowy evening, it took the Ramers 3 hours to travel 77 miles home from a closing service at Bible school. What was their average rate of speed? (Express any remainder as a fraction.) $25\frac{2}{3}$ miles per hour

26. On the way to Bible school before the snow began, the Ramers had traveled a slightly different route in $1\frac{1}{2}$ hours at an average rate of 50 miles per hour. What was the total distance they traveled on the way to Bible school? 75 miles

REVIEW EXERCISES

E. Expand each fraction by multiplying both terms by the number in parentheses. *(Lesson 36)*

27. $\frac{2}{9}$ (5) $\frac{10}{45}$ 28. $\frac{4}{5}$ (7) $\frac{28}{35}$

F. Reduce these fractions to lowest terms. *(Lesson 36)*

29. $\frac{10}{12}$ $\frac{5}{6}$ 30. $\frac{25}{60}$ $\frac{5}{12}$

G. Find the greatest common factor of each pair. *(Lesson 35)*

31. 45, 60   15   32. 18, 54   18

H. Find the lowest common multiple of each pair. *(Lesson 35)*

33. 12, 18   36   34. 63, 36   252

I. Find the missing parts. Write any remainder as a fraction. *(Lesson 29)*

Distance	Rate	Time
35. 123 m	6 m/s	$20\frac{1}{2}$ sec.
36. 1,800 km	75 km/h	24 hr.

J. Find these equivalents. *(Lesson 21)*

37. 12 km = ____ cm   1,200,000   38. 2,340 mm = 2.34 m

K. Do these additions mentally. *(Lesson 5)*

39. 23 + 65   88   40. 657 + 846   1,503

(5) Change the answer to simplest form.

a. $\begin{aligned} \frac{5}{8} &= \frac{10}{16} \\ -\frac{3}{16} &= \frac{3}{16} \\ \hline &\frac{7}{16} \end{aligned}$

b. $\begin{aligned} 9\frac{1}{5} &= 9\frac{6}{30} = 8\frac{36}{30} \\ -4\frac{5}{6} &= 4\frac{25}{30} = 4\frac{25}{30} \\ \hline &\phantom{9\frac{6}{30} =\,} 4\frac{11}{30} \end{aligned}$

c. $\begin{aligned} 7\frac{7}{9} &= 7\frac{28}{36} = 6\frac{64}{36} \\ -3\frac{11}{12} &= 3\frac{33}{36} = 3\frac{33}{36} \\ \hline &\phantom{7\frac{28}{36} =\,} 3\frac{31}{36} \end{aligned}$

---

23. $34\frac{1}{2} - (5\frac{1}{4} + 10\frac{1}{2} + 2\frac{1}{2} + \frac{3}{4})$

24. $15 - (3\frac{1}{4} + 4\frac{1}{4} + 2\frac{3}{4})$

25. $77 \div 3$

26. $1\frac{1}{2} \times 50$

T-127   Chapter 3   Factoring and Fractions

LESSON 38

Objectives

- To review mental multiplication of a whole number by a fraction when the denominator of the fraction is a factor of the whole number.
- To review multiplication problems containing fractions. (This lesson covers all the different kinds of multiplication with fractions.)
- To teach *multiplying fractions when there are more than two factors.
- To review cancellation in multiplying fractions.

Review

1. *Solve these addition and subtraction problems.* (Lesson 37)

   a. $9\frac{2}{5}$
      $+ 3\frac{7}{8}$
      $(13\frac{11}{40})$

   b. $21\frac{3}{4}$
      $+ 4\frac{5}{7}$
      $(26\frac{13}{28})$

   c. $10\frac{1}{5}$
      $- 5\frac{7}{9}$
      $(4\frac{19}{45})$

2. *Reduce these fractions to lowest terms.* (Lesson 36)

   a. $\frac{16}{32}$ $(\frac{1}{2})$
   b. $\frac{64}{96}$ $(\frac{2}{3})$
   c. $\frac{19}{76}$ $(\frac{1}{4})$
   d. $\frac{49}{63}$ $(\frac{7}{9})$

3. *Find the prime factors of each number. Use exponents to show repeating factors.* (Lesson 34)

   a. 28 $(28 = 2^2 \times 7)$
   b. 45 $(45 = 3^2 \times 5)$

4. *Find these equivalents.* (Lessons 27, 30)

   a. 300 lb. = ___ kg   (135)
   b. 145 kg = ___ lb.   (319)
   c. 21 in. = ___ cm   (53.34)
   d. 43 cm = ___ in.   (16.77)
   e. 4 hins = ___ cups $(85\frac{1}{3})$
   f. 25 Roman pounds = ___ g   (8,500)

38. Multiplying Fractions

Mentally Multiplying Whole Numbers by Fractions

To multiply a whole number by a unit fraction (a fraction with the numerator 1), divide the whole number by the denominator of the unit fraction. You can often do this mentally (Example A). Note that "$\frac{1}{4} \times 20$" means the same as "$\frac{1}{4}$ of 20."

Example A	**Example B**
$\frac{1}{4}$ of $20 = \frac{1}{4} \times 20 = 20 \div 4 = 5$	$\frac{3}{4} \times 20 = 20 \div 4 \times 3 = 15$

To multiply a whole number by a fraction other than a unit fraction, first divide the whole number by the denominator of the fraction. Then multiply that answer by the numerator of the fraction (Example B). Again, you may be able to do this mentally.

Multiplying Fractions and Mixed Numbers

Multiplication of fractions often requires written calculation. To multiply fractions, follow the steps below. Use the same steps for problems that have more than two factors.

1. Express all whole numbers and mixed numbers as improper fractions.
2. If possible, use cancellation to simplify the problem.
3. Multiply numerators by numerators and denominators by denominators.
4. Write the answer in simplest form.

Cancellation is a way to reduce fractions before multiplying. To cancel, divide any one numerator and one denominator by a common factor. **Always cancel only one numerator–denominator pair at a time.** If all possible cancellations are done, the answer will be in lowest terms.

Example C	**Example D**	**Example E**
$\frac{3}{5} \times \frac{\overset{1}{\cancel{5}}}{8} = \frac{3}{8}$	$17 \times \frac{5}{7} = \frac{17}{1} \times \frac{5}{7} = \frac{85}{7} = 12\frac{1}{7}$	$\frac{1}{2} \times 3\frac{3}{4} \times 2\frac{2}{3} = \frac{1}{2} \times \frac{\overset{5}{\cancel{15}}}{\underset{1}{\cancel{4}}} \times \frac{\overset{\overset{4}{\cancel{8}}}{\cancel{8}}}{\underset{1}{\cancel{3}}} = \frac{5}{1} = 5$

Example E illustrates multiplication of more than two fractions. In that problem, the same numerator is canceled twice. First the 2 of $\frac{1}{2}$ and the 8 of $\frac{8}{3}$ are canceled, leaving 1 and 4. This 4 is then canceled with the 4 of $\frac{15}{4}$, leaving 1 and 1. But note: **only one numerator–denominator pair at a time** is canceled.

128 Chapter 3 *Factoring and Fractions*

Vertical multiplication of mixed numbers and whole numbers is most efficient when you can mentally multiply the fraction by the whole number. See the steps below and the examples that follow.

1. Copy the problem, using the mixed number as the multiplier.
2. Multiply the multiplicand by the fraction in the multiplier.
3. Multiply the multiplicand by the ones' digit in the multiplier. Write the ones' digit of this partial product directly below the ones' digit of the first partial product.
4. If the multiplier has two digits, multiply the multiplicand by the tens' digit. Begin writing this partial product in the tens' place.
5. Add the partial products to obtain the final product.

Example F	**Example G**
25	32
$\times 18\frac{2}{3}$ ($\frac{2}{3} \times \frac{25}{1} = \frac{50}{3} = 16\frac{2}{3}$)	$\times 12\frac{3}{4}$ ($\frac{3}{4} \times 32 = 24$)
$16\frac{2}{3}$	24
200	64
25	32
$466\frac{2}{3}$	408

CLASS PRACTICE

Mentally do these multiplications containing unit fractions.

a. $\frac{1}{2}$ of 28   14   b. $\frac{1}{8}$ of 48   6   c. $\frac{1}{6} \times 30$ 5   d. $\frac{1}{5} \times 55$ 11

Find these answers mentally.

e. $\frac{5}{8}$ of 64   40   f. $\frac{3}{4}$ of 36   27   g. $\frac{7}{10} \times 70$ 49   h. $\frac{5}{9} \times 63$ 35

Copy and solve these problems.

i. $14 \times \frac{5}{6}$ $11\frac{2}{3}$ j. $\frac{7}{8}$ of 21 $18\frac{3}{8}$ k. $12 \times 1\frac{1}{4}$ 15   l. $\frac{1}{2} \times \frac{2}{5} \times \frac{3}{4}$ $\frac{3}{20}$

Copy and solve by vertical multiplication.

m. 25 $\times 1\frac{2}{5}$ 35 n. 18 $\times 3\frac{1}{9}$ 56 o. 33 $\times 3\frac{1}{3}$ 110 p. 42 $\times 5\frac{3}{7}$ 228

q. 15 $\times 12\frac{4}{5}$ 192 r. 38 $\times 16\frac{1}{3}$ $620\frac{2}{3}$ s. 51 $\times 22\frac{1}{2}$ $1{,}147\frac{1}{2}$ t. 40 $\times 13\frac{3}{4}$ 550

Lesson 38 T–128

Introduction

If a property has an area of 9 acres, and $\frac{1}{3}$ of the area is pasture, how much land is used for pasture? ($\frac{1}{3} \times 9 = 3$ acres. This is the same as dividing 9 by 3.)

If a property has an area of $\frac{3}{4}$ acre and $\frac{1}{3}$ of the area is used as a garden, what is the area of the garden? (Again, multiply $\frac{1}{3}$ times the number of acres: $\frac{1}{3} \times \frac{3}{4} = \frac{1}{4}$ acre.)

In this lesson we will multiply fractions, whole numbers, and mixed numbers.

Teaching Guide

1. **To mentally multiply a whole number by a unit fraction, divide the whole number by the denominator of the fraction.**
   a. $\frac{1}{2}$ of $16 = 16 \div 2 = 8$
   b. $\frac{1}{3}$ of $12 = 12 \div 3 = 4$
   c. $\frac{1}{8} \times 32 = 32 \div 8 = 4$
   d. $\frac{1}{5} \times 60 = 60 \div 5 = 12$

2. **To multiply a whole number by a fraction other than a unit fraction, first divide the whole number by the denominator of the fraction. Then multiply that answer by the numerator of the fraction.**
   a. $\frac{2}{3}$ of $12 = 12 \div 3 \times 2 = 8$
   b. $\frac{7}{10}$ of $60 = 60 \div 10 \times 7 = 42$
   c. $\frac{5}{6} \times 30 = 30 \div 6 \times 5 = 25$
   d. $\frac{3}{4} \times 32 = 32 \div 4 \times 3 = 24$

3. **To solve any multiplication problem involving fractions, follow the steps below.**
   (1) Express all whole numbers and mixed numbers as improper fractions.
   (2) If possible, use cancellation to simplify the problem.
   (3) Multiply numerators by numerators and denominators by denominators.
   (4) Write the answer in simplest form.

4. **Cancellation is a way to reduce fractions before multiplying.** To cancel, divide any one numerator and one denominator by a common factor. Always cancel only one numerator–denominator pair at a time. If all possible cancellations are done, the answer will be in lowest terms.

   Use the following problems to give practice with cancellation.

T-129   Chapter 3   *Factoring and Fractions*

a. $\frac{\cancel{3}^1}{4} \times \frac{5}{\cancel{6}_2}$ $(\frac{5}{8})$

b. $\frac{\cancel{6}^2}{7} \times \frac{\cancel{7}^1}{\cancel{9}_3}$ $(\frac{2}{3})$

c. $\frac{\cancel{5}^1}{\cancel{8}_2} \times \frac{\cancel{4}^1}{\cancel{5}_1}$ $(\frac{1}{2})$

d. $1\frac{7}{8} \times 3\frac{3}{5} = \frac{\cancel{15}^3}{\cancel{8}_4} \times \frac{\cancel{18}^9}{\cancel{5}_1} = 6\frac{3}{4}$

e. $4\frac{1}{2} \times 3\frac{3}{4} = \frac{9}{2} \times \frac{15}{4} = 16\frac{7}{8}$

5. **Multiplication of fractions may be done with more than two factors.** In such a problem, the same numerator or denominator may be canceled more than once. But only one numerator–denominator pair at a time is canceled.

   a. $\frac{1}{2} \times \frac{3}{4} \times \frac{3}{5}$ $(\frac{9}{40})$

   b. $\frac{\cancel{2}^1}{3} \times \frac{1}{\cancel{2}_1} \times \frac{\cancel{3}^1}{4}$ $(\frac{1}{4})$

   c. $\frac{1}{\cancel{3}_1} \times \frac{\cancel{3}^1}{\cancel{4}_1} \times \frac{1}{\cancel{2}_1} \times \frac{\cancel{8}^{2^1}}{9}$ $(\frac{1}{9})$

6. **Vertical multiplication of mixed numbers and whole numbers is most efficient when you can mentally multiply the whole number by the fraction.** Follow the steps below.

   (1) Copy the problem, using the mixed number as the multiplier.

   (2) Multiply the multiplicand by the fraction in the multiplier.

   (3) Multiply the multiplicand by the ones' digit in the multiplier. Write the ones' digit of this partial product directly below that of the first partial product.

   (4) If the multiplier has two digits, multiply the multiplicand by the tens' digit. Begin writing this partial product in the tens' place.

   (5) Add the partial products to obtain the final product.

   Be sure to emphasize that two partial products belong in the ones' place: the product of the fraction times the multiplicand and the product of the ones' digit in the multiplier times the multiplicand.

   a.
   $$\begin{array}{r} 14 \\ \times\, 4\frac{1}{2} \\ \hline 7 \\ 56 \\ \hline 63 \end{array}$$

   b.
   $$\begin{array}{r} 16 \\ \times\, 23\frac{1}{3} \\ \hline 5\frac{1}{3} \\ 48 \\ 32 \\ \hline 373\frac{1}{3} \end{array}$$

   c.
   $$\begin{array}{r} 16 \\ \times\, 12\frac{1}{4} \\ \hline 4 \\ 32 \\ 16 \\ \hline 196 \end{array}$$

Solutions for Part E

33. $\frac{1}{2} \times \frac{1}{10}$

34. $12 \times \frac{2}{10}$

35. $18 \times 1\frac{1}{2}$

WRITTEN EXERCISES

A. Mentally do these multiplications containing unit fractions.

1. $\frac{1}{4}$ of 36   9
2. $\frac{1}{3}$ of 39   13
3. $\frac{1}{8} \times 64$ 8
4. $\frac{1}{7} \times 84$ 12
5. $\frac{1}{20} \times 180$ 9
6. $\frac{1}{50} \times 800$ 16

B. Find these answers mentally.

7. $\frac{3}{4}$ of 24   18
8. $\frac{5}{6}$ of 30   25
9. $\frac{4}{9}$ of 45   20
10. $\frac{4}{7} \times 63$ 36
11. $\frac{3}{8} \times 56$ 21
12. $\frac{11}{12} \times 72$ 66

C. Copy and solve these problems.

13. $25 \times \frac{1}{4}$ $6\frac{1}{4}$
14. $27 \times \frac{3}{5}$ $16\frac{1}{5}$
15. $44 \times \frac{5}{9}$ $24\frac{4}{9}$
16. $\frac{7}{12}$ of 38 $22\frac{1}{6}$
17. $\frac{5}{7}$ of 55 $39\frac{2}{7}$
18. $\frac{1}{6}$ of 95 $15\frac{5}{6}$
19. $35 \times 1\frac{1}{4}$ $43\frac{3}{4}$
20. $32 \times 1\frac{3}{5}$ $51\frac{1}{5}$
21. $22 \times 1\frac{7}{8}$ $41\frac{1}{4}$
22. $3\frac{1}{2} \times 2\frac{3}{4}$ $9\frac{5}{8}$
23. $\frac{1}{5} \times \frac{2}{3} \times \frac{3}{4}$ $\frac{1}{10}$
24. $\frac{4}{9} \times \frac{1}{2} \times \frac{3}{5}$ $\frac{2}{15}$

D. Copy and solve by vertical multiplication.

25. $24 \times 2\frac{1}{2}$ = 60
26. $36 \times 6\frac{5}{6}$ = 246
27. $32 \times 9\frac{3}{4}$ = 312
28. $45 \times 2\frac{3}{5}$ = 117
29. $46 \times 14\frac{7}{8}$ = $684\frac{1}{4}$
30. $20 \times 11\frac{1}{6}$ = $223\frac{1}{3}$
31. $34 \times 27\frac{1}{16}$ = $920\frac{1}{8}$
32. $50 \times 31\frac{1}{12}$ = $1{,}554\frac{1}{6}$

E. Solve these reading problems.

33. In Leviticus 6:20, Aaron was instructed to offer $\frac{1}{10}$ ephah of fine flour as a meat offering every day. He was to offer $\frac{1}{2}$ of it in the morning and $\frac{1}{2}$ of it in the evening. What fractional part of an ephah of fine flour was Aaron to offer in the morning? $\frac{1}{20}$ ephah

34. The table of showbread in the tabernacle held 12 loaves of bread, each made with $\frac{2}{10}$ deal of fine flour (Leviticus 24:5, 6). How many deals in all were used for the 12 loaves of showbread? $2\frac{2}{5}$ deals

35. A veterinarian treated a sick horse with $1\frac{1}{2}$ milliliters of antibiotic per 100 pounds of body weight. Calculate the amount of antibiotic he used to treat this 1,800-pound draft horse.

130 Chapter 3 *Factoring and Fractions*

36. The Bosticks had $6\frac{1}{4}$ acres of cherry trees with 136 trees per acre. They applied 2 pounds of fertilizer per tree. How much fertilizer did they use?   1,700 pounds

37. The Balmers picked $5\frac{1}{2}$ bushels of apples one week. After using $3\frac{3}{4}$ bushels to make applesauce, how many bushels did they have left? $1\frac{3}{4}$ bushels

38. One afternoon Marlin planted corn for $2\frac{1}{2}$ hours and weeded the garden for $1\frac{3}{4}$ hours. How many hours did he work that afternoon? $4\frac{1}{4}$ hours

REVIEW EXERCISES

F. Solve these addition and subtraction problems. *(Lesson 37)*

39. $4\frac{5}{8}$
      $+3\frac{5}{6}$
      $8\frac{11}{24}$

40. $12\frac{1}{8}$
      $-9\frac{3}{4}$
      $2\frac{3}{8}$

G. Reduce these fractions to lowest terms. *(Lessons 36)*

41. $\frac{16}{36}$ $\frac{4}{9}$ **42.** $\frac{33}{44}$ $\frac{3}{4}$

H. Find the prime factors of each number. Use exponents to show repeating factors. *(Lesson 34)*

43. 48 $48 = 2^4 \times 3$ **44.** 57 $57 = 3 \times 19$

I. Find these equivalents. *(Lessons 27, 30)*

45. 218 lb. = 98.1 kg
46. 100 kg = 220 lb.
47. 35 cm = 13.65 in.
48. 19 in. = 48.26 cm
49. 3 handbreadths = 9 in.
50. 15 reeds = 48.75 m

Lesson 38 T–130

An Ounce of Prevention

Be sure to solve several problems by the vertical method. Placement of the partial products requires emphasis because it is different from the normal procedure.

Further Study

The mental math that is taught in this lesson, which students often know through practical experience, is actually another proof that multiplication and division are inverse operations. The next two lessons teach that division problems can be solved by **multiplying the dividend by the reciprocal of the divisor.** In the same way, multiplication problems can be solved by **dividing one factor by the reciprocal of the other.**

a. $7 \times 6 = 7 \div \frac{1}{6} = 42$
b. $8 \times \frac{1}{2} = 8 \div 2 = 4$
c. $4 \times 1\frac{1}{2} = 4 \div \frac{2}{3} = 2\frac{2}{3}$

Multiplying by unit fractions is one of the few cases in which it is more efficient to divide by the reciprocal. Another example is certain computations with calculators. It is more accurate, for instance, to divide by $\frac{3}{4}$ (0.75) than to multiply by $\frac{4}{3}$ (1.3333 . . .).

---

36. $2 \times 136 \times 6\frac{1}{4}$
37. $5\frac{1}{2} - 3\frac{3}{4}$
38. $2\frac{1}{2} + 1\frac{3}{4}$

T–131   Chapter 3   *Factoring and Fractions*

LESSON 39

Objectives

- To review finding reciprocals of whole numbers, fractions, and mixed numbers.
- To review dividing whole numbers by fractions by multiplying the whole number by the reciprocal of the fraction.

Review

1. Give Lesson 39 Quiz (Adding and Subtracting Fractions).

2. *Solve these problems containing fractions.* (Lessons 37, 38)

   a. $5\frac{1}{7}$
       $+ 8\frac{5}{14}$
       $(13\frac{1}{2})$

   b. $11\frac{1}{2}$
       $- 6\frac{7}{10}$
       $(4\frac{4}{5})$

   c. $1\frac{3}{5} \times 5\frac{1}{6}$ $(8\frac{4}{15})$

   d. $4\frac{2}{7} \times 4\frac{2}{3}$ (20)

3. *Find the greatest common factor of each pair.* (Lesson 35)

   a. 18, 45   (9)

   b. 32, 48   (16)

4. *Find the lowest common multiple of each pair.* (Lesson 35)

   a. 14, 21   (42)

   b. 18, 27   (54)

5. *Find these equivalents.* (Lesson 23)

   a. 24 ml = ___ l (0.024)

   b. 45 l = ___ qt.   (47.7)

6. *Solve by mental subtraction.* (Lesson 7)

   a. 352 − 164   (188)

   b. 730 − 492   (238)

39. Dividing Fractions, Part 1

Every number except 0 has a **reciprocal**. The reciprocal of a number is the factor that produces 1 when multiplied by the number. Since $\frac{3}{5} \times \frac{5}{3} = 1$, $\frac{3}{5}$ is the reciprocal of $\frac{5}{3}$, and $\frac{5}{3}$ is the reciprocal of $\frac{3}{5}$.

To find the reciprocal of any number, express it as a fraction and invert the fraction.

Numbers:	6	$\frac{5}{6}$	$2\frac{3}{4}$
Reciprocals:	$\frac{1}{6}$	$\frac{6}{5}$	$\frac{4}{11}$

Any division problem can be solved by changing the divisor to its reciprocal and multiplying. For example, $6 \div 3$ gives the same result as $6 \times \frac{1}{3}$. The answer to both problems is 2.

To divide with fractions and whole numbers, use the following steps.

1. Change any whole number to an improper fraction by writing it as a numerator over the denominator 1.

2. Change the **divisor** to its reciprocal by inverting it, and change the division sign to a multiplication sign.

3. Follow the normal rules for multiplying fractions. Use cancellation when possible.

4. Write the answer in simplest form.

Example A	Example B
$\frac{3}{4} \div \frac{3}{8} = \frac{\overset{1}{\cancel{3}}}{\underset{1}{\cancel{4}}} \times \frac{\overset{2}{\cancel{8}}}{\underset{1}{\cancel{3}}} = \frac{2}{1} = 2$	$9 \div \frac{6}{7} = \frac{9}{1} \div \frac{6}{7} = \frac{\overset{3}{\cancel{9}}}{1} \times \frac{7}{\underset{2}{\cancel{6}}} = \frac{21}{2} = 10\frac{1}{2}$

Remember that division tells how many times the divisor is contained in the dividend. Consider the problem $15 \div 3 = 5$. This means that the number of 3s in 15 is 5. Likewise, $4 \div \frac{1}{2} = 8$ means that the number of $\frac{1}{2}$s in 4 is 8.

$4 \div \frac{1}{2} = 8$
The number of $\frac{1}{2}$s in 4 is 8.

In reading problems, division is often indicated by phrases such as *how many times as much* or *how many times farther*. Be alert for phrases like this in part D.

132 Chapter 3 *Factoring and Fractions*

CLASS PRACTICE

Find the reciprocal of each number.

a. 11 $\frac{1}{11}$ b. $\frac{4}{5}$ $\frac{5}{4}$ c. $3\frac{1}{2}$ $\frac{2}{7}$ d. $6\frac{5}{6}$ $\frac{6}{41}$

Solve these division problems.

e. $15 \div \frac{5}{6}$ 18   f. $22 \div \frac{4}{7}$ $38\frac{1}{2}$ g. $34 \div \frac{3}{5}$ $56\frac{2}{3}$ h. $42 \div \frac{5}{9}$ $75\frac{3}{5}$

i. $\frac{5}{8} \div \frac{11}{16}$ $\frac{10}{11}$ j. $\frac{2}{3} \div \frac{1}{4}$ $2\frac{2}{3}$ k. $\frac{7}{12} \div \frac{2}{5}$ $1\frac{11}{24}$ l. $\frac{9}{10} \div \frac{3}{8}$ $2\frac{2}{5}$

WRITTEN EXERCISES

A. *Find the reciprocal of each number.*

1. 7 $\frac{1}{7}$ 2. $\frac{3}{4}$ $\frac{4}{3}$ 3. $2\frac{2}{3}$ $\frac{3}{8}$ 4. $5\frac{5}{8}$ $\frac{8}{45}$

B. *For each question, write a division problem and solve it.*

5. How many $\frac{1}{2}$s are in 4? $4 \div \frac{1}{2} = 8$

6. How many $\frac{1}{8}$s are in 3? $3 \div \frac{1}{8} = 24$

7. How many $\frac{3}{5}$s are in 6? $6 \div \frac{3}{5} = 10$

8. How many $\frac{3}{4}$s are in 9? $9 \div \frac{3}{4} = 12$

C. *Solve these division problems.*

9. $\frac{3}{4} \div \frac{1}{2}$ $1\frac{1}{2}$ 10. $\frac{4}{5} \div \frac{5}{6}$ $\frac{24}{25}$ 11. $\frac{5}{6} \div \frac{1}{3}$ $2\frac{1}{2}$ 12. $\frac{2}{3} \div \frac{7}{8}$ $\frac{16}{21}$

13. $15 \div \frac{3}{5}$ 25   14. $23 \div \frac{5}{8}$ $36\frac{4}{5}$ 15. $17 \div \frac{3}{11}$ $62\frac{1}{3}$ 16. $25 \div \frac{5}{7}$ 35

17. $26 \div \frac{5}{9}$ $46\frac{4}{5}$ 18. $18 \div \frac{3}{4}$ 24   19. $20 \div \frac{5}{7}$ 28   20. $35 \div \frac{6}{7}$ $40\frac{5}{6}$

D. *Solve these reading problems.*

21. Mother uses 4 yards of material to make a dress for herself and $\frac{4}{5}$ yard to make a dress for little Rachel. How many times as much material does it take for Mother's dress than for Rachel's?   5 times

22. The Seibel family lives $\frac{7}{10}$ mile from their school and 7 miles from their church. How many times greater is the distance to church than to school?   10 times

23. One morning David picked 2 bushels of peas, and his younger sister Ruth picked $\frac{7}{8}$ bushel. How many times more did David pick than his younger sister did? $2\frac{2}{7}$ times

24. An old mill has a water wheel with a diameter of 12 feet. The sprocket on the axle of the wheel has a diameter of $\frac{3}{4}$ foot. How many times larger is the water wheel than the sprocket?   16 times

25. Four boxes of cereal weigh $1\frac{1}{8}$ pounds, $1\frac{1}{4}$ pounds, $\frac{5}{8}$ pound, and $1\frac{1}{2}$ pounds. What is their combined weight? $4\frac{1}{2}$ pounds

26. From a 6-foot board, Mervin cut 3 pieces that were each $1\frac{5}{8}$ feet long. How long was the remaining piece of board? (Ignore the waste from cutting.) $1\frac{1}{8}$ feet

Introduction

How many ways can the students think of to express the problem "8 divided by 4 equals 2"?

$4\overline{)8}$ $8 \div 4 = 2$ $\frac{8}{4} = 2$ $\frac{1}{4} \times 8 = 2$

What relationship is there between the divisor (4) in the first three problems and the multiplier ($\frac{1}{4}$) in the fourth problem? It is that $\frac{1}{4}$ is the reciprocal of 4.

Teaching Guide

1. **Every number except 0 has a reciprocal.** To find the reciprocal of any number, express it as a fraction and invert the fraction. Any division problem can be solved by multiplying the dividend by the reciprocal of the divisor.

   Numbers: $\frac{1}{8}$ $\frac{2}{5}$ $1\frac{3}{4}$ $3\frac{5}{6}$

   Reciprocals: $\frac{8}{1}$ $\frac{5}{2}$ $\frac{4}{7}$ $\frac{6}{23}$

2. **To divide with fractions and whole numbers, use the following steps.**

   (1) Change any whole number to an improper fraction by writing it as a numerator over the denominator 1.

   (2) Change the divisor to its reciprocal by inverting it, and change the division sign to a multiplication sign.

   (3) Follow the normal rules for multiplying fractions. Use cancellation when possible.

   (4) Write the answer in simplest form.

   As you teach, stress that the quotient in a division problem tells how many times the divisor is contained in the dividend.

   a. How many $\frac{1}{4}$s are in 4?
   $4 \div \frac{1}{4} = \frac{4}{1} \times \frac{4}{1} = 16$
   (There are sixteen $\frac{1}{4}$s in 4.)

   b. How many $\frac{1}{10}$s are in 3?
   $3 \div \frac{1}{10} = \frac{3}{1} \times \frac{10}{1} = 30$
   (There are thirty $\frac{1}{10}$s in 3.)

   c. $8 \div \frac{3}{5} = \frac{8}{1} \times \frac{5}{3} = \frac{40}{3} = 13\frac{1}{3}$

   d. $9 \div \frac{3}{4} = \frac{\overset{3}{\cancel{9}}}{1} \times \frac{4}{\underset{1}{\cancel{3}}} = 12$

3. **In reading problems, division is often indicated by phrases such as *how many times as much* or *how many times farther*.** Tell students to be alert for such phrases in part D.

Solutions for Part D

21. $4 \div \frac{4}{5}$
22. $7 \div \frac{7}{10}$
23. $2 \div \frac{7}{8}$
24. $12 \div \frac{3}{4}$
25. $1\frac{1}{8} + 1\frac{1}{4} + \frac{5}{8} + 1\frac{1}{2}$
26. $6 - (3 \times 1\frac{5}{8})$

An Ounce of Prevention

1. Be sure the students understand that they must invert the divisor and not the dividend. If a student's answer is the reciprocal of the correct answer, he is probably making this mistake.

2. Reading problems involving division by proper fractions are among the more difficult ones to understand. You can help to overcome this by stressing what is being found in division by fractions. When we divide 8 by $\frac{1}{2}$, we are answering the question, "How many $\frac{1}{2}$s are in 8?"

REVIEW EXERCISES

E. Solve these problems containing fractions. *(Lessons 37, 38)*

27. $4\frac{5}{6}$
    $+9\frac{2}{9}$
    $\overline{14\frac{1}{18}}$

28. $12\frac{1}{2}$
    $-10\frac{5}{9}$
    $\overline{1\frac{17}{18}}$

29. $3\frac{2}{3} \times 2\frac{1}{5}$ $8\frac{1}{15}$

30. $\frac{1}{3} \times \frac{3}{8} \times \frac{1}{2}$ $\frac{1}{16}$

F. Find the greatest common factor of each pair. *(Lesson 35)*

31. 36, 54   18
32. 48, 72   24

G. Find the lowest common multiple of each pair. *(Lesson 35)*

33. 8, 20   40
34. 56, 64   448

H. Find these equivalents. *(Lesson 23)*

35. 186 qt. = 176.7 *l*
36. 950 *l* = 1,007 qt.

I. Solve by mental subtraction. *(Lesson 7)*

37. 7,837 − 3,587   4,250
38. $519.55 − $321.38 $198.17

134   Chapter 3   *Factoring and Fractions*

40. Dividing Fractions, Part 2

Consider this reading problem: Norman's father needs wooden blocks $\frac{1}{2}$ foot long. How many blocks can Norman cut from a board $2\frac{1}{2}$ feet long?

The question is really asking this: How many $\frac{1}{2}$s are in $2\frac{1}{2}$? As illustrated here, division gives the answer.

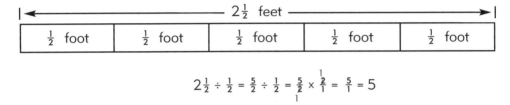

$$2\tfrac{1}{2} \div \tfrac{1}{2} = \tfrac{5}{2} \div \tfrac{1}{2} = \tfrac{5}{2} \times \tfrac{2}{1} = \tfrac{5}{1} = 5$$

Division with mixed numbers is done in the same way as division with proper fractions and whole numbers. First change all whole or mixed numbers to improper fractions. Then invert the divisor, follow the rules for multiplying, and express the answer in simplest form.

Example A

$$6\tfrac{2}{5} \div 1\tfrac{3}{5} = \tfrac{32}{5} \div \tfrac{8}{5} = \tfrac{32}{5} \times \tfrac{5}{8} = \tfrac{4}{1} = 4$$

Example B

$$1\tfrac{1}{8} \div 3\tfrac{1}{2} = \tfrac{9}{8} \div \tfrac{7}{2} = \tfrac{9}{8} \times \tfrac{2}{7} = \tfrac{9}{28}$$

Remember that a division problem answers the question, "How many times is the divisor contained in the dividend?" In Example A the question is, "How many $1\tfrac{3}{5}$s are in $6\tfrac{2}{5}$?" Notice in Example B that the divisor is larger than the dividend; therefore, the quotient is less than 1. The answer means that there is only $\tfrac{9}{28}$ of $3\tfrac{1}{2}$ in $1\tfrac{1}{8}$.

A complex fraction includes a fraction in the numerator, in the denominator, or in both terms. How can you simplify such a fraction? Remember that a fraction is one way of indicating division. That is, the complex fraction in Example C means $\tfrac{7}{8} \div \tfrac{3}{4}$. So you can simplify any complex fraction by dividing the numerator by the denominator. The answer should be expressed in the simplest form, as usual.

Example C

Simplify $\dfrac{\frac{7}{8}}{\frac{3}{4}}$.

Think: Divide $\tfrac{7}{8}$ by $\tfrac{3}{4}$.

$$\tfrac{7}{8} \div \tfrac{3}{4} = \tfrac{7}{8} \times \tfrac{4}{3} = \tfrac{7}{6} = 1\tfrac{1}{6}$$

Example D

Simplify $\dfrac{\frac{3}{5}}{6}$.

Think: Divide $\tfrac{3}{5}$ by 6.

$$\tfrac{3}{5} \div 6 = \tfrac{3}{5} \div \tfrac{6}{1} = \tfrac{3}{5} \times \tfrac{1}{6} = \tfrac{1}{10}$$

LESSON 40

Objectives

- To review division with fractions and mixed numbers.
- To teach *changing complex fractions to the simplest form.

Review

1. *Find the reciprocal of each number.* (Lesson 39)

    a. $\frac{19}{22}$ ($\frac{22}{19}$)

    b. $\frac{4}{5}$ ($\frac{5}{4}$)

    c. $4\frac{1}{2}$ ($\frac{2}{9}$)

    d. $12\frac{5}{6}$ ($\frac{6}{77}$)

2. *Solve these multiplication problems.* (Lesson 38)

    a. $1\frac{1}{6} \times 6\frac{2}{3}$ ($7\frac{7}{9}$)

    b. $2\frac{1}{4} \times 4\frac{2}{5}$ ($9\frac{9}{10}$)

3. *Solve by vertical multiplication.* (Lesson 38)

    a. $\begin{array}{r} 32 \\ \times\, 2\frac{5}{6} \\ \hline (90\frac{2}{3}) \end{array}$

    b. $\begin{array}{r} 56 \\ \times\, 4\frac{3}{8} \\ \hline (245) \end{array}$

4. *Find these equivalents.* (Lesson 24)

    a. 3,543 mm² = ___ cm² (35.43)

    b. 42 ha = ___ m² (420,000)

5. *Solve these multiplication problems, and check by casting out nines.* (Lesson 8)

    a. $\begin{array}{r} 154 \\ \times\, 857 \\ \hline (131{,}978) \end{array}$

    b. $\begin{array}{r} 826 \\ \times\, 205 \\ \hline (169{,}330) \end{array}$

T–135   *Chapter 3   Factoring and Fractions*

Introduction

Ask, "How many quarters are in $2.50?" There are 4 quarters in 1 dollar, so there are 10 quarters in $2\frac{1}{2}$ dollars ($2\frac{1}{2} \times 4 = 10$).

Now state the original question with fractions: How many $\frac{1}{4}$s are in $2\frac{1}{2}$? The problem would be written as $2\frac{1}{2} \div \frac{1}{4} = 10$. Have the students solve it by multiplying the dividend by the reciprocal of the divisor.

$$2\frac{1}{2} \div \frac{1}{4} = \frac{5}{2} \div \frac{1}{4} = \frac{5}{2} \times \frac{4}{1} = \frac{10}{1} = 10$$

Teaching Guide

1. Review the following steps for dividing fractions.

   (1) Change any whole or mixed number to an improper fraction.

   (2) Change the divisor to its reciprocal by inverting it, and change the division sign to a multiplication sign.

   (3) Follow the normal rules for multiplying fractions. Use cancellation when possible.

   (4) Write the answer in simplest form.

   Apply these principles to all the kinds of division problems containing fractions.

   a. $6 \div 1\frac{1}{2} = \frac{6}{1} \div \frac{3}{2} = \frac{\overset{2}{\cancel{6}}}{1} \times \frac{2}{\underset{1}{\cancel{3}}} = 4$

   (There are four $1\frac{1}{2}$s in 6. You may want to prove this on a number line.)

   b. $10\frac{1}{2} \div 1\frac{1}{2} = \frac{21}{2} \div \frac{3}{2} = \frac{\overset{7}{\cancel{21}}}{\underset{1}{\cancel{2}}} \times \frac{\overset{1}{\cancel{2}}}{\underset{1}{\cancel{3}}} = 7$

   (There are seven $1\frac{1}{2}$s in $10\frac{1}{2}$.)

   c. $6\frac{1}{2} \div 2\frac{3}{4} = \frac{13}{2} \div \frac{11}{4} = \frac{13}{\underset{1}{\cancel{2}}} \times \frac{\overset{2}{\cancel{4}}}{11} = 2\frac{4}{11}$

   ($2\frac{3}{4}$ is contained $2\frac{4}{11}$ times in $6\frac{1}{2}$.)

2. **A complex fraction can be simplified by dividing the numerator by the denominator.** In fact, a complex fraction is actually an unsolved division problem involving fractions.

   a. $\dfrac{\frac{3}{4}}{\frac{5}{6}} = \frac{3}{4} \div \frac{5}{6} = \frac{3}{\underset{2}{\cancel{4}}} \times \frac{\overset{3}{\cancel{6}}}{5} = \frac{9}{10}$

   b. $\dfrac{4}{\frac{5}{8}} = 4 \div \frac{5}{8} = \frac{4}{1} \times \frac{8}{5} = 6\frac{2}{5}$

   c. $\dfrac{1\frac{1}{4}}{3\frac{1}{3}} = 1\frac{1}{4} \div 3\frac{1}{3} = \frac{\overset{1}{\cancel{5}}}{4} \times \frac{3}{\underset{2}{\cancel{10}}} = \frac{3}{8}$

Solutions for Part C

21. $1\frac{4}{5} \div \frac{3}{10}$

22. $16 \div 1\frac{1}{3}$

23. $(10\frac{1}{2} - 1\frac{1}{4}) \div \frac{3}{4}$

24. $\frac{3}{4} \div \frac{3}{16}$

26. $\frac{7}{32} - \frac{1}{8}$

Lesson 40   135

CLASS PRACTICE

Solve these division problems.

a. $4 \div \frac{2}{7}$ 14
b. $6 \div 2\frac{1}{3}$ $2\frac{4}{7}$
c. $3\frac{3}{8} \div \frac{4}{5}$ $4\frac{7}{32}$
d. $2\frac{4}{9} \div 2\frac{1}{5}$ $1\frac{1}{9}$

e. $3\frac{3}{4} \div 2\frac{1}{3}$ $1\frac{17}{28}$
f. $1\frac{5}{9} \div 4\frac{1}{2}$ $\frac{28}{81}$
g. $7\frac{1}{6} \div 2\frac{3}{4}$ $2\frac{20}{33}$
h. $8\frac{2}{5} \div 3\frac{3}{4}$ $2\frac{6}{25}$

Simplify these complex fractions.

i. $\frac{5}{\frac{4}{7}}$ $8\frac{3}{4}$
j. $\frac{\frac{2}{3}}{\frac{7}{8}}$ $\frac{16}{21}$
k. $\frac{\frac{5}{8}}{3\frac{3}{4}}$ $\frac{1}{6}$
l. $\frac{\frac{6}{7}}{\frac{3}{5}}$ $1\frac{3}{7}$

WRITTEN EXERCISES

A. Solve these division problems.

1. $6 \div \frac{1}{5}$ 30
2. $9 \div \frac{1}{7}$ 63
3. $5 \div \frac{3}{4}$ $6\frac{2}{3}$
4. $7 \div \frac{3}{7}$ $16\frac{1}{3}$

5. $\frac{3}{4} \div \frac{1}{4}$ 3
6. $\frac{7}{8} \div \frac{1}{8}$ 7
7. $\frac{3}{5} \div \frac{1}{3}$ $1\frac{4}{5}$
8. $\frac{2}{3} \div \frac{1}{4}$ $2\frac{2}{3}$

9. $1\frac{3}{4} \div \frac{3}{4}$ $2\frac{1}{3}$
10. $3\frac{1}{2} \div \frac{1}{4}$ 14
11. $4\frac{1}{5} \div \frac{3}{10}$ 14
12. $3\frac{1}{6} \div \frac{5}{8}$ $5\frac{1}{15}$

13. $5\frac{1}{2} \div 2\frac{1}{4}$ $2\frac{4}{9}$
14. $3\frac{3}{5} \div 2\frac{3}{4}$ $1\frac{17}{55}$
15. $1\frac{7}{8} \div 2\frac{1}{4}$ $\frac{5}{6}$
16. $3\frac{4}{5} \div 6\frac{1}{3}$ $\frac{3}{5}$

B. Simplify these complex fractions.

17. $\frac{\frac{3}{4}}{\frac{5}{8}}$ $1\frac{1}{5}$
18. $\frac{\frac{7}{8}}{3}$ $\frac{7}{24}$
19. $\frac{6}{\frac{4}{5}}$ $7\frac{1}{2}$
20. $\frac{\frac{4}{9}}{\frac{2}{3}}$ $\frac{2}{3}$

C. Solve these reading problems.

21. Mother purchased $1\frac{4}{5}$ yards of material to make aprons for the girls. Each apron requires $\frac{3}{10}$ yard. How many aprons can she make?   6 aprons

22. Rhonda's recipe for pickle relish calls for $1\frac{1}{3}$ cups of vinegar. How many batches could she make with 1 gallon of vinegar?   12 batches

23. A builder purchased and developed a $10\frac{1}{2}$-acre field. He used $1\frac{1}{4}$ acres for a roadway and divided the rest of the land into $\frac{3}{4}$-acre lots. How many lots could he make? (Drop any fraction in your answer.)   12 lots

24. One family who purchased a $\frac{3}{4}$-acre lot used $\frac{3}{16}$ of an acre for a garden. How many times larger than the garden was the entire lot?   4 times

25. Father replaced a $2\frac{1}{2}$-inch sprocket with a $3\frac{1}{2}$-inch sprocket on a drive shaft in order to slow down the speed of the shaft. Write a complex fraction comparing the size of the smaller sprocket (the numerator) with the size of the larger sprocket (the denominator). $\frac{2\frac{1}{2}}{3\frac{1}{2}}$

26. According to the manual for a lawn mower, the length of the choke spring should be between $1\frac{1}{8}$ inches and $1\frac{7}{32}$ inches. What is the difference between these two lengths? $\frac{3}{32}$ inches

136   Chapter 3   Factoring and Fractions

REVIEW EXERCISES

D. Find the reciprocal of each number. *(Lesson 39)*

27. $\frac{5}{12}$ $\frac{12}{5}$ 28. $3\frac{4}{5}$ $\frac{5}{19}$

E. Solve these multiplication problems. *(Lesson 38)*

29. $\frac{1}{2} \times \frac{3}{5} \times \frac{4}{9}$ $\frac{2}{15}$ 30. $1\frac{7}{8} \times 9\frac{2}{3}$ $18\frac{1}{8}$

F. Solve by vertical multiplication. *(Lesson 38)*

31. $\begin{array}{r} 36 \\ \times\ 3\frac{8}{9} \\ \hline 140 \end{array}$ 32. $\begin{array}{r} 42 \\ \times\ 5\frac{5}{6} \\ \hline 245 \end{array}$

G. Label each item as *proper fraction, improper fraction, whole number, mixed number,* **or** *complex fraction.* *(Lesson 36)*

33. $7\frac{5}{9}$ mixed number   34. $\frac{25}{12}$ improper fraction

H. Find these equivalents. *(Lesson 24)*

35. 28,000 cm² = ___ ha   0.00028   36. 3.9 km² = ___ m²   3,900,000

I. Solve these multiplication problems, and check by casting out nines. *(Lesson 8)*

37. $\begin{array}{r} 407 \\ \times\ 389 \\ \hline 158{,}323 \end{array}$ 38. $\begin{array}{r} 486 \\ \times\ 901 \\ \hline 437{,}886 \end{array}$

LESSON 41

Objective

- To review finding a number when a fractional part of it is known.

Review

1. *Give the reciprocal of each number.* (Lesson 39)

   a. $\frac{3}{10}$ $\left(\frac{10}{3}\right)$

   b. $\frac{14}{33}$ $\left(\frac{33}{14}\right)$

   c. $3\frac{5}{12}$ $\left(\frac{12}{41}\right)$

2. *Solve these division problems.* (Lessons 39, 40)

   a. $14 \div 2\frac{4}{5}$ (5)

   b. $46 \div 3\frac{5}{6}$ (12)

   c. $2\frac{1}{5} \div 8\frac{1}{4}$ $\left(\frac{4}{15}\right)$

3. *Solve these addition and subtraction problems.* (Lesson 37)

   a. $12\frac{3}{4}$ b. 21
       $+ 10\frac{1}{6}$ $- 8\frac{7}{9}$
       $\left(22\frac{11}{22}\right)$ $\left(12\frac{2}{9}\right)$

4. *Add or subtract these compound measures.* (Lesson 25)

   a.   12 yd. 2 ft.
       + 7 yd. 1 ft.
        (20 yd.)

   b.   5 mi. 528 ft.
       – 1 mi. 950 ft.
        (3 mi. 4,858 ft.)

5. *Solve by mental multiplication.* (Lesson 9)

   a. 7×436 (3,052)

   b. 52×50 (2,600)

41. Finding a Number When a Fractional Part of It Is Known

After Marla hoed 90 feet of a row of beans, she had finished $\frac{3}{4}$ of the row. What was the total length of the row?

In the problem above, 90 feet is a fractional part ($\frac{3}{4}$) of the total length of the row. The numbers in this problem can be written in the expression "90 is $\frac{3}{4}$ of a certain number." That number can be found by dividing, as explained below.

> To find a number when a fractional part of it is known, divide the number by the fraction.

Example A

$$90 \text{ is } \tfrac{3}{4} \text{ of what number?}$$
Think: Divide 90 by $\frac{3}{4}$.
$$90 \div \tfrac{3}{4} = \tfrac{90}{1} \div \tfrac{3}{4} = \tfrac{\overset{30}{\cancel{90}}}{1} \times \tfrac{4}{\underset{1}{\cancel{3}}} = \tfrac{120}{1} = 120$$

Example B

12 is $\frac{4}{5}$ of ___

$$12 \div \tfrac{4}{5} = \tfrac{12}{1} \div \tfrac{4}{5} = \tfrac{\overset{3}{\cancel{12}}}{1} \times \tfrac{5}{\underset{1}{\cancel{4}}} = \tfrac{15}{1} = 15$$

Example C

$3\frac{1}{2}$ is $\frac{7}{8}$ of ___

$$3\tfrac{1}{2} \div \tfrac{7}{8} = \tfrac{7}{2} \div \tfrac{7}{8} = \tfrac{\cancel{7}}{\underset{1}{\cancel{2}}} \times \tfrac{\overset{4}{\cancel{8}}}{\underset{1}{\cancel{7}}} = \tfrac{4}{1} = 4$$

CLASS PRACTICE

Find the unknown numbers.

a. 14 is $\frac{1}{4}$ of <u>56</u>
b. 3 is $\frac{1}{8}$ of <u>24</u>
c. 5 is $\frac{5}{6}$ of <u>6</u>
d. 16 is $\frac{2}{5}$ of <u>40</u>
e. $\frac{2}{3}$ is $\frac{3}{8}$ of <u>$1\frac{7}{9}$</u>
f. $\frac{5}{9}$ is $\frac{1}{2}$ of <u>$1\frac{1}{9}$</u>
g. $2\frac{1}{5}$ is $\frac{5}{6}$ of <u>$2\frac{16}{25}$</u>
h. $4\frac{5}{7}$ is $\frac{3}{5}$ of <u>$7\frac{6}{7}$</u>
i. $1\frac{1}{2}$ is $\frac{5}{8}$ of <u>$2\frac{2}{5}$</u>

WRITTEN EXERCISES

A. *Find the unknown numbers.*

1. 12 is $\frac{1}{3}$ of <u>36</u>
2. 9 is $\frac{1}{5}$ of <u>45</u>
3. 7 is $\frac{1}{9}$ of <u>63</u>
4. 11 is $\frac{1}{7}$ of <u>77</u>
5. $\frac{3}{4}$ is $\frac{1}{4}$ of <u>3</u>
6. $\frac{5}{8}$ is $\frac{1}{8}$ of <u>5</u>
7. $\frac{5}{6}$ is $\frac{3}{4}$ of <u>$1\frac{1}{9}$</u>
8. $\frac{7}{8}$ is $\frac{5}{12}$ of <u>$2\frac{1}{10}$</u>
9. $3\frac{1}{5}$ is $\frac{4}{5}$ of <u>4</u>
10. $4\frac{1}{2}$ is $\frac{3}{8}$ of <u>12</u>
11. $6\frac{3}{4}$ is $\frac{5}{8}$ of <u>$10\frac{4}{5}$</u>
12. $5\frac{7}{12}$ is $\frac{5}{6}$ of <u>$6\frac{7}{10}$</u>

138   Chapter 3   Factoring and Fractions

B. Solve these reading problems. Set up numbers 13–16 as shown.

13. It took Marcus $2\frac{1}{2}$ hours to plow $\frac{3}{4}$ of a field. At that rate, how long should it take to plow the entire field?

    $\underline{2\frac{1}{2}}$ is $\underline{\frac{3}{4}}$ of $\underline{?}$ $3\frac{1}{3}$ hours

14. After a great victory over the Ethiopians, the people of Judah gathered at Jerusalem in the fifteenth year of King Asa's reign (2 Chronicles 15:10). They offered 700 oxen and 7,000 sheep from the spoil they had taken. If this represented $\frac{1}{10}$ of the spoil, how many sheep had they captured?

    $\underline{7{,}000}$ is $\underline{\frac{1}{10}}$ of $\underline{?}$ 70,000 sheep

15. The Garmans used 60 tons of gravel to cover $\frac{3}{4}$ of the length of their lane. At that rate, how many tons of gravel would it take to cover the entire lane?

    $\underline{60}$ is $\underline{\frac{3}{4}}$ of $\underline{?}$ 80 tons

16. The Groves have used 32 gallons of gasoline to travel $\frac{3}{8}$ of the distance to a Bible conference. At that rate, how many gallons of gasoline will they have used by the time they reach their destination?

    $\underline{32}$ is $\underline{\frac{3}{8}}$ of $\underline{?}$ $85\frac{1}{3}$ gallons

17. Mowing the orchard usually takes $2\frac{1}{2}$ hours; but when Edgar had finished $\frac{2}{3}$ of the job, he had to stop because of a storm. How long had he mowed in the orchard?

    $1\frac{2}{3}$ hours

18. Each year Pablo plants $\frac{2}{5}$ hectare of his land with yams. This year he also rented a $\frac{3}{10}$-hectare plot, where he planted more yams. How much land did Pablo plant with yams this year?

    $\frac{7}{10}$ hectare

REVIEW EXERCISES

C. Write the reciprocal of each number. *(Lesson 39)*

19. $\frac{7}{12}$ $\frac{12}{7}$ 20. $2\frac{1}{12}$ $\frac{12}{25}$

D. Solve these division problems. *(Lessons 39, 40)*

21. $24 \div 3\frac{1}{3}$ $7\frac{1}{5}$ 22. $2\frac{3}{8} \div 9\frac{1}{2}$ $\frac{1}{4}$

E. Solve these addition and subtraction problems. *(Lessons 25, 37)*

23. $4\frac{3}{5}$
    $+5\frac{5}{7}$
    $\overline{10\frac{11}{35}}$

24. 10
    $-8\frac{4}{5}$
    $\overline{1\frac{1}{5}}$

25. 3 ft.   9 in.
    +6 ft. 11 in.
    10 ft.   8 in.

26. 14 gal. 2 qt.
    − 8 gal. 3 qt.
    5 gal. 3 qt.

F. Solve by mental multiplication. *(Lesson 9)*

27. 6 × 435   2,610   28. 4 × 482   1,928   29. 40 × 34   1,360
30. 60 × 72   4,320   31. 6 × 7 × 15 630   32. 4 × 6 × 35 840

Lesson 41   T–138

Introduction

Ask the following questions.

> 8 is 2 times what number?   (4)
>
> 8 is $\frac{1}{2}$ times what number?   (16)
>
> 8 is $\frac{1}{4}$ of what number?   (32)

How did the students find the answers? They will likely tell you that they used division for the first problem and multiplication for the other two. Point out that division could have been used to solve all three problems.

> $8 \div 2 = 4$
>
> $8 \div \frac{1}{2} = \frac{8}{1} \times \frac{2}{1} = 16$
>
> $8 \div \frac{1}{4} = \frac{8}{1} \times \frac{4}{1} = 32$

Teaching Guide

To find a number when a fractional part of it is known, divide the known number by the fraction. This applies the principle that because factor times factor equals product, therefore product divided by one factor equals the other factor. In this case, the known factor is a fraction. (Note that *of* in these problems means "times.")

> a. 7 is $\frac{1}{3}$ of ___   (21)
>
> b. 19 is $\frac{1}{2}$ of ___   (38)
>
> c. 20 is $\frac{2}{3}$ of ___   (30)
>
> d. 56 is $\frac{7}{8}$ of ___   (64)

An Ounce of Prevention

The main idea of this lesson (finding a number when a fractional part of it is known) is fairly simple, yet it is one of the more difficult concepts to convey to students. Because this is the only concept presented in the lesson, you should have time to teach it well.

Solutions for Exercises 17 and 18

17. $\frac{2}{3} \times 2\frac{1}{2}$

18. $\frac{2}{5} + \frac{3}{10}$

T–139 Chapter 3 *Factoring and Fractions*

LESSON 42

Objective

- To teach *mentally dividing whole numbers by fractions by multiplying the whole number by the denominator of the fraction and dividing the result by the numerator. (Previous experience limited to unit fractions.)

Review

1. Give Lesson 42 Quiz (Calculating With Fractions).

2. *Find the answers.* (Lesson 41)
   a. $5\frac{1}{2}$ is $\frac{1}{8}$ of ___ (44)
   b. 21 is $\frac{3}{4}$ of ___ (28)
   c. $7\frac{1}{7}$ is $\frac{5}{7}$ of ___ (10)

3. *Solve these multiplication and division problems.* (Lessons 38–40)
   a. $6\frac{1}{4} \times 4\frac{3}{5}$ ($28\frac{3}{4}$)
   b. $2\frac{3}{4} \times 7\frac{1}{3}$ ($20\frac{1}{6}$)
   c. $3\frac{3}{4} \div 4\frac{1}{5}$ ($\frac{25}{28}$)
   d. $8\frac{3}{4} \div 6\frac{1}{2}$ ($1\frac{9}{26}$)

4. *Find the prime factors of these numbers.* (Lesson 34)
   a. 88 ($88 = 2^3 \times 11$)
   b. 132 ($132 = 2^2 \times 3 \times 11$)

5. *Solve these problems with compound measures.* (Lesson 26)
   a. 14 gal. 2 qt.
       × 7
       (101 gal. 2 qt.)
   b. 68 l 280 ml ÷ 6 = ___ l (11.38)

6. *Use the distributive law to find the missing numbers.* (Lesson 10)
   a. $12 \times 33 = (12 \times 30) + (\_ \times \_)$
       [12, 3]
   b. $15 \times 12 = (15 \times \_) + (15 \times \_)$
       [10, 2]

42. Mental Division by Proper Fractions

In Lesson 38 you practiced mentally multiplying whole numbers by fractions. It is also practical to mentally divide whole numbers by fractions if the numbers are simple enough. Dividing whole numbers by unit fractions is especially easy. **To mentally divide a whole number by a unit fraction, multiply the whole number by the denominator of the fraction.**

Example A	**Example B**
$7 \div \frac{1}{5} = 7 \times 5 = 35$	$4 \div \frac{1}{7} = 4 \times 7 = 28$

Mentally dividing a whole number by a fraction other than a unit fraction is often possible because it requires just two steps. **To mentally divide a whole number by a fraction, divide the whole number by the numerator of the fraction and multiply that quotient by the denominator of the fraction.**

Example C	**Example D**
$18 \div \frac{3}{8} = 18 \div 3 \times 8 = 48$	$35 \div \frac{5}{6} = 35 \div 5 \times 6 = 42$

Not all division problems involving whole numbers and fractions can be solved as easily as the examples above. However, if the whole number is a multiple of the numerator, you should be able to find the answer mentally.

CLASS PRACTICE

Mentally solve these problems, which have unit fractions as divisors.

a. $4 \div \frac{1}{5}$ 20
b. $8 \div \frac{1}{3}$ 24
c. $6 \div \frac{1}{11}$ 66
d. $9 \div \frac{1}{6}$ 54

Mentally solve these problems, which have other proper fractions as divisors.

e. $8 \div \frac{2}{7}$ 28
f. $15 \div \frac{5}{6}$ 18
g. $12 \div \frac{4}{9}$ 27
h. $28 \div \frac{7}{8}$ 32

WRITTEN EXERCISES

A. *Mentally solve these problems, which have unit fractions as divisors.*

1. $8 \div \frac{1}{4}$ 32
2. $9 \div \frac{1}{6}$ 54
3. $5 \div \frac{1}{12}$ 60
4. $7 \div \frac{1}{8}$ 56

5. $6 \div \frac{1}{3}$ 18
6. $12 \div \frac{1}{9}$ 108
7. $15 \div \frac{1}{2}$ 30
8. $22 \div \frac{1}{4}$ 88

140 Chapter 3 *Factoring and Fractions*

B. Mentally solve these problems, which have other proper fractions as divisors.

9. $6 \div \frac{2}{3}$ 9
10. $8 \div \frac{2}{5}$ 20
11. $10 \div \frac{5}{8}$ 16
12. $15 \div \frac{3}{7}$ 35
13. $16 \div \frac{4}{9}$ 36
14. $12 \div \frac{2}{3}$ 18
15. $12 \div \frac{6}{7}$ 14
16. $16 \div \frac{8}{9}$ 18
17. $18 \div \frac{3}{5}$ 30
18. $35 \div \frac{5}{8}$ 56
19. $42 \div \frac{7}{8}$ 48
20. $54 \div \frac{6}{7}$ 63
21. $22 \div \frac{11}{22}$ 44
22. $24 \div \frac{8}{9}$ 27
23. $36 \div \frac{18}{25}$ 50
24. $32 \div \frac{8}{15}$ 60

C. Solve these reading problems. Solve problems 25–28 mentally.

25. One icy morning 12 students arrived late at school. If that was $\frac{1}{5}$ of the total number, how many students were in the school?   60 students

26. During the famine in Egypt, Joseph gave the people seed to plant, and they were to give $\frac{1}{5}$ of their harvest to Pharaoh (Genesis 47:24). If a farmer brought 25 bushels of grain to Pharaoh, what was the amount of his harvest?   125 bushels

27. The Zimmermans have already stacked $\frac{3}{4}$ of the bales of hay from their fields, or 330 bales. How many bales of hay were there in all?   440 bales

28. Mark and Brent have dug $\frac{5}{6}$ of the potatoes in the garden. If they have filled 10 baskets, how many baskets of potatoes can they expect from the entire garden?   12 baskets

29. One year Mary and her mother canned 152 jars of fruit. Mary figured that this was $\frac{4}{7}$ of all the vegetables and fruit they canned that year. How many jars of vegetables and fruit did they can in all?   266 jars

30. On a trip to visit Grandfather, Arlin's family traveled 540 miles on the first day. If this was $\frac{3}{4}$ of the total distance, how far does Grandfather live from Arlin's home?   720 miles

REVIEW EXERCISES

D. Find the answers. *(Lesson 41)*

31. $2\frac{1}{5}$ is $\frac{1}{4}$ of $8\frac{4}{5}$
32. $3\frac{2}{3}$ is $\frac{5}{9}$ of $6\frac{3}{5}$

E. Solve these multiplication and division problems. *(Lessons 38–40)*

33. $\frac{4}{7} \times \frac{3}{8} \times \frac{1}{6}$ $\frac{1}{28}$
34. $5\frac{1}{4} \times 6\frac{2}{3}$ 35
35. $8\frac{2}{3} \div 6\frac{1}{2}$ $1\frac{1}{3}$
36. $5\frac{3}{4} \div 3\frac{2}{7}$ $1\frac{3}{4}$

F. Find the prime factors of these numbers. *(Lesson 34)*

37. 99 $99 = 3^2 \times 11$
38. 102 $102 = 2 \times 3 \times 17$

G. Solve these problems with compound measures. *(Lesson 26)*

39. 8×85 cm 23 mm = 698.4 cm
40. 5 kl 285 $l \div 7$ = 755 l

H. Use the distributive law to find the missing numbers. *(Lesson 10)*

41. $12 \times 41 = (12 \times 40) + (12 \times \underline{1})$
42. $2 \cdot 77 = (2 \cdot 70) + (\underline{2} \cdot \underline{7})$

Lesson 42   T–140

Introduction

Ask, "If you have 8 dollars' worth of quarters, how many quarters do you have?" (32) When you ask the students to explain how they found the answer, you will probably find that most of them "just know" it works because it makes sense. Some may mention that there are 4 quarters in 1 dollar, so there are 8 times 4 (or 32) quarters in 8 dollars.

Use a written example to show why the problem can be solved in that way.

$8 \div \frac{1}{4} = \frac{8}{1} \times \frac{4}{1} = 32$

Dividing a whole number by a unit fraction results in two factors that are actually whole numbers. The result is the same as multiplying the whole number by the denominator of the unit fraction.

Solutions for Part C

25. $12 = \frac{1}{5} \times \underline{}$; $12 \div \frac{1}{5}$
26. $25 = \frac{1}{5} \times \underline{}$; $25 \div \frac{1}{5}$
27. $330 = \frac{3}{4} \times \underline{}$; $330 \div \frac{3}{4}$
28. $10 = \frac{5}{6} \times \underline{}$; $10 \div \frac{5}{6}$
29. $152 = \frac{4}{7} \times \underline{}$; $152 \div \frac{4}{7}$
30. $540 = \frac{3}{4} \times \underline{}$; $540 \div \frac{3}{4}$

Teaching Guide

1. **To mentally divide a whole number by a unit fraction, multiply the whole number by the denominator of the fraction.** This puts into words what many students already know through experience.

    a. $6 \div \frac{1}{2} = \underline{}$ ($6 \times 2 = 12$; there are twelve $\frac{1}{2}$s in 6.)
    b. $9 \div \frac{1}{4} = \underline{}$ ($9 \times 4 = 36$; there are thirty-six $\frac{1}{4}$s in 9.)
    c. $12 \div \frac{1}{5}$ (60)
    d. $7 \div \frac{1}{8}$ (56)
    e. $20 \div \frac{1}{9}$ (180)
    f. $55 \div \frac{1}{3}$ (165)
    g. $100 \div \frac{1}{20}$ (2,000)
    h. $1,000 \div \frac{1}{90}$ (90,000)

2. **To mentally divide a whole number by a fraction, divide the whole number by the numerator of the fraction and multiply that quotient by the denominator of the fraction.** This is the opposite of mentally multiplying a whole number by a fraction. It works best when the numerator of the divisor is a factor of the whole number.

    a. $21 \div \frac{3}{4} = 21 \div 3 \times 4 = 28$
    b. $15 \div \frac{5}{8} = 15 \div 5 \times 8 = 24$
    c. $16 \div \frac{4}{9} = 16 \div 4 \times 9 = 36$
    d. $25 \div \frac{5}{11} = 25 \div 5 \times 11 = 55$
    e. $45 \div \frac{3}{5}$ (75)
    f. $27 \div \frac{9}{11}$ (33)
    g. $60 \div \frac{5}{12}$ (144)
    h. $56 \div \frac{7}{9}$ (72)

T-141   Chapter 3   Factoring and Fractions

LESSON 43

Objective

- To give help with distinguishing between multiplication and division in reading problems involving fractions.

Review

1. *Solve by mental division.* (Lesson 42)
   a. $6 \div \frac{1}{12}$ (72)
   b. $22 \div \frac{1}{3}$ (66)
   c. $14 \div \frac{7}{8}$ (16)
   d. $25 \div \frac{5}{7}$ (35)

2. *Find the answers.* (Lesson 41)
   a. $\frac{3}{4}$ is $\frac{6}{7}$ of ___ ($\frac{7}{8}$)
   b. $4\frac{2}{7}$ is $\frac{2}{3}$ of ___ ($6\frac{3}{7}$)

3. *Solve these division problems.* (Lessons 39, 40)
   a. $6\frac{2}{3} \div 4\frac{4}{5}$ ($1\frac{7}{18}$)
   b. $9\frac{1}{2} \div 5\frac{3}{7}$ ($1\frac{3}{4}$)

4. *Find the greatest common factor of each pair.* (Lesson 35)
   a. 72, 120 (24)
   b. 72, 80 (8)

5. *Find the lowest common multiple of each pair.* (Lesson 35)
   a. 35, 14 (70)
   b. 24, 36 (72)

6. *Find these equivalents.* (Lesson 27)
   a. 34 cm = ___ in. (13.26)
   b. 55 ml = ___ tbsp. (3.85)

7. *Solve these division problems.* (Lesson 11)
   a. $27\overline{)5{,}481}$ (203)
   b. $537\overline{)54{,}774}$ (102)

43. Reading Problems Containing Fractions

In working with whole numbers, the product is always larger than either of the factors in multiplication, and the quotient is always smaller than the dividend in division. In calculations with fractions, however, these patterns do not hold true. This can make it challenging to decide whether you should multiply or divide to solve a reading problem.

Another pattern of whole numbers is that we multiply to find a total or whole amount. But with fractions these terms sometimes indicate division. For example, if you know what $\frac{1}{2}$ of a number is, you can divide by $\frac{1}{2}$ to find the whole amount.

Multiplication and division problems always include three numbers, such as in the statement "9 is $\frac{3}{4}$ of 12." Twelve is the whole amount. Nine is the fractional part of the whole amount. Three-fourths tells us what fractional part 9 is of 12.

To decide whether you should multiply or divide to solve a reading problem, ask these questions:

1. Must I find part of the whole amount? Then multiply (Example A). Remember that *of* means "times."

2. Must I find the whole amount? Then divide the part by the fraction that it is of the whole amount (Example B).

3. Must I find how many times larger one number is than another? (The question may ask how many times longer, heavier, or faster one item is than another.) Then divide the larger number by the smaller.

Remember these principles.

Multiply to find **part of the whole amount.**

Divide to find **the whole amount**.

Divide to find **how many times larger** one number is than another.

Example A	**Example B**
Rachel peeled $\frac{3}{4}$ of the 12 potatoes in her bowl. How many potatoes did she peel?	Rachel peeled 9 potatoes. That was $\frac{3}{4}$ of the number in her bowl. How many potatoes were in her bowl?
___ is $\frac{3}{4}$ of 12	9 is $\frac{3}{4}$ of ___
___ = $\frac{3}{4} \times 12$	$9 = \frac{3}{4} \times$ ___
$\frac{3}{4} \times \frac{\cancel{12}^{3}}{1} = \frac{9}{1} = 9$ potatoes	$\frac{9}{1} \div \frac{3}{4} = \frac{\cancel{9}^{3}}{1} \times \frac{4}{\cancel{3}} = \frac{12}{1} = 12$ potatoes

142   Chapter 3   Factoring and Fractions

CLASS PRACTICE

a. The Lance family purchased a $\frac{3}{4}$-acre lot. They decided to use $\frac{1}{3}$ of the lot for a garden. What was the size of their garden? $\frac{1}{4}$ acre

b. Clifford calculated that the building area of his family's farm covers $\frac{1}{32}$ of the entire farm. How large is the farm if the building area contains $1\frac{1}{4}$ acres? 40 acres

c. The Gehmans moved to a 98-acre farm. How many times larger is this farm than the $1\frac{3}{4}$-acre lot where they lived before? 56 times

d. At Lowell's school, $\frac{11}{16}$ of the students have had chicken pox. How many students are in the school if 33 have had chicken pox? 48 students

WRITTEN EXERCISES

A. Use multiplication and division to solve these problems.

1. The Weavers own a car that weighs 3,000 pounds. This is $\frac{5}{8}$ as much as their van weighs. What is the weight of the van? 4,800 pounds

2. The Stauffers planted 9 rows of strawberries. This is $\frac{3}{8}$ of the total number of rows in their produce patch. How many rows are in the patch? 24 rows

3. An excavator removed a $\frac{3}{4}$-ton granite rock from a field. Later he dug up a rock with an estimated weight of $5\frac{1}{2}$ tons. How many times heavier was the second rock than the first? $7\frac{1}{3}$ times

4. Mother put $2\frac{1}{2}$ cups of noodles into a casserole. Then she added $\frac{1}{5}$ of that amount of water. How much water did she put into the casserole? $\frac{1}{2}$ cup

5. When Cheryl began mixing muffins, she found that the sugar canister had only $\frac{1}{2}$ cup of sugar. That was $\frac{2}{3}$ of the amount indicated in the recipe. How much sugar did the recipe call for? $\frac{3}{4}$ cup

6. Four years after Father started dairy farming, he said that only about $\frac{1}{8}$ of his present herd had been part of his original herd. How many of his 62 cows were part of the original herd? (Round to the nearest whole number.) 8 cows

7. The Macks planted $5\frac{1}{2}$ acres of sweet corn, but they could irrigate only $\frac{3}{4}$ of the patch. How many acres did not receive irrigation? $1\frac{3}{8}$ acres

8. A 3-inch water pipe has a circular area of $7\frac{1}{14}$ square inches. A 4-inch pipe has a circular area of $12\frac{4}{7}$ square inches. How many times larger is a 4-inch pipe than a 3-inch pipe? $1\frac{7}{9}$ times

9. Peter and Sylvia weeded 10 rows of peas, which was $\frac{5}{8}$ of the total number. How many rows of peas were there? 16 rows

10. On Saturday morning, Merle worked $\frac{7}{10}$ hour on a 2-horsepower motor and $1\frac{4}{5}$ hours on a 5-horsepower motor. How many times longer did he work on the larger motor than on the smaller motor? $2\frac{4}{7}$ times

Lesson 43   T–142

Introduction

What words or phrases in reading problems indicate multiplication and division?

Multiplication
—times
—total amount
—whole amount

Division
—how many times more
—how many times larger
—what is the average

These terms are good indicators of the operation if the problem contains whole numbers, but not all of them apply to problems that involve fractions.

Solutions for Part A

1. $3,000 = \frac{5}{8} \times \underline{}$; $3,000 \div \frac{5}{8}$
2. $9 = \frac{3}{8} \times \underline{}$; $9 \div \frac{3}{8}$
3. $5\frac{1}{2} \div \frac{3}{4}$
4. $\frac{1}{5} \times 2\frac{1}{2}$
5. $\frac{1}{2} = \frac{2}{3} \times \underline{}$; $\frac{1}{2} \div \frac{2}{3}$
6. $\frac{1}{8} \times 62$
7. $\frac{1}{4} \times 5\frac{1}{2}$
8. $12\frac{4}{7} \div 7\frac{1}{14}$
9. $10 = \frac{5}{8} \times \underline{}$; $10 \div \frac{5}{8}$
10. $1\frac{4}{5} \div \frac{7}{10}$

Teaching Guide

To decide whether you should multiply or divide to solve a reading problem, ask the following questions.

(1) Must I find part of the whole amount? Then multiply. Remember that *of* means "times."

(2) Must I find the whole amount? Then divide the part by the fraction that it is of the whole amount.

(3) Must I find how many times larger one number is than another? Then divide the larger number by the smaller.

Spend time discussing the problems in *Class Practice*. Help the students analyze which of the three parts are given: the part, the whole amount, or the fraction. Two of these three will always be present. They must then find the missing fact.

Remind them that *of* means "times" when working with fractions. The phrase "$\frac{3}{4}$ of 12" means "$\frac{3}{4} \times 12$," so we multiply to find the missing number. The phrase "9 is $\frac{3}{4}$ of ___" means "9 is $\frac{3}{4} \times$ ___." Since we know only one factor, $\frac{3}{4}$, we must divide to find the other factor.

Analyzing *Class Practice*:

a. whole: $\frac{3}{4}$ acre
   part: multiply to find
   fraction: $\frac{1}{3}$

b. whole: divide to find
   part: $1\frac{1}{4}$ acres
   fraction: $\frac{1}{32}$

c. whole: 98 acres
   part: $1\frac{3}{4}$ acres
   fraction: divide to find

d. whole: divide to find
   part: 33
   fraction: $\frac{11}{16}$

T-143 Chapter 3 Factoring and Fractions

An Ounce of Prevention

Help the students distinguish between the *part* and the *whole amount*. If the facts are set up in the following form, the whole amount always follows *of* and the part always precedes *is*.

16 is $\frac{4}{5}$ of 20

$16 = \frac{4}{5} \times 20$

$\frac{5}{8}$ is $\frac{1}{2}$ of $1\frac{1}{4}$

$\frac{5}{8} = \frac{1}{2} \times 1\frac{1}{4}$

The principles taught in this lesson apply to a number of concepts taught in later chapters: base, rate, and percentage; sales, rate, and commission; sales, rate, and sales tax; and principal, rate, and interest.

Solutions for Part B

11. $7 \times 2 \times (2\frac{1}{2} + 1\frac{1}{4} + \frac{3}{4})$
12. $16 - (3\frac{1}{2} + 4\frac{1}{4} + 1\frac{1}{2})$
13. $290 \div 9\frac{1}{2}$
14. $(260 - 20) \times 1\frac{1}{2}$
15. $90 \times 5\frac{1}{2}$
16. $100 \div 4\frac{1}{2}$
17. $455 \div 3\frac{1}{2}$
18. $8\frac{3}{4} - (3\frac{1}{6} + 2\frac{7}{8})$

B. Solve these reading problems.

11. While Father feeds the milk cows, David gives $2\frac{1}{2}$ bales of hay to the dry cows, $1\frac{1}{4}$ bales to the young heifers, and $\frac{3}{4}$ bale to the calves. If he feeds this much hay twice a day, how many bales of hay will David use in a week? 63 bales

12. One morning Sarah used $3\frac{1}{2}$ cups of flour to bake one cake, $4\frac{1}{4}$ cups to bake another cake, and $1\frac{1}{2}$ cups to make a pie crust. The canister held 16 cups of flour when she started. How much flour was left when Sarah finished? $6\frac{3}{4}$ cups

13. An excavator worked $9\frac{1}{2}$ hours to cut a driveway 290 feet into a wood lot. On the average, how many feet of driveway did he cut per hour? (Round your answer to the nearest foot.) 31 feet

14. A well driller dug a well 260 feet deep. The water came within 20 feet of the top of the well, and each foot of depth contained $1\frac{1}{2}$ gallons of water. How much water did the well contain? 360 gallons

15. The well was used for drip irrigation. If water was pumped at the rate of $5\frac{1}{2}$ gallons per minute for 90 minutes, how many gallons of water were pumped in all? 495 gallons

16. If water is pumped out of a well at the rate of $4\frac{1}{2}$ gallons per minute, how many minutes will it take to pump 100 gallons? (Round your answer to the nearest whole minute.) 22 minutes

17. A strip of black plastic used for weed control covers a total of 455 square feet. If the width of the row covered is $3\frac{1}{2}$ feet, what is its length? 130 feet

18. Sister Alice bought $8\frac{3}{4}$ yards of fabric to sew dresses at the sewing circle. She used $3\frac{1}{6}$ yards for one dress and $2\frac{7}{8}$ yards for another. How much fabric did she have left? $2\frac{17}{24}$ yards

144 Chapter 3 Factoring and Fractions

REVIEW EXERCISES

C. Solve by mental division. *(Lesson 42)*

19. $5 \div \frac{1}{11}$ 55
20. $19 \div \frac{1}{2}$ 38
21. $18 \div \frac{3}{5}$ 30
22. $12 \div \frac{3}{8}$ 32

D. Find the answers. *(Lesson 41)*

23. 21 is $\frac{3}{7}$ of 49
24. $\frac{5}{8}$ is $\frac{3}{4}$ of $\frac{5}{6}$
25. $2\frac{1}{3}$ is $\frac{1}{9}$ of 21
26. $3\frac{3}{5}$ is $\frac{2}{9}$ of $16\frac{1}{5}$

E. Solve these division problems. *(Lessons 39, 40)*

27. $3\frac{1}{7} \div 2\frac{1}{5}$ $1\frac{3}{7}$
28. $8\frac{4}{7} \div 3\frac{1}{3}$ $2\frac{4}{7}$

F. Find the greatest common factor of each pair. *(Lesson 35)*

29. 24, 72   24
30. 32, 48   16

G. Find the lowest common multiple of each pair. *(Lesson 35)*

31. 15, 25   75
32. 36, 48   144

H. Find these equivalents. *(Lesson 27)*

33. 34 cm = 13.26 in.
34. 25 tons = 22.75 MT

I. Solve these division problems. *(Lesson 11)*

35. $34 \overline{)6{,}834}$ 201
36. $643 \overline{)20{,}576}$ 32

T–145 Chapter 3 Factoring and Fractions

LESSON 44

Objective

- To give practice with recognizing what additional information is needed to solve a given reading problem.

Review

1. Give Lesson 44 Quiz (Using Fractions).
2. *Solve by mental division.* (Lesson 42)
   a. $5 \div \frac{1}{6}$ (30)
   b. $21 \div \frac{1}{2}$ (42)
   c. $8 \div \frac{1}{12}$ (96)
3. *Solve these division problems.* (Lessons 39, 40)
   a. $3\frac{1}{8} \div \frac{5}{7}$ $(4\frac{3}{8})$
   b. $2\frac{1}{6} \div \frac{1}{3}$ $(6\frac{1}{2})$
   c. $3\frac{2}{9} \div 2\frac{5}{12}$ $(1\frac{1}{3})$
4. *Write the greatest common factor of the numerator and denominator of each fraction. Then reduce the fraction to lowest terms.* (Lesson 36)
   a. $\frac{16}{24}$ $(8; \frac{2}{3})$
   b. $\frac{48}{80}$ $(16; \frac{3}{5})$
5. *Find these temperature equivalents to the nearest degree.* (Lesson 28)
   a. $195°F = \_\_°C$ (91)
   b. $52°C = \_\_°F$ (126)
6. *Use the divide-and-divide method to solve these problems mentally.* (Lesson 12)
   a. $192 \div 16$ (12)
   b. $126 \div 18$ (7)

44. Reading Problems: Deciding What Information Is Missing

In everyday life, math problems do not come to us as neat paragraphs including all the information we need to solve them. Rather, we usually gather facts from several different places and perhaps sort through much unnecessary information to find what we need. Then we must decide what operations will give the answer.

The reading problems in this lesson, like many situations in everyday life, contain some of the facts necessary for solution. However, not all the needed information is present. You will study these problems and decide what additional information is needed to solve them. The following steps should be helpful.

1. Read the problem carefully. Think, "What is the question really asking?"
2. Decide what additional information is needed to solve the problem.

Example A

On the planet Venus, a day is longer than a year. One day on Venus is 243 earth days long, and one year on Venus is 225 earth days long. The temperatures during that long day rise as high as 900°F. What is the difference between the highest daytime temperature and the lowest nighttime temperature on Venus?

Information needed: Coldest temperature on Venus

Example B

More than 1,300 earths could fit inside Jupiter if it were hollow. On its surface is a red spot, which scientists say is a huge storm that has been raging for at least 300 earth years. How many times longer is the duration of a 300-year storm on Jupiter than that of a typical hurricane on the earth?

Information needed: Duration of a typical hurricane on the earth

CLASS PRACTICE

Tell what missing information is needed to solve each problem.

a. A greenhouse owner buys pine-bark mulch by the cubic yard. What is the price of the mulch per cubic foot if 1 cubic yard equals 27 cubic feet? *price per cubic yard*

b. Gettysburg, Pennsylvania, received 11 inches of rain one evening. An inch of rain equals 5.2 pounds per square foot of land; 226,512 pounds per acre; and 144,967,680 pounds per square mile. What part of the average annual precipitation fell at Gettysburg that evening? *average annual precipitation*

146   Chapter 3   Factoring and Fractions

WRITTEN EXERCISES

A. Tell what missing information is needed to solve each problem.

1. Marlin plans to use a 3-ton hydraulic jack to lift a sagging beam in the crawl space under the kitchen floor. If the top of the jack is $10\frac{1}{4}$ inches above the ground, how far is the bottom of the floor joist from the top of the jack? *distance from the ground to the floor joist*

2. David balanced a car tire by replacing a 2-ounce weight that had been on it. How much less was the new weight than the one used formerly? *size of the new weight*

3. Uncle Henry is mixing pesticide to spray part of his 40-acre field of corn. He wants to apply $\frac{3}{4}$ cup of pesticide per acre. How much pesticide will he use? *number of acres to be sprayed*

4. The Masts' telephone bill usually arrives on the tenth day of the month following the billing period. What was the average cost per day of their telephone service in November? *amount of the telephone bill*

5. Father calculated that the annual food bill for the family was $5,037 and the electric bill was $1,150. What was the average cost of food per person? *number of persons in the family*

6. Father also calculated that the family drove a total of 18,000 miles with the family van. The gasoline for the van cost $1,434.00. What was the average number of miles traveled for each gallon of fuel consumed? *gallons of gasoline used (or price per gallon of gasoline)*

7. Brother Harold bought some used furniture at an auction. He paid $120 for a bed, $67 for a dresser, $162 for a table, and $138 for chairs. What was his total bill, including sales tax? *sales tax rate*

8. One summer Mother canned 175 quarts of peaches, 150 quarts of pears, 75 quarts of cherries, and 90 quarts of applesauce. On the average, how many quarts of fruit is that per family member? *number of family members*

9. The Lehmans soaked and canned dry navy beans. Each pound of dry beans, when soaked, swells to $1\frac{2}{5}$ quarts or $5\frac{3}{5}$ cups of beans. How many quarts of beans did the Lehmans can? *number of pounds of dry beans*

10. The cashier at the hardware store rang up Father's items for a total of $12.95 plus $0.65 tax. Just then Mother brought another item that cost $1.25 plus $0.06 tax. How much change did they receive? *amount of money given*

11. For a number of years, the Harveys provided a home for foster children. Their invalid grandmother also lived with them during one-third of that time. How long did their grandmother stay with them? *how long they cared for foster children*

He did good, and gave us rain from heaven, and fruitful seasons, filling our hearts with food and gladness. Acts 14:17

Introduction

Give the students the following information. After the information are some possible questions with which to make reading problems. Ask the students what facts would be needed to answer the questions.

The facts needed are given in parentheses. You may want to give these facts to the students afterward.

Some of the most skillful climbers in North America are the mountain goat and the bighorn sheep. The more skillful of the two is the mountain goat. There are few mountains that this goat cannot climb, and it seldom is injured or killed from falling.

The bighorn sheep is nearly as skillful in climbing. Bighorn sheep can walk on ledges only 2 inches wide.

a. How much longer is a bighorn sheep than a mountain goat?
Information needed: the lengths of both animals
(bighorn sheep, 72 in.; mountain goat, 70 in.; answer: 2 in.)

b. How much taller does the mountain goat grow than the bighorn sheep?
Information needed: the heights of both animals
(bighorn sheep, 36 in.; mountain goat, 40 in.; answer: 4 in.)

Teaching Guide

If a reading problem does not give all the information needed to solve it, use the following steps.

(1) Read the problem carefully. Think, "What is the question really asking?"

(2) Decide what additional information is needed to solve the problem.

a. One year there were 95,500 dairy cattle in Lancaster County, Pennsylvania, and they produced a total of 1,560,000,000 pounds of milk. To the nearest whole number, what was the average number of cows per square mile in Lancaster County?
(Information needed: Square miles in Lancaster County)

b. One year the United States produced 1,100,000 gallons of maple syrup, of which 365,000 gallons came from the state of Vermont. What was the total value of the maple syrup produced in Vermont that year?
(Information needed: Value per gallon of maple syrup)

12. The Sensenigs are planning to visit their grandparents, who live 550 miles away. How long should it take to reach their destination?   average rate (speed) of travel
13. Ashville is 81 miles from Newton. Caldwall is between Ashville and Newton. How far is it from Newton to Caldwall?   distance from Ashville to Caldwall
   (or what fraction of the total distance Caldwall is from one of the cities)
14. Brother Horton spread lime on three fields. He put 4 tons per acre on the north field, $\frac{1}{2}$ as many tons per acre on the south field as on the north field, and $\frac{3}{8}$ as many tons per acre on the southwest field as on the north field. How many tons of lime did he put on the south field?   acres in the south field

REVIEW EXERCISES

B. Solve by mental division. *(Lesson 42)*

15. $4 \div \frac{1}{12}$ 48
16. $13 \div \frac{1}{4}$ 52
17. $15 \div \frac{3}{7}$ 35
18. $36 \div \frac{4}{5}$ 45

C. Solve these division problems. *(Lessons 39, 40)*

19. $2\frac{1}{4} \div \frac{3}{8}$ 6
20. $3\frac{1}{6} \div 2\frac{2}{3}$ $1\frac{3}{16}$
21. $3\frac{3}{8} \div \frac{3}{4}$ $4\frac{1}{2}$
22. $6\frac{2}{5} \div 4\frac{4}{7}$ $1\frac{2}{5}$

D. Write the greatest common factor of the numerator and denominator of each fraction. Then reduce the fraction to lowest terms. *(Lesson 36)*

23. $\frac{12}{20}$ 4 $\frac{3}{5}$
24. $\frac{64}{96}$ 32 $\frac{2}{3}$

E. Find these temperature equivalents to the nearest degree. *(Lesson 28)*

25. 77°F = __25__ °C
26. 41°C = __106__ °F
27. 28°C = __82__ °F
28. 91°F = __33__ °C

F. Use the divide-and-divide method to solve these problems mentally. *(Lesson 12)*

29. $216 \div 24$ 9
30. $144 \div 18$ 8

148   Chapter 3   Factoring and Fractions

45. Chapter 3 Review

A. Solve these fraction problems. *(Lessons 37–41)*

1. $\frac{1}{4} + \frac{3}{8}$ $\frac{5}{8}$
2. $\frac{3}{4} + \frac{5}{6}$ $1\frac{7}{12}$
3. $\frac{7}{8} + \frac{4}{5}$ $1\frac{27}{40}$
4. $\frac{7}{9} + \frac{3}{8}$ $1\frac{11}{72}$
5. $\frac{2}{3} - \frac{5}{8}$ $\frac{1}{24}$
6. $\frac{7}{8} - \frac{3}{4}$ $\frac{1}{8}$
7. $\frac{7}{12} - \frac{3}{8}$ $\frac{5}{24}$
8. $\frac{8}{15} - \frac{1}{5}$ $\frac{1}{3}$
9. $\frac{1}{5}$ of 35   7
10. $\frac{1}{6}$ of 24   4
11. $43 \times \frac{5}{9}$ $23\frac{8}{9}$
12. $17 \times \frac{3}{7}$ $7\frac{2}{7}$
13. $11 \times 2\frac{1}{8}$ $23\frac{3}{8}$
14. $15 \times 1\frac{5}{6}$ $27\frac{1}{2}$
15. $\frac{2}{3} \times \frac{3}{5} \times \frac{1}{4}$ $\frac{1}{10}$
16. $\frac{5}{6} \times \frac{3}{8} \times \frac{4}{5}$ $\frac{1}{4}$
17. $21 \div \frac{1}{8}$ 168
18. $26 \div \frac{1}{2}$ 52
19. $\frac{7}{9} \div \frac{7}{8}$ $\frac{8}{9}$
20. $\frac{2}{5} \div \frac{4}{5}$ $\frac{1}{2}$
21. $24 \div \frac{6}{7}$ 28
22. $35 \div \frac{7}{8}$ 40
23. $2\frac{1}{6} \div 2\frac{1}{4}$ $\frac{26}{27}$
24. $4\frac{1}{8} \div 3\frac{2}{3}$ $1\frac{1}{8}$
25. 9 is $\frac{3}{8}$ of __24__
26. 15 is $\frac{3}{4}$ of __20__
27. $3\frac{1}{8}$ is $\frac{5}{8}$ of __5__
28. $4\frac{3}{4}$ is $\frac{1}{2}$ of __$9\frac{1}{2}$__

B. Copy and solve by vertical multiplication. *(Lesson 38)*

29. $16 \times 3\frac{1}{2}$ 56
30. $18 \times 2\frac{2}{3}$ 48
31. $20 \times 3\frac{2}{5}$ 68
32. $25 \times 2\frac{3}{5}$ 65

C. Simplify these complex fractions. *(Lesson 40)*

33. $\dfrac{\frac{1}{6}}{\frac{1}{4}}$ $\frac{2}{3}$
34. $\dfrac{2\frac{3}{4}}{\frac{7}{8}}$ $3\frac{1}{7}$

D. Find these answers mentally. *(Lessons 38, 42)*

35. $\frac{3}{4}$ of 16   12
36. $\frac{3}{8}$ of 40   15
37. $\frac{5}{9}$ of 45   25
38. $\frac{4}{7}$ of 42   24
39. $5 \div \frac{1}{11}$ 55
40. $7 \div \frac{1}{8}$ 56
41. $5 \div \frac{5}{7}$ 7
42. $9 \div \frac{3}{4}$ 12

LESSON 45

Objective

- To review the material taught in Chapter 3 (Lessons 34–44).

Teaching Guide

1. Lesson 45 reviews the material taught in Lessons 34–44. For pointers on using these review lessons, see *Teaching Guide* for Lesson 15.

2. Review the following concepts. Although most of these items are not new, plenty of drill is needed to master them.

   a. Finding greatest common factors and lowest common multiples (Lesson 35).

   b. Dividing fractions (Lessons 39, 40).

   c. Complex fractions (Lessons 36, 40).

   d. Finding a number when a fractional part of it is known (Lesson 41).

3. Be sure to review the following new concepts taught in this chapter.

Lesson number and new concept	Exercises in Lesson 45
34—Memorizing the prime factors from 2 to 19.	43
34—Using exponents to express prime factors.	45, 46
36—Unit fractions.	53
36—Complex fractions.	54
40—Simplifying complex fractions.	33, 34
42—Mentally dividing whole numbers by fractions.	39–42

T–149   Chapter 3   Factoring and Fractions

Solutions for Part J

61. $38\frac{1}{4} - 13\frac{1}{2}$

62. $\frac{1}{4} \times 13\frac{1}{2}$

63. $\frac{1}{4} \times \$12{,}500$

64. $3\frac{1}{2} + 2 \times \frac{3}{8} + 2 \times \frac{3}{8}$

E. Do these exercises on prime numbers and factors. *(Lesson 34)*

43. List the prime numbers from 1 to 20.   2, 3, 5, 7, 11, 13, 17, 19
44. Is 51 prime or composite?   composite
45. Find the prime factors of 36. Use exponents for repeating factors. $36 = 2^2 \times 3^2$
46. Find the prime factors of 52. Use exponents for repeating factors. $52 = 2^2 \times 13$

F. Find the greatest common factor of each pair. *(Lesson 35)*

47. 28, 42   14   48. 54, 72   18

G. Find the lowest common multiple of each pair. *(Lesson 35)*

49. 15, 25   75   50. 16, 20   80

H. Do these fraction exercises. *(Lesson 36)*

51. Compare $\frac{5}{8}$ and $\frac{3}{5}$, and write < or > between them. $\frac{25}{40} > \frac{24}{40}$
52. Compare $\frac{1}{3}$ and $\frac{3}{16}$, and write < or > between them. $\frac{16}{48} > \frac{9}{48}$
53. True or false: $\frac{1}{4}$ is a unit fraction.   true
54. True or false: $\frac{103}{121}$ is a complex fraction.   false
55. Expand $\frac{5}{8}$ by multiplying both parts by 9. $\frac{45}{72}$
56. Expand $\frac{6}{7}$ by multiplying both parts by 12. $\frac{72}{84}$
57. Reduce $\frac{36}{63}$ to lowest terms. $\frac{4}{7}$
58. Reduce $\frac{48}{72}$ to lowest terms. $\frac{2}{3}$

I. Write the reciprocal of each expression. *(Lesson 39)*

59. $3\frac{3}{5}$ $\frac{5}{18}$ 60. $\frac{4}{29}$ $\frac{29}{4}$

J. Solve these reading problems. *(Lesson 43)*

61. The honeycombs from a certain beehive weighed $38\frac{1}{4}$ pounds. After the honey was extracted, the combs weighed $13\frac{1}{2}$ pounds. What was the weight of the honey extracted? $24\frac{3}{4}$ pounds

62. In the $13\frac{1}{2}$ pounds of honeycombs that were left, $\frac{1}{4}$ of the weight was honey that could not be extracted. What was the weight of the honey left in the combs? $3\frac{3}{8}$ pounds

63. Father contracted to build a garage for $12,500. He asked the customer to pay $\frac{1}{4}$ of the money as a down payment before starting the job, $\frac{1}{4}$ when it was half finished, and the remaining $\frac{1}{2}$ when it was completed. What was the amount of the down payment that he requested? $3,125

64. Inside a wall 8 feet high are studs $3\frac{1}{2}$ inches thick. Both sides of the studs are covered with $\frac{3}{8}$-inch plaster lath and $\frac{3}{8}$ inch of plaster. What is the total thickness of the wall?   5 inches

150 Chapter 3 Factoring and Fractions

65. Deborah's goat is producing $1\frac{1}{2}$ gallons of milk per day. This is only $\frac{3}{4}$ of the amount that she produced a few weeks ago. How much did the goat formerly produce each day?   2 gallons

66. Delmar usually pays $12 per hundred pounds of regular rabbit feed. When he needed to buy medicated feed, he found that the cost of regular feed was $\frac{4}{5}$ the cost of medicated feed. What was the cost of medicated feed per hundred pounds? $15

65. $1\frac{1}{2} = \frac{3}{4} \times$ ____ ; $1\frac{1}{2} \div \frac{3}{4}$

66. $12 = \frac{4}{5} \times$ ____ ; $12 \div \frac{4}{5}$

46. Chapter 3 Test

LESSON 46

Objective

- To test the students' mastery of the concepts in Chapter 3.

Teaching Guide

1. Correct Lesson 45.
2. Review any areas of special difficulty.
3. Administer the test. For pointers on giving tests, see *Teaching Guide* for Lesson 16.

"Having then gifts differing according to the grace that is given to us, whether prophecy, let us prophesy according to the proportion of faith; or ministry, let us wait on our ministering: or he that teacheth, on teaching; or he that exhorteth, on exhortation: he that giveth, let him do it with simplicity; he that ruleth, with diligence; he that sheweth mercy, with cheerfulness. Let love be without dissimulation. Abhor that which is evil; cleave to that which is good" (Romans 12:6–9).

Chapter 4
Decimals, Ratios, and Proportions

Decimal fractions, commonly called decimals, are written by extending the place value system of Arabic numeration to the right of the ones' place. Only the numerator of a decimal fraction is written. The denominator is 10 or a power of 10. A decimal point separates the fraction from the whole number in a decimal. As with whole numbers, the value of each decimal place is one-tenth as great as the next place to its left, and ten times as great as the next place to its right.

Decimals, ratios, and proportions can all be written in the form of common fractions. These mathematical expressions are useful in solving many practical problems.

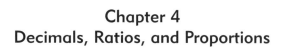

$$\frac{heavens}{earth} = \frac{God's\ ways}{our\ ways}$$

For as the heavens are higher than the earth, so are my ways higher than your ways, and my thoughts than your thoughts.
Isaiah 55:9

47. Using Decimals to Express Fractional Parts

The Arabic number system is based on tens. Moving to the right in a numeral, each place has a value $\frac{1}{10}$ as great as the place before: thousands, hundreds, tens, ones. After the ones' place, you can write a decimal point and continue the same pattern: tenths, hundredths, thousandths, and so on. The interval between any two numbers can be divided into 10 parts, and the number of those parts is expressed by the digit in the next place to the right. Study the following illustration.

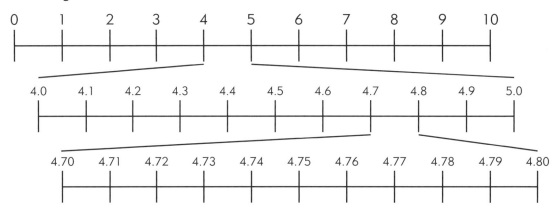

The interval between 4 and 5 above is enlarged and divided to show the ten parts from 4.0 to 5.0. The interval between 4.7 and 4.8 is enlarged and divided to show the ten parts from 4.70 to 4.80. This subdividing can go on and on without end. Each time the intervals are divided again, another decimal place is added.

The digits to the right of the decimal point indicate values less than 1 and are therefore a type of fraction. Decimal fractions, often simply called decimals, are fractions written with a decimal point.

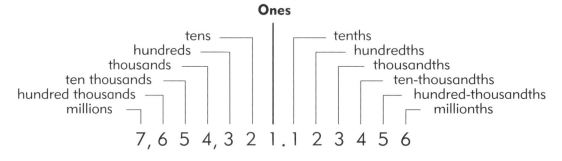

The number above is read, "Seven million, six hundred fifty-four thousand, three hundred twenty-one **and** one hundred twenty-three thousand four hundred fifty-six millionths."

LESSON 47

Objectives

- To review place values to millionths.
- To practice reading and writing decimals.
- To practice comparing numbers by annexing zeroes to give an equal number of decimal places.
- To practice adding and subtracting decimals, including more difficult subtraction problems.

Review

1. *Find the reciprocal of each number.* (Lesson 39)

   a. $\frac{4}{9}$ ($\frac{9}{4}$)   b. 14 ($\frac{1}{14}$)

2. *Choose the facts that you will need to solve these reading problems. Then find the answers.* (Lesson 31)

   a. A box of 12 birthday cards marked 20% off cost Maria $3.95. The cards individually would have cost $0.65. How much did Maria save by buying the box of cards?
   (12 cards, $3.95, $0.65; Answer: $3.85)

   b. The number of students attending Sunday school one morning was as follows: preschool, 3; primary, 2; junior, 3; intermediate, 5; youth, 3; and adult, 11. The total attendance including those not yet in Sunday school was 55. How many students went to the basement for Sunday school if only the intermediate and adult classes were held on the main floor?
   (preschool, 3; primary, 2; junior, 3; youth, 3; Answer: 11 students)

Chapter 4   Decimals, Ratios, and Proportions

Introduction

Write the number 1,000 on the board. Ask the students to divide it by 10. Repeat the process until the quotient is 1. Why is dropping a zero in this case the same as dividing by 10? It is because we use a tens-based system. The value of each place in a number is ten times that of the next place to the right and $\frac{1}{10}$ the value of the next place to the left.

Now write the number 1.000 on the board and ask the students to divide it by 10. Does dropping a zero change its value in the same way? No, it does not. The decimal point is what determines the value of a number. Each time a zero was dropped in the preceding example, the decimal point was actually being moved one place to the left. Thus the value of the number was divided by 10 each time the decimal point was moved. Illustrate this by writing the number 1,000 on the board again and doing the division by moving the decimal point.

Teaching Guide

1. **The Arabic number system is based on tens; therefore, the value of each place in a number is determined by its relationship to the decimal point.** If no decimal point is written, it is understood to be after the last digit on the right of the number. Have the students tell where the decimal point belongs in the following numbers.

   a. 1 = (1.)   b. 1,005 = (1,005.)

2. **The principles of place value apply to both sides of the decimal point.** Moving one place to the right divides the place value by 10. Moving one place to the left multiplies the place value by 10. Have the students give the values of the decimal places up to millionths. (See the place-value diagram in the lesson.)

   Give the place value of each underlined digit.

   a. 0.002<u>3</u>4   (ten-thousandths)

   b. 0.00011<u>1</u>   (millionths)

   c. 0.12345<u>6</u>   (hundred-thousandths)

   d. 0.987<u>6</u>54   (ten-thousandths)

3. **A decimal is read with the same words used for a common fraction of the same value.**

   a. 0.078 = $\frac{78}{1,000}$ = seventy-eight thousandths

Answers for CLASS PRACTICE e–h

e. Eight and sixty-four hundredths

f. Five and eight hundred forty-eight ten-thousandths

g. One and six hundred two hundred-thousandths

h. Eleven and fifty-five millionths

Lesson 47

To write or read a decimal, remember the following points.

(1) The number to the right of the decimal point is the numerator of the fraction. It should be written in the same way as a number to the left of the decimal point, except that there are no commas between periods.

(2) In reading a decimal, determine the denominator by counting the number of places to the right of the decimal point. Read the decimal just as you would any other number, and then say the correct denominator.
0.14 = fourteen hundredths
0.00256 = two hundred fifty-six hundred-thousandths

Study the following examples. Example A shows the relationship between decimals and common fractions. Example B shows how to compare decimals.

Example A	**Example B**
Change 4.35 to a mixed number in simplest form. $4.35 = 4\frac{35}{100} = 4\frac{7}{20}$	Compare 0.11 and 0.109. (1) Annex zeroes so that both decimals have the same number of decimal places. $\quad$ 0.11 = 0.110 $\quad$ 0.109 = 0.109 (2) Write < or > between the decimals. $\quad$ 0.110 > 0.109 $\quad$ 0.11 > 0.109

To add or subtract decimals, align the digits vertically by place value. (Keep the decimal points in a straight line.) Annex zeroes so that all the decimals have the same number of decimal places; then add or subtract as usual. Place the decimal point in the answer directly below the decimal points in the problem.

Example C	**Example D**
1.008 + 0.2537 = _____ $\quad$ 1.0080 $\quad$ + 0.2537 $\quad$ 1.2617	3 − 0.8976 = _____ $\quad$ 3.0000 $\quad$ − 0.8976 $\quad$ 2.1024

CLASS PRACTICE

Name the place farthest to the right in each decimal.
- a. 2.158 — thousandths
- b. 5.12957 — hundred-thousandths
- c. 5.3489 — ten-thousandths
- d. 1.982546 — millionths

Read these numbers, using words. (See facing page.)
- e. 8.64
- f. 5.0848
- g. 1.00602
- h. 11.000055

154   Chapter 4   *Decimals, Ratios, and Proportions*

Change each number to a common fraction or mixed number in simplest form.

i. 0.4 $\frac{2}{5}$ j. 0.51 $\frac{51}{100}$ k. 5.648 $5\frac{81}{125}$ l. 1.000245 $1\frac{49}{200,000}$

Compare each set of decimals, and place < or > between them.

m. 9.3 _>_ 9.23   n. 2.054 _<_ 2.06

o. 0.1025 _>_ 0.0991   p. 0.5493 _<_ 0.5601

WRITTEN EXERCISES

A. Name the place farthest to the right in each decimal.

1. 3.876   2. 2.0991   3. 4.90983   4. 2.123111
   thousandths   ten-thousandths   hundred-thousandths   millionths

B. Write these numbers, using words. (See facing page.)

5. 4.211   6. 3.0871   7. 2.00007   8. 5.000199

C. Change each number to a common fraction or mixed number in simplest form.

9. 0.9 $\frac{9}{10}$ 10. 0.77 $\frac{77}{100}$ 11. 3.45 $3\frac{9}{20}$ 12. 4.375 $4\frac{3}{8}$

D. Compare each set of decimals, and place < or > between them.

13. 4.2 _>_ 4.12   14. 3.096 _<_ 3.1001

15. 0.009 _<_ 0.011   16. 0.348 _>_ 0.3051

E. Solve these addition and subtraction problems.

17.   4.2   18.   5.77   19.   7.0078   20.   32.05
    + 3.7   + 3.908   + 2.109   + 14.202
    ─────   ──────   ──────   ───────
      7.9   9.678   9.1168   46.252

21. 16 + 21.11 + 16.8   53.91   22. 2.002 + 3.7 + 1.66 + 8.4309   15.7929

23. 0.0035 + 24 + 3.70009   27.70359   24. 4.561 + 57 + 3.2 + 0.06067   64.82167

25.   7.6   26.   8.287   27.   8.1   28.   6
    - 2.8   - 2.567   - 3.0012   - 4.6788
    ─────   ──────   ───────   ───────
      4.8   5.72   5.0988   1.3212

29. 15 - 3.5676   11.4324   30. 27.1 - 2.015   25.085

31. 0.46 - 0.4578   0.0022   32. 0.5 - 0.00456   0.49544

33. 0.0125 - 0.0079   0.0046   34. 0.0068 - 0.0009   0.0059

b. $1.1507 = 1\frac{1,507}{10,000}$ = one and one thousand five hundred seven ten-thousandths

c. $3.111111 = 3\frac{111,111}{1,000,000}$ = three and one hundred eleven thousand one hundred eleven millionths

4. **To compare two decimals, first annex zeroes until both numbers have the same number of decimal places.** Then compare the decimals, and write < or > between them.

   a. Compare 0.7 and 0.695.
   Think: 0.7 = 0.700
   0.700 > 0.695, so 0.7 > 0.695

   b. Compare 3.078 and 3.0807.
   Think: 3.078 = 3.0780
   3.0780 < 3.0807, so 3.078 < 3.0807

5. **Addition and subtraction of decimals is much like that of whole numbers.** Align the digits vertically by place value. (Keep the decimal points in a straight line.) Annex zeroes so that all the decimals have the same number of decimal places; then add or subtract as usual. Place the decimal point in the answer directly below the decimal points in the problem.

   Be sure to review subtracting a decimal from a whole number as in problem *d* below.

   a. 2.08   b. 5.7823
     + 0.077   + 3.09
      (2.157)   (8.8723)

   c. 3.07
     – 1.9
      (1.17)

   d. 7 – 2.6187   (4.3813)

Answers for Part B

5. Four and two hundred eleven thousandths
6. Three and eight hundred seventy-one ten-thousandths
7. Two and seven hundred-thousandths
8. Five and one hundred ninety-nine millionths

T–155   Chapter 4   Decimals, Ratios, and Proportions

Solutions for Part F

35. 0.00208 − 0.00125
36. (2 × 0.002) + 0.016
37. 1 − (0.51 + 0.23)
38. 2.0625 − 0.0015
39. 48 = $\frac{4}{5}$ × ___ ; 48 ÷ $\frac{4}{5}$
40. $\frac{9}{16}$ × 80

Solutions for Part H

45. 750 ÷ 8
46. 6 ÷ $1\frac{1}{2}$

F. Solve these reading problems.

35. Harold measured a section of pages in his Bible and divided that measurement by the number of sheets. He concluded that each sheet was 0.00125 inch thick. He did the same with his concordance and found one sheet to be 0.00208 inch thick. How much thicker is the paper in his concordance than in his Bible? 0.00083 inch

36. A roller on a printing press is covered with a metal jacket 0.016 inch thick. Sheets of paper 0.002 inch thick are used for packing under this jacket. What is the total thickness of two packing sheets and the metal jacket? 0.02 inch

37. The weather forecast predicted 1 inch of rainfall. This morning 0.51 inch of rain fell, and an afternoon shower brought 0.23 inch. How much more rain is needed to make exactly 1 inch? 0.26 inch

38. An engine part is specified to be 2.0625 inches long. There is a tolerance of 0.0015 inch over or under the given measurement. What is the smallest acceptable length for this part? 2.061 inches

39. The Landis family is milking 48 cows at present. That is $\frac{4}{5}$ of their entire herd. How many cows do they have? 60 cows

40. The Landises live on an 80-acre farm. Only $\frac{9}{16}$ of the land is tillable. How many acres are tillable? 45 acres

REVIEW EXERCISES

G. Write the reciprocal of each number. *(Lesson 39)*

41. $\frac{5}{8}$ $\frac{8}{5}$ 42. $\frac{7}{16}$ $\frac{16}{7}$ 43. 81 $\frac{1}{81}$ 44. $7\frac{5}{6}$ $\frac{6}{47}$

H. Solve these reading problems containing extra information. *(Lesson 31)*

45. A 75-watt light bulb, rated for 750 hours, burned out after only 450 hours of service. How many 8-hour days should the bulb have lasted before burning out? $93\frac{3}{4}$ days

46. Gilbert spent a total of 9 hours building a bird feeder for Mother's birthday. He worked 6 hours at sawing and sanding the wood pieces, $1\frac{1}{2}$ hours at assembling the feeder, and $1\frac{1}{2}$ hours at applying two coats of finish. How many times longer did it take to saw and sand than to apply the finish? 4 times

48. Decimals in Multiplication

Multiplying decimals is exactly the same process as multiplying whole numbers, with the additional step of placing the decimal point. To multiply when one or both of the factors are decimals, follow these steps.

1. Multiply as with whole numbers.
2. Count the total number of decimal places in both factors together.
3. Put an equal number of decimal places in the product. Annex zeroes to the left of the product if necessary.
4. Drop any final zeroes in the product.

Example A

```
    0.0 3 4
  × 0.2 5
    1 7 0
    6 8
  0.0 0 8 5 0   = 0.0085
```

Because each place in the Arabic number system represents a power of 10 (10; 100; 1,000; and so on), multiplying by a power of 10 is very simple. Just move the decimal point as many places to the right as there are zeroes in the multiplier (Example B). Annex zeroes if necessary to make enough decimal places.

Example B

```
10 × 23.42      = 234.2
100 × 23.42     = 2,342
1,000 × 23.42   = 23,420
10,000 × 23.42  = 234,200
```

CLASS PRACTICE

Solve these multiplication problems. Drop any final zeroes in the products.

a. 5.56 × 0.03 = 0.1668

b. 25.15 × 1.046 = 26.3069

c. 2,102.5 × 0.1504 = 316.216

d. 0.011 × 0.042 = 0.000462

e. 19.6 × 0.15   2.94

f. 126.1 × 1.104   139.2144

Do these multiplications mentally.

g. 100 × 1.54   154

h. 10 × 6.008   60.08

i. 1,000 × 16.04   16,040

j. 100 × 81.4   8,140

LESSON 48

Objectives

- To review multiplying decimals, including those for which zeroes must be annexed to the left of the product.
- To review that when decimals are multiplied, final zeroes in the product may be dropped. (Students are instructed to drop final zeroes in the products in some exercises.)
- To review multiplying by powers of 10 by moving the decimal point to the right as many places as there are zeroes in the multiplier.

Review

1. *Change each decimal to a common fraction in lowest terms.* (Lesson 47)

    a. 0.8 ($\frac{4}{5}$)   b. 0.56 ($\frac{14}{25}$)

2. *Write these decimals, using words.* (Lesson 47)

    a. 0.28 (twenty-eight hundredths)

    b. 15.000095 (fifteen and ninety-five millionths)

3. *Compare each set of decimals, and place < or > between them.* (Lesson 47)

    a. 0.26 (>) 0.258

    b. 2.9 (<) 3.0001

4. *Solve these division problems.* (Lesson 40)

    a. $2\frac{4}{5} \div \frac{2}{5}$ (7)

    b. $\frac{8}{9} \div 7\frac{1}{3}$ ($\frac{4}{33}$)

Introduction

Solve the following multiplication problem on the board. Do not place a decimal point in the product.

$$\begin{array}{r} 5.12 \\ \times\ 3.5 \\ \hline 17920 \end{array}$$

Use estimation to get an idea of what the answer should be (4 × 5 = 20). Ask, "Where is the only sensible place to put the decimal point?" (Between the 7 and the 9. The whole number definitely has two digits.)

Review the rule for determining where to place a decimal point in the product of a multiplication problem. Then count the total number of decimal places in both factors, and place the decimal point correctly in the product.

The students probably do not need this demonstration to remember where to put the decimal point. But it does help them to think through what is happening when they place it.

Teaching Guide

1. **To multiply when one or both of the factors are decimals, follow these steps.**

   (1) Multiply as with whole numbers.

   (2) Count the total number of decimal places in both factors together.

   (3) Put an equal number of decimal places in the product. Annex zeroes to the left of the product if necessary.

   (4) Drop any final zeroes in the product.

   a. 2.5
      × 3.4
      (8.5)

   b. 5.26
      × 1.15
      (6.049)

   c. 0.03
      × 0.7
      (0.021)

   d. 0.015
      × 0.011
      (0.000165)

   e. 0.525
      × 0.004
      (0.0021)

   f. 0.5025
      × 0.008
      (0.00402)

2. **To multiply by a power of 10, move the decimal point as many places to the right as there are zeroes in the multiplier.** Annex zeroes if necessary to make enough decimal places (see *d*, *e*, and *f* below)

   a. 10 × 1.17 (11.7)

   b. 100 × 0.0171 (1.71)

   c. 1,000 × 3.261 (3,261)

   d. 100 × 4.1 (410)

   e. 1,000 × 0.6 (600)

   f. 1,000 × 2.71 (2,710)

Solutions for Part C

25. 5,000 × 0.4

26. 9.408 × $1.249

Lesson 48

WRITTEN EXERCISES

A. Solve these multiplication problems. Drop any final zeroes in the products.

1. 4.9 × 3.7 = 18.13
2. 2.7 × 3.6 = 9.72
3. 4.06 × 3.15 = 12.789
4. 4.71 × 3.09 = 14.5539
5. 2.7178 × 0.03 = 0.081534
6. 3.1412 × 0.015 = 0.047118
7. 0.978 × 0.056 = 0.054768
8. 0.098 × 0.054 = 0.005292
9. 4.5675 × 0.032 = 0.14616
10. 34.518 × 0.076 = 2.623368
11. 1,187.2 × 0.003 = 3.5616
12. 2,357.5 × 0.004 = 9.43
13. 16.3 × 0.23   3.749
14. 15.7 × 1.002   15.7314
15. 15.2 × 3.025   45.98
16. 2.62 × 2.015   5.2793

B. Do these multiplications mentally.

17. 10 × 3.45   34.5
18. 10 × 7.7761   77.761
19. 100 × 3.515   351.5
20. 100 × 0.0231   2.31
21. 100 × 456.124   45,612.4
22. 10 × 4.616   46.16
23. 1,000 × 0.05   50
24. 1,000 × 53.2   53,200

C. Solve these reading problems.

25. Goliath's coat weighed 5,000 shekels of brass (1 Samuel 17:5). If a shekel was equal to 0.4 ounce, what was the weight of Goliath's coat in pounds?   125 pounds

26. Sister Sarah filled her fuel tank on the way home from school. She bought 9.408 gallons of 87-octane gasoline at the price of $1.249 per gallon. What was her bill, to the nearest cent? $11.75

Them that honour me I will honour, and they that despise me shall be lightly esteemed.

1 Samuel 2:30

158   Chapter 4   Decimals, Ratios, and Proportions

27. The Musser family's house was assessed at $25,300. Property tax was charged at the rate of 29.75 mills ($0.02975 per dollar). To the nearest cent, find the tax due on this property. $752.68

28. The Friesen family's house was assessed at $85,600. The rate of their property tax was 13.41 mills ($0.01341 per dollar). To the nearest cent, find the tax due on this property. $1,147.90

29. A 6-acre plot of land is being divided into $\frac{3}{4}$-acre building lots. How many lots can be made from 6 acres? 8 lots

30. In some areas, the minimum size of a building lot is $1\frac{1}{2}$ acres. How many full lots can be made from an 11-acre field? 7 lots

REVIEW EXERCISES

D. Change each decimal to a common fraction in lowest terms. *(Lesson 47)*

31. 0.75 $\frac{3}{4}$
32. 0.52 $\frac{13}{25}$

E. Write these decimals, using words. *(Lesson 47)*

33. 2.69
    Two and sixty-nine hundredths
34. 11.015
    Eleven and fifteen thousandths

F. Compare each set of decimals, and write < or > between them. *(Lesson 47)*

35. 0.1 _<_ 0.11
36. 5.908 _<_ 5.91

G. Solve these division problems. *(Lesson 40)*

37. $1\frac{11}{15} \div \frac{2}{3}$ $2\frac{3}{5}$
38. $6\frac{3}{5} \div \frac{3}{10}$ 22

27. 0.02975 × $25,300
28. 0.01341 × $85,600
29. $6 \div \frac{3}{4}$
30. $11 \div 1\frac{1}{2}$

Chapter 4 Decimals, Ratios, and Proportions

LESSON 49

Objectives

- To review all the kinds of division problems involving decimals. (All answers in this lesson divide evenly by the ten-thousandths' place.)
- To review moving the decimal point to divide decimals by powers of ten.

Review

1. *Solve these multiplication problems. Remember to drop any final zeroes.* (Lesson 48)

    a. 10×0.098 (0.98)

    b. $100{,}000 \times 0.000567$ (56.7)

    c. 2.069
        $\times$ 1.4
        (2.8966)

    d. 0.052
        $\times$ 0.025
        (0.0013)

2. *Write these decimals, using words.* (Lesson 47)

    a. 0.543 (five hundred forty-three thousandths)

    b. 8.00976 (eight and nine hundred seventy-six hundred-thousandths)

3. *Compare each set of decimals, and place < or > between them.* (Lesson 47)

    a. 3.00697 (<) 3.00701

    b. 0.9087 (>) 0.8907

4. *Find the unknown numbers.* (Lesson 41)

    a. $\frac{7}{12}$ is $\frac{7}{9}$ of ___ ($\frac{3}{4}$)

    b. $\frac{1}{10}$ is $\frac{2}{5}$ of ___ ($\frac{1}{4}$)

5. *Find these equivalents.* (Lessons 17, 18)

    a. 20 ft. = ___ in. (240)

    b. 23,760 ft. = ___ mi. ($4\frac{1}{2}$)

    c. 9 fl. oz. ___ tbsp. (18)

    d. 32 pt. = ___ gal. (4)

49. Decimals in Division

Dividing a decimal by a whole number is done in the same way as dividing whole numbers (see Example A). The decimal point in the quotient goes directly above the one in the dividend. If the answer is to be rounded to the nearest cent, annex a zero and divide one place further than the place to which you will round. Then round the answer as indicated.

Example A
Find $276.30 ÷ 42 to the nearest cent.

$$\begin{array}{r} \$6.578 \\ 42\overline{)\$276.300} \end{array} = \$6.58$$

Example B
Divide 0.012 by 1.5, and check by casting out nines.

$$(6)\ \ 1.5_\wedge\overline{)0.0_\wedge120}^{\ 0.008}\ \to\ 3\ \overset{(8)}{\underset{1}{}}$$

Check: 6 × 8 = 48 → 12 → 3

To divide a decimal by a decimal as in Example B, follow these steps.

1. Move the decimal point in the divisor to the far right, and mark its new position with a caret ($_\wedge$).

2. Move the decimal point in the dividend the same number of places as in the divisor, and also mark its new position with a caret. Annex zeroes to the right if needed.

3. Divide as usual. Annex zeroes if they are needed to continue dividing.

4. Place the decimal point in the quotient directly above the caret in the dividend.

Casting out nines can be used to check division problems involving decimals. Simply disregard the decimal points, and proceed as usual. In Example B, the numbers in parentheses are check numbers for casting out nines.

To divide by a power of 10, simply move the decimal point to the left the same number of places as there are zeroes in the divisor. Annex zeroes if necessary.

$$1.5 ÷ 10 = 0.15$$
$$1.5 ÷ 100 = 0.015$$
$$1.5 ÷ 1{,}000 = 0.0015$$
$$1.5 ÷ 10{,}000 = 0.00015$$

Chapter 4 Decimals, Ratios, and Proportions

CLASS PRACTICE

Solve these division problems; all answers divide evenly by the ten-thousandths' place. If an answer is a money amount, round it to the nearest cent.

a. 4)4.32 → 1.08
b. $2.50)$16.75 → 6.7
c. 1.6)$15.42 → $9.64
d. 0.011)115.5 → 10,500

Solve these division problems mentally.

e. 18.6 ÷ 10 → 1.86
f. 0.248 ÷ 100 → 0.00248
g. 1.59 ÷ 1,000 → 0.00159
h. 184.05 ÷ 1,000 → 0.18405

WRITTEN EXERCISES

A. *Solve these division problems; all answers divide evenly by the ten-thousandths' place. If an answer is a money amount, round it to the nearest cent.*

1. 8)3.56 → 0.445
2. 6)3.21 → 0.535
3. 4)0.414 → 0.1035
4. 16)$0.28 → $0.02
5. 15)$12.48 → $0.83
6. 2.5)6 → 2.4
7. 0.6)2.73 → 4.55
8. $0.40)$2.68 → 6.7
9. $0.08)$4.45 → 55.625
10. 0.028)0.14 → 5
11. 0.015)0.066 → 4.4
12. 0.03)0.045 → 1.5

B. *Solve these division problems mentally.*

13. 23.1 ÷ 10 → 2.31
14. 45.66 ÷ 100 → 0.4566
15. 1.575 ÷ 10 → 0.1575
16. 476.543 ÷ 1,000 → 0.476543
17. 6.657 ÷ 100 → 0.06657
18. 0.3 ÷ 100 → 0.003
19. 0.45 ÷ 100 → 0.0045
20. 7.3 ÷ 1,000 → 0.0073

C. *Solve these reading problems.*

21. A tool wholesaler advertised an assortment of 1,000 nuts, bolts, screws, and washers for $13.00. Calculate mentally the price per item. (Do not round your answer.) $0.013

22. A school supply catalog advertised 100 school tablets for $50.95. Calculate mentally the price per tablet, to the nearest cent. $0.51

23. In one year a kestrel (sparrow hawk) weighing 0.44 pound ate 12.32 pounds of food. How many times its weight did the kestrel eat in one year?   28 times

24. A dog weighing 20 pounds ate 182.5 pounds of food in one year. How many times its own weight did the dog eat in one year?   9.125 times

Introduction

Write the problem 10 ÷ 0.25 on the board. See if the students can give examples for which this problem would be the solution. Here are some suggestions.

a. How many quarters are in 10 dollars?

b. How many $\frac{1}{4}$-yard strips of cloth can be cut from 10 yards of cloth?

c. How many sheep can be grazed in a 10-acre meadow if $\frac{1}{4}$ acre is allowed for each sheep?

d. If a certain jet travels an average of $\frac{1}{4}$ mile per gallon of jet fuel, how many gallons does it need to fly 10 miles?

Ask the students to consider the first question: "How many quarters are in 10 dollars?" Will the answer be more or less than 10 quarters? (more) How do they know? (A quarter dollar is less than a whole dollar, so there will be more quarter dollars than whole dollars in any amount of money.)

Have the students tell how many quarters are in 10 dollars. Because there are 4 quarters in 1 dollar, there are 40 quarters in 10 dollars.

Now set up and solve the problem below. How should the decimal point be placed? Because the answer is 40, obviously two zeroes must be annexed to the right of the dividend, and the decimal point placed after it.

$$0.25 \overline{)10}$$

Solutions for Part C

21. $13.00 ÷ 1,000
22. $50.95 ÷ 100
23. 12.32 ÷ 0.44
24. 182.5 ÷ 20

Teaching Guide

1. **Dividing a decimal by a whole number is done in the same way as dividing whole numbers.** The decimal point in the quotient goes directly above the one in the dividend. If the answer is to be rounded to the nearest cent, annex a zero and divide one place further than the place to which you will round. Then round the answer as indicated.

    a. $24 \overline{)0.036}$ (0.0015)

    b. $7 \overline{)\$6.07}$ ($0.867 = $0.87)

2. **To divide a decimal by a decimal, follow these steps.**

    (1) Move the decimal point to the far right of the divisor and mark its new position with a caret ($_\wedge$).

    (2) Move the decimal point in the dividend the same number of places as in the divisor, and also mark its new position with a caret. Annex zeroes to the right if needed.

    (3) Divide as usual. Annex zeroes if they are needed to continue dividing.

    (4) Place the decimal point in the quotient directly above the caret in the dividend.

    Show that moving the decimal point an equal number of places in the dividend and in the divisor does not change the quotient. This has the effect of multiplying both dividend and divisor by a power of 10; and as with the double-and-double shortcut, the quotient stays the same.

    $8 \overline{)40}$ $80 \overline{)400}$

    $800 \overline{)4,000}$ $8,000 \overline{)40,000}$

    (each with quotient 5)

    Tell students to be alert for

T–161   Chapter 4   Decimals, Ratios, and Proportions

problems in which zeroes must be annexed to the dividend (example *a* below) and for problems in which the quotient has one or more zeroes to the immediate right of the decimal point (example *b*).

a. $2.2\overline{)11}^{(5)}$ b. $1.2\overline{)0.078}^{(0.065)}$

3. **To divide by a power of 10, move the decimal point to the left the same number of places as there are zeroes in the divisor.** Annex zeroes if necessary.

   a. $1.117 \div 10$ (0.1117)
   b. $131.17 \div 100$ (1.3117)
   c. $4.2 \div 1{,}000$ (0.0042)

---

25. $1\frac{3}{4} = \frac{2}{3} \times \underline{}$; $1\frac{3}{4} \div \frac{2}{3}$

26. $1\frac{1}{2} = \frac{3}{4} \times \underline{}$; $1\frac{1}{2} \div \frac{3}{4}$

Answers for Part F

31. Six and thirty-seven ten-thousandths

32. Eight and two thousand six hundred eight ten-thousandths

25. After reading a total of $1\frac{3}{4}$ hours in his spare time, David had read $\frac{2}{3}$ of a 100-page book. At that rate, how long should it take him to read the entire book? $2\frac{5}{8}$ hours

26. Richard weeded $\frac{3}{4}$ of the garden in $1\frac{1}{2}$ hours. At that rate, how long would it take to weed the whole garden?   2 hours

REVIEW EXERCISES

D. *Solve by mental multiplication. (Lesson 48)*

27. 10 × 0.589   5.89   **28.** 100 × 0.0054   0.54

E. *Solve these multiplication problems. Remember to drop any final zeroes. (Lesson 48)*

29. 2.008
      × 1.25
      ────
       2.51

30. 8.75
      × 5.64
      ────
      49.35

F. *Write these decimals, using words. (Lesson 47)* (See facing page.)

31. 6.0037   **32.** 8.2608

G. *Compare each set of decimals, and write < or > between them. (Lesson 47)*

33. 0.509 __<__ 0.59   **34.** 3.801 __>__ 3.8007

H. *Find the unknown numbers. (Lesson 41)*

35. $\frac{5}{12}$ is $\frac{3}{10}$ of $1\frac{7}{18}$ **36.** $1\frac{3}{4}$ is $\frac{6}{7}$ of $2\frac{1}{24}$

I. *Find these equivalents. (Lessons 17, 18)*

37. 150 in. = __$12\frac{1}{2}$__ ft.   **38.** 219 yd. = __657__ ft.

39. 3 bu. = __12__ pk.   **40.** 84 fl. oz. = __$10\frac{1}{2}$__ cups

50. Rounding Decimals and Expressing Fractions as Decimals

Rounding Decimals

To round a decimal, follow the same basic rules as when rounding a whole number. The steps are given below.

1. Choose the place to which you will round the number.

2. Look at the digit to the right of the place being rounded. If the digit to the right has a value of 5 or more, add 1 to the digit being rounded. If the digit to the right has a value of 4 or less, do not change the digit being rounded.

3. Drop all digits to the right of the place being rounded. If the digit you are rounding is zero, keep the zero to show that the decimal was rounded to that place.

> Round each number to the place indicated.
> **Example A:** 57.34 to the nearest tenth = 57.3
> **Example B:** 2.5358 to the nearest hundredth = 2.54
> **Example C:** 15.6597 to the nearest thousandth = 15.660

Expressing Fractions as Decimals

Decimal fractions and common fractions are two ways to express parts of a number. A decimal may be a mixed number as well as a value less than 1. Changing a decimal to a fraction or a fraction to a decimal can often simplify math problems.

A fraction indicates division; therefore, to change a fraction to its decimal equivalent, divide the numerator by the denominator. If a fraction divides evenly, the result is a **terminating decimal**. Such a decimal is the exact equivalent of the fraction.

> **Example D**
> Change $\frac{11}{40}$ to its decimal equivalent.
>
> $$\begin{array}{r} 0.275 \\ 40\overline{)11.000} \\ \underline{80} \\ 300 \\ \underline{280} \\ 200 \\ \underline{200} \\ 0 \end{array}$$

LESSON 50

Objectives

- To review rounding decimals.
- To review using division to express fractions in decimal form—both rounding off remainders and using a vinculum to indicate repeating decimals.
- To review memorization of the decimal equivalents of common fractions (one-half and the thirds, fourths, fifths, eighths, and tenths); and to memorize *the sixths. (Students will need to know these equivalents by memory for the test.)

Review

1. Give Lesson 50 Quiz (Multiplying Decimals).

2. *Solve these problems mentally.* (Lessons 48, 49)
   a. 100×1.008 (100.8)
   b. $1{,}000 \times 154.02$ (154,020)
   c. $0.05 \div 100$ (0.0005)
   d. $2.15 \div 1{,}000$ (0.00215)

3. *Do these divisions with decimals.* (Lesson 49)
   a. $1.2\overline{)0.09}$ (0.075)
   b. $2.6\overline{)0.5408}$ (0.208)

4. *Solve by mental division.* (Lesson 42)
   a. $21 \div \frac{1}{2}$ (42)
   b. $13 \div \frac{1}{3}$ (39)
   c. $15 \div \frac{5}{6}$ (18)
   d. $21 \div \frac{3}{7}$ (49)

5. *Find these equivalents.* (Lessons 19, 20)
   a. 65,340 sq. ft. = ___ a. ($1\frac{1}{2}$)
   b. 1,296 sq. in. = ___ sq. ft. (9)
   c. 60 decades = ___ centuries (6)
   d. 15 min. = ___ sec. (900)

T-163 Chapter 4 Decimals, Ratios, and Proportions

Introduction

Have the students get out their metric rulers and show you about how long 6.935 centimeters is. Which of the following measures is it closest to? If they seem confused, ask them how many tenths 6.935 is nearest to. What is $\frac{1}{100}$ of a centimeter?

a. 6.9 centimeters (69 millimeters)

b. 7 centimeters (70 millimeters)

Are these measures exactly 6.935 centimeters? Then how did the students know that the length was closer to 6.9 centimeters? Those who gave this answer rounded the length to the nearest millimeter (tenth of a centimeter).

Have the students look at their rulers again, and find $\frac{3}{4}$ decimeter. This time they should all point to 7.5 centimeters (75 millimeters). If the students have problems thinking this through, ask the following questions.

a. How many millimeters are in a decimeter? (100)

b. How many millimeters is $\frac{3}{4}$ of 100? (75)

In these exercises with metric units, the students have worked with both parts of today's lesson. In the first problem they rounded a decimal fraction. In the second problem they changed a fraction to a decimal.

Teaching Guide

1. **To round a decimal, follow the same basic rules as when rounding a whole number.**

   (1) Choose the place to which you will round the number.

   (2) Look at the digit to the right of the place being rounded. If the digit to the right has a value of 5 or more, add 1 to the digit being rounded. If the digit to the right has a value of 4 or less, do not change the digit being rounded.

   (3) Drop all digits to the right of the place being rounded. If the digit you are rounding is zero, keep the zero to show that the decimal was rounded to that place.

   Have students round the following numbers as indicated. Make special mention of exercises c and d, where a zero is written in the place to which the decimals are rounded.

   a. 1.2134 to the nearest thousandth (1.213)

   b. 2.189 to the nearest hundredth (1.19)

   c. 5.13012 to the nearest thousandth (5.130)

   d. 2.619934 to the nearest ten-thousandth (2.6200)

2. **To change a fraction to its decimal equivalent, divide the numerator by the denominator.** If the fraction divides evenly, the result is a terminating decimal. Such a decimal is the exact equivalent of the fraction. All the examples below terminate by the hundred-thousandths' place.

   a. $\frac{1}{16}$ = ___ (0.0625)

   b. $\frac{1}{8}$ = ___ (0.125)

   c. $\frac{7}{16}$ = ___ (0.4375)

   d. $\frac{3}{32}$ = ___ (0.09375)

If a fraction does not divide evenly, the result is a **nonterminating decimal**. Such a division can never be completed; rather, a certain digit or series of digits in the quotient begins to repeat. For this reason, nonterminating decimals are also called **repeating decimals**. These decimal equivalents are expressed by one of the following methods.
1. Express the remainder as a fraction, as in $0.33\frac{1}{3}$.
2. Round the quotient to a certain place. See Example E.
3. Use a vinculum (bar) to indicate the repeating digits. See Example F.

Example E	**Example F**
Express $\frac{4}{15}$ as a decimal rounded to the nearest hundredth.	Express $\frac{5}{33}$ as a decimal. Indicate repeating digits with a vinculum.
$$\begin{array}{r} 0.266 \\ 15\overline{)4.000} \\ \underline{30} \\ 100 \\ \underline{90} \\ 100 \\ \underline{90} \\ 10 \end{array}$$ = 0.27 (to the nearest hundredth)	$$\begin{array}{r} 0.151 \\ 33\overline{)5.000} \\ \underline{33} \\ 170 \\ \underline{165} \\ 50 \\ \underline{33} \\ 17 \end{array}$$ = $0.\overline{15}$

In Example F, the subtraction 170–165 leaves a remainder of 5. The original dividend was 5; so when we obtain a remainder of 5 (which becomes the dividend for the next step), we know that the quotient figures will repeat (0.151515 . . .).

Some of the most commonly used fractions and their decimal equivalents are shown on the bars below. Memorize these equivalents.

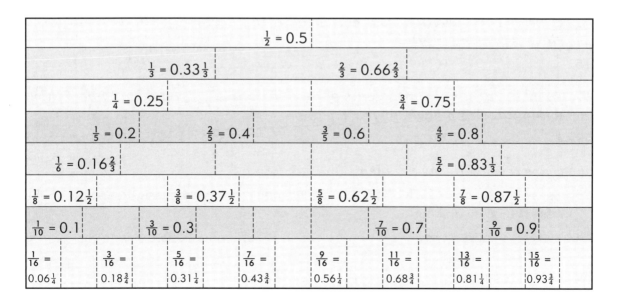

164 Chapter 4 Decimals, Ratios, and Proportions

CLASS PRACTICE

Round each decimal to the place indicated.

a. 2.861 (tenths)   2.9
b. 0.1532 (hundredths)   0.15
c. 3.9968 (hundredths)   4.00
d. 5.159951 (ten-thousandths)   5.1600

Express each fraction as a decimal. If an answer does not divide evenly by the ten-thousandths' place, round it to the nearest thousandth.

e. $\frac{3}{40}$ 0.075
f. $\frac{79}{80}$ 0.9875
g. $\frac{23}{30}$ 0.767
h. $\frac{10}{17}$ 0.588

Express each fraction as a nonterminating decimal, with a vinculum to show repeating digits.

i. $\frac{8}{9}$ $0.\overline{8}$
j. $\frac{7}{11}$ $0.\overline{63}$
k. $\frac{9}{22}$ $0.4\overline{09}$
l. $\frac{14}{99}$ $0.\overline{14}$

WRITTEN EXERCISES

A. *Round each decimal to the place indicated.*

1. 0.345 (tenths)   0.3
2. 0.999 (tenths)   1.0
3. 0.467 (hundredths)   0.47
4. 0.07438 (thousandths)   0.074
5. 3.26811 (thousandths)   3.268
6. 5.6199 (thousandths)   5.620
7. 5.27996 (ten-thousandths)   5.2800
8. 2.36996 (ten-thousandths)   2.3700

B. *Express each fraction as a decimal. If an answer does not divide evenly by the thousandths' place, round it to the nearest hundredth.*

9. $\frac{3}{16}$ 0.19
10. $\frac{19}{400}$ 0.05
11. $\frac{11}{15}$ 0.73
12. $\frac{17}{18}$ 0.94
13. $\frac{22}{7}$ 3.14
14. $\frac{31}{11}$ 2.82

C. *Express each fraction as a nonterminating decimal, with a vinculum to show repeating digits.*

15. $\frac{7}{9}$ $0.\overline{7}$
16. $\frac{2}{15}$ $0.1\overline{3}$
17. $\frac{4}{33}$ $0.\overline{12}$
18. $\frac{9}{55}$ $0.1\overline{63}$

D. *Write the decimal equivalents of these fractions. If you do not know them without looking back, memorize them.*

19. $\frac{5}{8}$ $0.62\frac{1}{2}$
20. $\frac{3}{8}$ $0.37\frac{1}{2}$
21. $\frac{5}{6}$ $0.83\frac{1}{3}$
22. $\frac{1}{8}$ $0.12\frac{1}{2}$

E. *Solve these reading problems.*

23. The earth requires 23.934 hours to complete one rotation on its axis. Round this to the nearest tenth of an hour.   23.9 hours

24. Jupiter completes one rotation on its axis in 9.925 hours. Round this to the nearest hundredth of an hour.   9.93 hours

25. Mercury orbits the sun in $\frac{6}{25}$ of an earth year. Convert this number to a decimal.   0.24

If a fraction does not divide evenly, the result is a nonterminating decimal, also called a repeating decimal. In the decimal equivalent of $\frac{1}{3}$, the digit 3 repeats endlessly. In the decimal equivalent of $\frac{1}{7}$, the digits 142857 repeat endlessly. Nonterminating decimals are expressed in one of the following ways.

(1) By expressing the remainder as a fraction, as in $0.33\frac{1}{3}$.

(2) By rounding the quotient to a certain place.

(3) By using a vinculum (bar) to indicate the repeating digits.

Emphasize that only when the remainder is equal to the dividend in a previous division can one be certain that the repetition has been found. Illustrate with the examples below and in the pupil's lesson.

Express each fraction as a decimal with two places. Show the remainder as a fraction.

e. $\frac{7}{15}$ $(0.46\frac{2}{3})$ f. $\frac{5}{9}$ $(0.55\frac{5}{9})$

Express each fraction as a decimal rounded to the nearest thousandth.

g. $\frac{1}{13}$ (0.077) h. $\frac{17}{24}$ (0.708)

Express each fraction as a decimal. Indicate repeating digits with a vinculum.

i. $\frac{4}{15}$ $(0.2\overline{6})$ j. $\frac{5}{7}$ $(0.\overline{714285})$

(All the sevenths repeat the sequence 142857, but each starts at a different place. For example, $\frac{1}{7} = 0.\overline{142857}$, $\frac{3}{7} = 0.\overline{428571}$, and $\frac{4}{7} = 0.\overline{571428}$.)

3. **It is useful to know the decimal equivalents of commonly used fractions.** Introduce the equivalents for the sixths, and direct the students to thoroughly memorize all the equivalents in the box at the end of the lesson text.

Note that each decimal equivalent in the box is either an exact decimal with one or two places, or a two place decimal with a fraction. This is for the sake of simplicity as well as convenience, since the equivalents are often expressed as percents. Point out that an equivalent like $0.18\frac{3}{4}$ can also be expressed as 0.1875; in fact, that is the required form when using a calculator. Since the decimal equivalent of $\frac{3}{4}$ is 0.75, one simply replaces the fraction with "75."

Further Study

Concerning problem 23, students may logically ask this question: "If each rotation takes less than 24 hours, how do our 24-hour clocks keep correct time? Shouldn't noon come earlier and earlier by about 4 minutes every day?"

The reason is that while the earth is rotating on its axis, it is also revolving about the sun. At noon for a given location on earth, the earth is at a slightly different place in its orbit from where it was at noon on the previous day. Since its rotation and revolution are both in the same direction, the earth must turn slightly more than 360 degrees from one noon to the next (when the sun is highest in the sky). Such a rotation, from one noon to another, takes exactly 24 hours. But it takes only 23.934 hours for the earth to rotate 360 degrees.

T-165　Chapter 4　Decimals, Ratios, and Proportions

Solutions for Exercises 27 and 28

27. $250 \div 0.125$

28. $23.75 \div 0.25$

26. The orbit of Saturn requires $29\frac{23}{100}$ earth years. Convert this number to a decimal.
29.23

27. At nesting time, many birds claim a certain area as their own. The nesting area of a kestrel (sparrow hawk) is 250 acres and that of a mallard duck is 0.125 acre. How many times larger is the nesting territory of the kestrel than that of the mallard duck?
2,000 times

28. A pair of hairy woodpeckers will claim about 0.25 acre as their nesting territory. How many pairs of hairy woodpeckers could nest in a woodland of 23.75 acres?   95 pairs

REVIEW EXERCISES

F. Solve these problems mentally. *(Lessons 48, 49)*

29. 10 × 2.45   24.5   **30.** 5.745 ÷ 100   0.05745

G. Solve these division problems. *(Lesson 49)*

31. 7)1.456   0.208   **32.** 1.1)0.1078   0.098

H. Solve each multiplication problem. Drop any final zero in the product. *(Lesson 48)*

33. 3.6
       × 1.9
        6.84

34. 0.746
       × 0.045
        0.03357

I. Solve by mental division. *(Lesson 42)*

35. $15 \div \frac{1}{5}$ 75   **36.** $12 \div \frac{2}{3}$ 18

J. Identify each number as *prime* or *composite*. *(Lesson 34)*

37. 52   composite   **38.** 54   composite

K. Find the prime factors of these composite numbers. Use exponents to express repeating factors. *(Lesson 34)*

39. 24 $24 = 2^3 \times 3$ **40.** 46 $46 = 2 \times 23$

L. Find these equivalents. *(Lessons 19, 20)*

41. 4 a. = _____ sq. ft.   174,240   **42.** 63 sq. ft. = _7_ sq. yd.

43. 5 millennia = _50_ centuries   **44.** 54,000 sec. = _15_ hr.

166 Chapter 4 *Decimals, Ratios, and Proportions*

51. Multiplying by the Simpler Method

Because you have worked with decimals for a few years, you may wonder why anyone would calculate with fractions. After all, decimal calculations follow the same rules as those for whole numbers except for placing the decimal point.

However, sometimes calculating with fractions is much simpler than calculating with equivalent decimals. This is especially true when the fractional equivalent is a unit fraction (Example A) or an easy-to-calculate fraction (Examples B, C, and D). It is much easier to multiply by $\frac{1}{3}$ (divide by 3) than to multiply by $0.33\frac{1}{3}$. And it is often simpler to multiply by $\frac{3}{8}$ than to multiply by 0.375.

Example A
  0.125 of 72 = ___
  Think: 0.125 = $\frac{1}{8}$
  72 ÷ 8 = 9

Example B
  0.375 of 32 = ___
  Think: 0.375 = $\frac{3}{8}$
  32 ÷ 8 = 4, × 3 = 12

Example C
  0.7 of 70 = ___
  Think: 0.7 = $\frac{7}{10}$
  70 ÷ 10 = 7, × 7 = 49

Example D
  $0.66\frac{2}{3}$ of 27 = ___
  Think: $0.66\frac{2}{3}$ = $\frac{2}{3}$
  27 ÷ 3 = 9, × 2 = 18

Each multiplication above is especially simple because the denominator of the fraction is a factor of the whole number. To simplify multiplications in this way, you must thoroughly memorize the equivalents on the chart in Lesson 50.

CLASS PRACTICE

For each pair, give the number of the problem that is easier to solve mentally.

a. (1) 0.8 of 24   (2) $\frac{4}{5}$ of 24   b. (1) $0.33\frac{1}{3}$ of 27   (2) $\frac{1}{3}$ of 27
c. (1) 0.875 of 32   (2) $\frac{7}{8}$ of 32   d. (1) 0.4 of 43   (2) $\frac{2}{5}$ of 43

(Circled answers: a. (1), b. (2), c. (2), d. (1))

Solve mentally by using the fractional equivalent of each decimal.

e. 0.8 of 40   32   f. $0.83\frac{1}{3}$ of 42   35   g. 64 × 0.625   40
h. 0.375 of 16   6   i. $0.66\frac{2}{3}$ of 18   12   j. 56 × 0.125   7

Solve by using decimals.

k. 0.9 of 21   18.9   l. 37 × 0.2   7.4   m. 0.4 of 28   11.2
n. 0.625 of 25   15.625   o. 62 × 0.7   43.4   p. 0.75 of 33   24.75

LESSON 51

Objective

- To practice simplifying multiplication by changing decimals to common fractions.

Review

1. *Express each fractions as a decimal. If an answer does not divide evenly by the ten-thousandth's place, round it to the nearest thousandth.* (Lesson 50)

   a. $\frac{25}{32}$ (0.781)

   b. $\frac{23}{34}$ (0.676)

2. *Round each decimal as indicated.* (Lesson 50)

   a. 1.049 to the nearest tenth (1.0)

   b. 4.10971 to the nearest thousandth (4.110)

3. *Solve these problems. All answers divide evenly by the ten-thousandths' place.* (Lesson 49)

   a. $3.2 \overline{)0.04}$ (0.0125)

   b. $0.6 \overline{)0.003}$ (0.005)

4. *Find the greatest common factor and the lowest common multiple of each pair.* (Lesson 35)

   a. 56, 70 (g.c.f. = 14; l.c.m. = 280)

   b. 18, 54 (g.c.f. = 18; l.c.m. = 54)

5. *Find these equivalents.* (Lesson 21)

   a. 15 dkm = ___ dm (1,500)

   b. 1,080 cm = ___ m (10.8)

Introduction

Ask, "Do you prefer multiplying with decimals or with fractions?" Typically, the students will choose multiplying with decimals.

Write the problems 48 × 0.875 and 48 × $\frac{7}{8}$ on the board. Assign half the class the decimal problem and the other half the fraction problem to see which half gets the answer first. It should take only a few seconds for the ones using the fraction to solve the problem. They will probably have the answer before the others have the problem copied.

Sometimes it is much easier to calculate with fractions than to use equivalent decimals. This lesson challenges the students not to be fixed on using only one method to solve problems, but rather to be flexible in choosing the simpler method of solution.

Solutions for Part E

23. 0.8 × 45
24. $\frac{3}{8}$ × 24
25. $\frac{7}{8}$ × 32
26. $\frac{2}{3}$ × 21
27. 3 × 0.65
28. $3.95 ÷ 25

Teaching Guide

1. **Multiplying with a fraction is sometimes much simpler than multiplying with an equivalent decimal.** This is especially true in the following cases.

   (1) *If the fraction is a unit fraction.*

   a. 38 × 0.5 = 38 × $\frac{1}{2}$ = 38 ÷ 2 = 19
   b. 44 × 0.25 = 44 × $\frac{1}{4}$ = 44 ÷ 4 = 11

   (2) *If the denominator of the fraction is a factor of the whole number.*

   c. 48 × 0.75 = 48 × $\frac{3}{4}$
      = 48 ÷ 4 × 3 = 36
   d. 35 × 0.8 = 35 × $\frac{4}{5}$
      = 35 ÷ 5 × 4 = 28

   (3) *If the decimal equivalent includes a fraction.* This applies especially to the equivalents of the thirds and sixths.

   e. 36 × 0.33$\frac{1}{3}$ = 36 × $\frac{1}{3}$
      = 36 ÷ 3 = 12
   f. 24 × 0.83$\frac{1}{3}$ = 24 × $\frac{5}{6}$
      = 24 ÷ 6 × 5 = 20

   For a person trying to avoid division, it does work sometimes to multiply with decimals. (See problems *a–d* above.) But division cannot be avoided with a fraction like $\frac{1}{3}$. (See problem *e*.) If you multiply by 0.33$\frac{1}{3}$, the first thing you must do is divide by 3 to find $\frac{1}{3}$. So you may as well divide by 3 at the start!

2. **To multiply a whole number by a fraction, remember to divide the whole number by the denominator and multiply the quotient by the numerator.**

3. **To simplify multiplications by using fractions, one must thoroughly memorize the decimal–fraction equivalents.** Review the equivalents in Lesson 50. Are the students memorizing them?

Solve by using fractions or decimals, whichever is simpler.

q. 0.6 of 35   21
r. 0.6 of 39   23.4
s. $42 \times 0.16\frac{2}{3}$ 7
t. 0.125 of 30   3.75
u. 0.375 of 32   12
v. $36 \times 0.83\frac{1}{3}$ 30

WRITTEN EXERCISES

A. *For each pair, write the letter of the problem that is easier to solve mentally.*

1. a. 0.5 of 20   ⓑ. $\frac{1}{2}$ of 20
2. ⓐ. 0.6 of 47   b. $\frac{3}{5}$ of 47
3. a. 0.625 of 48   ⓑ. $\frac{5}{8}$ of 48
4. a. 0.8 of 55   ⓑ. $\frac{4}{5}$ of 55

B. *Solve mentally by using the fractional equivalent of each decimal.*

5. 0.2 of 60   12
6. 0.125 of 24   3
7. 0.5 of 74   37
8. 0.6 of 45   27
9. 0.8 of 55   44
10. 0.375 of 48   18

C. *Solve by using decimals.*

11. 0.8 of 27   21.6
12. 0.75 of 39   29.25
13. 0.375 of 26   9.75
14. 0.4 of 28   11.2
15. 0.25 of 78   19.5
16. 0.6 of 57   34.2

D. *Solve by using fractions or decimals, whichever is simpler.*

17. 0.25 of 36   9
18. 0.75 of 40   30
19. 0.75 of 31   23.25
20. 0.4 of 33   13.2
21. 0.6 of 35   21
22. 0.375 of 64   24

E. *Solve these reading problems.*

23. Neal washed 0.8 of the 45 milk buckets in the calf barn before breakfast. How many buckets did he wash?   36 buckets

24. On a long trip, the Kauffmans traveled for 0.375 of a whole day. How many hours was that?   9 hours

25. One month Father completed 0.875 of the 32 pieces of furniture for which he had orders. How many pieces of furniture did he complete that month?   28 pieces

26. Dwight is $0.66\frac{2}{3}$ as old as his brother Luke, who is 21 years old. What is Dwight's age?   14 years

27. William tightened a V-belt by removing 3 shims between the halves of one pulley. The thickness of each shim was 0.65 millimeter. What was the total thickness of the shims he removed?   1.95 mm

28. Father purchased a box of 25 carriage bolts for $3.95. Find the cost of each bolt, to the nearest tenth of a cent. $0.158 (15.8¢)

168   Chapter 4   Decimals, Ratios, and Proportions

REVIEW EXERCISES

F. Change each fraction to a decimal. If an answer does not divide evenly by the ten-thousandths' place, round it to the nearest thousandth. *(Lesson 50)*

29. $\frac{37}{40}$ 0.925   30. $\frac{13}{46}$ 0.283

G. Round each decimal to the place indicated. *(Lesson 50)*

31. 0.0954 (hundredths)   0.10   32. 11.05071 (thousandths)   11.051

H. Solve these problems. All answers divide evenly by the ten-thousandths' place. *(Lesson 49)*

33. 4.5)0.0945   0.021   34. 0.08)0.025   0.3125

I. Write these decimals, using words. *(Lesson 47)*

35. 1.805   36. 2.0082
    One and eight hundred five thousandths   Two and eighty-two ten-thousandths

J. Find the greatest common factor of each pair. *(Lesson 35)*

37. 56, 84   28   38. 36, 63   9

K. Find the lowest common multiple of each pair. *(Lesson 35)*

39. 12, 28   84   40. 16, 24   48

L. Find these equivalents. *(Lesson 21)*

41. 12 dkm = ____ cm   12,000   42. 1,580 km = ____ dkm   158,000
43. 500 hm = ____ dm   500,000   44. 49 m = 0.49 hm

Lesson 51 T-168

T-169 Chapter 4 Decimals, Ratios, and Proportions

LESSON 52

Objectives

- To review writing ratios and reducing them to lowest terms.
- To review the terms *antecedent* and *consequent*.
- To teach *expressing a ratio with either the antecedent or the consequent as 1 and the other part in fraction or decimal form.

Review

1. *Solve mentally by using the fractional equivalent of each decimal.* (Lesson 51)
   a. 0.2 of 75 (15)
   b. $0.33\frac{1}{3}$ of 81 (27)
   c. $0.83\frac{1}{3}$ of 60 (50)
   d. 0.875 of 72 (63)

2. *Express these fractions as decimals. If an answer does not divide evenly by the ten-thousandth's place, round it to the nearest thousandth.* (Lesson 50)
   a. $\frac{5}{14}$ (0.357)
   b. $\frac{3}{17}$ (0.176)

3. *Reduce these fractions to lowest terms.* (Lesson 36)
   a. $\frac{20}{48}$ ($\frac{5}{12}$)
   b. $\frac{13}{39}$ ($\frac{1}{3}$)

4. *Find these equivalents.* (Lessons 22, 23)
   a. 8.14 kg = ___ g (8,140)
   b. 150 kg = ___ MT (0.15)
   c. 1.5 *l* ___ ml (1,500)
   d. 12 *l* = ___ kl (0.012)

52. Ratios: Tools of Comparison

A **ratio** is an expression that shows the relationship between two numbers. For example, the ratio of shaded blocks to unshaded blocks in the following illustration is 6 to 9. Ratios can be reduced to lowest terms in the same way as fractions; the ratio 6 to 9 in lowest terms is 2 to 3. This means there are 2 shaded blocks for every 3 unshaded blocks.

The first number in a ratio is the **antecedent**, and the second number is the **consequent**. Ratios can be written in three forms: as a fraction, with a colon, or with the word *to*.

$$\frac{6}{9} \begin{array}{l} \text{antecedent} \\ \text{consequent} \end{array}$$

Reduced to lowest terms: $\frac{6}{9} = \frac{2}{3}$

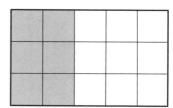

Three forms of ratios: $\frac{2}{3}$ 2:3   2 to 3   All three forms are read "2 to 3."

Ratios are often written like proper fractions. However, if the antecedent is larger than the consequent, a ratio looks like an improper fraction. The ratio of unshaded blocks to shaded blocks above is $\frac{9}{6}$ or $\frac{3}{2}$.

Ratios are sometimes reduced so that either the antecedent or the consequent is 1. The other part may then be in fraction or decimal form. For example, the ratio of shaded blocks to unshaded blocks above could be written as 1 to $1\frac{1}{2}$, or 1 to 1.5.

The following steps explain how to reduce a ratio so that one of the parts is 1.

1. Write the ratio in fraction form.

2. If the antecedent is to be expressed as 1, divide it by itself to obtain 1. Divide the consequent by the same number.

   If the consequent is to be expressed as 1, divide it by itself to obtain 1. Divide the antecedent by the same number.

Example A

Reduce 4:7 so that the antecedent is 1. Express the consequent in fraction form.

$$\frac{4 \div 4 = 1}{7 \div 4 = 1\frac{3}{4}} \quad \text{Answer: 1 to } 1\frac{3}{4}$$

Example B

Reduce 12:5 so that the consequent is 1. Express the antecedent in decimal form.

$$\frac{12 \div 5 = 2.4}{5 \div 5 = 1} \quad \text{Answer: 2.4 to 1}$$

170 Chapter 4 Decimals, Ratios, and Proportions

CLASS PRACTICE

Identify the antecedent and consequent of each ratio. Then reduce the ratio to lowest terms.

a. 19 to 76   1 to 4   b. 36 to 42   6 to 7   c. 18:48   3:8   d. $\frac{45}{54}$ $\frac{5}{6}$

Give a ratio for each statement. Then reduce the ratio to lowest terms.

e. Kevin read 75 pages in a book with 125 pages.   75 to 125; 3 to 5

f. Jonathan ran 160 feet in 10 seconds.   160 to 10; 16 to 1

g. The output of an electric transformer is 115 volts compared to an input of 460 volts.
   115 to 460; 1 to 4

Reduce each ratio as indicated.

h. 7 to 8 (consequent = 1; antecedent in decimal form)   0.875 to 1

i. 5 to 7 (antecedent = 1; consequent in decimal form)   1 to 1.4

j. 2 to 3 (consequent = 1; antecedent in fraction form) $\frac{2}{3}$ to 1

k. 9 to 5 (antecedent = 1; consequent in fraction form)   1 to $\frac{5}{9}$

WRITTEN EXERCISES

A. *Reduce each ratio to lowest terms, and write your answer in the same form as in the exercise. Label each part of the ratio with a (antecedent) or c (consequent).*

1. 15 to 21   5 to 7
2. 18:24   3:4
3. 25:60   5:12
4. $\frac{7}{14}$ $\frac{1}{2}$
5. 51 to 85   3 to 5
6. 65:91   5:7

B. *Give a ratio in lowest terms for each statement.*

7. Brent has memorized 6 of the 15 verses assigned for Bible memory.   2 to 5

8. Dwight is riding his bicycle at the rate of 125 feet in 5 seconds.   25 to 1

9. Marcus walked 22 feet in 5 seconds.   22 to 5

10. A driven sprocket has 18 teeth compared with a drive sprocket having 40 teeth.   9 to 20

11. A driven sprocket has 60 teeth compared with a drive sprocket having 20 teeth.   3 to 1

12. The output of an electric transformer is 2,000 volts compared with an input of 50,000 volts.   1 to 25

13. The output of an electric transformer is 120 volts compared with an input of 2,000 volts.   3 to 50

14. The output of an electric transformer is 9 volts compared with an input of 120 volts.   3 to 40

Lesson 52   T–170

Introduction

Ask the students to give a fraction stating how the number of math lessons already studied compares with the number of lessons yet to be studied. Because 51 lessons have been studied and there are 129 more, that fraction is $\frac{51}{129}$.

Now reduce the fraction to lowest terms ($\frac{17}{43}$). The fractions $\frac{51}{129}$ and $\frac{17}{43}$ are also ratios. What is the relationship of lessons studied to lessons not yet studied? It is about 1 to 2.5.

Ratios are a special kind of fractions that we use specifically to compare numbers.

Teaching Guide

1. **A ratio is an expression that shows the relationship between two numbers.** Ratios can be reduced to lowest terms in the same way that fractions are reduced.

    a. 6 to 3 = 2 to 1

    b. 20 to 5 = 4 to 1

    c. 21 to 9 = 7 to 3

2. **The first number in a ratio is the antecedent, and the second number is the consequent.** The antecedent is the number being compared, and the consequent is the number with which the antecedent is compared.

    antecedent → 3 to 5 ← consequent

3. **A ratio may be expressed in three forms.** It may be written as a fraction, with a colon between the numbers, and with *to* between the numbers. All three forms are read with *to* between the numbers.

    a. $\frac{2}{7}$ = 2:7 = 2 to 7

    b. $\frac{9}{13}$ = 9:13 = 9 to 13

    In fraction form, a ratio looks like an improper fraction if the antecedent is larger than the consequent. Such a ratio is not changed to a mixed number.

    a. 7 to 5 = $\frac{7}{5}$

    b. 15 to 6 = $\frac{15}{6}$ = $\frac{5}{2}$

4. **Ratios are sometimes reduced so that either the antecedent or the consequent is 1.** The other part may then be in fraction or decimal form. For example, the ratio of a quarter to a dime could be expressed as $2\frac{1}{2}$ to 1 or 2.5 to 1.

    The following steps explain how to reduce a ratio so that one of the parts is 1.

    (1) Write the ratio in fraction form.

T–171   Chapter 4 Decimals, Ratios, and Proportions

(2) If the antecedent is to be expressed as 1, divide it by itself to obtain 1. Divide the consequent by the same number.

If the consequent is to be expressed as 1, divide it by itself to obtain 1. Divide the antecedent by the same number.

a. Reduce 9 to 12 so that the antecedent is 1. Express the consequent in fraction form.

$$\frac{9 \div 9 = 1}{12 \div 9 = 1\frac{1}{3}} \quad \text{Answer: 1 to } 1\frac{1}{3}$$

b. Reduce 5 to 7 so that the antecedent is 1. Express the consequent in decimal form.

$$\frac{5 \div 5 = 1}{7 \div 5 = 1.4} \quad \text{Answer: 1 to 1.4}$$

c. Reduce 4 to 3 so that the consequent is 1. Express the antecedent in fraction form.

$$\frac{4 \div 3 = 1\frac{1}{3}}{3 \div 3 = 1} \quad \text{Answer: } 1\frac{1}{3} \text{ to 1}$$

d. The dimensions of the United States flag form a ratio of 19 (length) to 10 (width). Express this ratio with a consequent of 1 and the antecedent in decimal form.

$$\frac{19 \div 10 = 1.9}{10 \div 10 = 1} \quad \text{Answer: 1.9 to 1}$$

Further Study

1. Using the colon to express a ratio is an older method of expressing both fractions and ratios. Thus, 3:4 at one time represented both the common fraction $\frac{3}{4}$ and the ratio 3 to 4.

2. Ratios were formerly expressed with the words "is to." Over time the word *is* was dropped so now 3:4 is generally read "three to four" rather than "three is to four."

Solutions for Exercises 27 and 28

27. $346 \div 6.6$

28. $425 \div 16$

15. The sun's circumference is 2,700,000 miles compared with the earth's circumference of 25,000 miles. 108 to 1

16. Methuselah's age was 969 years compared with the average life span today of about 72 years. 323 to 24

C. Reduce these ratios as indicated.

17. 6 to 4 (consequent = 1; antecedent in decimal form) 1.5 to 1

18. 8 to 5 (consequent = 1; antecedent in decimal form) 1.6 to 1

19. 9 to 8 (consequent = 1; antecedent in fraction form) $1\frac{1}{8}$ to 1

20. 8 to 11 (antecedent = 1; consequent in fraction form) 1 to $1\frac{3}{8}$

21. 2 to 3 (antecedent = 1; consequent in decimal form) 1 to 1.5

22. 9 to 7 (antecedent = 1; consequent in fraction form) 1 to $\frac{7}{9}$

D. Solve these reading problems.

23. In Luke 17, Jesus met 10 lepers and commanded them to show themselves to the priest. As they went, they were healed. Only 1 returned to thank Jesus; 9 did not. Write a ratio comparing the number that did not return with the number that did return. 9 to 1

24. Christ spoke of three different harvest yields in His parable of the sower. Some ground yielded thirtyfold, some sixtyfold, and some a hundredfold. Write a ratio in lowest terms to compare the lowest yield with the highest. 3 to 10

25. David reigned as king for 40 years: 7 years at Hebron and then 33 years at Jerusalem. Write a ratio to compare the number of years he reigned at Hebron with the total years in his reign. (Express the antecedent of the ratio as 1 and the consequent as a mixed number.) 1 to $5\frac{5}{7}$

26. Twelve spies went into the land of Canaan, but only Joshua and Caleb returned with a good report. Write a ratio to compare those who gave a bad report with those who gave a good report. (Express the antecedent of the ratio as 1 and the consequent as a decimal.) 1 to 0.2

27. A car traveled 346 miles in 6.6 hours. What was its average speed in miles per hour? (Round your answer to the nearest tenth.) 52.4 m.p.h.

28. A car used 16 gallons of gasoline to go 425 miles. How many miles per gallon did it average? (Round your answer to the nearest tenth.) 26.6 mi. per gal.

REVIEW EXERCISES

E. Solve mentally by using the fractional equivalent of each decimal. (Lesson 51)

29. 0.5 of 78   39
30. 0.625 of 32   20
31. $0.33\frac{1}{3}$ of 36   12
32. $0.83\frac{1}{3}$ of 36   30

172   Chapter 4   Decimals, Ratios, and Proportions

F. Round each decimal to the place indicated. *(Lesson 50)*

33. 1.2058 (thousandths)   1.206   34. 18.2519 (hundredths)   18.25

G. Express each fraction as a decimal. If it does not divide evenly by the ten-thousandths' place, round to the nearest thousandth. *(Lesson 50)*

35. $\frac{5}{7}$ 0.714   36. $\frac{5}{11}$ 0.455

H. Solve these multiplication problems involving decimals. *(Lesson 48)*

37. $\begin{array}{r} 1.008 \\ \times\, 0.09 \\ \hline \end{array}$
    0.09072

38. $\begin{array}{r} 1.0521 \\ \times\, 0.02 \\ \hline \end{array}$
    0.021042

I. Reduce these fractions to lowest terms. *(Lesson 36)*

39. $\frac{31}{93}$ $\frac{1}{3}$ 40. $\frac{24}{54}$ $\frac{4}{9}$

J. Find these equivalents. *(Lessons 22, 23)*

41. 4.7 kg = <u>4,700</u> g   42. 2,600 kg = <u>2.6</u> MT
43. 120 *l* = <u>0.12</u> kl   44. 120 *l* = ____ ml   120,000

*Other fell into good ground,
and brought forth fruit.*

Matthew 13:8

100:1

60:1

30:1

T–173   Chapter 4 Decimals, Ratios, and Proportions

LESSON 53

Objectives

- To review proportions and proportion terminology.
- To review that the product of the means equals the product of the extremes.

Review

1. Give Lesson 53 Quiz (Using Decimals).
2. *Give the ratios indicated.* (Lesson 52)
   a. Compare 6 teachers with 123 students. Express the antecedent as 1 and the consequent in decimal form. (1 to 20.5)
   b. Compare an 80-pound load lifted with 25 pounds of force exerted. Express the consequent as 1 and the antecedent in fraction form. ($3\frac{1}{5}$ to 1)
3. *Solve mentally by using the fractional equivalent of each decimal.* (Lesson 51)
   a. 0.8 of 45 (36)
   b. 0.25 of 48 (12)
4. *Solve these division problems. If an answer does not divide evenly by the ten-thousandth's place, round it to the nearest thousandth.* (Lesson 50)
   a. $8\overline{)5.4}$ (0.675)
   b. $1.8\overline{)0.82}$ (0.456)
5. *Solve these addition and subtraction problems.* (Lesson 37)

   a. $9\frac{1}{5}$
      $+6\frac{3}{7}$
      ($15\frac{22}{35}$)

   b. 11
      $-6\frac{11}{16}$
      ($4\frac{5}{16}$)

53. Direct Proportions

A proportion consists of two equivalent ratios joined by an equal sign. In a **direct proportion**, changing one part of a ratio requires a similar change in the other ratio. Direct proportions are useful for solving many kinds of math problems in which an increase or a decrease in one amount is matched by a similar increase or decrease in another amount. (You will study another kind of proportion in Lesson 54.)

A proportion is correct if the product of the **means** (middle numbers) is equal to the product of the **extremes** (outer numbers). The means and extremes are easy to identify if the ratios are written in horizontal form with colons or the word *to*. With the ratios written in fraction form, cross multiply to find the product of the means or the extremes.

$$\underset{\text{extremes}}{\overset{\text{means}}{2 \text{ to } 3 = 6 \text{ to } 9}} \qquad \frac{2}{3} = \frac{6}{9}$$

$$3 \times 6 = 18, \text{ and } 2 \times 9 = 18$$

Because the products of the means and the extremes are always equal, knowing three parts of a proportion makes it possible to calculate the fourth part. The steps are given below.

1. Write the proportion as two ratios in fraction form. Multiply the two numbers that are located diagonally from each other.

2. Divide that product by the third number in the proportion. The result is the missing part of the proportion.

3. Check the answer by cross multiplying to make sure the products of the means and the extremes are equal.

> **Example A**
> Find the missing part of the following proportion.
> $$\frac{8}{6} = \frac{12}{n}$$
> $6 \times 12 = 72$; $72 \div 8 = 9$ $n = 9$
> Check: $6 \times 12 = 72$, and $8 \times 9 = 72$

174 Chapter 4 *Decimals, Ratios, and Proportions*

Proportions are helpful in solving many reading problems. The steps for using them are shown in the following example.

Example B

Father traveled 141 miles in 3 hours. At that rate, how long will it take him to travel 235 miles?

(1) Choose two labels from the problem. Write them for both ratios.

$$\frac{\text{hours}}{\text{miles}} = \frac{\text{hours}}{\text{miles}}$$

(2) Write the primary ratio to match the labels.

$$\frac{\text{hours}}{\text{miles}} \; \frac{3}{141} = \frac{\text{hours}}{\text{miles}}$$

(3) Place the third fact beside its label in the second ratio. Write n for the unknown part.

$$\frac{\text{hours}}{\text{miles}} \; \frac{3}{141} = \frac{n}{235} \; \frac{\text{hours}}{\text{miles}}$$

(4) Find the missing part of the proportion.

$3 \times 235 = 705$; $705 \div 141 = 5$
$n = 5$ hours
Check: $3 \times 235 = 705$
$141 \times 5 = 705$

Sometimes one part of a proportion needs to be calculated. This becomes evident when a problem contains three labels instead of two. Consider this problem.

> One day Mother baked 120 cookies. She made 3 oatmeal cookies for every 2 sugar cookies. How many oatmeal cookies did she make?

Could this proportion be used to solve the problem? Can you tell why it will not work?

$$\frac{\text{oatmeal cookies}}{\text{sugar cookies}} \; \frac{3}{2} = \frac{n}{120} \; \frac{\text{oatmeal cookies}}{\text{total cookies}}$$

Look at the labels in the proportion above. The two ratios represent different relationships. Notice what happens if we try to find the missing part by cross multiplication: $3 \times 120 = 360$, and $360 \div 2 = 180$. That answer cannot be right, for Mother baked only 120 cookies in all!

Both ratios in a direct proportion must represent the same relationship. If the second ratio compares oatmeal cookies with total cookies, the primary ratio must make the same comparison. Writing labels while setting up a proportion will help you to avoid the mistake illustrated above.

To find a total number of cookies for the primary ratio, add 3 (oatmeal cookies) plus 2 (sugar cookies) to obtain 5. The result means that out of every 5 cookies, 3 were oatmeal cookies and 2 were sugar cookies.

Lesson 53   T-174

Introduction

Ask the students how they would write a ratio comparing 70 pounds of hay with 14 cattle (70:14). Now suppose the herd is increased to 25 head of cattle. At the same amount of hay per head, how much hay would be needed?

Of course, we could divide 70 by 14 to find the amount of hay per head, and then multiply the result by 25 to find the amount needed for 25 head of cattle. However, those calculations can all be incorporated into a proportion to find the answer.

Write the original ratio in fraction form and label its parts. Then write another ratio, fill in the numbers you know, and write n for the unknown part. Because we do not know the amount of hay needed, the antecedent for the new ratio is n. Because the new number of cattle to be fed is 25, that is the consequent of the second ratio. Notice that the relationship of hay to cattle in the new ratio is stated in the same way as in the first. Hay is the antecedent and cattle is the consequent.

$$\text{hay} \quad \frac{70}{14} = \frac{n}{25} \quad \text{hay}$$
$$\text{cattle} \qquad\qquad\qquad \text{cattle}$$

Teaching Guide

1. **A proportion consists of two equivalent ratios joined by an equal sign.** In a direct proportion, changing one part of a ratio requires a similar change in the other ratio. Direct proportions are useful for solving many kinds of math problems in which an increase or a decrease in one amount is matched by a similar increase or decrease in another amount.

2. **A proportion is correct if the product of the extremes (outer numbers) is equal to the product of the means (middle numbers).** This can be shown by writing the ratios in fraction form and cross multiplying.

    a. $\dfrac{6}{5} = \dfrac{12}{10}$

    extremes = 6 and 10; 6 × 10 = 60

    means = 5 and 12; 5 × 12 = 60

    b. $\dfrac{9}{6} = \dfrac{21}{14}$

    extremes = 9 and 14; 9 × 14 = 126

    means = 6 and 21; 6 × 21 = 126

3. **Because the products of the extremes and the means are always equal, knowing three parts of a proportion makes it possible to calculate the fourth part.** The steps are given below.

    (1) Write the proportion as two ratios in fraction form. Multiply the two numbers that are located diagonally from each other.

    (2) Divide that product by the third number in the proportion. The result is the missing part of the proportion.

    (3) Check the answer by cross multiplying to make sure the products of the extremes and the

T–175   Chapter 4   Decimals, Ratios, and Proportions

means are equal.

a. $\dfrac{6}{14} = \dfrac{27}{n}$

$14 \times 27 = 378$
$378 \div 6 = 63$
Check: $14 \times 27 = 378$
$6 \times 63 = 378$

b. $\dfrac{15}{50} = \dfrac{21}{n}$

$50 \times 21 = 1{,}050$
$1{,}050 \div 15 = 70$
Check: $50 \times 21 = 1{,}050$
$15 \times 70 = 1{,}050$

4. **Proportions are helpful in solving many reading problems.** The steps for using them are given below.

   (1) Choose two labels from the problem. Write them for both ratios.

   (2) Write the primary ratio to match the labels.

   (3) Place the third fact beside its label in the second ratio. Write n for the unknown part.

(4) Find the missing part of the proportion.

   a. A 30-foot tree casts a 21-foot shadow. How tall is the communications tower standing nearby if it casts a 126-foot shadow at the same time?

   $\dfrac{\text{height}}{\text{shadow}} \ \dfrac{30}{21} = \dfrac{n}{126} \ \dfrac{\text{height}}{\text{shadow}}$

   $30 \times 126 = 3{,}780$
   $3{,}780 \div 21 = 180$ ft.
   Check: $30 \times 126 = 3{,}780$
   $21 \times 180 = 3{,}780$

   b. A certain kind of rabbit feed contains 420 pounds of alfalfa meal per 2,000 pounds of finished feed. At that rate, how many pounds of alfalfa meal are needed for 3,500 pounds of rabbit feed?

   $\dfrac{\text{alfalfa}}{\text{feed}} \ \dfrac{420}{2{,}000} = \dfrac{n}{3{,}500} \ \dfrac{\text{alfalfa}}{\text{feed}}$

   $420 \times 3{,}500 = 1{,}470{,}000$
   $1{,}470{,}000 \div 2{,}000 = 735$ lb.
   Check: $420 \times 3{,}500 = 1{,}470{,}000$
   $2{,}000 \times 735 = 1{,}470{,}000$

Proportions for CLASS PRACTICE d–i

d. $\dfrac{\text{bushels}}{\text{cost}} \ \dfrac{3}{\$25.50} = \dfrac{10}{n} \ \dfrac{\text{bushels}}{\text{cost}}$ $n = \$85.00$

e. $\dfrac{\text{height}}{\text{shadow}} \ \dfrac{5\frac{1}{2}}{2} = \dfrac{n}{24} \ \dfrac{\text{height}}{\text{shadow}}$ $n = 66$ feet

f. $\dfrac{\text{gallons}}{\text{miles}} \ \dfrac{15}{240} = \dfrac{n}{360} \ \dfrac{\text{gallons}}{\text{miles}}$ $n = 22\frac{1}{2}$ gallons

g. $\dfrac{\text{hours}}{\text{rows}} \ \dfrac{1\frac{3}{4}}{7} = \dfrac{3}{n} \ \dfrac{\text{hours}}{\text{rows}}$ $n = 12$ rows

h. $\dfrac{\text{finished}}{\text{total}} \ \dfrac{2}{3} = \dfrac{n}{18} \ \dfrac{\text{finished}}{\text{total}}$ $n = 12$ drawings

i. $\dfrac{\text{Marcus}}{\text{total}} \ \dfrac{2}{5} = \dfrac{n}{35} \ \dfrac{\text{Marcus}}{\text{total}}$ $n = 14$ hours

$$\frac{\text{oatmeal cookies}}{\text{total cookies (3 + 2)}} \quad \frac{3}{5} = \frac{n}{120} \quad \frac{\text{oatmeal cookies}}{\text{total cookies}}$$

$3 \times 120 = 360; \; 360 \div 5 = 72$

$n = 72$ oatmeal cookies

Check: $3 \times 120 = 360$, and $5 \times 72 = 360$

CLASS PRACTICE

Find the missing part of each proportion. Show your work and the check.

a. $\dfrac{11}{33} = \dfrac{8}{n}$ $n = 24$ b. $\dfrac{10}{35} = \dfrac{n}{14}$ $n = 4$ c. $\dfrac{3}{14} = \dfrac{42}{n}$ $n = 196$

Use proportions to solve these problems. (See facing page for proportions.)

d. If 3 bushels of peaches cost $25.50, what is the cost of 10 bushels? $85.00

e. A $5\frac{1}{2}$-foot clothesline post casts a 2-foot shadow when a nearby oak tree casts a 24-foot shadow. How tall is the tree?   66 feet

f. The Amstutzes' van used 15 gallons of gasoline to travel 135 miles to visit friends and then 105 miles in driving to and from school. At that rate, how many gallons will it take to drive 360 miles? $22\frac{1}{2}$ gallons

g. In $1\frac{3}{4}$ hours, Susanna hoed 4 rows of strawberries, and Sarah hoed 3 rows. If they spent 3 hours at hoeing strawberries, how many rows did they finish?   12 rows

h. Darlene sorted the 18 drawings from art class. She found that 2 out of every 3 drawings were finished. How many of the drawings were finished?   12 drawings

i. On a certain job, Marcus worked 2 hours for every 3 hours that Jacob worked. Together they worked a total of 35 hours. How many hours did Marcus work?   14 hours

176   Chapter 4   Decimals, Ratios, and Proportions

WRITTEN EXERCISES

A. Find the missing part of each proportion. Show your work and the check.

1. $\dfrac{6}{12} = \dfrac{15}{n}$ $n = 30$
2. $\dfrac{42}{18} = \dfrac{n}{24}$ $n = 56$
3. $\dfrac{8}{14} = \dfrac{20}{n}$ $n = 35$
4. $\dfrac{10}{15} = \dfrac{n}{24}$ $n = 16$
5. $\dfrac{22}{10} = \dfrac{n}{15}$ $n = 33$
6. $\dfrac{14}{30} = \dfrac{35}{n}$ $n = 75$

B. Use proportions to solve these problems. (See facing page for proportions.)

7. A 4-foot fence post casts a 3-foot shadow when a nearby utility pole casts a 33-foot shadow. How high is the utility pole?   44 feet

8. A 7-foot clothesline post casts a 2-foot shadow when the house casts a 7-foot shadow. How high is the house? $24\frac{1}{2}$ feet

9. Six yards of fabric cost $22.50. Find the cost of $8\frac{1}{2}$ yards of the same fabric, to the nearest cent. $31.88

10. If 3 bushels of apples cost $18.75, what is the cost of 8 bushels? $50.00

11. A pig-grower ration contains 50 grams of antibiotic per ton. Find the amount of antibiotic in an 80-pound bag of this feed.   2 grams

12. A feeder-pig ration contains 175 pounds of wheat per 1,000 pounds of feed. How many pounds of this ration can be mixed when 700 pounds of wheat is available?   4,000 pounds

C. Use proportions to solve these problems. Be careful, for you need to calculate one part of each proportion.

13. The eighth grade classroom at the Valley View Mennonite School has 24 students. The ratio of girls to boys is 5:3. How many girls are in the room?   15 girls

14. Father and Daniel picked sweet corn one day in August. Daniel picked 5 ears for every 7 that Father picked. Together they picked 228 dozen ears. How many dozen did Daniel pick?   95 dozen

15. Mother planted irises at the rate of 8 purple ones for every 6 white ones. She planted 70 irises in all. How many white irises did she plant?   30 white irises

16. The Baumans have 1 dry cow for every 5 cows they are milking. If they are milking 55 cows, how many cows do they own?   66 cows

D. Use proportions to solve these problems. Be careful, for you may need to calculate one part of a proportion.

17. One kind of rabbit feed has $1\frac{1}{4}$ pounds of molasses per 50-pound bag. How much molasses is in 1 ton of this feed?   50 pounds

18. Brother Mark planted 5 acres of corn for every 2 acres of soybeans. In all, he planted 140 acres of corn and soybeans. How many acres of soybeans did he plant?   40 acres

5. **Sometimes one part of a proportion needs to be calculated.** This becomes evident when the problem includes three labels instead of two. When you write a direct proportion, you must always form both ratios by the same logic. For example, if the number given for the second ratio is a total, the corresponding number in the primary ratio must also give a total.

a. One winter the Masts counted 5 male cardinals for every 2 female cardinals at their bird feeder. They counted 70 cardinals in all. How many female cardinals did they count?

Proportions for Reading Problems

7. $\dfrac{\text{height}}{\text{shadow}} \; \dfrac{4}{3} = \dfrac{n}{33} \; \dfrac{\text{height}}{\text{shadow}}$ $n = 44$ feet

8. $\dfrac{\text{height}}{\text{shadow}} \; \dfrac{7}{2} = \dfrac{n}{7} \; \dfrac{\text{height}}{\text{shadow}}$ $n = 24\tfrac{1}{2}$ feet

9. $\dfrac{\text{yards}}{\text{cost}} \; \dfrac{6}{\$22.50} = \dfrac{8\tfrac{1}{2}}{n} \; \dfrac{\text{yards}}{\text{cost}}$ $n = \$31.88$

10. $\dfrac{\text{bushels}}{\text{cost}} \; \dfrac{3}{\$18.75} = \dfrac{8}{n} \; \dfrac{\text{bushels}}{\text{cost}}$ $n = \$50.00$

11. $\dfrac{\text{grams}}{\text{pounds}} \; \dfrac{50}{2{,}000} = \dfrac{n}{80} \; \dfrac{\text{grams}}{\text{pounds}}$ $n = 2$ grams

12. $\dfrac{\text{wheat}}{\text{feed}} \; \dfrac{175}{1{,}000} = \dfrac{700}{n} \; \dfrac{\text{wheat}}{\text{feed}}$ $n = 4{,}000$ pounds

13. $\dfrac{\text{girls}}{\text{students}} \; \dfrac{5}{8} = \dfrac{n}{24} \; \dfrac{\text{girls}}{\text{students}}$ $n = 15$ girls

14. $\dfrac{\text{Daniel}}{\text{total}} \; \dfrac{5}{12} = \dfrac{n}{228} \; \dfrac{\text{Daniel}}{\text{total}}$ $n = 95$ dozen

15. $\dfrac{\text{white}}{\text{total}} \; \dfrac{6}{14} = \dfrac{n}{70} \; \dfrac{\text{white}}{\text{total}}$ $n = 30$ white irises

16. $\dfrac{\text{milking}}{\text{total}} \; \dfrac{5}{6} = \dfrac{55}{n} \; \dfrac{\text{milking}}{\text{total}}$ $n = 66$ cows

17. $\dfrac{\text{molasses}}{\text{feed}} \; \dfrac{1\tfrac{1}{4}}{50} = \dfrac{n}{2{,}000} \; \dfrac{\text{molasses}}{\text{feed}}$ $n = 50$ pounds

18. $\dfrac{\text{soybeans}}{\text{total}} \; \dfrac{2}{7} = \dfrac{n}{140} \; \dfrac{\text{soybeans}}{\text{total}}$ $n = 40$ acres

(The ratio needed to find the answer is female cardinals to all cardinals. That ratio is not 2 to 5 but rather 2 to 7, obtained by adding the numbers of male and female cardinals.)

$\dfrac{\text{female}}{\text{total}} \dfrac{2}{7} = \dfrac{n}{70} \dfrac{\text{female}}{\text{total}}$

$2 \times 70 = 140; 140 \div 7 = 20$

$n = 20$ female cardinals

b. The grades 8–10 classroom at the Lowell Valley Mennonite School has 2 boys for every 3 girls. There are 25 students in the class. How many of them are boys?
(The ratio needed to find the answer is boys to all students. That ratio is not 2 to 3 but rather 2 to 5, obtained by adding the numbers of boys and girls.)

$\dfrac{\text{boys}}{\text{total}} \dfrac{2}{5} = \dfrac{n}{25} \dfrac{\text{boys}}{\text{total}}$

$2 \times 25 = 50; 50 \div 5 = 10$

$n = 10$ boys

An Ounce of Prevention

1. A proportion can usually be set up in several different ways. If a student uses an arrangement different from that in the answer key, give him credit as long as the products of the extremes and the means are equal.

2. Stress to the students that if a total is given for the secondary ratio or if the question asks for a total, they must be sure to have a total for both the primary and the secondary ratios.

---

19. $\dfrac{\text{corn}}{\text{feed}} \dfrac{4}{5} = \dfrac{n}{360} \dfrac{\text{corn}}{\text{feed}}$ $n = 288$ pounds

20. $\dfrac{\text{cups}}{\text{dozen}} \dfrac{4}{6} = \dfrac{10}{n} \dfrac{\text{cups}}{\text{dozen}}$ $n = 15$ dozen

19. A finishing ration for feeder pigs contains 4 pounds of corn to every pound of other ingredients. If a typical feeder pig consumes 360 pounds of finishing ration, how much corn does it eat? 288 pounds

20. A recipe calls for 4 cups of flour to make 6 dozen cookies. If Grandmother has 10 cups of flour, how many cookies can she make? 15 dozen

REVIEW EXERCISES

E. Write the ratios indicated. *(Lesson 52)*

21. Compare 700 bushels with 4 acres. Express the consequent as 1. 175 to 1

22. Compare 128 ounces of gasoline with 2.6 ounces of oil to be mixed in. Express the consequent as 1, and round the antecedent to the nearest tenth. 49.2 to 1

F. Solve mentally by using the fractional equivalent of each decimal. *(Lesson 51)*

23. 0.7 of 40   28

24. 0.5 of 46   23

G. Solve these division problems. If an answer does not divide evenly by the ten-thousandths' place, round it to the nearest thousandth. *(Lesson 50)*

25. $1.6 \overline{)0.82}$ 0.5125

26. $1.2 \overline{)0.46}$ 0.383

H. Solve these addition and subtraction problems. *(Lesson 37)*

27. $16 \tfrac{1}{4} + 1 \tfrac{1}{3} = 17 \tfrac{7}{12}$

28. $15 - 2 \tfrac{5}{6} = 12 \tfrac{1}{6}$

I. Find these equivalents. *(Lesson 24)*

29. 12 cm² = 1,200 mm²

30. 1,585 m² = _____ cm²   15,850,000

31. 10 ha = 25 a.

32. 50 a. = 20 ha

54. Inverse Proportions

In a direct proportion, both quantities in each ratio increase or decrease at the same rate. For example, if the cost of material is $3 for 1 yard, it will be $6 for 2 yards and $1.50 for $\frac{1}{2}$ yard.

The opposite of a direct proportion is an **inverse proportion**. In this kind of proportion, the ratios also change at the same rate, but they change in opposite directions. For example, if a bin holds enough grain to feed 5 cows for 8 days, it will feed 10 cows for only 4 days. Thus, 2 times as many cows will eat the feed in $\frac{1}{2}$ the original time, and $\frac{1}{2}$ as many cows will eat the feed in 2 times the original time.

Inverse proportions are set up differently from direct proportions. For a direct proportion, each ratio contains two numbers in the same relationship (such as the 3 and 1 in "$3 for 1 yard").

Direct proportion

$$\frac{\text{cost}}{\text{yards}} \quad \frac{\$3}{1} = \frac{\$6}{2} \quad \frac{\text{cost}}{\text{yards}} \qquad 3 \times 2 = 6, \text{ and } 1 \times 6 = 6$$

For an inverse proportion, the two numbers with one relationship are placed diagonally from each other, and not in the same ratio. (See the 5 and 8 below for the relationship "5 cows for 8 days".) This is because one quantity decreases when the other increases.

Inverse proportion

$$\frac{\text{cows (1)}}{\text{cows (2)}} \quad \frac{5}{10} = \frac{4}{8} \quad \frac{\text{days (2)}}{\text{days (1)}} \qquad 5 \times 8 = 40, \text{ and } 10 \times 4 = 40$$

Notice the special way in which inverse proportions are written. Both quantities of cows are on the left side. Both quantities of days are on the right side, but each part is diagonal from the number that it relates to. (The numbers 1 and 2 in parentheses show the related facts.) Cross multiplication yields two equal products, the same as with direct proportions.

Follow these steps to write an inverse proportion.

1. Keep similar parts of each relationship on the same side of the proportion, such as both speeds and both hours.
2. Place the values of each relationship in positions diagonal to each other. It is helpful to use (1) and (2) in the labels to show which facts are related.

LESSON 54

Objectives

- To teach that *some facts are related by inverse proportion.
- To teach *how to write inverse proportions.

Review

1. *Write a ratio in fractional form for each statement, and reduce it to lowest terms.* (Lesson 52)

    a. Janette has read 72 pages of a 156-page book. ($\frac{72}{156} = \frac{6}{13}$)

    b. The output of an electric transformer is 24 volts compared with an input of 440 volts. ($\frac{24}{440} = \frac{3}{55}$)

2. *Solve by mental multiplication.* (Lesson 38)

    a. $\frac{5}{9}$ of 108 (60)

    b. $\frac{7}{12}$ of 72 (42)

3. *Solve these multiplication problems.* (Lesson 38)

    a. $\frac{5}{7} \times \frac{2}{5}$ ($\frac{2}{7}$)

    b. $6\frac{3}{4} \times \frac{2}{3}$ ($4\frac{1}{2}$)

4. *Solve these problems involving compound measures.* (Lesson 25)

    a. 6 tbsp. 1 tsp.
        + 3 tbsp. 2 tsp.
        (10 tbsp.)

    b. 20 min.
        − 9 min. 16 sec.
        (10 min. 44 sec.)

    c. 2.4 kl + 815 l = ___ kl (3.215)

    d. 0.9 MT − 210 kg = ___ MT (4.69)

T–179 Chapter 4 *Decimals, Ratios, and Proportions*

Introduction

Can the students think of any situation in which something becomes smaller as another thing becomes larger or vice versa? They may think of examples such as the following.

a. The faster the rate of speed, the less time is needed to travel a given distance.

b. The more animals being fed, the less time a given amount of feed will last.

c. The more people helping with a project, the less time is needed to complete it.

d. The smaller the unit of measure, the larger the number of units is in a certain quantity.

These are examples of inverse proportions, the topic of this lesson.

Teaching Guide

1. **In an inverse proportion, two quantities change at the same rate but in opposite directions.** For example, if it takes 2 boys 20 minutes to unload hay from a truck, 4 boys should be able to unload the hay in 10 minutes.

2. **Inverse proportions are set up differently from direct proportions.** In a direct proportion, each ratio contains two numbers in the same relationship. But in an inverse proportion, the two numbers in each relationship are located diagonally from each other. As in direct proportions, the products obtained by cross multiplication are the same.

Here are the steps for setting up inverse proportions.

(1) Keep similar parts of each relationship on the same side of the proportion, such as both lengths and both prices.

(2) Place the values of each relationship in positions diagonal to each other. It is helpful to use (1) and (2) in the labels to show which facts are related.

First solve the following examples and the proportions in *Class Practice*. Then solve the reading problems.

Proportions for **CLASS PRACTICE d–f**

d. $\dfrac{\text{diameter (1)}}{\text{diameter (2)}}$ $\dfrac{5}{8} = \dfrac{n}{240}$ $\dfrac{\text{r.p.m. (2)}}{\text{r.p.m. (1)}}$ $n = 150$ r.p.m.

e. $\dfrac{\text{diameter (1)}}{\text{diameter (2)}}$ $\dfrac{12}{5} = \dfrac{n}{30}$ $\dfrac{\text{r.p.m. (2)}}{\text{r.p.m. (1)}}$ $n = 72$ r.p.m.

f. $\dfrac{\text{pounds on side 1}}{\text{pounds on side 2}}$ $\dfrac{80}{n} = \dfrac{2}{5}$ $\dfrac{\text{feet from fulcrum (2)}}{\text{feet from fulcrum (1)}}$ $n = 200$ pounds

Study these examples of reading problems solved by inverse proportions.

Example A
At a rate of 50 miles per hour, it takes 6 hours to reach Creston. How long does it take at 60 miles per hour?

$$\frac{\text{m.p.h. (1)}}{\text{m.p.h. (2)}} \quad \frac{50}{60} = \frac{n}{6} \quad \frac{\text{hours (2)}}{\text{hours (1)}}$$

50 × 6 = 300; 300 ÷ 60 = 5
n = 5 hr.
Check: 50 × 6 = 300; 60 × 5 = 300

Example B
A gear with a diameter of 8 inches is driving a gear with a diameter of 5 inches. When the speed of the drive gear is 90 revolutions per minute (r.p.m.), what is the speed of the driven gear?

$$\frac{\text{diameter (1)}}{\text{diameter (2)}} \quad \frac{8}{5} = \frac{n}{90} \quad \frac{\text{r.p.m. (2)}}{\text{r.p.m. (1)}}$$

8 × 90 = 720; 720 ÷ 5 = 144
n = 144 r.p.m.
Check: 8 × 90 = 720; 5 × 144 = 720

Inverse proportions have various practical applications in everyday life. The following list names a few common ones.
- rate of travel and time of travel
- number of workers and hours to complete a task
- size and speed of a drive gear and size and speed of a driven gear
- weight and distance from fulcrum on side 1 of a lever, and weight and distance from fulcrum on side 2 of a lever

CLASS PRACTICE

Find the missing parts of these inverse proportions.

a. $\frac{\text{workers (1)}}{\text{workers (2)}} \quad \frac{8}{6} = \frac{n}{12} \quad \frac{\text{hours (2)}}{\text{hours (1)}}$ n = 16 hours

b. $\frac{\text{pounds on side 1 of seesaw}}{\text{pounds on side 2 of seesaw}} \quad \frac{75}{125} = \frac{n}{60} \quad \frac{\text{inches from fulcrum (2)}}{\text{inches from fulcrum (1)}}$ n = 36 inches

c. $\frac{\text{m.p.h. (1)}}{\text{m.p.h. (2)}} \quad \frac{45}{50} = \frac{9}{n} \quad \frac{\text{hours (2)}}{\text{hours (1)}}$ n = 10 hours

Use inverse proportions to solve these problems. (See facing page for proportions.)

d. A pulley with a 5-inch diameter is turning at 240 revolutions per minute as it drives an 8-inch pulley. At what speed is the 8-inch pulley turning?   150 r.p.m.

e. A pulley with a 12-inch diameter is turning at 30 revolutions per minute and driving a 5-inch pulley. At what speed is the 5-inch pulley rotating?   72 r.p.m.

f. Keith and Edward are trying to lift a rock by using a plank as a lever. They are pushing down with 80 pounds of force at a point 5 feet from the fulcrum. If the rock rests on the plank at a point 2 feet from the fulcrum, how many pounds of force are the boys exerting on the rock?   200 pounds

180 Chapter 4 Decimals, Ratios, and Proportions

g. A boy weighing 40 pounds is sitting 8 feet from the fulcrum of a seesaw. His older brother, who weighs 100 pounds, is sitting on the opposite side. How far from the fulcrum must the older brother sit for the seesaw to balance?   3.2 feet

WRITTEN EXERCISES

A. Find the missing parts of these inverse proportions.

1. $\dfrac{\text{workmen (1)}}{\text{workmen (2)}} \quad \dfrac{3}{5} = \dfrac{n}{15} \quad \dfrac{\text{hours (2)}}{\text{hours (1)}}$ $n = 9$ hours

2. $\dfrac{\text{children shelling peas (1)}}{\text{children shelling peas (2)}} \quad \dfrac{2}{3} = \dfrac{n}{3} \quad \dfrac{\text{hours (2)}}{\text{hours (1)}}$ $n = 2$ hours

3. $\dfrac{\text{pounds on side 1}}{\text{pounds on side 2}} \quad \dfrac{50}{60} = \dfrac{n}{66} \quad \dfrac{\text{inches from fulcrum (2)}}{\text{inches from fulcrum (1)}}$ $n = 55$ inches

4. $\dfrac{\text{pounds on side 1}}{\text{pounds on side 2}} \quad \dfrac{45}{120} = \dfrac{n}{96} \quad \dfrac{\text{inches from fulcrum (2)}}{\text{inches from fulcrum (1)}}$ $n = 36$ inches

5. $\dfrac{\text{m.p.h. (1)}}{\text{m.p.h. (2)}} \quad \dfrac{80}{50} = \dfrac{8}{n} \quad \dfrac{\text{hours (2)}}{\text{hours (1)}}$ $n = 5$ hours

6. $\dfrac{\text{diameter of gear 1}}{\text{diameter of gear 2}} \quad \dfrac{6}{8} = \dfrac{n}{76} \quad \dfrac{\text{r.p.m. (2)}}{\text{r.p.m. (1)}}$ $n = 57$ r.p.m.

B. Use inverse proportions to solve these problems. (See facing page for proportions.)

7. Several women were helping Sister Louise with her canning because her children had been sick. In 7 hours 2 women were able to prepare 10 bushels of snapped green beans for canning. How long would it take 6 women? $2\frac{1}{3}$ hours

8. It took 3 people 8 hours to pick strawberries one day. Father calculated that there will be about the same amount of strawberries to pick the next day. If 5 people pick strawberries the next day at the same rate, how long should it take to finish picking? $4\frac{4}{5}$ hours

9. A bicycle has a drive sprocket with 45 teeth and a driven sprocket with 18 teeth. When the drive sprocket is pedaled at 60 revolutions per minute, how fast does the driven sprocket turn?   150 r.p.m.

10. A drive pulley with a 3-inch diameter is rotating at 20 revolutions per second (r.p.s.) and is driving a 2-inch pulley. At what speed is the 2-inch pulley rotating?   30 r.p.s.

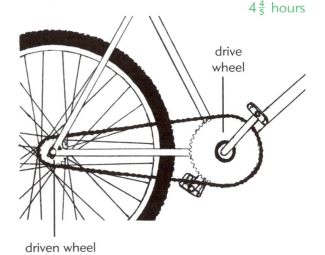

drive wheel

driven wheel

a. $\dfrac{\text{women quilting (1)}}{\text{women quilting (2)}}\ \dfrac{5}{9} = \dfrac{n}{6}\ \dfrac{\text{hours (2)}}{\text{hours (1)}}$

$5 \times 6 = 30$

$30 \div 9 = 3\frac{1}{3}$

$n = 3\frac{1}{3}$ hr.

Check: $5 \times 6 = 30$
$\qquad\quad 9 \times 3\frac{1}{3} = 30$

b. $\dfrac{\text{pounds on side 1}}{\text{pounds on side 2}}\ \dfrac{120}{300} = \dfrac{n}{90}\ \dfrac{\text{inches from fulcrum (2)}}{\text{inches from fulcrum (1)}}$

$120 \times 90 = 10{,}800$

$10{,}800 \div 300 = 36$

$n = 36$ in.

Check: $120 \times 90 = 10{,}800$
$\qquad\quad 300 \times 36 = 10{,}800$

The students might enjoy writing a reading problem for the following proportion.

$\dfrac{5}{n} = \dfrac{60}{300}$

(Sample: It takes 5 hours for a plane to make a certain trip at 300 miles per hour. How long would it take a car to make the same trip at 60 miles per hour? Answer: 25 hr.)

---

g. $\dfrac{\text{pounds on side 1}}{\text{pounds on side 2}}\ \dfrac{40}{100} = \dfrac{n}{8}\ \dfrac{\text{feet from fulcrum (2)}}{\text{feet from fulcrum (1)}}\qquad n = 3.2$ feet

Proportions for Part B

7. $\dfrac{\text{workers (1)}}{\text{workers (2)}}\ \dfrac{2}{6} = \dfrac{n}{7}\ \dfrac{\text{hours (2)}}{\text{hours (1)}} \qquad n = 2\frac{1}{3}$ hours

8. $\dfrac{\text{workers (1)}}{\text{workers (2)}}\ \dfrac{3}{5} = \dfrac{n}{8}\ \dfrac{\text{hours (2)}}{\text{hours (1)}} \qquad n = 4\frac{4}{5}$ hours

9. $\dfrac{\text{teeth (1)}}{\text{teeth (2)}}\ \dfrac{45}{18} = \dfrac{n}{60}\ \dfrac{\text{r.p.m. (2)}}{\text{r.p.m. (1)}} \qquad n = 150$ r.p.m.

10. $\dfrac{\text{diameter (1)}}{\text{diameter (2)}}\ \dfrac{3}{2} = \dfrac{n}{20}\ \dfrac{\text{r.p.s. (2)}}{\text{r.p.s. (1)}} \qquad n = 30$ r.p.s.

An Ounce of Prevention

Inverse proportions can be set up in several different ways. However, do not confuse your students. Stick to the method taught in the lesson; and if a student discovers another arrangement that works, give him credit. Be careful, though, because writing an inverse proportion is not a matter of just inverting the second ratio of a direct proportion. See the following example.

If a feed bin holds enough to feed 5 cows for 8 days, it will feed 10 cows only 4 days.

a. $\dfrac{\text{cows (1)}}{\text{cows (2)}}\ \dfrac{5}{10} = \dfrac{4}{8}\ \dfrac{\text{days (2)}}{\text{days (1)}}$

$5 \times 8 = 40;\ 10 \times 4 = 40$ (correct)

b. $\dfrac{\text{cows (1)}}{\text{days (2)}}\ \dfrac{5}{4} = \dfrac{10}{8}\ \dfrac{\text{cows (2)}}{\text{days (1)}}$

$5 \times 8 = 40;\ 4 \times 10 = 40$ (correct)

c. $\dfrac{\text{cows (1)}}{\text{days (1)}}\ \dfrac{5}{8} = \dfrac{4}{10}\ \dfrac{\text{days (2)}}{\text{cows (2)}}$

$5 \times 10 = 50;\ 8 \times 4 = 32$ (incorrect)

---

11. $\dfrac{\text{pounds on side 1}}{\text{pounds on side 2}}\ \dfrac{90}{75} = \dfrac{n}{5}\ \dfrac{\text{feet from fulcrum (2)}}{\text{feet from fulcrum (1)}}$ $n = 6$ feet

12. $\dfrac{\text{pounds on side 1}}{\text{pounds on side 2}}\ \dfrac{120}{80} = \dfrac{n}{5}\ \dfrac{\text{feet from fulcrum (2)}}{\text{feet from fulcrum (1)}}$ $n = 7\tfrac{1}{2}$ feet

13. $\dfrac{\text{diameter (1)}}{\text{diameter (2)}}\ \dfrac{8}{3\tfrac{1}{2}} = \dfrac{400}{n}\ \dfrac{\text{r.p.m. (2)}}{\text{r.p.m. (1)}}$ $n = 175$ r.p.m.

14. $\dfrac{\text{diameter (1)}}{\text{diameter (2)}}\ \dfrac{5}{12} = \dfrac{1{,}750}{n}\ \dfrac{\text{r.p.m. (2)}}{\text{r.p.m. (1)}}$ $n = 4{,}200$ r.p.m.

15. $\dfrac{\text{diameter (1)}}{\text{diameter (2)}}\ \dfrac{7}{3} = \dfrac{n}{60}\ \dfrac{\text{r.p.m. (2)}}{\text{r.p.m. (1)}}$ $n = 140$ r.p.m.

16. $\dfrac{\text{teeth (1)}}{\text{teeth (2)}}\ \dfrac{44}{20} = \dfrac{n}{100}\ \dfrac{\text{r.p.m. (2)}}{\text{r.p.m. (1)}}$ $n = 220$ r.p.m.

Proportions for Part C

17. $\dfrac{\text{height}}{\text{shadow}}\ \dfrac{3}{4\tfrac{1}{2}} = \dfrac{n}{96}\ \dfrac{\text{height}}{\text{shadow}}$ $n = 64$ feet

18. $\dfrac{\text{minutes}}{\text{miles}}\ \dfrac{8}{1} = \dfrac{n}{4\tfrac{1}{2}}\ \dfrac{\text{minutes}}{\text{miles}}$ $n = 36$ minutes

19. $\dfrac{\text{corn}}{\text{feed}}\ \dfrac{3}{5} = \dfrac{n}{150}\ \dfrac{\text{corn}}{\text{feed}}$ $n = 90$ pounds

20. $\dfrac{\text{loaves}}{\text{fish}}\ \dfrac{5}{2} = \dfrac{n}{153}\ \dfrac{\text{loaves}}{\text{fish}}$ $n = 382\tfrac{1}{2}$ loaves

11. On the school playground, a boy weighing 90 pounds is sitting 5 feet from the fulcrum, seesawing with a lower-grade boy weighing 75 pounds. How far from the fulcrum must the lighter boy sit for the seesaw to balance? 6 feet

12. An upper-grade boy weighing 120 pounds is sitting 5 feet from the fulcrum, seesawing with a lower-grade boy weighing 80 pounds. How far is the smaller boy sitting from the fulcrum? $7\frac{1}{2}$ feet

13. A large gear on a corn picker is 8 inches in diameter. It is driven by a gear $3\frac{1}{2}$ inches in diameter that turns at 400 revolutions per minute. How fast does the large gear turn? 175 r.p.m.

14. On a lumber planer, a 5-inch pulley is driven by a 12-inch pulley that turns at 1,750 revolutions per minute. What is the speed of the 5-inch pulley? 4,200 r.p.m.

15. A bicycle has a 7-inch drive sprocket and a 3-inch sprocket on the driven wheel. If Daniel is pedaling at the rate of 60 revolutions per minute, how fast is the bicycle wheel turning? 140 r.p.m.

16. Dwight is pedaling the drive sprocket, which has 44 teeth, at a speed of 100 revolutions per minute. The driven sprocket has 20 teeth. How fast is the driven wheel turning? 220 r.p.m.

C. Solve these reading problems by using direct proportions. Remember, these are set up differently from inverse proportions. *(Lesson 53)*

17. Phoebe and Joanna wanted to calculate the height of the rock cliff on Uncle Arlin's farm. The girls measured a 96-foot shadow cast by the cliff and a $4\frac{1}{2}$-foot shadow cast by their yardstick. How high was the cliff? 64 feet

18. It is $4\frac{1}{2}$ miles from Wayne's home to Grandfather's house. If it takes him 8 minutes to drive his bicycle 1 mile, how long will it take to get to his grandfather's house? 36 minutes

19. Jason mixed 150 pounds of chicken feed at the rate of 3 pounds of corn for every 5 pounds of feed. How many pounds of corn did he use? 90 pounds

20. Jesus used 5 loaves and 2 fish to feed a multitude. Later, Peter caught 153 fish in his net. Using the same ratio of loaves to fish, how many loaves would Peter have needed to complement his catch of fish? $382\frac{1}{2}$ loaves

182 Chapter 4 Decimals, Ratios, and Proportions

REVIEW EXERCISES

D. Write ratios for these facts. *(Lesson 52)*

21. The Landis family is raising corn on 55 of their 80 acres of farmland. Write a ratio as a fraction in lowest terms to show what part of their acreage the Landises have planted in corn. $\frac{11}{16}$

22. The Landises have 15 Jerseys in their herd of 75 cows. Write a ratio as a fraction in lowest terms to show what part of the herd is Jerseys. $\frac{1}{5}$

E. Solve by mental multiplication. *(Lesson 38)*

23. $\frac{2}{7}$ of 35   10
24. $\frac{5}{8}$ of 48   30

F. Solve these multiplication problems. *(Lesson 38)*

25. $\frac{2}{9} \times \frac{3}{4}$ $\frac{1}{6}$
26. $3\frac{3}{4} \times \frac{1}{12}$ $\frac{5}{16}$

G. Solve these problems involving compound measures. *(Lesson 25)*

27. 12 gal. 3 qt.
   + 5 gal. 2 qt.
   ─────────────
    18 gal. 1 qt.

28. 9 hr. 12 min.
   − 4 hr. 50 min.
   ─────────────
    4 hr. 22 min.

*Every beast of the forest is mine,
and the cattle upon a thousand hills.*

Psalm 50:10

Further Study

Why do direct and inverse proportions work as they do? In mathematical terms, direct variation (which occurs in direct proportions) is change in which the *quotient* of two variables is a constant. That is, the two parts of a direct proportion are the dividend and the divisor in a division problem; and if both numbers are multiplied by the same factor, the result stays the same. For example, 15 ÷ 5 = 3, 30 ÷ 10 = 3, and 60 ÷ 20 = 3. (Remember the double-and-double shortcut for division.) In the setup of direct proportions, the numbers are positioned in such a way that cross multiplication yields the same product.

By contrast, inverse variation (which occurs in inverse proportions) is change in which the *product* of two variables is a constant. In other words, the two parts of an inverse proportion are the factors in a multiplication problem; and an increase in one factor is matched by a corresponding decrease in the other factor. For example, 5 × 12 = 60 and 10 × 6 = 60; 2 × 30 = 60 and 4 × 15 = 60. (Remember the double-and-divide shortcut for multiplication.) In the setup of inverse proportions, the numbers are again positioned in such a way that cross multiplication yields the same product.

The connection between inverse proportions and multiplication factors is especially evident in distance-rate-time problems. A product like 1,500 (the distance) can be obtained by multiplying 5 hours times 300 miles per hour, 10 hours times 150 miles per hour, or 25 hours times 60 miles per hour.

T–183   Chapter 4   Decimals, Ratios, and Proportions

LESSON 55

Objectives

- To review the difference between direct and inverse proportions.
- To introduce *the expressions *vary directly* and *vary inversely*.
- To practice writing proportions to solve reading problems.

Review

1. *Solve mentally by using the fractional equivalent of each decimal.* (Lesson 51)

   a. 0.5 of 54   (27)

   b. $0.83\frac{1}{3}$ of 54   (45)

2. *Write these decimals, using words.* (Lesson 47)

   a. 0.829   (eight hundred twenty-nine thousandths)

   b. 3.5034   (three and five thousand thirty-four ten-thousandths)

3. *Find the reciprocal of each number.* (Lesson 39)

   a. $\frac{7}{16}$ ($\frac{16}{7}$)

   b. $2\frac{3}{8}$ ($\frac{8}{19}$)

4. *Solve these division problems.* (Lesson 39)

   a. $\frac{1}{6} \div \frac{3}{14}$ ($\frac{7}{9}$)

   b. $4\frac{1}{6} \div \frac{5}{7}$ ($5\frac{5}{6}$)

5. *Solve these problems involving compound measures.* (Lesson 26)

   a. 3 hr. 45 min.
       $\times$ 7
       ―――――――――
       (26 hr. 15 min.)

   b. (2 tons $866\frac{2}{3}$ lb.)
       3)7 tons 600 lb.

55. Reading Problems: Using Direct and Inverse Proportions

In the last two lessons you worked with direct proportions and inverse proportions. If two facts are in direct proportion to each other, they both increase or decrease at the same rate. For example, doubling the speed of an airplane also doubles the distance traveled in a given time. Because of this relationship, the two parts are said to **vary directly** in relation to each other.

The facts of a direct proportion are arranged in a way that shows this direct relationship. Matching subjects are placed in the same position in both ratios, directly across from each other in the proportion.

Example A

The Millers had an 8-acre field that produced 1,200 bushels of corn. At that rate, how many acres must they harvest to have 9,000 bushels of corn?

As the number of acres increases, the number of bushels also increases. These facts are in direct proportion.

$$\frac{\text{bushels}}{\text{acres}} \ \frac{1,200}{8} = \frac{9,000}{n} \ \frac{\text{bushels}}{\text{acres}}$$

$8 \times 9,000 = 72,000$

$72,000 \div 1,200 = 60$

$n = 60$ acres

Check: $8 \times 9,000 = 72,000$; $1,200 \times 60 = 72,000$

If two facts are in inverse proportion to each other, an increase in one item means a decrease at the same rate in the other item. For example, doubling the speed of an airplane cuts in half the amount of time needed to travel a given distance. The two facts are said to **vary inversely** in relation to each other.

The facts of an inverse proportion are arranged in a way that shows this inverse relationship. Matching subjects are placed in opposite positions in the two ratios. They are located diagonally from each other in the proportion.

$$\frac{\text{miles per hour (1)}}{\text{miles per hour (2)}} = \frac{\text{time (2)}}{\text{time (1)}}$$

184   Chapter 4   Decimals, Ratios, and Proportions

Example B

Last year the Millers' 60-acre field yielded 12,000 bushels of corn. This year it yielded 10,000 bushels. At that rate, how many acres will it take this year to yield the same amount of corn that the field produced last year?

As the yield per acre decreases, the number of acres must increase.

These facts are in inverse proportion.

$$\frac{\text{bushels (1)}}{\text{bushels (2)}} \frac{12{,}000}{10{,}000} = \frac{n}{60} \frac{\text{acres (2)}}{\text{acres (1)}}$$

12,000 × 60 = 720,000

720,000 ÷ 10,000 = 72

n = 72 acres

Check: 12,000 × 60 = 720,000; 10,000 × 72 = 720,000

CLASS PRACTICE

Tell whether the facts in each relationship vary directly or inversely.

a. The amount of time it takes Arthur to make one weld in relation to the number of welds he can make in one hour. (Does more time per weld mean more or fewer welds per hour?)   inversely

b. The speed of a vehicle in relation to the number of miles traveled. (Does more speed mean more or less distance traveled?)   directly

c. The speed of a vehicle in relation to the amount of time it takes to travel a certain distance. (Does more speed mean more or less travel time?)   inversely

d. The weight of a load in relation to the distance it is placed from the fulcrum of a lever in order to balance. (Does more weight mean more or less distance from the fulcrum?)
                                                                                         inversely

Solve by using direct or inverse proportions. (See facing page.)

e. It took 3 men 45 hours to complete a small construction project. At the same rate, how long should it take 5 men to complete a similar project?   27 hours

f. If 3 men built 12 pieces of furniture in 30 hours, how many pieces could 5 men build in the same amount of time?   20 pieces

g. Dwight picked 20 dozen ears of sweet corn in 45 minutes. How long should it take him to pick 12 dozen ears?   27 minutes

h. Mary Lou and one of her sisters picked 30 quarts of strawberries in $1\frac{1}{2}$ hours. If another sister helps them, how long should it take to pick the same amount of strawberries?
                                                                                         1 hour

Introduction

Ask, "If you increase speed from 50 to 60 miles per hour, will you travel more or fewer miles in 3 hours?" You will travel more miles. When one number increases at the same rate that another increases, we say the two numbers vary directly.

Have the students give a few more examples of direct proportions. The amount of harvest is directly related to the number of acres and the bushels per acre. The number of miles driven on a tankful of gasoline is directly related to the number of miles per gallon and the number of gallons in the tank.

Now ask the students for some examples of inverse proportions. Rate and time are inversely related. The greater the speed, the less time is required to travel a given distance. In using a lever, the number of pounds and the distance between the fulcrum and load are related by inverse proportion. When one number decreases at the same rate that another increases, we say the two numbers vary inversely.

This is a review of the last two lessons. Use the *Class Practice* exercises to review proportions and the *Teaching Guide* points below to reinforce the skills exercised in the last two lessons.

Teaching Guide

1. **In a direct proportion, one quantity increases at the same rate that another quantity increases. The two quantities vary directly in relation to each other.**

2. **In an inverse proportion, one quantity increases at the same rate that another item decreases. The two parts vary inversely in relation to each other.**

Use the exercises below and in the first part of *Class Practice* to review direct and inverse proportions.

Tell whether the facts in each relationship vary directly or inversely.

a. The number of quarts of peas that Mother freezes in relation to the bushels of peas that she picks. (Do more bushels picked mean more or fewer quarts of peas?) (directly)

b. The time required to travel from Philadelphia to Washington, D.C., in relation to the rate of speed. (Does more speed mean more or less time spent in travel?) (inversely)

Proportions for CLASS PRACTICE e–h

e. $\dfrac{\text{workers (1)}}{\text{workers (2)}} \quad \dfrac{3}{5} = \dfrac{n}{45} \quad \dfrac{\text{hours (2)}}{\text{hours (1)}}$ $n = 27$ hours

f. $\dfrac{\text{workers}}{\text{pieces}} \quad \dfrac{3}{12} = \dfrac{5}{n} \quad \dfrac{\text{workers}}{\text{pieces}}$ $n = 20$ pieces

g. $\dfrac{\text{dozens}}{\text{minutes}} \quad \dfrac{20}{45} = \dfrac{12}{n} \quad \dfrac{\text{dozens}}{\text{minutes}}$ $n = 27$ minutes

h. $\dfrac{\text{workers (1)}}{\text{workers (2)}} \quad \dfrac{2}{3} = \dfrac{n}{1\frac{1}{2}} \quad \dfrac{\text{hours (2)}}{\text{hours (1)}}$ $n = 1$ hour

T–185   Chapter 4   Decimals, Ratios, and Proportions

c. The weight of water in relation to the number of gallons of water. (Do more gallons of water mean more or less weight of water?)
(directly)

d. The number of men shingling a roof in relation to the time required to complete the roofing job. (Do more men require more or less time to complete the job?)
(inversely)

Use proportions to solve these problems.

e. A 20-foot building casts a 12-foot shadow. How tall is a nearby tree if it casts a 51-foot shadow?

$$\frac{\text{height}}{\text{shadow}} \quad \frac{20}{12} = \frac{n}{51} \quad \frac{\text{height}}{\text{shadow}}$$

$20 \times 51 = 1{,}020$
$1{,}020 \div 12 = 85$
($n = 85$ ft.)

f. A truck traveled 336 miles on 42 gallons of fuel. At that rate, how many miles can it travel on 55 gallons of fuel?

$$\frac{\text{miles}}{\text{gallons}} \quad \frac{336}{42} = \frac{n}{55} \quad \frac{\text{miles}}{\text{gallons}}$$

$336 \times 55 = 18{,}480$
$18{,}480 \div 42 = 440$
($n = 440$ mi.)

Proportions for Part B

7. $\dfrac{\text{miles}}{\text{hours}} \quad \dfrac{125}{2\frac{1}{2}} = \dfrac{n}{3\frac{1}{2}} \quad \dfrac{\text{miles}}{\text{hours}}$ $n = 175$ miles

8. $\dfrac{\text{miles}}{\text{minutes}} \quad \dfrac{2}{14} = \dfrac{3\frac{1}{2}}{n} \quad \dfrac{\text{miles}}{\text{minutes}}$ $n = 24\frac{1}{2}$ minutes

9. $\dfrac{\text{height}}{\text{shadow}} \quad \dfrac{7}{12} = \dfrac{n}{78} \quad \dfrac{\text{height}}{\text{shadow}}$ $n = 45\frac{1}{2}$ feet

10. $\dfrac{\text{pounds on side 1}}{\text{pounds on side 2}} \quad \dfrac{90}{75} = \dfrac{4}{n} \quad \dfrac{\text{feet from fulcrum (2)}}{\text{feet from fulcrum (1)}}$ $n = 3\frac{1}{3}$ feet

11. $\dfrac{\text{workers (1)}}{\text{workers (2)}} \quad \dfrac{6}{8} = \dfrac{n}{5} \quad \dfrac{\text{days (2)}}{\text{days (1)}}$ $n = 3\frac{3}{4}$ days

12. $\dfrac{\text{men (1)}}{\text{men (2)}} \quad \dfrac{4}{7} = \dfrac{n}{2} \quad \dfrac{\text{hours (2)}}{\text{hours (1)}}$ $n = 1\frac{1}{7}$ hours

13. $\dfrac{\text{oil}}{\text{smelt}} \quad \dfrac{\frac{3}{4}}{6} = \dfrac{n}{44} \quad \dfrac{\text{oil}}{\text{smelt}}$ $n = 5\frac{1}{2}$ cups

14. $\dfrac{\text{salt}}{\text{okra}} \quad \dfrac{\frac{1}{3}}{3\frac{1}{2}} = \dfrac{n}{28} \quad \dfrac{\text{salt}}{\text{okra}}$ $n = 2\frac{2}{3}$ cups

15. $\dfrac{\text{quarts per hour (1)}}{\text{quarts per hour (2)}} \quad \dfrac{10}{12} = \dfrac{n}{5\frac{1}{2}} \quad \dfrac{\text{hours (2)}}{\text{hours (1)}}$ $n = 4\frac{7}{12}$ hours

Lesson 55

WRITTEN EXERCISES

A. Write whether the facts in each relationship vary *directly* **or** *inversely*.

1. The number of bushels of peas picked in relation to the number of quarts of shelled peas. (Do more bushels of peas result in more or fewer quarts of shelled peas?)
   directly

2. Greater speed in relation to the time needed to complete a trip. (Does more speed result in more or less travel time?)
   inversely

3. The distance traveled in relation to the speed of travel. (Does more distance require more or less speed?)
   directly

4. The height of a stick in relation to the length of the shadow it casts. (Does a taller stick cast a longer or shorter shadow?)
   directly

5. The number of men working on a job in relation to the time required to finish the job. (Do more workers result in more or less time required?)
   inversely

6. The weight of a person in relation to his distance away from the fulcrum of a seesaw in order to balance. (Does more weight mean more or less distance from the fulcrum?)
   inversely

B. Solve by using direct or inverse proportions. (See facing page.)

7. The Millers drove 125 miles in $2\frac{1}{2}$ hours. At that rate, how far can they drive in $3\frac{1}{2}$ hours?
   175 miles

8. Marvin rode his bicycle 2 miles in 14 minutes. At that rate, how long will it take him to drive $3\frac{1}{2}$ miles?
   $24\frac{1}{2}$ minutes

9. If a clothesline post is 7 feet tall and casts a 12-foot shadow, how tall is a tree that casts a 78-foot shadow?
   $45\frac{1}{2}$ feet

10. Stephen weighs 90 pounds and Dwight weighs 75 pounds. How far must Stephen sit from the fulcrum of the seesaw in order to balance Dwight, who sits 4 feet from the fulcrum?
    $3\frac{1}{3}$ feet

11. If it normally takes 6 men 5 days to complete a given job, how long should it take 8 men to complete the same job?
    $3\frac{3}{4}$ days

12. If it took 4 men 2 hours to distribute a certain number of tracts, how long should it take 7 men to distribute the same number of tracts?
    $1\frac{1}{7}$ hours

13. Jean's family netted some little fish called smelt. To can the smelt, Mother is using $\frac{3}{4}$ cup of vegetable oil per 6 pounds of smelt. How much oil is needed to can 44 pounds of smelt?
    $5\frac{1}{2}$ cups

14. To make okra pickles, Mother uses $\frac{1}{3}$ cup of salt for $3\frac{1}{2}$ pounds of okra pods. How much salt is needed to pickle 28 pounds of okra pods?
    $2\frac{2}{3}$ cups

15. One morning Dale picked strawberries at the rate of 10 quarts per hour for $5\frac{1}{2}$ hours. If he increases his rate to 12 quarts per hour, how long will it take to pick the same amount?
    $4\frac{7}{12}$ hours

Chapter 4 Decimals, Ratios, and Proportions

16. Another morning, Dale picked 55 quarts in 5 hours. At that rate, how many quarts should he be able to pick in 6 hours? 66 quarts

17. Mother has a piece of material large enough to make 3 dresses that take $2\frac{1}{2}$ yards each. How many dresses can Mother make out of the same piece of material if she makes smaller girls' dresses that take $1\frac{3}{4}$ yards each? (Drop any fraction in your answer.) 4 dresses

18. If Mother can make 3 dresses out of an 8-yard piece of material, how many yards will she need to make 5 dresses of the same size? $13\frac{1}{3}$ yards

REVIEW EXERCISES

C. Solve mentally by using the fractional equivalent of each decimal. *(Lesson 51)*

19. $0.33\frac{1}{3}$ of 39   13

20. $0.16\frac{2}{3}$ of 42   7

D. Write these decimals, using words. *(Lesson 47)*

21. 0.0803
    Eight hundred three ten-thousandths

22. 12.000411   Twelve and four hundred eleven millionths

E. Find the reciprocal of each number. *(Lesson 39)*

23. $\frac{5}{12}$ $\frac{12}{5}$

24. $1\frac{5}{6}$ $\frac{6}{11}$

F. Solve these division problems. *(Lesson 39)*

25. $\frac{4}{5} \div \frac{8}{15}$ $1\frac{1}{2}$

26. $\frac{13}{24} \div \frac{1}{8}$ $4\frac{1}{3}$

G. Solve these problems involving compound measures. *(Lesson 26)*

27. 1 hr. 38 min.
    5)8 hr. 10 min.

28. 15 yd. 2 ft.
     $\times$ 4
     62 yd. 2 ft.

g. An automobile traveled a certain distance in $5\frac{1}{2}$ hours at an average rate of 50 miles per hour. After the speed limit was increased, the automobile averaged 60 miles per hour for the same distance. How long did the second trip take?

$$\frac{\text{hours (1)}}{\text{hours (2)}} \quad \frac{5\frac{1}{2}}{n} = \frac{60}{50} \quad \frac{\text{m.p.h. (2)}}{\text{m.p.h. (1)}}$$

$5\frac{1}{2} \times 50 = 275$
$275 \div 60 = 4\frac{7}{12}$
$(n = 4\frac{7}{12} \text{ hr.})$

h. A 5-inch pulley is driving a 3-inch pulley. If the 3-inch pulley is turning at a rate of 2,000 revolutions per minute, how fast is the 5-inch pulley turning?

$$\frac{\text{diameter (1)}}{\text{diameter (2)}} \quad \frac{5}{3} = \frac{2,000}{n} \quad \frac{\text{r.p.m. (2)}}{\text{r.p.m. (1)}}$$

$3 \times 2,000 = 6,000$
$6,000 \div 5 = 1,200$
$(n = 1,200 \text{ r.p.m.})$

---

16. $\frac{\text{quarts}}{\text{hours}} \quad \frac{55}{5} = \frac{n}{6} \quad \frac{\text{quarts}}{\text{hours}}$ $n = 66$ quarts

17. $\frac{\text{dresses (1)}}{\text{dresses (2)}} \quad \frac{3}{n} = \frac{1\frac{3}{4}}{2\frac{1}{2}} \quad \frac{\text{yards (2)}}{\text{yards (1)}}$ $n = 4$ dresses

18. $\frac{\text{dresses}}{\text{yards}} \quad \frac{3}{8} = \frac{5}{n} \quad \frac{\text{dresses}}{\text{yards}}$ $n = 13\frac{1}{3}$ yards

T–187   Chapter 4   Decimals, Ratios, and Proportions

LESSON 56

Objectives

- To review the proportion method of using the scale of miles to find actual distances on a map.
- To review changing actual distances to scale distances on a map.

Review

1. Give Lesson 56 Quiz (Proportions).

2. *Solve by using inverse proportions.* (Lesson 54)

   a. Lamar is riding his bicycle up a hill. The diameter of the drive sprocket is 8 inches, and the diameter of the sprocket on the wheel is $3\frac{1}{4}$ inches. How many times must Lamar turn his pedals to make the wheel turn 16 times?

   diameter (1) $\dfrac{8}{3\frac{1}{4}} = \dfrac{16}{n}$ rotations (2)
   diameter (2)   rotations (1)
   ($n = 6\frac{1}{2}$ times)

   b. Father is using a lever to lift a 200-pound rock, which is 14 inches from the fulcrum. If Father pushes down 40 inches from the fulcrum, how much force must exert to lift the rock?

   lb. on side 1 $\dfrac{n}{200} = \dfrac{14}{40}$ in. from fulcrum (2)
   lb. on side 2   in. from fulcrum (1)
   ($n = 70$ lb.)

3. *Write ratios in lowest terms to make the following comparisons.* (Lesson 52)

   a. Compare the 27 chapters in Leviticus with the 36 chapters in Numbers. ($\frac{27}{36} = \frac{3}{4}$)

   b. The load lifted by a lever is 55 pounds compared with the applied force of 25 pounds. ($\frac{55}{25} = \frac{11}{5}$)

4. *Multiply these decimals. Do the first two mentally.* (Lesson 48)

   a. 10×1.806 (18.06)

   b. 100×2.5011 (250.11)

   c. $\begin{array}{r} 0.078 \\ \times\, 0.9 \\ \hline \end{array}$ (0.0702)

   d. $\begin{array}{r} 1.025 \\ \times\, 0.06 \\ \hline \end{array}$ (0.0615)

5. *Write these equivalents by memory.* (Lesson 27)

   a. 1 cm = ___ in.   (0.39)
   b. 1 m = ___ ft.   (3.28)
   c. 1 kg = ___ lb.   (2.2)
   d. 1 l = ___ qt.   (1.06)
   e. 1 ha = ___ a.   (2.5)

56. Applying Direct Proportions to Maps

Maps are small drawings that show the locations of places and the distances between them. A map drawn to scale shows each place in direct proportion to its actual location. Suppose the distance between point A and point B on a map is twice as many inches as from point B to point C. Then the distance from the real place labeled A to the real place labeled B is actually twice as far as from the real place B to the real place C.

In making maps, a cartographer (mapmaker) chooses a ratio by which to reduce all the actual distances to a size that is convenient for his map. This ratio is commonly stated as a scale indicating that 1 inch or 1 centimeter equals a certain number of miles or kilometers. The antecedent of the ratio represents the drawing, and the consequent represents the actual distance.

Because maps use proportion, they provide a simple way to calculate actual distances. The following steps explain how to do this.

1. Write the scale as the primary ratio of the proportion. Label the two ratios as usual for a direct proportion.

2. Carefully measure the distance on the map, and write the measurement beside the correct label in the second ratio. Write n for the missing part of the second ratio.

3. Calculate the missing part of the proportion.

All proportions can be checked by cross multiplication. The product of the means equals the product of the extremes.

Example A

Two cities are 3.5 inches apart on a map with a scale of 1 inch = 25 miles. Find the actual distance between the cities.

$$\text{scale in.} \quad \frac{1}{25} = \frac{3.5}{n} \quad \text{scale in.}$$
$$\text{actual mi.} \qquad \qquad \qquad \text{actual mi.}$$

$25 \times 3.5 = 87.5; \quad 87.5 \div 1 = 87.5$

$n = 87.5$ miles (Division is actually not needed since the divisor is 1.)

Check: $25 \times 3.5 = 87.5; \quad 1 \times 87.5 = 87.5$

Example B

Find the distance between two villages that are $4\frac{1}{4}$ inches apart on a map with a scale of $\frac{3}{4}$ inch = 2 miles.

$$\text{scale in.} \quad \frac{\frac{3}{4}}{2} = \frac{4\frac{1}{4}}{n} \quad \text{scale in.}$$
$$\text{actual mi.} \qquad \qquad \qquad \text{actual mi.}$$

$2 \times 4\frac{1}{4} = 8\frac{1}{2}; \quad 8\frac{1}{2} \div \frac{3}{4} = 11\frac{1}{3}$

$n = 11\frac{1}{3}$ miles

Check: $2 \times 4\frac{1}{4} = 8\frac{1}{2}; \quad \frac{3}{4} \times 11\frac{1}{3} = 8\frac{1}{2}$

188 Chapter 4 *Decimals, Ratios, and Proportions*

The same method can be used when making a map to decide how far apart two places should be. The only difference is in writing the second ratio.

1. Write the scale as the primary ratio of the proportion. Label the two ratios as usual for a direct proportion.

2. Write the distance between the two locations beside the correct label in the second ratio. Write n for the missing part of the second ratio.

3. Calculate the missing part of the proportion.

Example C

Two cities are 40 miles apart. Find the distance that should be between them on a map with a scale of 1 inch = 15 miles.

$$\text{scale in.} \quad \frac{1}{15} = \frac{n}{40} \quad \text{scale in.}$$
$$\text{actual mi.} \qquad\qquad\qquad \text{actual mi.}$$

$1 \times 40 = 40; \quad 40 \div 15 = 2\frac{2}{3}$

$n = 2\frac{2}{3}$ inches (Multiplication is actually not needed since one factor is 1.)

Check: $1 \times 40 = 40; \quad 15 \times 2\frac{2}{3} = 40$

Example D

Two cities are 288 miles apart. How far apart should they be on a map with a scale of $\frac{3}{8}$ inch = 100 miles?

$$\text{scale in.} \quad \frac{\frac{3}{8}}{100} = \frac{n}{288} \quad \text{scale in.}$$
$$\text{actual mi.} \qquad\qquad\qquad \text{actual mi.}$$

$\frac{3}{8} \times 288 = 108; \quad 108 \div 100 = 1.08$

$n = 1.08$ inches

Check: $\frac{3}{8} \times 288 = 108; \quad 100 \times 1.08 = 108$

CLASS PRACTICE

Use proportions to find these distances.

a. scale: 1 in. = 24 mi.; measurement on map: $2\frac{5}{8}$ in. $n = 63$ miles

b. scale: 1 in. = 35 mi.; measurement on map: $5\frac{1}{2}$ in. $n = 192\frac{1}{2}$ miles

c. scale: $\frac{1}{2}$ in. = 8 mi.; measurement on map: $8\frac{2}{3}$ in. $n = 138\frac{2}{3}$ miles

d. scale: 1 in. = $2\frac{1}{2}$ mi.; measurement on map: $2\frac{4}{5}$ in. $n = 7$ miles

Use proportions to find the correct measurements.

e. scale: 1 in. = 28 mi.; actual distance: 168 mi. $n = 6$ inches

f. scale: 1 in. = $2\frac{1}{2}$ mi.; actual distance: $8\frac{1}{2}$ mi. $n = 3\frac{2}{5}$ inches

g. scale: $\frac{3}{4}$ in. = 15 mi.; actual distance: 42 mi. $n = 2\frac{1}{10}$ inches

h. scale: $\frac{1}{5}$ in. = 40 mi.; actual distance: 220 mi. $n = 1\frac{1}{10}$ inches

Introduction

Sometimes the scale of a map is printed in such a manner as 1:13,939,000. What does this ratio mean? It means that if the map were made 13,939,000 times longer and wider, it would be the same size as the land it represents. The cartographers shrank 13,939,000 inches to 1 inch on the map.

Perhaps the globe in your classroom shows a scale. Multiplying the size of a 12-inch globe by 41,851,400 would make it about the same size as the actual earth.

Although no map is perfect, maps are drawn in direct proportion to the geographic areas they represent. For this reason, the map provides a way to find actual distances by using proportions. If two locations are 1 inch apart on a 12-inch globe, they are 41,851,400 inches apart on the earth. This translates to a scale of about 1 inch = 661 miles.

Teaching Guide

1. **Since a map shows each place in direct proportion to its actual location on the earth, proportions can be used to find distances on a map.** Follow the steps below.

    (1) Write the scale as the primary ratio of the proportion. Label the two ratios as usual for a direct proportion.

    (2) Carefully measure the distance on the map, and write the measurement beside the correct label in the second ratio. Write n for the missing part of the second ratio.

    (3) Calculate the missing part of the proportion.

    All proportions can be checked by cross multiplication.

Proportions for CLASS PRACTICE

a. $\dfrac{\text{scale in.}}{\text{actual mi.}} \quad \dfrac{1}{24} = \dfrac{2\frac{5}{8}}{n} \quad \dfrac{\text{scale in.}}{\text{actual mi.}} \qquad n = 63 \text{ mi.}$

b. $\dfrac{\text{scale in.}}{\text{actual mi.}} \quad \dfrac{1}{35} = \dfrac{5\frac{1}{2}}{n} \quad \dfrac{\text{scale in.}}{\text{actual mi.}} \qquad n = 192\frac{1}{2} \text{ mi.}$

c. $\dfrac{\text{scale in.}}{\text{actual mi.}} \quad \dfrac{\frac{1}{2}}{8} = \dfrac{8\frac{2}{3}}{n} \quad \dfrac{\text{scale in.}}{\text{actual mi.}} \qquad n = 138\frac{2}{3} \text{ mi.}$

d. $\dfrac{\text{scale in.}}{\text{actual mi.}} \quad \dfrac{1}{2\frac{1}{2}} = \dfrac{2\frac{4}{5}}{n} \quad \dfrac{\text{scale in.}}{\text{actual mi.}} \qquad n = 7 \text{ mi.}$

e. $\dfrac{\text{scale in.}}{\text{actual mi.}} \quad \dfrac{1}{28} = \dfrac{n}{168} \quad \dfrac{\text{scale in.}}{\text{actual mi.}} \qquad n = 6 \text{ in.}$

f. $\dfrac{\text{scale in.}}{\text{actual mi.}} \quad \dfrac{1}{2\frac{1}{2}} = \dfrac{n}{8\frac{1}{2}} \quad \dfrac{\text{scale in.}}{\text{actual mi.}} \qquad n = 3\frac{2}{5} \text{ in.}$

g. $\dfrac{\text{scale in.}}{\text{actual mi.}} \quad \dfrac{\frac{3}{4}}{15} = \dfrac{n}{42} \quad \dfrac{\text{scale in.}}{\text{actual mi.}} \qquad n = 2\frac{1}{10} \text{ in.}$

h. $\dfrac{\text{scale in.}}{\text{actual mi.}} \quad \dfrac{\frac{1}{5}}{40} = \dfrac{n}{220} \quad \dfrac{\text{scale in.}}{\text{actual mi.}} \qquad n = 1\frac{1}{10} \text{ in.}$

T-189 Chapter 4 Decimals, Ratios, and Proportions

a. scale: 1 in. = 376 mi.
distance on map: $12\frac{3}{8}$ in.

$$\begin{array}{c}\text{inches}\\ \text{miles}\end{array}\ \frac{1}{376}=\frac{12\frac{3}{8}}{n}\ \begin{array}{c}\text{inches}\\ \text{miles}\end{array}$$

$376 \times 12\frac{3}{8} = 4{,}653;\ 4{,}653 \div 1 = 4{,}653$
($n = 4{,}653$ mi.)

Note: Division is actually not needed since the divisor is 1.

b. scale: $\frac{3}{4}$ in. = 3 mi.
distance on map: $12\frac{1}{4}$ in.

$$\begin{array}{c}\text{inches}\\ \text{miles}\end{array}\ \frac{\frac{3}{4}}{3}=\frac{12\frac{1}{4}}{n}\ \begin{array}{c}\text{inches}\\ \text{miles}\end{array}$$

$3 \times 12\frac{1}{4} = 36\frac{3}{4};\ 36\frac{3}{4} \div \frac{3}{4} = 49$
($n = 49$ mi.)

Proportion Arrangement for Part A
(Given numbers go in asterisk position.)

1–4. $\quad\begin{array}{c}\text{scale in.}\\ \text{actual mi.}\end{array}\ \dfrac{1}{40}=\dfrac{*}{n}\ \begin{array}{c}\text{scale in.}\\ \text{actual mi.}\end{array}$

5–8. $\quad\begin{array}{c}\text{scale in.}\\ \text{actual mi.}\end{array}\ \dfrac{1}{24}=\dfrac{*}{n}\ \begin{array}{c}\text{scale in.}\\ \text{actual mi.}\end{array}$

9–12. $\quad\begin{array}{c}\text{scale in.}\\ \text{actual mi.}\end{array}\ \dfrac{1}{3\frac{1}{2}}=\dfrac{*}{n}\ \begin{array}{c}\text{scale in.}\\ \text{actual mi.}\end{array}$

13–16. $\quad\begin{array}{c}\text{scale in.}\\ \text{actual mi.}\end{array}\ \dfrac{\frac{5}{8}}{12}=\dfrac{*}{n}\ \begin{array}{c}\text{scale in.}\\ \text{actual mi.}\end{array}$

Proportions for Part B

17. $\quad\begin{array}{c}\text{scale in.}\\ \text{actual mi.}\end{array}\ \dfrac{1}{16}=\dfrac{n}{40}\ \begin{array}{c}\text{scale in.}\\ \text{actual mi.}\end{array}\qquad n = 2\frac{1}{2}$ in.

18. $\quad\begin{array}{c}\text{scale in.}\\ \text{actual mi.}\end{array}\ \dfrac{1}{32}=\dfrac{n}{88}\ \begin{array}{c}\text{scale in.}\\ \text{actual mi.}\end{array}\qquad n = 2\frac{3}{4}$ in.

19. $\quad\begin{array}{c}\text{scale in.}\\ \text{actual mi.}\end{array}\ \dfrac{1}{64}=\dfrac{n}{328}\ \begin{array}{c}\text{scale in.}\\ \text{actual mi.}\end{array}\qquad n = 5\frac{1}{8}$ in.

20. $\quad\begin{array}{c}\text{scale in.}\\ \text{actual mi.}\end{array}\ \dfrac{1}{24}=\dfrac{n}{87}\ \begin{array}{c}\text{scale in.}\\ \text{actual mi.}\end{array}\qquad n = 3\frac{5}{8}$ in.

Proportions for Part C

21. $\quad\begin{array}{c}\text{scale in.}\\ \text{actual mi.}\end{array}\ \dfrac{1}{\frac{8}{9}}=\dfrac{8\frac{3}{8}}{n}\ \begin{array}{c}\text{scale in.}\\ \text{actual mi.}\end{array}\qquad n = 7\frac{4}{9}$ mi.

22. $\quad\begin{array}{c}\text{scale in.}\\ \text{actual mi.}\end{array}\ \dfrac{1}{0.4}=\dfrac{1\frac{1}{4}}{n}\ \begin{array}{c}\text{scale in.}\\ \text{actual mi.}\end{array}\qquad n = \frac{1}{2}$ mi.

23. $\quad\begin{array}{c}\text{scale in.}\\ \text{actual mi.}\end{array}\ \dfrac{1}{12}=\dfrac{1\frac{1}{2}}{n}\ \begin{array}{c}\text{scale in. }(5\frac{1}{4}-3\frac{3}{4})\\ \text{actual mi.}\end{array}\qquad n = 18$ mi.

WRITTEN EXERCISES

A. Use proportions to find these distances.

Scale for numbers 1–4: 1 in. = 40 mi.

1. $3\frac{1}{2}$ in. $n = 140$ mi.
2. $7\frac{1}{4}$ in. $n = 290$ mi.
3. $5\frac{3}{8}$ in. $n = 215$ mi.
4. $6\frac{7}{8}$ in. $n = 275$ mi.

Scale for numbers 5–8: 1 in. = 24 mi.

5. $2\frac{1}{2}$ in. $n = 60$ mi.
6. $6\frac{3}{4}$ in. $n = 162$ mi.
7. $3\frac{7}{8}$ in. $n = 93$ mi.
8. $7\frac{5}{8}$ in. $n = 183$ mi.

Scale for numbers 9–12: 1 in. = $3\frac{1}{2}$ mi.

9. $4\frac{1}{2}$ in. $n = 15\frac{3}{4}$ mi.
10. $5\frac{1}{4}$ in. $n = 18\frac{3}{8}$ mi.
11. $4\frac{3}{8}$ in. $n = 15\frac{5}{16}$ mi.
12. $8\frac{1}{8}$ in. $n = 28\frac{7}{16}$ mi.

Scale for numbers 13–16: $\frac{5}{8}$ in. = 12 mi.

13. $1\frac{1}{2}$ in. $n = 28\frac{4}{5}$ mi.
14. $2\frac{3}{4}$ in. $n = 52\frac{4}{5}$ mi.
15. $3\frac{5}{8}$ in. $n = 69\frac{3}{5}$ mi.
16. $6\frac{7}{8}$ in. $n = 132$ mi.

B. Use proportions to find the correct measurements.

17. scale: 1 in. = 16 mi.; actual distance: 40 mi. $n = 2\frac{1}{2}$ in.
18. scale: 1 in. = 32 mi.; actual distance: 88 mi. $n = 2\frac{3}{4}$ in.
19. scale: 1 in. = 64 mi.; actual distance: 328 mi. $n = 5\frac{1}{8}$ in.
20. scale: 1 in. = 24 mi.; actual distance: 87 mi. $n = 3\frac{5}{8}$ in.

C. Solve these problems by using direct or inverse proportions. (See facing page.)

21. On a map with a scale of 1 inch = $\frac{8}{9}$ mile, the distance between the Readville and the Ruggles railroad stations is $8\frac{3}{8}$ inches. How many miles is it by railroad from Readville to Ruggles? $7\frac{4}{9}$ miles

22. On a map with a scale of 1 inch = 0.4 mile, the subway line between South Station and Park Street Station is $1\frac{1}{4}$ inches long. What is the distance from South Station to Park Street Station? $\frac{1}{2}$ mile

23. On a map with a scale of 1 inch = 12 miles, the distance from Caryville to Brownsville is $3\frac{3}{4}$ inches and from Caryville to Villa is $5\frac{1}{4}$ inches. From Caryville, how many miles farther is it to Villa than to Brownsville?   18 miles

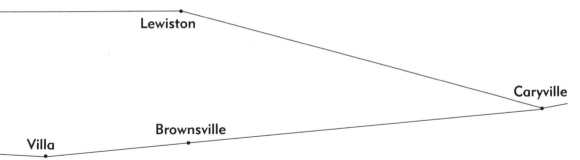

190   Chapter 4   Decimals, Ratios, and Proportions

24. A map having a scale of 1 inch = 20 miles shows a road from Smithsburg to Keysport passing through Lewiston and Caryville. On the map, the distance from Smithsburg to Lewiston is $1\frac{5}{8}$ inches, from Lewiston to Caryville is $2\frac{3}{8}$ inches, and from Caryville to Keysport is $1\frac{1}{4}$ inches. How many miles long is the route from Smithsburg to Keysport?
   105 miles

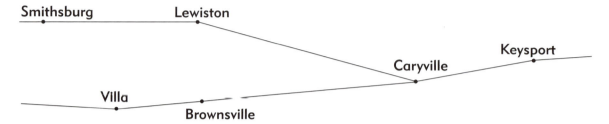

25. Father is planning to work on a remodeling project. On a similar project, it took 3 men 50 hours to finish the job. If Father has a fourth man on this project, how many hours should it take to complete the project? $37\frac{1}{2}$ hours

26. Five years ago, it took Father 20 minutes to drive to his job at an average speed of 45 miles per hour. Because of new housing in the area, he now averages 40 miles per hour. How many minutes does it take Father to drive to his job? $22\frac{1}{2}$ minutes

REVIEW EXERCISES

D. Write ratios in lowest terms for the following comparisons. (Lesson 52)

27. 50 chapters in Genesis compared with 40 chapters in Exodus   5 to 4
28. 27 books in the New Testament compared with 39 books in the Old Testament   9 to 13

E. Multiply these decimals. Do numbers 29 and 30 mentally. (Lesson 48)

29. 10 × 2.085   20.85
30. 1,000 × 11.9052   11,905.2

31. 0.092
    × 1.5
    ―――
    0.138

32. 3.012
    × 0.05
    ―――
    0.1506

F. Write these equivalents. Do as many as you can by memory. (Lesson 27)

33. 1 cm = 0.39 in.
34. 1 in. = 2.54 cm
35. 1 lb. = 0.45 kg
36. 1 kg = 2.2 lb.
37. 1 l = 1.06 qt.
38. 1 qt. = 0.95 l
39. 1 ft. = 0.3 m
40. 1 m = 3.28 ft.
41. 1 ha = 2.5 a.
42. 1 ton = 907 kg

Lesson 56 T-190

2. **A proportion can also be used when making a map to determine how far apart the places should be.** The only difference is in writing the second ratio.

   (1) Write the scale as the primary ratio of the proportion. Label the two ratios as usual for a direct proportion.

   (2) Write the distance between the two locations beside the correct label in the second ratio. Write n for the missing part of the second ratio.

   (3) Calculate the missing part of the proportion.

   a. Two cities are 21 miles apart. Find the distance that should be between them on a map with a scale of 1 inch = 12 miles.

   $$\frac{\text{inches}}{\text{miles}} \quad \frac{1}{12} = \frac{n}{21} \quad \frac{\text{inches}}{\text{miles}}$$

   $1 \times 21 = 21; 21 \div 12 = 1\frac{3}{4}$
   ($n = 1\frac{3}{4}$ mi.)

   Note: Multiplication is actually not needed since one factor is 1.

   b. Two cities are 261 miles apart. How far apart should they be on a map with a scale of $\frac{3}{8}$ inch = 27 miles?

   $$\frac{\text{inches}}{\text{miles}} \quad \frac{\frac{3}{8}}{27} = \frac{n}{261} \quad \frac{\text{inches}}{\text{miles}}$$

   $\frac{3}{8} \times 261 = 97\frac{7}{8}; 97\frac{7}{8} \div 27 = 3\frac{5}{8}$
   ($n = 3\frac{5}{8}$ in.)

---

24. $\frac{\text{inch}}{\text{miles}} \quad \frac{1}{20} = \frac{5\frac{1}{4}}{n} \quad \frac{\text{inches}}{\text{miles}} \quad (1\frac{5}{8} + 2\frac{3}{8} + 1\frac{1}{4})$ $n = 105$ miles

25. $\frac{\text{workers (1)}}{\text{workers (2)}} \quad \frac{3}{4} = \frac{n}{50} \quad \frac{\text{hours (2)}}{\text{hours (1)}}$ $n = 37\frac{1}{2}$ hours

26. $\frac{\text{m.p.h. (1)}}{\text{m.p.h. (2)}} \quad \frac{45}{40} = \frac{n}{20} \quad \frac{\text{minutes (2)}}{\text{minutes (1)}}$ $n = 22\frac{1}{2}$ minutes

Chapter 4 Decimals, Ratios, and Proportions

LESSON 57

Objectives

- To review the proportion method of using the scale on a blueprint to find actual distances.
- To review changing actual distances to scale distances on a blueprint.
- To teach *using a ratio to determine lengths on a blueprint.

Review

1. *Solve by mental division.* (Lesson 49)
    a. $106.4 \div 100$ (1.064)
    b. $1{,}804 \div 1{,}000$ (1.804)

2. *Solve these division problems. All answers divide evenly by the ten-thousandths' place.* (Lesson 49)
    a. $0.8 \overline{)0.02}$ (0.025)
    b. $8 \overline{)0.7}$ (0.0875)

3. *Find the unknown numbers.* (Lesson 41)
    a. 15 is $\frac{5}{7}$ of ___ (21)
    b. 21 is $\frac{7}{9}$ of ___ (27)

4. *Find these equivalents.* (Lesson 27)
    a. 15 qt. = ___ l (14.25)
    b. 7 m = ___ ft. (22.96)
    c. 37 cm = ___ in. (14.43)
    d. 5 a. = ___ ha. (2)

57. Applying Direct Proportions to Scale Drawings

A map is a scale drawing of a geographic area that may be millions of times larger than the map itself. Blueprints and other technical drawings are scale drawings of objects such as houses, barns, equipment, and even animals. A scale drawing may be larger or smaller than the actual size of the object. The relationship between the size of the scale drawing and the actual object is stated as a ratio.

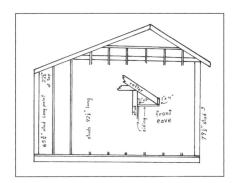

A blueprint of a barn is much smaller than the actual building. **Therefore, the antecedent of the ratio will be smaller than the consequent.** A drawing smaller than the actual object is called a *reduction*.

A detailed diagram of a small creature or a microscopic cross section of a leaf will be larger than the actual object. **Therefore, the antecedent of the ratio will be larger than the consequent.** A drawing larger than the actual object is called an *enlargement*.

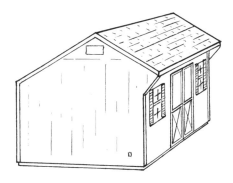

Reduction ratio: 1:24, or 1 inch = 24 inches
Enlargement ratio: 2:1, or 2 inches = 1 inch

When a ratio is given without labels, both parts represent the same unit of measure. For example, a ratio of 1:20 means 1 inch = 20 inches, 1 centimeter = 20 centimeters, and so on. The antecedent of the ratio always represents the drawing, and the consequent always represents the actual object.

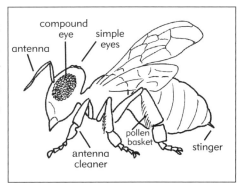

Terms of ratio for scale drawing

antecedent	1	units on drawing
consequent	50	units on actual object

Sometimes a map is labeled with a ratio such as 1:633,600, which means 1 inch equals 633,600 inches (52,800 feet). But it is more practical to express that ratio as 1 inch = 10 miles.

192   Chapter 4   Decimals, Ratios, and Proportions

The steps for calculating distances on scale drawings are the same as for distances on maps. They are given below.

1. Write the scale as the primary ratio of the proportion. Label the two ratios as usual for a direct proportion.
2. Carefully measure the distance on the scale drawing, and write the measurement beside the correct label in the second ratio. Write n for the missing part of the second ratio.
3. Calculate the missing part of the proportion. Check by cross multiplication.

Notice that the enlargement in Example B is calculated in the same way as the reduction in Example A. Be careful not to confuse the parts of the ratio when working with enlargements.

Example A
Find the distance represented by 35 centimeters if the scale is 1:40.

$$\frac{\text{scale}}{\text{actual}} \quad \frac{1}{40} = \frac{35}{n} \quad \frac{\text{scale cm}}{\text{actual cm}}$$

$40 \times 35 = 1{,}400$ cm

Example B
Find the distance represented by $1\frac{3}{4}$ inches if the scale is 2 inches = 1 inch.

$$\frac{\text{scale in.}}{\text{actual in.}} \quad \frac{2}{1} = \frac{1\frac{3}{4}}{n} \quad \frac{\text{scale in.}}{\text{actual in.}}$$

$1 \times 1\frac{3}{4} \div 2 = \frac{7}{8}$ in.

When making a scale drawing, you can use a direct proportion to calculate the distances on your paper. The only difference is that n, the unknown part of the proportion, represents the size of the drawn part rather than the actual part. Be sure to use correct labels, especially if one part of a ratio is inches and the other part is feet as in Example D.

Example C
Find the distance needed to show 12 feet on a blueprint if the scale is 1 foot = 8 feet.

$$\frac{\text{scale ft.}}{\text{actual ft.}} \quad \frac{1}{8} = \frac{n}{12} \quad \frac{\text{scale ft.}}{\text{actual ft.}}$$

$1 \times 12 \div 8 = 1\frac{1}{2}$ ft.

Example D
Find the distance needed to show 18 feet on a scale drawing if the scale is 1 inch = 4 feet.

$$\frac{\text{scale in.}}{\text{actual ft.}} \quad \frac{1}{4} = \frac{n}{18} \quad \frac{\text{scale in.}}{\text{actual ft.}}$$

$1 \times 18 \div 4 = 4\frac{1}{2}$ in.

CLASS PRACTICE

Use the scale drawing on the next page to find the lengths of the following items. Give all answers to the nearest foot.

a. How long is side 6 of the house (outside measure), not counting the porch?   25 feet
b. How long is the wall in the sewing room (inside measure) along side 2?   7 feet
c. How long is the porch on side 4?   15 feet
d. How wide is the porch on side 4?   5 feet

Introduction

If a blueprint is available, bring it along to class. Show it to the students, and ask them how it differs from a map. The main difference is that it is not reduced nearly as much as a map.

Blueprints need to be drawn large enough so that many details can be included. Show the students some of the details on the blueprint, such as doors, windows, closets, and even the stove and the refrigerator in the kitchen.

Teaching Guide

1. **A scale drawing is based on a ratio, just as a map is.** The antecedent of the scale ratio always represents the drawing, and the consequent always represents the actual object. Sometimes instead of having labels, a scale simply shows the ratio between the scale drawing and the actual object. In that case, both parts represent the same unit of measure.

   In the following ratios, have the students identify which part represents the scale drawing and which represents the actual object.

   a. (scale) → 1 in. = 8 ft. ← (actual)

   b. (scale) → 15:1 ← (actual)

2. **A scale drawing may be larger or smaller than the actual size of the object.** If the antecedent of the ratio is smaller than the consequent, the drawing is smaller than the actual object and is called a reduction. If the antecedent is larger than the consequent, the drawing is larger than the actual object and is called an enlargement.

   Have the students tell whether each ratio below represents an enlargement or a reduction.

   a. 1 in. = 4 ft. (reduction)

   b. 1 ft. = 1 in. (enlargement)

   c. 15 mm = 10 mm (enlargement)

   d. 1 in. = 1.5 in. (reduction)

   e. 3:1 (enlargement)

   f. 1:12 (reduction)

3. **The steps for calculating distances on scale drawings are the same as for distances on maps.** There may also be the added step of simplifying the answer.

T–193   Chapter 4   Decimals, Ratios, and Proportions

(1) Write the scale as the primary ratio of the proportion. Label the two ratios as usual for a direct proportion. Be sure the labels are correct, especially if one part of a ratio is inches and the other part is feet.

(2) Carefully measure the distance on the scale drawing, and write the measurement beside the correct label in the second ratio. Write n for the missing part of the second ratio.

(3) Calculate the missing part of the proportion.

(4) Simplify the answer. If possible, change inches to feet, or centimeters to meters. This will often be needed if the scale is a ratio with no labels.

As with all proportions, the solution can be checked by cross multiplication.

a. scale: 1 in. = 4 ft.
   measurement: $12\frac{1}{2}$ in.

   $$\frac{\text{scale in.}}{\text{actual ft.}} \quad \frac{1}{4} = \frac{12\frac{1}{2}}{n} \quad \frac{\text{scale in.}}{\text{actual ft.}}$$

   $4 \times 12\frac{1}{2} = 50; 50 \div 1 = 50$
   ($n = 50$ ft.)

b. scale: 5 cm = 1 m
   measurement: 12 cm

   $$\frac{\text{scale cm}}{\text{actual m}} \quad \frac{5}{1} = \frac{12}{n} \quad \frac{\text{scale cm}}{\text{actual m}}$$

   $1 \times 12 = 12; 12 \div 5 = 2.4$
   ($n = 2.4$ m)

4. **For making a scale drawing, a direct proportion can be used to calculate the distances on the paper.** The only difference is that n, the unknown part of the proportion, will be the size of the drawn part rather than the actual part. Emphasize that in proportions, any one of the four numbers can be found if the other three are known.

   a. Find the distance needed on a blueprint to show a 15-foot length if the scale is 1 in. = 4 ft.

   $$\frac{\text{scale in.}}{\text{actual ft.}} \quad \frac{1}{4} = \frac{n}{15} \quad \frac{\text{scale in.}}{\text{actual ft.}}$$

   $1 \times 15 = 15; 15 \div 4 = 3\frac{3}{4}$
   ($n = 3\frac{3}{4}$ in.)

   b. Find the distance needed on a scale drawing to show a 14-millimeter line if the scale is 5:2. Express your answer in centimeters.

   $$\frac{\text{scale}}{\text{actual}} \quad \frac{5}{2} = \frac{n}{14} \quad \frac{\text{scale mm}}{\text{actual mm}}$$

   $5 \times 14 = 70$
   $70 \div 2 = 35; 35$ mm $= 3.5$ cm
   ($n = 3.5$ cm)

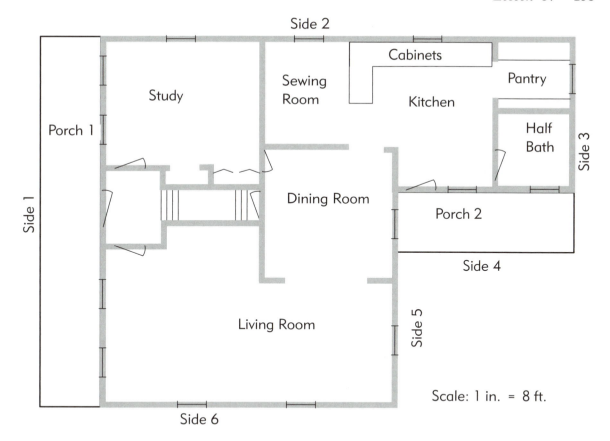

Tell whether each scale indicates enlargement or reduction.

e. 3 to 2
   enlargement
f. 2 to 3
   reduction
g. 1 cm to 4 m
   reduction
h. 7 in. to 1 in.
   enlargement

Find the distance represented by each of these lengths on a scale drawing.

i. $3\frac{1}{2}$ inches (scale: 1 in. = 4 ft.)   14 feet
j. $4\frac{7}{8}$ inches (scale: 1 in. = 8 ft.)   39 feet
k. $10\frac{1}{2}$ inches (scale: 3 in. = 1 in.) $3\frac{1}{2}$ inches
l. $2\frac{1}{2}$ inches (scale: 5:3) $1\frac{1}{2}$ inches
m. 45 centimeters (scale: 1:80) Answer in meters.   36 meters
n. 5.5 centimeters (scale: 2.5 to 1)   2.2 centimeters

Find the correct length on a scale drawing for each distance.

o. 22 feet (scale: 1 in. = 4 ft.) $5\frac{1}{2}$ inches
p. 14 feet (scale: 1 in. = 8 ft.) $1\frac{3}{4}$ inches
q. 4 inches (scale: 3 in. = 1 in.)   12 inches
r. $3\frac{3}{4}$ inches (scale: 5:3) $6\frac{1}{4}$ inches

194 Chapter 4 Decimals, Ratios, and Proportions

WRITTEN EXERCISES

A. Use the scale drawing in the lesson to find the lengths of the following items. Give all answers to the nearest foot.

1. How long is the wall in the living room (inside measure) along side 6? 24 feet
2. What is the outside length of side 1? 31 feet
3. How long are the cabinets along side 2 in the kitchen? 12 feet
4. How long is the wall between the sewing room and the dining room? 7 feet
5. What is the outside length of side 5, including the part at the end of porch 2? 18 feet
6. What is the outside length of side 2, not including the porch? 40 feet
7. How long is the sewing room along side 2 (from the cabinets to the corner)? 7 feet
8. How wide is the door that opens onto porch 1 from side 1? 3 feet

B. Write whether each scale indicates *enlargement* or *reduction*.

9. 5 to 7   10. 6 to 1   11. 4 m = 1 m   12. 1 in. = 6 in.
   reduction   enlargement   enlargement   reduction

C. Find the actual distance represented by each of these lengths on a scale drawing.

Scale for numbers 13 and 14: 1 in. = 4 ft.

13. $2\frac{1}{2}$ in.   10 feet   14. $3\frac{3}{8}$ in. $13\frac{1}{2}$ feet

Scale for numbers 15 and 16: 1:8

15. $3\frac{1}{4}$ in.   26 inches   16. $4\frac{5}{8}$ in.   37 inches

Scale for numbers 17 and 18: 1:90

17. 20 cm   1,800 cm   18. 12.6 cm   1,134 cm

Scale for numbers 19 and 20: 5 in. = 2 in.

19. $1\frac{1}{4}$ in. $\frac{1}{2}$ inch   20. $3\frac{1}{8}$ in. $1\frac{1}{4}$ inches

D. Find the correct length on a scale drawing for each distance. Use the scale $\frac{1}{4}$ inch = 1 foot.

21. 12 ft.   3 inches   22. $3\frac{1}{2}$ ft. $\frac{7}{8}$ inch

E. Solve these reading problems. Complete the proportions for numbers 23–26.

23. Dwight was helping his father build a corner cupboard. Dwight found the height of the cupboard on the drawing to be 13 inches. If the scale is 2 inches = 1 foot, tell how high the cupboard will be.

$$\text{scale in.} \quad \frac{2}{1} = \frac{13}{n} \quad \text{scale in.} \qquad n = 6\frac{1}{2} \text{ feet}$$
$$\text{actual ft.} \qquad \qquad \text{actual ft.}$$

T–195   Chapter 4   Decimals, Ratios, and Proportions

Solutions for Exercises 27 and 28

27. $779.00 − ($366.00 + $195.75 + $88.25)

28. $95.52 ÷ 16

24. The depth of the cupboard on the drawing (problem 23) was $3\frac{1}{2}$ inches. What was the actual depth of the cupboard?

$$\frac{\text{scale in.}}{\text{actual ft.}} \quad \frac{2}{1} = \frac{3\frac{1}{2}}{n} \quad \frac{\text{scale in.}}{\text{actual ft.}} \qquad n = 1\frac{3}{4} \text{ feet}$$

25. David worked on a model van just like the one his family has. The van was built to a scale of 1:30. If the model van was 7 inches long, what was the length of the family van in feet?

$$\frac{\text{scale}}{\text{actual}} \quad \frac{1}{30} = \frac{7}{n} \quad \frac{\text{scale in.}}{\text{actual in.}} \qquad n = 17\frac{1}{2} \text{ feet}$$

26. The students at the Woodvale Mennonite School are building a model of the Mayflower. The scale is 1:60, and the length of the actual Mayflower was 90 feet. How long will the model be?

$$\frac{\text{scale}}{\text{actual}} \quad \frac{1}{60} = \frac{n}{90} \quad \frac{\text{scale ft.}}{\text{actual ft.}} \qquad n = 1\frac{1}{2} \text{ feet}$$

27. Edward's father needs to replace the engine of the family van. The price of a rebuilt engine is $779.00. If he gets an engine from a salvage yard and rebuilds it, the costs will be as follows: engine—$366.00; parts—$195.75; and machine work—$88.25. How much less will the cost be if Father rebuilds the salvaged engine? $129.00

28. Father is replacing the tines on the family's garden tiller. The cost of 16 tines is $95.52. What is the cost of each tine? $5.97

REVIEW EXERCISES

F. Solve by mental division. *(Lesson 49)*

29. 1,005 ÷ 1,000   1.005

30. 506.01 ÷ 100   5.0601

G. Solve these division problems. All answers divide evenly by the thousandths' place. *(Lesson 49)*

31. 0.08)‾1.5   18.75

32. 40)‾2.6   0.065

H. Find the unknown numbers. *(Lesson 41)*

33. 16 is $\frac{8}{9}$ of 18

34. 12 is $\frac{2}{5}$ of 30

I. Find these equivalents. *(Lesson 27)*

35. 5 qt. = 4.75 l

36. 18 ft. = 5.4 m

37. 22 cm = 8.58 in.

38. 60 kg = 132 lb.

196 Chapter 4 Decimals, Ratios, and Proportions

58. Chapter 4 Review

A. Solve these problems. *(Lessons 47–51)*

1. 6.0398
    + 4.515
    10.5548

2. 15
    + 7.894
    22.894

3. 4.7
    − 2.9876
    1.7124

4. 7
    − 3.8687
    3.1313

5. 4.6125
    × 0.08
    0.369

6. 2.7735
    × 0.26
    0.72111

7. 0.03)2.76 **92**

8. 0.08)2.11 **26.375**

9. 14 + 16.75 + 15.9   **46.65**

10. 3.108 + 6.9 + 2.54 + 1.7779   **14.3259**

11. 0.81 − 0.7893   **0.0207**

12. 0.6 − 0.01765   **0.58235**

13. 14.7 × 0.45   **6.615**

14. 13.8 × 1.007   **13.8966**

B. Solve these problems mentally. Change decimals to fractions if it makes calculation easier. *(Lessons 48, 49, 51)*

15. 100 × 313.854   **31,385.4**

16. 1,000 × 2.31   **2,310**

17. 461.03 ÷ 1,000   **0.46103**

18. 11.43 ÷ 1,000   **0.01143**

19. 0.5 of 58   **29**

20. 0.75 of 88   **66**

21. 0.375 of 48   **18**

22. 0.625 of 32   **20**

C. Change these decimals to common fractions in lowest terms. *(Lesson 47)*

23. 0.64   **16/25**

24. 0.45   **9/20**

D. Do these exercises involving decimals. *(Lesson 47)*

25. Name the place farthest to the right in 0.3156.   **ten-thousandths**

26. Name the place farthest to the right in 0.217198.   **millionths**

27. Write 6.0085, using words.   **Six and eighty-five ten-thousandths**

28. Write 20.000175, using words.   **Twenty and one hundred seventy-five millionths**

29. Compare 0.0021 and 0.0199, and write < or > between them.   **0.0021 < 0.0199**

30. Compare 1.507 and 1.5071, and write < or > between them.   **1.507 < 1.5071**

E. Round each value to the place indicated. *(Lesson 50)*

31. 2.89456 (thousandth)   **2.895**

32. 3.17198 (ten-thousandth)   **3.1720**

LESSON 58

Objective

- To review the material taught in Chapter 4 (Lessons 47–57).

Teaching Guide

1. Give Lesson 58 Quiz (Proportions and Scale Drawings).
2. Lesson 58 reviews the material taught in Lessons 47–57. For pointers on using these review lessons, see *Teaching Guide* for Lesson 15.
3. Review the following concepts.
   a. Decimal places up to millionths. (Lesson 47)
   b. Multiplying decimals. (Lesson 48)
   c. Mentally multiplying and dividing by powers of ten by moving the decimal point. (Lessons 48, 49)
   d. Dividing decimals. (Lesson 49)
   e. Rounding decimals. (Lesson 50)
   f. Using division to change fractions to decimals. (Lesson 50)
   g. Know fraction–decimal equivalents by memory, including sixths. (Lesson 50)
   h. Simplifying multiplication by changing decimals to fractions. (Lesson 51)
   i. Ratios and the terms *antecedent* and the *consequent*. (Lesson 52)
   j. Using proportions in working with actual distances and scale distances on maps and scale drawings. (Lessons 56, 57)
   k. Using unlabeled ratios to determine lengths in scale drawings. (Lesson 57)
4. Be sure to review the following new concepts taught in this chapter.

Lesson number and new concept	Exercises in Lesson 58
52—Expressing ratios with either the antecedent or the consequent as 1 and the other part as a fraction or a decimal.	45, 46
54—Inverse proportions, and reading problems to be solved by them.	41, 42, 51–62
57—Using proportions to determine lengths on a blueprint.	47–50

F. Express these fractions as decimals. *(Lesson 50)*

33. $\frac{7}{16}$ 0.4375

34. $\frac{19}{40}$ 0.475

G. Solve each division. Round the quotients to the nearest ten-thousandth. *(Lesson 50)*

35. $45.7 \div 0.14$ 326.4286

36. $71.25 \div 6.7$ 10.6343

H. Write the letter of the problem that is easier to solve mentally. *(Lesson 51)*

37. a. 0.75 of 16   (b.) $\frac{3}{4}$ of 16

38. (a.) 0.4 of 12   b. $\frac{2}{5}$ of 12

I. Write whether each scale indicates *enlargement* **or** *reduction.* *(Lesson 57)*

39. 1:10   reduction

40. 3:1   enlargement

J. Label each proportion as *direct* **or** *inverse.* *(Lesson 54)*

41. $\frac{\text{size of drive gear}}{\text{size of driven gear}} \quad \frac{4}{5} = \frac{12}{n} \quad \frac{\text{r.p.m. of driven gear}}{\text{r.p.m. of drive gear}}$ inverse

42. $\frac{\text{number of items}}{\text{price}} \quad \frac{4}{\$5} = \frac{6}{n} \quad \frac{\text{number of items}}{\text{price}}$ direct

K. Write ratios as indicated for these statements. *(Lesson 52)*

43. One evening Sarah put 18 of the family's 21 spoons on the table. (Reduce to lowest terms.)   6 to 7

44. At the end of a trip, Father pumped 18 gallons of fuel into the car's 20-gallon tank. (Reduce to lowest terms.)   9 to 10

45. Dale has memorized 9 of the verses in John 14:1–15. (Write a ratio in which the consequent is 1 and the antecedent is in decimal form.)   0.6 to 1

46. Sanford raked 8,000 square feet of the lawn, compared with 5,000 square feet raked by his younger brother. (Write a ratio in which the antecedent is 1 and the consequent is a fraction in lowest terms.)   1 to $\frac{5}{8}$

L. Find the distance represented by each length on a map or scale drawing. *(Lessons 56, 57)*

Scale for numbers 47 and 48: 1:32

47. $2\frac{1}{4}$ in.   72 in.

48. $7\frac{7}{8}$ in.   252 in.

Scale for numbers 49 and 50: 2.5 cm = 1 cm

49. 3.5 cm   1.4 cm

50. 5.6 cm   2.24 cm

198   Chapter 4   *Decimals, Ratios, and Proportions*

(See facing page.)

M. Use direct or inverse proportions to solve these problems. Be careful, for one part of the proportion may need to be calculated. *(Lessons 53–55)*

51. Marcus, who is 60 inches tall, casts a 66-inch shadow when the school building casts a 28-foot shadow. How tall is the school building? $25\frac{5}{11}$ feet

52. An arborvitae bush in the Landises' lawn casts a 22-foot shadow. A nearby shrub 5 feet tall casts an 8-foot shadow. How tall is the arborvitae bush? $13\frac{3}{4}$ feet

53. A seesaw balanced when Douglas sat 5 feet from the fulcrum and Dwight sat 4 feet from the fulcrum. If Douglas weighed 80 pounds, how much did Dwight weigh? 100 pounds

54. Later Susan and Marilyn used the seesaw. Susan weighed 90 pounds and sat $4\frac{1}{2}$ feet from the fulcrum. If Marilyn's weight was 60 pounds, how far away did she need to sit in order to make the seesaw balance? $6\frac{3}{4}$ feet

55. One summer the Martins put away food at the ratio of 5 quarts canned for every 3 quarts frozen. If they canned 320 quarts, how many quarts did they freeze? 192 quarts

56. One week Father worked 5 hours on a building project for every 2 hours that Philip worked on the same project. If Philip worked 16 hours, what was the total number of hours that he and Father worked on the project that week? 56 hours

57. Father said that it once took 5 men 4 days to put up a building similar to one that the Smiths are starting. With only 2 men doing the job, how long can the project be expected to take? 10 days

58. An airplane traveling at an average of 350 miles per hour makes a certain flight in 7 hours. If a plane travels at an average of 560 miles per hour, how long will it take for the same trip? $4\frac{3}{8}$ hours

59. When the Witmers put a new roof on their house, they nailed 4 rows of shingles in 10 minutes. At that rate, how long should it take to nail the 54 rows of shingles on one side of the roof? (Give the answer in hours.) $2\frac{1}{4}$ hours

60. The Fields are insulating the attic of their old house. They have used 12 bales of insulation to cover 130 square feet of attic space. How many bales of insulation will the Fields need to insulate 660 square feet of attic space? (Round up to the next full bale.) 61 bales

61. One Sunday afternoon, Jonathan and Michael passed out 6 tracts for every 4 tracts that Father handed out. If Father distributed 124 tracts, how many tracts did they all distribute together? 310 tracts

62. One autumn, the Miller family plowed 3 acres of land for every 2 acres that they left unplowed. If they plowed 90 acres, how many acres did they leave unplowed? 60 acres

59. Chapter 4 Test

LESSON 59

Objective

- To test the students' mastery of the concepts in Chapter 4.

Teaching Guide

1. Correct Lesson 58.
2. Review any areas of special difficulty.
3. Administer the test. For pointers on giving tests, see *Teaching Guide* for Lesson 16.

Proportions for Part M

51. $\dfrac{\text{height}}{\text{shadow}} \quad \dfrac{60}{66} = \dfrac{n}{28} \quad \dfrac{\text{height}}{\text{shadow}} \qquad n = 25\tfrac{5}{11}$ feet

52. $\dfrac{\text{height}}{\text{shadow}} \quad \dfrac{n}{22} = \dfrac{5}{8} \quad \dfrac{\text{height}}{\text{shadow}} \qquad n = 13\tfrac{3}{4}$ feet

53. $\dfrac{\text{pounds on side 1}}{\text{pounds on side 2}} \quad \dfrac{80}{n} = \dfrac{4}{5} \quad \dfrac{\text{feet from fulcrum (2)}}{\text{feet from fulcrum (1)}} \qquad n = 100$ pounds

54. $\dfrac{\text{pounds on side 1}}{\text{pounds on side 2}} \quad \dfrac{90}{60} = \dfrac{n}{4\tfrac{1}{2}} \quad \dfrac{\text{feet from fulcrum (2)}}{\text{feet from fulcrum (1)}} \qquad n = 6\tfrac{3}{4}$ feet

55. $\dfrac{\text{canned}}{\text{frozen}} \quad \dfrac{5}{3} = \dfrac{320}{n} \quad \dfrac{\text{canned}}{\text{frozen}} \qquad n = 192$ quarts

56. $\dfrac{\text{Philip}}{\text{total}} \quad \dfrac{2}{7} = \dfrac{16}{n} \quad \dfrac{\text{Philip}}{\text{total}} \qquad n = 56$ hours

57. $\dfrac{\text{workers (1)}}{\text{workers (2)}} \quad \dfrac{5}{2} = \dfrac{n}{4} \quad \dfrac{\text{days (2)}}{\text{days (1)}} \qquad n = 10$ days

58. $\dfrac{\text{m.p.h. (1)}}{\text{m.p.h. (2)}} \quad \dfrac{350}{560} = \dfrac{n}{7} \quad \dfrac{\text{hours (2)}}{\text{hours (1)}} \qquad n = 4\tfrac{3}{8}$ hours

59. $\dfrac{\text{rows}}{\text{minutes}} \quad \dfrac{4}{10} = \dfrac{54}{n} \quad \dfrac{\text{rows}}{\text{minutes}} \qquad n = 2\tfrac{1}{4}$ hours

60. $\dfrac{\text{bales}}{\text{sq. ft.}} \quad \dfrac{12}{130} = \dfrac{n}{660} \quad \dfrac{\text{bales}}{\text{sq. ft.}} \qquad n = 61$ bales

61. $\dfrac{\text{Father}}{\text{total}} \quad \dfrac{4}{10} = \dfrac{124}{n} \quad \dfrac{\text{Father}}{\text{total}} \qquad n = 310$ tracts

62. $\dfrac{\text{plowed}}{\text{unplowed}} \quad \dfrac{3}{2} = \dfrac{90}{n} \quad \dfrac{\text{plowed}}{\text{unplowed}} \qquad n = 60$ acres

If a man is given one pound and accumulates ten pounds, is he a better servant than one who is given forty pounds and accumulates one hundred? Four times better!

Learn to think of gain and loss or savings on the percent scale. A percentage mentality helps us practice wise stewardship over that which God has entrusted to us.

Chapter 5
Mastering Percents

Percents are fractions in which numbers are written for the numerator alone. The denominator is indicated by the word or sign for percent. *Percent* comes from the Latin phrase *per centum*, which means "per hundred" or "by the hundred."

Percents are very common in business because they are easy to state and to grasp. They are used especially to express rates, such as for interest, discounts, and commissions. For calculation, percents are changed to decimals or fractions. Knowing how to work with percents is a useful skill with many applications in everyday life.

Lord, thy pound hath gained ten pounds.

And he said unto him,
Well, thou good servant:
because thou hast been faithful
in a very little, have thou authority
over ten cities.
Luke 19:16, 17

60. Expressing Rates as Percents

Decimals and percents are types of fractions. Decimal fractions have denominators that are powers of 10: tenths, hundredths, thousandths, and so on. Percents are limited to one denominator, 100, and they include the percent sign (%).

$$5\% = \frac{5}{100} \qquad 75\% = \frac{75}{100} \qquad 125\% = \frac{125}{100}$$

Percents, decimals, fractions, and ratios are four methods of expressing the same value, and their forms are interchangeable. The following points explain how to change percents to the other three forms.

(1) To change a percent to a decimal, drop the percent sign and move the decimal point two places to the left. Sometimes when doing this, you will need to annex a zero to the left as in Example B.

(2) To change a percent to a fraction, write the percent as the numerator and 100 as the denominator. Reduce to lowest terms.

(3) To change a percent to a ratio, write the percent as the antecedent and 100 as the consequent. Reduce to lowest terms.

	Percent	Decimal	Fraction	Ratio
Example A:	32%	= 0.32	= $\frac{32}{100}$ (or $\frac{8}{25}$)	= 32:100 (or 8:25)
Example B:	9%	= 0.09	= $\frac{9}{100}$	= 9:100
Example C:	135%	= 1.35	= $1\frac{35}{100}$ (or $1\frac{7}{20}$)	= 135:100 (or 27:20)

The process can also be reversed. The following steps will change ratios, fractions, and decimals to percents.

1. Write a ratio in fraction form.

2. Change the fraction to decimal form by dividing the numerator by the denominator. If the answer does not come out even in the hundredths' place, write the remainder as a fraction in lowest terms.

3. Change the decimal to a percent by moving the decimal point two places to the right and adding the percent sign.

	Ratio	Fraction	Decimal	Percent
Example D:	3:7	= $\frac{3}{7}$ (3 ÷ 7)	= $0.42\frac{6}{7}$	= $42\frac{6}{7}\%$
Example E:	1:20	= $\frac{1}{20}$ (1 ÷ 20)	= 0.05	= 5%
Example F:	3:2.2	= $\frac{3}{2.2}$ (3 ÷ 2.2)	= $1.36\frac{4}{11}$	= $136\frac{4}{11}\%$

LESSON 60

Objectives

- To review the meaning of percents.
- To review changing fractions to percents by using division.
- To review changing one- and two-place decimals to percents.
- To teach *changing percents to ratios and vice versa.

Review

1. *For each distance, use a proportion to find the correct measurement on a map.* (Lesson 56)

   a. scale: 1 in. = 32 mi.
      actual distance: 76 mi. ($2\frac{3}{8}$ in.)

   b. scale: $\frac{5}{8}$ in. = 14 mi.
      actual distance, 56 mi. ($2\frac{1}{2}$ in.)

2. *Write a ratio in lowest terms for each statement.* (Lesson 52)

   a. The load lifted by a lever is 55 pounds compared with the 20-pound force required to lift it. (11 to 4)

   b. It took 5 trailer trucks to transport the 40,000 chickens from the pullet house to the layer house. (1 to 8,000)

3. *Find these temperature equivalents to the nearest degree.* (Lesson 28)

   a. 39°F = ___°C (4)
   b. 142°F = ___°C (61)
   c. 25°C = ___°F (77)
   d. 111°C = ___°F (232)

Introduction

Ask the students how the quarter received its name. (It is a quarter of a dollar.)

How did the dime receive its name? (It comes from the Latin word *decima*, meaning "one-tenth.")

How did the cent receive its name? (It comes from the Latin word *centum*, meaning "hundred.")

The word *percent* comes from the same root. Percents express hundredth parts of a whole.

Can the students think of any other words with the *cent* root? (A *century* equals 100 years. A *centimeter* is $\frac{1}{100}$ of a meter. A *centipede* is supposed to have 100 feet. A *centenarian* is a person who is one hundred years old. A *centennial* is a hundredth anniversary.)

Teaching Guide

1. **Percents, like decimals, are a kind of fraction.** Decimals are fractions with denominators that are powers of 10 (tenths, hundredths, thousandths, and so on). Percents have only one denominator, 100. Percents include the percent sign (%) and show relationships in terms of hundredths.

   a. $65\% = 0.65 = \frac{65}{100}$

   b. $42\% = 0.42 = \frac{42}{100}$

   c. $3\% = 0.03 = \frac{3}{100}$

   d. $9\% = 0.09 = \frac{9}{100}$

2. **Percents can be changed to decimals, fractions, and ratios.**

   a. To change a percent to a decimal, drop the percent sign and move the decimal point two places to the left. Sometimes when doing this, you will need to annex a zero to the left.

   (1) $25\% = 0.25$

   (2) $32\% = 0.32$

   (3) $799\% = 7.99$

   (4) $7\% = 0.07$

   b. To change a percent to a fraction, write the percent as the numerator and 100 as the denominator. Reduce to lowest terms.

   (5) $31\% = \frac{31}{100}$

   (6) $49\% = \frac{49}{100}$

   (7) $32\% = \frac{32}{100} = \frac{8}{25}$

   (8) $96\% = \frac{96}{100} = \frac{24}{25}$

   c. To change a percent to a ratio, write the percent as the antecedent and 100 as the consequent. Reduce to lowest terms.

   (9) $68\% = 68:100 = 17:25$

   (10) $188\% = 188:100 = 47:25$

   (11) $91\% = 91:100$

   (12) $75\% = 75:100 = 3:4$

Lesson 60

Some of the most commonly used fractions and their percent equivalents are shown on the bars below. Note how similar they are to the decimal equivalents that you learned in Chapter 4. Memorize these equivalents well; you will need to know them for the chapter test.

10%	20%	30%	40%	50%	60%	70%	80%	90%	100%
				$\frac{1}{2}$ = 50%					
		$\frac{1}{3}$ = 33$\frac{1}{3}$%				$\frac{2}{3}$ = 66$\frac{2}{3}$%			
		$\frac{1}{4}$ = 25%					$\frac{3}{4}$ = 75%		
	$\frac{1}{5}$ = 20%		$\frac{2}{5}$ = 40%		$\frac{3}{5}$ = 60%		$\frac{4}{5}$ = 80%		
$\frac{1}{6}$ = 16$\frac{2}{3}$%							$\frac{5}{6}$ = 83$\frac{1}{3}$%		
$\frac{1}{8}$ = 12$\frac{1}{2}$%		$\frac{3}{8}$ = 37$\frac{1}{2}$%			$\frac{5}{8}$ = 62$\frac{1}{2}$%			$\frac{7}{8}$ = 87$\frac{1}{2}$%	
$\frac{1}{10}$ = 10%		$\frac{3}{10}$ = 30%				$\frac{7}{10}$ = 70%		$\frac{9}{10}$ = 90%	

CLASS PRACTICE

Express these percents as decimals.

a. 48%   0.48   b. 31%   0.31   c. 73%   0.73   d. 97%   0.97

Express these percents as fractions in lowest terms.

e. 64% $\frac{16}{25}$ f. 37% $\frac{37}{100}$ g. 15% $\frac{3}{20}$ h. 96% $\frac{24}{25}$

Express these percents as ratios in lowest terms.

i. 25%   1:4   j. 63%   63:100   k. 11%   11:100   l. 88%   22:25

m. 49%   49:100   n. 28%   7:25   o. 50%   1:2   p. 95%   19:20

Express these fractions, decimals, and ratios as percents.

q. $\frac{3}{4}$ 75%   r. $\frac{11}{20}$ 55%   s. $\frac{18}{25}$ 72%   t. $\frac{8}{9}$ 88$\frac{8}{9}$%

u. 9:25   36%   v. 19:50   38%   w. 0.6   60%   x. 0.08   8%

WRITTEN EXERCISES

A. *Express these percents as decimals.*

1. 61%   0.61   2. 5%   0.05   3. 40%   0.4

4. 8%   0.08   5. 1%   0.01   6. 86%   0.86

Chapter 5 Mastering Percents

B. Express these percents as fractions in lowest terms.

7. 31% $\frac{31}{100}$ 8. 3% $\frac{3}{100}$
9. 44% $\frac{11}{25}$ 10. 70% $\frac{7}{10}$
11. 7% $\frac{7}{100}$ 12. 2% $\frac{1}{50}$

C. Express these percents as ratios in lowest terms.

13. 19%   19:100   14. 30%   3:10
15. 77%   77:100   16. 16%   4:25
17. 48%   12:25   18. 6%   3:50

D. Express these fractions, decimals, and ratios as percents.

19. $\frac{3}{5}$ 60%   20. $\frac{17}{20}$ 85%
21. $\frac{19}{25}$ 76%   22. 0.2   20%
23. 0.3   30%   24. 13:25   52%

E. Express these fractions as percents. Try to do 25–27 by memory.

25. $\frac{3}{4}$ 75%   26. $\frac{7}{8}$ $87\frac{1}{2}$%
27. $\frac{1}{6}$ $16\frac{2}{3}$%   28. $\frac{7}{16}$ $43\frac{3}{4}$%
29. $\frac{13}{18}$ $72\frac{2}{9}$% 30. $\frac{17}{24}$ $70\frac{5}{6}$%

F. Solve these reading problems. Some problems contain extra information.

31. In Numbers 31, Israel was victorious in battle and took much spoil. A tribute was imposed on the people's half of the spoil. Of every 50 cattle and captives, 1 was to be given to the priests as an offering. Write 1 out of 50 as a percent.   2%

32. Christ took 3 of the 12 disciples with Him to the mountain where He was transfigured. Write as a percent the portion of the disciples who went with Jesus.   25%

33. During the 15-minute afternoon recess, David threw the basketball 8 times and scored a 2-point basket 3 times. On what percent of David's throws did he score? $37\frac{1}{2}$%

34. That afternoon on a 2-minute speed test, David's answers were correct for 7 out of 8 math problems. What percent of the problems did David solve correctly? $87\frac{1}{2}$%

35. One sultry afternoon during a severe thunderstorm, the temperature dropped from 38°C to 28°C. To the nearest whole degree, what was the temperature on the Fahrenheit scale before the thunderstorm?   100°F

36. That night, the temperature dropped to 65°F. To the nearest whole degree, what was the Celsius temperature?   18°C

3. **Ratios, fractions, and decimals can also be changed to percents.**

    a. Write a ratio in fraction form.

    b. Change a fraction to a decimal by dividing the numerator by the denominator. If the answer does not divide evenly to the hundredths' place, write the remainder as a fraction in lowest terms.

    c. Change a decimal to a percent by moving the decimal point two places to the right and adding the percent sign.

    (1) $7:12 = \frac{7}{12} = 7 \div 12 = 0.58\frac{1}{3} = 58\frac{1}{3}\%$

    (2) $\frac{11}{14} = 11 \div 14 = 0.78\frac{4}{7} = 78\frac{4}{7}\%$

    (3) $0.48 = 48\%$

    (4) $0.27 = 27\%$

4. Remind the students that they will need to memorize the fraction–percent equivalents on the chart by the end of the chapter.

Further Study

The word *percent* is actually an abbreviation for the Latin phrase *per centum*; the form *per cent.* would actually be the correct abbreviation. Through common usage, however, both the period and the space between the two words were eventually dropped.

It appears that percents were developed even before decimals; a form of them was possibly used as far back as the time of the Romans. The advent of decimals has made percents mathematically unnecessary; in fact, the percents in percent problems are changed to decimals for calculation. But percents are so common in business that they are not likely to pass out of use.

Solutions for Part F

31. $\frac{1}{50} = 2\%$

32. $\frac{3}{12} = \frac{1}{4} = 25\%$

33. $\frac{3}{8} = 37\frac{1}{2}\%$

34. $\frac{7}{8} = 87\frac{1}{2}\%$

35. $F = \frac{9}{5} \times 38 + 32$

36. $C = \frac{5}{9}(65 - 32)$

T–203   Chapter 5   Mastering Percents

Proportions for Part G

37. $\dfrac{\text{scale in.}}{\text{actual mi.}} \; \dfrac{1}{30} = \dfrac{n}{135} \; \dfrac{\text{scale in.}}{\text{actual mi.}}$ $n = 4\tfrac{1}{2}$ in.

38. $\dfrac{\text{scale in.}}{\text{actual mi.}} \; \dfrac{\frac{5}{8}}{13} = \dfrac{n}{52} \; \dfrac{\text{scale in.}}{\text{actual mi.}}$ $n = 2\tfrac{1}{2}$ in.

REVIEW EXERCISES

G. For each distance, use a proportion to find the correct measurement on a map. *(Lesson 56)*

37. scale: 1 in. = 30 mi.; actual distance: 135 mi. $4\frac{1}{2}$ in.
38. scale: $\frac{5}{8}$ in. = 13 mi.; actual distance: 52 mi. $2\frac{1}{2}$ in.

H. Write a ratio in lowest terms for each statement. *(Lesson 52)*

39. The Winters family passed out 120 copies of the *Star of Hope* in 45 minutes.   8:3
40. The drive sprocket has 36 teeth, and the driven sprocket has 16 teeth.   9:4

I. Find these temperature equivalents to the nearest degree. *(Lesson 28)*

41. 88°F = __31__ °C
42. 102°F = __39__ °C
43. 32°C = __90__ °F
44. 15°C = __59__ °F

Praise the LORD from the earth . . .

Fire, and hail; snow, and vapour; stormy wind fulfilling his word . . .

Let them praise the name of the LORD: for his name alone is excellent;

his glory is above the earth and heaven.

Psalm 148:7–13

61. Finding Percentages

One of the most frequent uses of percents is to find percentages, as illustrated here.

$$\begin{array}{ccc} \text{Base} & \text{Rate} & \text{Percentage} \\ 50 & \times \quad 20\% \quad = & 10 \end{array}$$

Three terms are important in percent problems: the base, the rate, and the percentage.

The **base** is the whole or original amount. The **rate** is the percent that shows what fractional part the percentage is of the base. The **percentage** is the fractional part of the whole or original amount. Observe how these terms are used above.

To find a percentage of a number, change the percent to a decimal and multiply. (Remember that *of* means "times.") Be sure to place the decimal point correctly in the answer. The answer will be a decimal rather than a percent, so you should not label the answer with the percent sign.

> To find the percentage when the base and the rate are known, multiply the base by the rate.
>
> Percentage = Base × Rate, or $P = B \times R$

Example A
38% of 96 = ___
38% = 0.38

$$\begin{array}{r} 96 \\ \times\, 0.38 \\ \hline 36.48 \end{array}$$

Example B
7% of 82 = ___
7% = 0.07

$$\begin{array}{r} 82 \\ \times\, 0.07 \\ \hline 5.74 \end{array}$$

CLASS PRACTICE

Use **base, rate,** *or* **percentage** *to complete each sentence.*

a. 20% of 50 = 10. In the problem at the left, 50 is the ___. base
b. 40% of 30 = 12. In the problem at the left, 12 is the ___. percentage
c. 25% of 36 = 9. In the problem at the left, 25% is the ___. rate
d. 36% of 50 = 18. In the problem at the left, 18 is the ___. percentage

Find these percentages.

e. 30% of 90   27
f. 50% of 76   38
g. 4% of 38   1.52
h. 90% of 21   18.9
i. 37% of 43   15.91
j. 79% of 155   122.45
k. 8% of 65   5.2
l. 3% of 18   0.54

LESSON 61

Objectives

- To review finding a percentage of a number.
- To review the terms *base, rate,* and *percentage.*
- To review the formula for finding the percentage: Percentage = Base × Rate.

Review

1. *Write these fractions as percents.* (Lesson 60)
   a. $\frac{3}{5}$ (60%)   b. $\frac{7}{20}$ (35%)
   c. $\frac{5}{8}$ (62$\frac{1}{2}$%)   d. $\frac{7}{10}$ (70%)
   e. $\frac{1}{5}$ (20%)   f. $\frac{3}{4}$ (75%)

2. *Write these decimals as percents.* (Lesson 60)
   a. 0.9 (90%)   b. 0.4 (40%)

3. *Find the correct length on a scale drawing for each distance.* (Lesson 57)
   a. scale: 1 in. = 3 ft.
      distance: 15 ft. (5 in.)
   b. scale: $\frac{5}{8}$ in. = 2 ft.
      distance: 14 ft. (4$\frac{3}{8}$ in.)

4. *Find the missing parts. Write any remainder as a fraction.* (Lesson 29)

Distance	Rate	Time	
a. 304 mi.	64 m.p.h.	___ hr.	(4$\frac{3}{4}$)
b. ___ km	88 km/h	6$\frac{1}{4}$ hr.	(550)
c. 300 mi.	___ m.p.h.	2$\frac{1}{2}$ hr.	(120)

5. *Use proportions to solve these problems.* (Lesson 53)
   a. If 11 out of every 12 hens lay one egg per day, how many eggs would 3,000 hens lay in one day? (2,750 eggs)
   $$\frac{\text{eggs}}{\text{hens}} \quad \frac{11}{12} = \frac{n}{3,000} \quad \frac{\text{eggs}}{\text{hens}}$$
   b. The Gingrichs use 10 gallons of water every 4 minutes as they water their garden. How much water would they use in 1 hour and 20 minutes? (200 gal.)
   $$\frac{\text{gallons}}{\text{minutes}} \quad \frac{10}{4} = \frac{n}{80} \quad \frac{\text{gallons}}{\text{minutes}}$$

T–205 Chapter 5 Mastering Percents

Introduction

Write the problem "60% of 50 = 30" on the board and ask the students, "Which of the three numbers is the percentage?"

It is easy to think that the 60% is the percentage because it has the percent sign. But the 30 is actually the percentage, or hundredth part of the whole.

It could be clarified this way. In the problem "60% of 50 = 30," which number is the whole amount? (50) Which number is the fractional part (percentage) of 50? The number 30 is $\frac{60}{100}$ of 50, or the part that is 60% of 50.

Teaching Guide

1. **Three terms are important in percent problems: the *base*, the *rate*, and the *percentage*.** Review these terms and their place in working with percents.

   a. The base is the whole or complete amount.
      (1) In the problem 25% of 40 = 10, 40 is the base.
      (2) In the problem 60% of 20 = 12, 20 is the base.
      (3) In the problem 120% of 30 = 36, 30 is the base.

   b. The rate shows in percent form what part the percentage is of the base.
      (4) In the problem 45% of 20 = 9, 45% is the rate.
      (5) In the problem 80% of 30 = 24, 80% is the rate.
      (6) In the problem 120% of 90 = 108, 120% is the rate.

   c. The percentage is the amount that is the given part of the whole.
      (7) In the problem 30% of 20 = 6, 6 is the percentage.
      (8) In the problem 90% of 50 = 45, 45 is the percentage.
      (9) In the problem 140% of 60 = 84, 84 is the percentage.

2. **To find a percentage of a number, change the percent to a decimal and multiply.** (Remember that *of* means "times.") Be sure to place the decimal point correctly in the answer. The answer will be a decimal rather than a percent, so you should not label the answer with the percent sign.

   a. 45% of 36 = 0.45 × 36 = 16.20 = 16.2
   b. 6% of 27 = 0.06 × 27 = 1.62
   c. 108% of 49 = 1.08 × 49 = 52.92

Solutions for Part D

27. 0.20 × 175
28. 1,000 × 8.34 × 0.25
29. 0.45 × 74
30. 0.10 × $39.97

WRITTEN EXERCISES

A. Use *base,* **rate,** *or* **percentage** *to complete each sentence.*

1. 60% of 60 = 36. In the problem at the left, 60% is the ___. rate
2. 40% of 20 = 8. In the problem at the left, 8 is the ___. percentage
3. 25% of 40 = 10. In the problem at the left, 10 is the ___. percentage
4. 46% of 50 = 23. In the problem at the left, 50 is the ___. base

B. Find these percentages.

5. 40% of 60   24
6. 50% of 65   32.5
7. 60% of 38   22.8
8. 70% of 76   53.2
9. 29% of 58   16.82
10. 43% of 28   12.04
11. 22% of 47   10.34
12. 34% of 69   23.46
13. 71 × 3%   2.13
14. 78 × 6%   4.68
15. 45 × 5%   2.25
16. 16 × 7%   1.12

C. *In these exercises, each number is the base in a percentage problem. Find the percentage if the rate is 12%.*

17. 16   1.92
18. 85   10.2
19. 45   5.4
20. 95   11.4
21. 128   15.36
22. 212   25.44
23. 402   48.24
24. 155   18.6
25. 570   68.4
26. 167   20.04

D. *Solve these reading problems. Use proportions for numbers 31 and 32.*

27. In Genesis 47, Joseph gave the Egyptians seed to sow and requested a 20% share of the harvest for Pharaoh. If a farmer harvested 175 bushels of grain, how much would he need to give to Pharaoh?   35 bushels

28. The water of the Dead Sea has a mineral content of almost 25%. If 1,000 gallons of water containing 25% minerals were evaporated, how many pounds of minerals would remain? (One gallon of water weighs 8.34 pounds.)   2,085 pounds

29. Marlin reported that due to a drought, the soybean yield was only 45% of last year's yield. The harvest last year was 74 bushels per acre. What was the yield this year?   33.3 bushels per acre

30. A bookstore offered the *Martyrs Mirror* at a 10% discount. The regular price was $39.97. By how much was the price reduced during the sale? $4.00

206   Chapter 5   Mastering Percents

31. The Conleys harvested 5 quarts of an early variety of strawberries for every 11 quarts of other varieties. In all, they harvested 640 quarts of strawberries. How many quarts of the early variety of strawberries did they harvest?   200 quarts

32. Of their 640 quarts of strawberries, the Conleys sold 7 quarts for every quart they kept for their own use. How many quarts of berries did they keep for their own use?   80 quarts

REVIEW EXERCISES

E. Write these fractions and decimals as percents. *(Lesson 60)*

33. $\frac{4}{5}$ 80%   **34.** $\frac{13}{20}$ 65%

35. 0.4   40%   **36.** 0.6   60%

F. Use proportions to find the correct length on a scale drawing for each distance. *(Lesson 57)*

37. scale: 1 in. = 4 ft.; distance: 12 ft.   3 in.

38. scale, 3 in. = 2 in.; distance: 15 in. $22\frac{1}{2}$ in.

G. Find the missing parts. Write any remainder as a fraction. *(Lesson 29)*

	Distance	Rate	Time
39.	231 mi.	42 m.p.h.	$5\frac{1}{2}$ hr.
40.	665 km	95 km/h	7 hr.

An Ounce of Prevention

1. Sometimes a student thinks that a whole number multiplied by a percent yields a percent answer. This probably comes from thinking of the percent sign as a label like *feet*. The percent sign is not a label, however, but a denominator (hundredths). In multiplying with percents, the percent is changed to a decimal and the decimal places (rather than the percent sign) carry through to the product. Stress to the students that the answers in this lesson should not be percents.

2. The terms *base*, *rate*, and *percentage* are not new. The calculations to find each one are very simple. But be sure to drill and review throughout this chapter the difference between the *base* and the *percentage*. The base is the whole or original amount, it is equal to 100%, and it always follows *of*. The percentage is *part* of the whole amount. After Lesson 62, some percentages (parts) will be larger than the bases (whole amounts), and that can make this concept especially confusing.

3. The mnemonic device below can be helpful for remembering the relationships among base, rate, and percentage. Cover the letter of the missing part, and find that part by using the operation indicated by the remaining letters. The dot means multiplication, and the fraction bar means division.

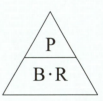

---

31. $\dfrac{\text{early}}{\text{total}} \quad \dfrac{5}{16} = \dfrac{n}{640} \quad \dfrac{\text{early}}{\text{total}}$ n = 200 quarts

32. $\dfrac{\text{kept}}{\text{total}} \quad \dfrac{1}{8} = \dfrac{n}{640} \quad \dfrac{\text{kept}}{\text{total}}$ n = 80 quarts

Proportions for Part F

37. $\dfrac{\text{scale in.}}{\text{actual ft.}} \quad \dfrac{1}{4} = \dfrac{n}{12} \quad \dfrac{\text{scale in.}}{\text{actual ft.}}$ n = 3 in.

38. $\dfrac{\text{scale in.}}{\text{actual in.}} \quad \dfrac{3}{2} = \dfrac{n}{15} \quad \dfrac{\text{scale in.}}{\text{actual in.}}$ n = $22\frac{1}{2}$ in.

T–207   Chapter 5   Mastering Percents

LESSON 62

Objectives

- To review working with rates over 100%.
- To review finding percentages when the rate is over 100%.
- To review working with percents containing fractions.
- To review finding percentages when the rate contains a fraction.

Review

1. Give Lesson 62 Speed Test (Percents, Fractions, and Decimals).

2. *Find these percentages.* (Lesson 61)
   a. 35% of 64   (22.4)
   b. 44% of 44   (19.36)

3. *Express these decimals as percents.* (Lesson 60)
   a. 0.15   (15%)
   b. 0.07   (7%)

4. *Find these equivalents.* (Lesson 30)
   a. 14 spans = ___ in.   (126)
   b. 9 cabs = ___ l (13.5)

5. *Use inverse proportions to solve these problems.* (Lesson 54)

   a. Father estimated that 2 men would be able to complete a certain job in 18 hours. How long should it take 3 men to complete the same job?   (12 hours)

   $$\begin{array}{c}\text{men (1)}\\ \text{men (2)}\end{array}\ \frac{2}{3} = \frac{n}{18}\ \begin{array}{c}\text{hours (2)}\\ \text{hours (1)}\end{array}$$

   b. The sewing circle is making dresses for a distant mission. One piece of fabric will make either 5 smaller dresses or 3 larger dresses. If the smaller dresses require $1\frac{4}{5}$ yards each, how much fabric will a larger dress take?   (3 yards)

   $$\begin{array}{c}\text{dresses (1)}\\ \text{dresses (2)}\end{array}\ \frac{5}{3} = \frac{n}{1\frac{4}{5}}\ \begin{array}{c}\text{yards (2)}\\ \text{yards (1)}\end{array}$$

62. Rates Greater Than 100% and Less Than 1%

One hundred percent means the whole amount. If your score on a spelling test is 100%, you spelled every word correctly. If a drink is 100% grape juice, it is all grape juice; nothing has been added.

When a percent refers to part of a whole, it cannot be more than 100%. It is not possible to have 125% of your spelling words correct, neither is it possible to have a drink that is 115% grape juice. However, when a percent compares two numbers, it may be more than 100%. For example, one cow can give 135% as much milk as another. A student can memorize 150% as many Bible verses this year as he did the previous year.

To find a percentage when the rate is more than 100%, use the same steps as for other percentages. Move the decimal point two places to the left, drop the percent sign, and multiply. Remember to place the decimal point correctly in the answer.

Percent		*Decimal*
225%	=	2.25
536%	=	5.36

Example A
225% of 426 = ___   (225% = 2.25)

$$\begin{array}{r} 426 \\ \times\, 2.25 \\ \hline 958.50 \end{array} = 958.5$$

Rates less than 1% are usually expressed as fractions. For example, $\frac{1}{2}\%$ is equal to $\frac{1}{2}$ of 1%, and $2\frac{1}{2}\%$ is halfway between 2% and 3%.

To express fractional percents as decimals, use these steps.

1. Change the fraction in the percent to a decimal. This number is still a percent.

2. Change the percent to a decimal by moving the decimal point two places to the left and dropping the percent sign.

Fractional Percent		*Decimal Percent*		*Decimal*
$\frac{1}{2}\%$	=	0.5%	=	0.005
$4\frac{1}{5}\%$	=	4.2%	=	0.042
$28\frac{3}{4}\%$	=	28.75%	=	0.2875

To find a percentage when the rate includes a fraction, change the percent to a decimal and multiply.

Example B
$3\frac{1}{2}\%$ of 94 = ___   ($3\frac{1}{2}\%$ = 0.035)

$$\begin{array}{r} 94 \\ \times\, 0.035 \\ \hline 3.290 \end{array} = 3.29$$

Example C
$5\frac{4}{5}\%$ of 175 = ___   ($5\frac{4}{5}\%$ = 0.058)

$$\begin{array}{r} 175 \\ \times\, 0.058 \\ \hline 10.150 \end{array} = 10.15$$

208 Chapter 5 Mastering Percents

CLASS PRACTICE

Change these percents to decimals.

a. 145%   1.45
b. 7½%   0.075
c. ⅗%   0.006
d. 420%   4.2
e. 305%   3.05
f. 27¾%   0.2775
g. 4 3/10 %   0.043
h. 135%   1.35

Find these percentages. Round money amounts to the nearest cent.

i. 105% of 50   52.5
j. 225% of 92   207
k. 145% of $8.75 $12.69
l. 350% of $125 $437.50
m. ½% of 490   2.45
n. 3/10 % of 175   0.525
o. ⅖% of $605 $2.42
p. 2¾% of $850 $23.38

Solve these reading problems.

q. Six months ago, the production rate of the Reinfords' laying hens peaked at 58,000 eggs per day. Now it has dropped by 9%. What is the present production per day?
   52,780 eggs

r. The living area in the Fishers' house was 2,500 square feet. Since they built an addition, the living area is 118% of the original area. What is their present living area?
   2,950 square feet

WRITTEN EXERCISES

A. *Change these percents to decimals.*

1. 145%   1.45
2. 325%   3.25
3. 980%   9.8
4. 600%   6 (6.0)
5. 900%   9 (9.0)
6. 1,175%   11.75

B. *Find these percentages. Round money amounts to the nearest cent.*

7. 110% of 70   77
8. 120% of 75   90
9. 135% of 90   121.5
10. 195% of $1.20 $2.34
11. 325% of $92.50 $300.63
12. 375% of $275 $1,031.25

C. *Change these percents to decimals.*

13. ½%   0.005
14. ⅖%   0.004
15. 1⅗%   0.016
16. 26¼%   0.2625
17. 101½%   1.015
18. 210¼%   2.1025

D. *Find these percentages.*

19. ½% of 200   1
20. ¾% of 400   3
21. 38⅘% of 500   194
22. 2⅕% of $500 $11
23. 105⅗% of $800 $844.80
24. 182 7/10 % of 175   319.725

Introduction

Ask the students to give the meaning of 100% in a single word. Words such as *everything* and *all* certainly apply.

If 100% means everything, is it possible to have more than 100%? Generally it is not. It is not possible to have more than 100% of the math problems done correctly, nor is it possible to have water that is more than 100% pure.

Sometimes 100% means "equals" or "as many" as in the following case. The sentence "David picked 100% as many bushels of apples as Donald picked" could be reworded, "The number of bushels that David picked *equals* the number that Donald picked."

Would it be possible for David to pick more than 100% as many bushels as Donald did? Yes; if David picked 15 bushels and Donald picked 10 bushels, we could say David picked 150% as many bushels of apples as Donald did.

It is not possible to have more than 100% when considering parts of a whole. However, it is possible to have more than 100% when comparing two numbers.

Teaching Guide

1. **When a percent compares two numbers, it may be greater than 100%.** For example, it is possible for a field to produce 110% of the harvest that it produced last year.

2. **To find a percentage when the rate is more than 100%, use the same steps as for other percentages.** Move the decimal point two places to the left, drop the percent sign, and multiply. Remember to place the decimal point correctly in the answer.

   Change these percents to decimals.

   a. 729% (7.29)

   b. 198% (1.98)

   c. 987% (9.87)

   d. 1,190% (11.90 = 11.9)

   Find these percentages.

   e. 145% of 70 ($1.45 \times 70 = 101.5$)

   f. 475% of 750 ($4.75 \times 750 = 3,562.50 = 3,562.5$)

3. **To express fractional percents as decimals, use the following steps.**

   (1) Change the fraction in the percent to a decimal. This number is still a percent.

   (2) Change the percent to a decimal by moving the decimal point two places to the left and dropping the percent sign.

   a. $\frac{1}{2}\% = 0.5\% = 0.005$

   b. $\frac{3}{4}\% = 0.75\% = 0.0075$

   c. $\frac{3}{8}\% = 0.375\% = 0.00375$

   d. $\frac{7}{8}\% = 0.875\% = 0.00875$

   e. $1\frac{1}{2}\% = 1.5\% = 0.015$

   f. $3\frac{3}{4}\% = 3.75\% = 0.0375$

4. **To find a percentage when the rate includes a fraction, change the percent to a decimal and multiply.**

   a. $\frac{1}{4}$% of 240 = 0.25% of 240 = 0.0025 of 240 = 0.6

   b. $\frac{4}{5}$% of 275 = 0.8% of 275 = 0.008 of 275 = 2.2

   c. $1\frac{3}{4}$% of 360 = 1.75% of 360 = 0.0175 of 360 = 6.3

   d. $2\frac{7}{8}$% of 400 = 2.875% of 400 = 0.02875 of 400 = 11.5

An Ounce of Prevention

Working with fractions in percents is no different from working with other percents. However, students may easily be confused by the decimal point in a form like 1.5%, and multiply by 1.5 rather than 0.015. Take time to review the logic behind the steps in making these conversions.

Solutions for Part E

25. 2 × 10
26. 1.3943 × $120
27. 0.055 × 4,800
28. 0.0425 × $92
29. 0.06 × $32.75
30. 0.045 × 120

E. Solve these reading problems.

25. In Solomon's temple, each dimension of the most holy place was 200% as large as its counterpart in the tabernacle. If the most holy place was 10 cubits wide in the tabernacle, how wide was it in Solomon's temple? 20 cubits

26. A man exchanged $120 United States money for Canadian money. He received 139.43% as many dollars as he turned in. How many Canadian dollars did he receive, to the nearest cent? $167.32

27. The Masts had a supply of coal weighing 4,800 pounds. One week they burned $5\frac{1}{2}$% of this amount. How many pounds of coal did they burn that week? 264 pounds

28. When Brother James went to purchase a load of coal, he discovered that the price had risen $4\frac{1}{4}$% over the previous price of $92.00 per ton. What was the amount of increase in price? $3.91

29. At the dry goods store, Mother's bill was $32.75 plus 6% sales tax. Find the amount of tax to the nearest cent. $1.97

30. One year Darrell's weight increased by $4\frac{1}{2}$%. At the beginning of the year, he weighed 120 pounds. How much did he gain during the year? 5.4 pounds

REVIEW EXERCISES

F. Find these percentages. (Lesson 61)

31. 25% of 62   15.5
32. 84% of 19   15.96
33. 51% of 94   47.94
34. 33% of 228   75.24

G. Express these decimals as percents. (Lesson 60)

35. 0.09   9%
36. 0.5   50%

H. Find these equivalents. (Lesson 30)

37. 4 spans = 92 cm
38. 6 ephahs = $19\frac{1}{2}$ pk.

63. Calculating Increase and Decrease (Method 1)

Price changes and other changes are often expressed as percents based on the original amount. For example, a price reduction may be stated as a 10% discount; that is, 10% of the regular price is subtracted. A newspaper article may say that the population of a certain city has increased by 25%.

To find the amount of increase or decrease and the new amount, use the following steps:

1. Multiply the original amount by the percent of change.

2. Round the answer to the nearest whole number. If the amount is money, round to the nearest cent unless specified otherwise.

3. If the change is an increase, add the result to the original amount. If the change is a decrease, subtract the result from the original amount.

Example A

Brother Alvin buys circuit beakers for $3.68 each. He adds a 35% markup before reselling them. What is his selling price? Find both the amount of increase and the new price.

```
   $ 3.68              $3.68
   × 0.35              + 1.29
   1.2880 = $1.29 (increase)    $4.97 (new price)
```

Example B

Due to a summer heat wave, the Witmers suffered a 3% mortality rate on their last batch of 60,000 chicks. Find the number of chicks that died and the number that survived.

```
   60,000              60,000
   × 0.03              - 1,800
   1,800 (decrease)    58,200 (surviving chicks)
```

CLASS PRACTICE   (See facing page.)

Find the amount of each change, to the nearest cent. Also find the new price.

a. $55.00 increased by 25%

b. $49.00 decreased by 14%

c. $30.25 increased by 85%

d. $23.50 increased by 32%

e. $6.45 decreased by 45%

f. $11.75 decreased by 65%

LESSON 63

Objectives

- To review finding a percent more or less than a number.
- To apply this skill in finding new prices at a given rate of increase or decrease, especially when discounts are given.

Review

1. *Find these percentages to the nearest whole number.* (Lessons 61, 62)
   a. 52% of 218 (113)
   b. 15% of 415 (62)
   c. $7\frac{3}{5}$% of 45 (3)
   d. $\frac{3}{4}$% of 280 (2)

2. *Express these decimals as percents.* (Lesson 60)
   a. 0.34 (34%)
   b. 0.09 (9%)

3. *Use proportions to find the actual distances.* Scale: $\frac{3}{4}$ in. = 16 mi. (Lesson 56)
   a. $2\frac{1}{4}$ in. (48 mi.)
   $$\frac{\text{scale in.}}{\text{actual mi.}} \quad \frac{\frac{3}{4}}{16} = \frac{2\frac{1}{4}}{n} \quad \frac{\text{scale in.}}{\text{actual mi.}}$$
   b. $4\frac{7}{8}$ in. (104 mi.)
   $$\frac{\text{scale in.}}{\text{actual mi.}} \quad \frac{\frac{3}{4}}{16} = \frac{4\frac{7}{8}}{n} \quad \frac{\text{scale in.}}{\text{actual mi.}}$$

4. *Multiply these decimals.* (Lesson 48)
   a. 6.4
      × 1.2
      (7.68)
   b. 0.007
      × 1.02
      (0.00714)

5. *Use direct or inverse proportions to solve these problems.* (Lesson 55)
   a. Sister Katrina has 1 bushel of apples, and she figures that she will get 18 quarts of applesauce from them. How many jars will she need if she uses the 3-cup jars that her neighbor gave her? (24 jars)
   $$\frac{\text{jars (1)}}{\text{jars (2)}} \quad \frac{18}{n} = \frac{3}{4} \quad \frac{\text{cups (2)}}{\text{cups (1)}}$$
   b. Yesterday Mother canned 3 baskets of peaches and got 36 quarts. Today she has 5 more baskets of peaches to can. How many quarts can she expect to have in all? (96 qt.)
   $$\frac{\text{baskets}}{\text{quarts}} \quad \frac{3}{36} = \frac{8}{n} \quad \frac{\text{baskets in all}}{\text{quarts in all}}$$

Answers for CLASS PRACTICE a–f

	change	new price
a.	$13.75	$68.75
b.	$6.86	$42.14
c.	$25.71	$55.96
d.	$7.52	$31.02
e.	$2.90	$3.55
f.	$7.64	$4.11

T–211   Chapter 5   Mastering Percents

Introduction

If the population of a town is 4,100 and it increases by 5%, how would you find the number of additional people? That increase would be calculated by finding 5% of 4,100 (0.05 × 4,100 = 205).

How would you calculate the population of the town after the increase? You would add the increase to the previous population.

If the students understand this concept, they have grasped the main idea of this lesson.

Teaching Guide

1. **Price changes and other changes are often expressed as percents based on the original amount.** For example, sale prices may be advertised as "20% off regular prices." When there is a general price increase throughout the country, it may be described as a "5% increase in prices."

2. **To find the amount of increase or decrease and the new amount, use the following steps.**

    (1) Multiply the original amount by the percent of change.

    (2) Round the answer to the nearest whole number. If the amount is money, round to the nearest cent unless specified otherwise.

    (3) If the change is an increase, add the result to the original amount. If the change is a decrease, subtract the result from the original amount.

    a. $2.75 increased by 8%
        (increase: $0.22
        new price: $2.97)

    b. $4.68 increased by 13%
        (increase: $0.61
        new price: $5.29)

    c. $5.98 decreased by 15%
        (decrease: $0.90
        new price: $5.08)

    d. $15.95 decreased by 45%
        (decrease: $7.18
        new price: $8.77)

Solutions for Part D

19. 0.20 × 5 = 1; 5 + 1

20. 0.02 × $347.68 = $6.95; $347.68 − $6.95

Find the new amount after each change, to the nearest whole number.

g. 475 increased by 30%   618
h. 290 decreased by 18%   238
i. 924 decreased by 35%   601
j. 525 increased by 70%   893

WRITTEN EXERCISES

A. Find the amount of each increase, to the nearest cent. Also find the new price.

		increase	*new price*
1.	$25.00 increased by 15%	$3.75	$28.75
2.	$30.00 increased by 25%	$7.50	$37.50
3.	$45.00 increased by 65%	$29.25	$74.25
4.	$23.00 increased by 40%	$9.20	$32.20
5.	$2.85 increased by 85%	$2.42	$5.27
6.	$14.75 increased by 30%	$4.43	$19.18

B. Find the amount of each discount, to the nearest cent. Also find the discount price.

		discount	*discount price*
7.	$26.75 decreased by 12%	$3.21	$23.54
8.	$38.40 decreased by 16%	$6.14	$32.26
9.	$52.75 decreased by 25%	$13.19	$39.56
10.	$38.65 decreased by 28%	$10.82	$27.83
11.	$199.95 decreased by 8%	$16.00	$183.95
12.	$215.50 decreased by 18%	$38.79	$176.71

C. Find the new amount after each change, to the nearest whole number.

13. 450 increased by 25%   563
14. 560 decreased by 16%   470
15. 675 decreased by 40%   405
16. 395 decreased by 60%   158
17. 1,500 decreased by 4%   1,440
18. 4,000 increased by 12%   4,480

D. Solve these reading problems. Use proportions for numbers 23 and 24.

19. Under the Law (Leviticus 6), a 20% penalty was to be added in several cases where a man wronged his neighbor. If a farmer was keeping his neighbor's sheep and stole 5 of them, how many sheep would he need to give back to his neighbor?   6 sheep

20. Many companies allow a 2% discount for prompt payment. If a customer has a bill of $347.68 and a 2% discount is allowed, how much will he need to pay? $340.73

212 Chapter 5 Mastering Percents

21. Rod and Staff Publishers allowed a 12% discount on a book normally selling for $15.95. What was the reduced price of the book? $14.04

22. Willow Valley Woodworking gives special discounts to its larger customers. On one order for $1,475, a 7% discount was given. What was the amount due after the discount? $1,371.75

23. It normally takes 2 people 8 days to plant Uncle Mervin's produce. If 5 people do the planting this year, how long should it take to complete the work? $3\frac{1}{5}$ days

24. The Robinsons formerly traveled 21 hours at an average speed of 50 miles per hour to visit their grandparents. With the speed limit increased by 10 miles per hour, they should now be able to average 60 miles per hour. How long can they expect the journey to take? $17\frac{1}{2}$ hours

REVIEW EXERCISES

E. Find these percentages to the nearest whole number. *(Lessons 61, 62)*

25. $4\frac{2}{5}$% of 86 4
26. $8\frac{3}{4}$% of 35 3
27. $\frac{7}{10}$% of 738 5
28. $36\frac{1}{2}$% of 149 54

F. Express these decimals as percents. *(Lesson 60)*

29. 0.2 20%
30. 0.12 12%

G. Use proportions to find the actual distances. *(Lesson 56)*

Scale: 1 in. = 32 mi.

31. $3\frac{3}{4}$ in. 120 mi.
32. $4\frac{5}{8}$ in. 148 mi.

H. Multiply these decimals. *(Lesson 48)*

33. 9.5
    × 1.6
    ─────
    15.2

34. 4.007
    × 1.8
    ─────
    7.2126

An Ounce of Prevention

Insist that the students use method 1, especially in Written Exercises, part C.

Further Study

Two different methods can be used to find an amount after a percent of increase or decrease. The method taught in this lesson is to find the percentage of the base and add it to or subtract it from the base. The other method is to first find what percent the new amount is of the original amount by adding the rate of increase to 100% or subtracting the rate of decrease from 100% and then to multiply the base by that rate. This second method is taught in Lesson 64.

---

21. $0.12 \times \$15.95 = \$1.91$; $\$15.95 - \$1.91$

22. $0.07 \times \$1,475 = \$103.25$; $\$1,475.00 - \$103.25$

23. $\dfrac{\text{workers (1)}}{\text{workers (2)}} \ \dfrac{2}{5} = \dfrac{n}{8} \ \dfrac{\text{days (2)}}{\text{days (1)}}$ $\quad n = 3\tfrac{1}{5}$ days

24. $\dfrac{\text{m.p.h. (1)}}{\text{m.p.h. (2)}} \ \dfrac{50}{60} = \dfrac{n}{21} \ \dfrac{\text{hours (2)}}{\text{hours (1)}}$ $\quad n = 17\tfrac{1}{2}$ hours

Proportions for Part G

31. $\dfrac{\text{scale in.}}{\text{actual mi.}} \ \dfrac{1}{32} = \dfrac{3\tfrac{3}{4}}{n} \ \dfrac{\text{scale in.}}{\text{actual mi.}}$ $\quad n = 120$ mi.

32. $\dfrac{\text{scale in.}}{\text{actual mi.}} \ \dfrac{1}{3} = \dfrac{4\tfrac{5}{8}}{n} \ \dfrac{\text{scale in.}}{\text{actual mi.}}$ $\quad n = 148$ mi.

T–213 Chapter 5 Mastering Percents

LESSON 64

Objectives

- To review adding percents to and subtracting percents from 100%.
- To review finding a percentage more or less than a number by first adding or subtracting the rate of increase or decrease.

Review

1. *Find the new amount after each change, to the nearest cent.* (Lesson 63)
   a. $62.00 decreased by 25% ($46.50)
   b. $2.53 increased by 85% ($4.68)
2. *Find these percentages. Round money amounts to the nearest cent.* (Lesson 62)
   a. 175% of 40 (70)
   b. 205% of $9.75 ($19.99)
   c. $\frac{3}{8}$% of $135 ($0.51)
   d. $\frac{4}{5}$% of 325 (2.6)
3. *Read these decimals.* (Lesson 47)
   a. 1.032
      (one and thirty-two thousandths)
   b. 3.00415 (three and four hundred fifteen hundred-thousandths)
4. *Read these large numbers.* (Lesson 1)
   a. 31,987,000,000 (thirty-one billion, nine hundred eighty-seven million)
   b. 1,000,080,000,000,000
         (one quadrillion, eighty billion)

64. Calculating Increase and Decrease (Method 2)

In Lesson 63, you found the new price by first multiplying to find the percentage and then adding or subtracting to find the new amount.

Sometimes the amount of increase or decrease is not needed; rather, all that is needed is the new amount. For example, a store may sell merchandise at a 15% discount, or it may add a certain percent to its purchase prices to find the selling prices. Then the new amount is all that is needed. In these cases, find what percent the new amount is of the original by subtracting the percent of change from 100% or adding it to 100%. A 12% increase means 112% of the original amount. A 12% decrease means 88% of the original amount. Multiply the original amount by the new percent. The following steps explain how to find the new amount by this method.

1. To find an increase, add the percent of increase to 100%. To find a decrease, subtract the percent of decrease from 100%.

2. Change the new percent to a decimal by dropping the percent sign and moving the decimal point two places to the left.

3. Multiply the price or other amount by the decimal found in step 2.

4. If the answer is a dollar amount, round it to the nearest cent. Otherwise, round to the nearest whole number.

Example A
   The milk production of Brother Daniel's dairy herd increased by 11%. What percent is the new production of the earlier production?
   100% + 11% = 111%

Example B
   During a week of high temperatures, milk production decreased by 13%. The production at the end of the week was what percent of that at the beginning?
   100% − 13% = 87%

Example C
   To find its selling prices, the Hogarth General Store adds 35% to the cost of its merchandise. If the cost for a set of plates is $9.50, what is the selling price?

   100% + 35% = 135% = 1.35

   $9.50
   × 1.35
   $12.8250 = $12.83 (rounded)

Example D
   One month the Hogarth General Store had a clearance sale in which all its prices were reduced by 15%. What was the sale price for the set of plates regularly priced at $12.83?

   100% − 15% = 85% = 0.85

   $12.83
   × 0.85
   10.9055 = $10.91 (rounded)

214   Chapter 5   *Mastering Percents*

CLASS PRACTICE

Solve these problems by adding or subtracting percents.

a. The population of Belmont decreased by 4%. What percent is the new population of the former population?   96%

b. The landlord increased the rent by 5%. What percent is the new rent of the previous rent?   105%

c. The price of diesel fuel decreased by 7%. What percent is the new price of the former price?   93%

d. Brother Daniel increased the size of his pullet house by 12%. What percent is the new size of the former size?   112%

Find the new price after each change, to the nearest cent.

e. $32.45 increased by 11% $36.02
f. $20.95 decreased by 46% $11.31
g. $18.95 increased by 28% $24.26
h. $54.12 decreased by 15% $46.00

Find the new amount after each change, to the nearest whole number.

i. 29 increased by 63%   47
j. 85 decreased by 31%   59
k. 342 decreased by 82%   62
l. 821 increased by 24%   1,018

WRITTEN EXERCISES

A. *Solve these problems by adding or subtracting percents.*

1. The price of gasoline increased by 15%. What percent is the new price of the former price?   115%

2. The price of milk decreased by 23%. What percent is the new price of the former price?   77%

3. The price of an electronic component decreased by 25%. What percent is the new price of the former price?   75%

4. The price of sugar increased by 19%. What percent is the new price of the former price?   119%

B. *Find the new price after each increase, to the nearest cent.*

5. $35.00 increased by 20% $42.00
6. $40.00 increased by 30% $52.00
7. $6.85 increased by 9% $7.47
8. $9.75 increased by 8% $10.53

C. *Find the new price after each decrease, to the nearest cent.*

9. $31.00 decreased by 10% $27.90
10. $26.00 decreased by 30% $18.20
11. $159.95 decreased by 12% $140.76
12. $319.95 decreased by 16% $268.76

D. *Find the new price after each change, to the nearest cent.*

13. $35.30 increased by 25% $44.13
14. $22.75 decreased by 22% $17.75
15. $14.25 increased by 13% $16.10
16. $31.15 decreased by 41% $18.38
17. $122.75 decreased by 9% $111.70
18. $157.60 increased by 19% $187.54

Lesson 64 T–214

Introduction

One day in the grocery store, Mother found margarine priced at $0.89 per pound. The regular price of butter was $1.16 per pound, but the butter was on sale at a 20% discount. Which was cheaper, margarine at the regular price or butter at the discount price?

Since $1.16 is the original amount or base, it is equal to 100%. A discount of 20% is a decrease of 20%. Subtracting 20% from 100% and multiplying $1.16 by 80% will give the same answer as multiplying $1.16 by 20% and then subtracting.

Multiplying $1.16 by 80% yields $0.928. The margarine is still cheaper than the butter.

Teaching Guide

1. **Adding the rate of increase to 100% or subtracting the rate of decrease from 100% gives the percent that the new amount is of the original amount.**

    a. If a price increases by 9%, the new price is ___ of the original price.
    (109%)

    b. If a price increases by $4\frac{1}{2}$%, the new price is ___ of the original price. ($104\frac{1}{2}$%)

    c. If a price decreases by 8%, the new price is ___ of the original price.
    (92%)

    d. If a price decreases by $5\frac{1}{2}$%, the new price is ___ of the original price. ($94\frac{1}{2}$%)

2. **To find the new amount after an increase or a decrease, multiply the original amount by the rate that the new amount is of the original amount. Use the following steps.**

    (1) To find an increase, add the percent of increase to 100%. To find a decrease, subtract the percent of decrease from 100%.

    (2) Change the new percent to a decimal by dropping the percent sign and moving the decimal point two places to the left.

    (3) Multiply the price or other amount by the decimal found in step 2.

    (4) If the answer is a dollar amount, round it to the nearest cent. Otherwise, round to the nearest whole number.

    a. Find $1.75 increased by 12%.
    100% + 12% = 112% = 1.12
    1.12 × $1.75 = $1.96

T–215   Chapter 5   Mastering Percents

    b. Find $4.79 increased by 23%.
       100% + 23% = 123% = 1.23
       1.23 × $4.79 = $5.8917 = $5.89
                           (rounded)

    c. Find $12.51 decreased by 16%
       100% − 16% = 84% = 0.84
       0.84 × $12.51 = $10.5084
       = $10.51 (rounded)

    d. Find $12.79 decreased by 47%.
       100% − 47% = 53% = 0.53
       0.53 × $12.79 = $6.7787
       = $6.78 (rounded)

Solutions for Part F

25. 1.45 × $1.75

26. 0.65 × $1.45

27. 1.07 × 4,050

28. 0.40 × $49.95

29. $\dfrac{\text{Fred}}{\text{total}} \quad \dfrac{3}{5} = \dfrac{n}{80} \quad \dfrac{\text{Fred}}{\text{total}} \qquad n = 48 \text{ melons}$

30. $\dfrac{\text{cow \#62}}{\text{cow \#59}} \quad \dfrac{3}{2} = \dfrac{n}{44} \quad \dfrac{\text{cow \#62}}{\text{cow \#59}} \qquad n = 66 \text{ pounds}$

Lesson 64

E. Find the new amount after each change, to the nearest whole number.

19. 75 increased by 32%   99
20. 94 decreased by 16%   79
21. 795 increased by 99%   1,582
22. 677 decreased by 99%   7
23. 3,900 decreased by 5%   3,705
24. 6,500 increased by 11%   7,215

F. Solve these reading problems. Use proportions for numbers 29 and 30.

25. At his roadside stand, Mr. Shattock sold watermelons for 45% more than he paid for them. If his cost was $1.75 per melon, what was the selling price? $2.54

26. Mr. Shattock had some day-old baked goods that he discounted 35%. If a loaf of bread regularly sold for $1.45, what was the discount price? $0.94

27. The population of Newton increased by about 7% since the last census. If the last census figure was 4,050, what is the estimated population now? (Round to the nearest whole number.)   4,334

28. Dudley Hardware had a clearance sale when it went out of business. On the last day of the sale, all merchandise was offered at a 60% discount. What was the discount price for a pole lamp that had previously sold for $49.95? $19.98

29. One day after school, Fred picked 3 melons for every 2 melons picked by his younger brother. Together they picked 80 melons. How many melons did Fred pick?   48 melons

30. One morning cow #62 gave 3 pounds of milk for every 2 pounds produced by cow #59. If cow #59 gave 44 pounds of milk, what was the production of cow #62?   66 pounds

REVIEW EXERCISES

G. Find the amount of each change, to the nearest cent. Also find the new price. *(Lesson 63)*

31. $8.75 decreased by 35% $3.06 $5.69
32. $1.85 increased by 76% $1.41 $3.26

H. Find these percentages. Round money amounts to the nearest cent. *(Lesson 62)*

33. 285% of $63.50 $180.98
34. 125% of $34.15 $42.69
35. $\frac{1}{8}$% of 360   0.45
36. $\frac{7}{8}$% of 640   5.6

I. Write these numbers, using words. *(Lessons 1, 47)*

37. 1.205   One and two hundred five thousandths
38. 8.0631   Eight and six hundred thirty-one ten-thousandths
39. 10,240,000,010   Ten billion, two hundred forty million, ten
40. 2,000,150,000,900,000   Two quadrillion, one hundred fifty billion, nine hundred thousand

216   Chapter 5   Mastering Percents

65. Finding What Percent One Number Is of Another

Suppose one day you have 28 of 30 math problems solved correctly, and the next day you have 37 of 40 problems correct. How does your work compare for the two days? One good way to compare numbers is to use percents. What percent is 28 of 30? What percent is 37 of 40?

> To find what percent one number is of another, divide the percentage by the base.
>
> $$\text{Rate} = \text{Percentage} \div \text{Base, or } R = \frac{P}{B}$$

To apply the formula above, use the following steps.

1. Write the two numbers as a fraction with the percentage as the numerator and the base as the denominator. (The number following *of* is the denominator.)

2. Divide the percentage (numerator) by the base (denominator). Round the quotient to the nearest hundredth unless instructed otherwise.

3. Change the quotient to a percent by moving the decimal point two places to the right and adding the percent sign.

Example A

In the eighth grade, 8 out of the 11 students are over 14 years old. What percent of them are over age 14? Express any remainder as a fraction.

$$\frac{8}{11} = 11\overline{)8.00} \quad 0.72\tfrac{8}{11} = 72\tfrac{8}{11}\%$$

Example B

The Stauffers planted 7 acres of produce on their own land and 15 acres on rented land. The acres of rented land are what percent of the acres of their own land?

$$\frac{15}{7} = 7\overline{)15.00} \quad 2.14\,2 = 214\% \text{ (rounded)}$$

CLASS PRACTICE

Find these rates. They are all whole percents.

a. 21 is __75%__ of 28

b. 16 is __40%__ of 40

c. 60 is __15%__ of 400

Find each rate. Express any remainder as a fraction of a percent.

d. 5 is ____ of 12
    $41\tfrac{2}{3}\%$

e. 4 is ____ of 9
    $44\tfrac{4}{9}\%$

f. 23 is ____ of 27
    $85\tfrac{5}{27}\%$

LESSON 65

Objectives

- To review finding what percent one number is of another.
- To review the formula for finding the rate: Rate = Percentage ÷ Base.
- To review finding what percent one number is of another and expressing the remainder as a fraction or *rounding it to the nearest whole percent.

Review

1. Give Lesson 65 Quiz (Finding Percentages).

2. *Find the new amount after each change, to the nearest whole number.* (Lesson 64)
    a. 53 increased by 82% (96)
    b. 24 decreased by 35% (16)

3. *Find these percentages. Round money amounts to the nearest cent.* (Lessons 61, 62)
    a. $\frac{1}{8}$% of 320 (0.4)
    b. $1\frac{4}{5}$% of 275 (4.95)
    c. 45% of $77.58 ($34.91)
    d. 79% of $63.24 ($49.96)

4. *Use proportions to do these exercises on scale drawings. Scale:* 1 in. = 24 ft. (Lesson 57)
    a. $1\frac{5}{8}$ in. represents ___. (39 ft.)

    $$\frac{\text{inch}}{\text{feet}} \quad \frac{1}{24} = \frac{1\frac{5}{8}}{n} \quad \frac{\text{inches}}{\text{feet}}$$

    b. 18 ft. is represented by ___. ($\frac{3}{4}$ in.)

    $$\frac{\text{inch}}{\text{feet}} \quad \frac{1}{24} = \frac{n}{18} \quad \frac{\text{inches}}{\text{feet}}$$

5. *Solve these division problems. All answers divide evenly by the ten-thousandths' place.* (Lesson 49)
    a. $0.08 \overline{)0.003}$ (0.0375)
    b. $3.1 \overline{)6.448}$ (2.08)

Introduction

In the eighth grade, 3 of the 5 students are boys. What percent of the class is boys?

To find what percent one number is of another, divide the percentage by the base. The base is 5. The percentage is 3.

$3 \div 5 = 0.6 = 60\%$

Adjust these numbers to fit your class.

Solutions for Part C

15. $\frac{5}{39} = 5 \div 39$
16. $\frac{11}{120} = 11 \div 120$
17. $\frac{6}{14} = 6 \div 14$
18. $\frac{8}{43} = 8 \div 43$
19. $100\% + 12.3\%$
20. $100\% - 25\%$

Teaching Guide

1. **To find what percent one number is of another, divide the percentage by the base.** Write the formula on the board and tell the students to memorize it.

   Rate = Percentage ÷ Base, or $R = \frac{P}{B}$

   (If rate is missing in the mnemonic triangle, the rest indicates percentage divided by base.)

2. **To apply the formula above, use the following steps.**

   (1) Write the two numbers as a fraction with the percentage as the numerator and the base as the denominator. (The number following *of* is the denominator.)

   (2) Divide the percentage (numerator) by the base (denominator). Round the quotient to the nearest hundredth unless instructed otherwise.

   (3) Change the quotient to a percent by moving the decimal point two places to the right and adding the percent sign.

   Find each rate to the nearest whole percent.

   a. 12 is what percent of 26?
   $\frac{12}{26} = \frac{6}{13} = 6 \div 13 = 0.46$ (rounded) = 46%

   b. 21 is what percent of 16?
   $\frac{21}{16} = 21 \div 16 = 1.31$ (rounded) = 131%

   Find each rate, and express any remainder as a fraction.

   c. 8 is what percent of 15?
   $\frac{8}{15} = 8 \div 15 = 0.53\frac{1}{3} = 53\frac{1}{3}\%$

   d. 9 is what percent of 22?
   $\frac{9}{22} = 9 \div 22 = 0.40\frac{10}{11} = 40\frac{10}{11}\%$

Find each rate. Round to the nearest whole percent.

	Base	Rate	Percentage		Base	Rate	Percentage
g.	32	47%	15	h.	28	250%	70
i.	51	24%	12	j.	44	205%	90

WRITTEN EXERCISES

A. Find these rates.

1. 4 is 25% of 16
2. 5 is 25% of 20
3. 20 is 5% of 400
4. 90 is 15% of 600

B. Find each rate. Round to the nearest whole percent.

5. 4 is 67% of 6
6. 5 is 45% of 11
7. 20 is 61% of 33
8. 18 is 69% of 26
9. 25 is 147% of 17
10. 50 is 455% of 11
11. 60 is 429% of 14
12. 32 is 168% of 19
13. $27 is 55% of $49
14. $17 is 44% of $39

C. Solve these reading problems. Some of them contain extra information.

15. Of the 39 books in the Old Testament, 5 are books of poetry. To the nearest whole percent, what part of the Old Testament books are poetry? 13%

16. There were about 120 persons in the group when a man was ordained to replace Judas. What percent of the group was made up of the 11 disciples? (Express any remainder as a fraction.) $9\frac{1}{6}\%$

17. The Masts raised 6 more acres of sweet corn this year than the 14 acres they raised last year. What percent is the increase over the amount raised last year? (Express any remainder as a fraction of a percent.) $42\frac{6}{7}\%$

18. The attendance at the River Point Mennonite School increased by 8 over the previous total of 43. To the nearest whole number, what percent is the increase over the previous attendance? 19%

19. One year the population of Drexter was recorded as 4,500. During the next 10 years it increased 12.3%. The population at the end of the 10-year period was what percent of the population at the beginning? 112.3%

20. Gabel's Lawn and Garden Center had an end-of-season sale in which they reduced prices on all merchandise by 25%. At the sale they sold 35% of their stock. What percent were the discount prices of the original price? 75%

218 Chapter 5 Mastering Percents

REVIEW EXERCISES

D. Find the new amount after each change, to the nearest whole number. *(Lesson 64)*

21. 62 increased by 45%   90
22. 14 decreased by 14%   12

E. Find these percentages. Round money amounts to the nearest cent. *(Lessons 61, 62)*

23. $\frac{2}{5}$% of 260   1.04
24. $2\frac{3}{4}$% of 425   11.6875
25. 83% of $15.20 $12.62
26. 46% of $32.95 $15.16

F. Use proportions to find these answers. *(Lesson 57)*

Scale: 1:25   (See facing page for proportions.)

27. scale length = 14 cm; actual length = ___   350 cm
28. scale length = 19.4 cm; actual length = ___   485 cm

Scale: $\frac{1}{4}$ in. = 1 ft.

29. scale length = ___; actual length = 14 ft. $3\frac{1}{2}$ in.
30. scale length = ___; actual length = $8\frac{1}{2}$ ft. $2\frac{1}{8}$ in.

G. Solve these division problems. All answers divide evenly by the ten-thousandths' place. *(Lesson 49)*

31. 0.04) 0.001   0.025
32. 2.6) 0.0312   0.012

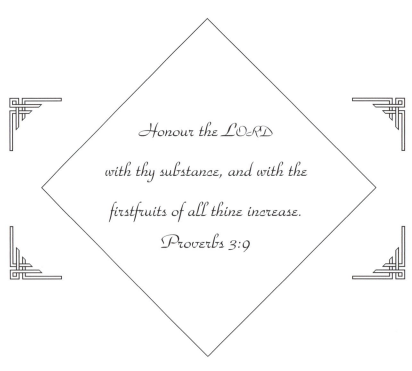

Honour the LORD with thy substance, and with the firstfruits of all thine increase.

Proverbs 3:9

Proportions for Part F

27. $\dfrac{\text{scale}}{\text{actual}} \quad \dfrac{1}{25} = \dfrac{14}{n} \quad \dfrac{\text{scale cm}}{\text{actual cm}}$ $\qquad n = 350$ cm

28. $\dfrac{\text{scale}}{\text{actual}} \quad \dfrac{1}{25} = \dfrac{19.4}{n} \quad \dfrac{\text{scale cm}}{\text{actual cm}}$ $\qquad n = 485$ cm

29. $\dfrac{\text{scale in.}}{\text{actual ft.}} \quad \dfrac{\frac{1}{4}}{1} = \dfrac{n}{14} \quad \dfrac{\text{scale in.}}{\text{actual ft.}}$ $\qquad n = 3\frac{1}{2}$ in.

30. $\dfrac{\text{scale in.}}{\text{actual ft.}} \quad \dfrac{\frac{1}{4}}{1} = \dfrac{n}{8\frac{1}{2}} \quad \dfrac{\text{scale in.}}{\text{actual ft.}}$ $\qquad n = 2\frac{1}{8}$ in.

T–219   Chapter 5   Mastering Percents

LESSON 66

Objectives

- To review finding the rate of increase or decrease, including the rounding of rates to the nearest whole percent.
- To teach *finding the rate of increase when the rate is greater than 100%.

Review

1. *Find each percent. Express any remainder as a fraction.* (Lesson 65)
   a. 6 is ___ of 11   ($54\frac{6}{11}\%$)
   b. 32 is ___ of 45   ($71\frac{1}{9}\%$)

2. *Find the new price after each change, to the nearest cent.* (Lesson 64)
   a. $15.32 increased by 44%   ($22.06)
   b $36.65 decreased by 32%   ($24.92)

3. *Find the amount of each change, to the nearest whole number. Also find the new amount.* (Lesson 63)
   a. 240 increased by 7%   (17; 257)
   b. 525 decreased by 26%   (137; 388)

4. *Solve these division problems.* (Lesson 49)
   a. $12\overline{)1.05}$ (0.0875)   b. $0.06\overline{)0.0714}$ (1.19)

5. *Find the prime factors of these composite numbers.* (Lesson 34)
   a. 38   (38 = 2 × 19)
   b. 96   (96 = 2^5 × 3)

6. *Write the place value of each underlined digit.* (Lesson 2)
   a. 302,4<u>1</u>9,914,800,000   (ten billions)
   b. 1<u>3</u>2,912,000,000,100,000   (ten quadrillions)

66. Finding the Percent of Increase or Decrease

During a spring sale at Line's Lawn Equipment, Father bought a used mower for $60. The regular price of the mower was $75. What was the percent of discount? (See Example A.)

With sales tax added to the $60, Father's bill was $64.20. What was the rate of sales tax? (See Example B.)

To find the percent (rate) of increase or decrease, divide the percentage by the base as in Lesson 65. The base is the original amount; it is $75 in the first problem above and $60 in the second problem. The percentage is the amount of increase or decrease; it is $15 in the first problem and $4.20 in the second problem.

To find the percent of increase or decrease, use the following steps.

1. Subtract to find the difference between the new and the original amounts.

2. Divide the difference by the original amount. Round the quotient to the nearest hundredth.

3. Change the quotient to a percent by moving the decimal point two places to the right and adding the percent sign.

Sometimes the increase is more than 100%, as illustrated in Example C.

Example A

Find the percent of decrease from $75 to $60.

$75 − $60 = $15

$$75\overline{)15.00} = 0.20 = 20\% \text{ decrease}$$

Example B

Find the percent of increase from $60 to $64.20.

$64.20 − $60 = $4.20

$$60\overline{)4.20} = 0.07 = 7\% \text{ increase}$$

Example C

At the end of October, Brendon had 15 sets of bookends ready to sell at his uncle's store. By the end of November, he had 42 sets ready. What was the percent of increase in his supply of bookends?

42 − 15 = 27

$$15\overline{)27.00} = 1.80 = 180\% \text{ increase}$$

CLASS PRACTICE

Find the rate of each change, to the nearest whole percent. Include the label increase *or* decrease.

a. year 1 enrollment, 25 48% increase
   year 2 enrollment, 37

b. year 5 enrollment, 45 16% decrease
   year 6 enrollment, 38

c. week 1 production, 15 47% increase
   week 2 production, 22

d. week 4 test score, 97% 3% decrease
   week 5 test score, 94%

e. year 6 weight, 46 lb. 13% increase
   year 7 weight, 52 lb.

f. 1985 population, 2,450 18% decrease
   1990 population, 2,000

WRITTEN EXERCISES

A. *Find the rate of each increase. The answers are all whole percents.*

	Original price	New price			Original price	New price	
1.	$2.00	$2.30	15%	2.	$3.00	$3.60	20%
3.	$5.50	$7.15	30%	4.	$6.40	$6.72	5%
5.	$7.90	$15.01	90%	6.	$5.80	$12.76	120%

B. *Find the rate of each decrease. The answers are all whole percents.*

	Original price	New price			Original price	New price	
7.	$4.00	$3.80	5%	8.	$1.20	$1.02	15%
9.	$7.00	$4.41	37%	10.	$8.50	$6.63	22%
11.	$9.20	$5.98	35%	12.	$8.50	$5.78	32%

C. *Find the rate of each change, to the nearest whole percent. Include the label* increase *or* decrease.

13. year 1 enrollment, 50 12% increase
    year 2 enrollment, 56

14. year 1 enrollment, 50 6% decrease
    year 2 enrollment, 47

15. year 1 weight, 22 lb. 36% increase
    year 2 weight, 30 lb.

16. week 1 weight, 3,400 g 6% decrease
    week 2 weight, 3,200 g

17. week 1 test score, 96% 4% decrease
    week 2 test score, 92%

18. week 3 test score, 90% 11% increase
    week 4 test score, 100%

19. 1980 population, 3,200 138% increase
    1990 population, 7,600

20. 1980 population, 4,100 137% increase
    1990 population, 9,700

D. *Solve these reading problems. Use proportions for numbers 25 and 26.*

21. On the day of Pentecost, about 3,000 disciples were added to the 120 that were already in the group. From 120 to 3,120 is an increase of what percent? 2,500%

Introduction

Ask the students if they know what inflation is. Inflation occurs when money loses value and prices rise. If a price of $1.00 increases to $1.05 and a price of $75.00 increases to $75.05, which one represents a higher rate of inflation?

Both items increased the same amount, but the increase in proportion to the price is not the same. The $1.00 price increased 5% while the $75.00 price increased only about 0.07%. An increase or decrease is usually given as a percent that shows the change in proportion to the original amount.

Teaching Guide

1. **To find the percent of increase or decrease, use the following steps.**

    (1) Subtract to find the difference between the new and the original amounts.

    (2) Divide the difference by the original amount. Round the quotient to the nearest hundredth.

    (3) Change the quotient to a percent by moving the decimal point two places to the right and adding the percent sign.

    a. From $2.50 to $2.85
    ($2.85 − 2.50 = $0.35
    $0.35 ÷ $2.50 = 0.14
    = 14% increase)

    b. From $1.66 to $0.99
    ($1.66 − $0.99 = $0.67
    $0.67 ÷ $1.66 = 0.40
    = 40% decrease)

    c. From 272 to 214
    (272 − 214 = 58
    58 ÷ 272 = 0.21
    = 21% decrease)

2. **The rate of increase may be more than 100%.** However, the calculations are no different.

    a. From $3.75 to $8.50
    ($8.50 − $3.75 = $4.75
    $4.75 ÷ $3.75 = 1.27 = 127%)

    b. From 215 to 975
    (975 − 215 = 760
    760 ÷ 215 = 3.53 = 353%)

Solutions for Part D

21. 3,120 − 120 = 3,000; 3,000 ÷ 120

T–221   Chapter 5   Mastering Percents

An Ounce of Prevention

Stress that the original number is not always the larger number; it is rather the starting number. Be sure to use a few examples to clinch this fact.

---

22. $120 - 70 = 50$; $50 \div 120$

23. $7 - 6 = 1$; $1 \div 7$

24. $4.6 - 3.6 = 1$; $1 \div 3.6$

25. $\dfrac{\text{below 600}}{\text{total}} \quad \dfrac{3}{4} = \dfrac{n}{60} \quad \dfrac{\text{below 600}}{\text{total}} \qquad n = 45 \text{ head}$

26. $\dfrac{\text{below 600}}{\text{total}} \quad \dfrac{5}{6} = \dfrac{45}{n} \quad \dfrac{\text{below 600}}{\text{total}} \qquad n = 54 \text{ head}$

22. In Genesis 6:3, God said the number of man's days would be 120 years. In Psalm 90:10, man's days are said to be 70 years. If these numbers represent a decrease in man's life span, what is the rate of decrease? (Answer to the nearest whole percent.)   42%

23. The Martins' truck patch yielded 7 bushels of beans on Monday and 6 bushels on Thursday. What was the percent of decrease in the yield? (Express any remainder as a fraction.) $14\frac{2}{7}\%$

24. In one month, Galen's new baby sister gained in weight from 3.6 kilograms to 4.6 kilograms. What was the rate of increase? (Express any remainder as a fraction of a percent.) $27\frac{7}{9}\%$

25. When the Steiners bought a group of beef cattle, 3 out of 4 weighed less than 600 pounds. How many of the 60 head of cattle weighed less than 600 pounds?   45 head

26. In an earlier group of cattle, 5 out of 6 weighed less than 600 pounds. If 45 weighed less than 600 pounds, how many head of cattle were in that group?   54 head

REVIEW EXERCISES

E. Find each rate to the nearest whole percent. *(Lesson 65)*

27. 18 is 72% of 25

28. 34 is 83% of 41

F. Find the new price after each change, to the nearest cent. *(Lesson 64)*

29. $315 decreased by 45% $173.25

30. $12.28 increased by 36% $16.70

G. Find the amount of each change, to the nearest whole number. Also find the new amount. *(Lesson 63)*

31. 360 increased by 8%   29   389

32. 475 decreased by 32%   152   323

H. Solve these division problems. *(Lesson 49)*

33. $2.5\overline{)0.214}$ 0.0856

34. $0.08\overline{)0.654}$ 8.175

I. Find the prime factors of these composite numbers. *(Lesson 34)*

35. 48 $48 = 2^4 \times 3$

36. 126 $126 = 2 \times 3^2 \times 7$

J. Write the place value of each underlined digit. *(Lesson 2)*

37. 159,015,687,000   hundred millions

38. 9,350,005,000,000   hundred billions

222 Chapter 5 Mastering Percents

67. Calculating the Base

David correctly solved 27 problems in his math assignment. His grade was 90%. How many problems was he assigned in all?

In the problem above, 90% is the rate. The number of problems correct, 27, is a percentage of a larger number (the whole assignment). This problem is solved in Example A below.

> To find the base when the percentage and the rate are known, divide the percentage by the rate.
>
> $$\text{Base} = \text{Percentage} \div \text{Rate, or } B = \frac{P}{R}$$

If the rate is less than 100%, the base will be larger than the percentage (Example A). If the rate is more than 100%, the base will be less than the percentage (Example B). Be especially careful in changing the percent to a decimal when it includes a fraction (Examples C and D).

Example A
27 is 90% of ___
90% = 0.9

$$0.9_\wedge \overline{)27.0_\wedge} = 30$$

Example B
25 is 125% of ___
125% = 1.25

$$1.25_\wedge \overline{)25.00_\wedge} = 20$$

Example C
9 is ½% of ___
½% = 0.005

$$0.005_\wedge \overline{)9.000_\wedge} = 1{,}800$$

Example D
60 is 1½% of ___
1½% = 0.015

$$0.015_\wedge \overline{)60.000_\wedge} = 4{,}000$$

CLASS PRACTICE

Find the base in each problem. All answers are whole numbers.

a. 16 is 25% of ___   64
b. 60 is 12% of ___   500
c. 15 is 3% of ___   500
d. 12 is 5% of ___   240
e. 56 is ⅞% of ___   6,400
f. 25 is ¼% of ___   10,000
g. 18 is 2¼% of ___   800
h. 35 is 12½% of ___   280
i. 12 is 240% of ___   5
j. 99 is 450% of ___   22

LESSON 67

Objectives

- To review the formula for finding the base: Base = Percentage ÷ Rate.
- To teach *finding the base when the rate is a fractional part of 1 percent.
- To teach *finding the base when the rate is more than 100 percent.

Review

1. *Find the rate of each change, to the nearest whole percent. Label your answer as an increase or a decrease.* (Lesson 66)

    a. From $3.40 to $2.72
    (20% decrease)

    b. From $4.15 to $4.70
    (13% increase)

2. *Find each percent to the nearest whole number.* (Lesson 65)

    a. 21 is ___ of 26 (81%)

    b. $36 is ___ of $40 (90%)

3. *Find these percentages.* (Lesson 62)

    a. 142% of 35 (49.7)

    b. $\frac{3}{4}$% of 620 (4.65)

4. *Solve mentally by changing the decimals to fractions.* (Lesson 51)

    a. $0.66\frac{2}{3}$ of 18 (12)

    b. 0.4 of 55 (22)

5. *Find the greatest common factor of each pair.* (Lesson 35)

    a. 18, 30 (6)

    b. 60, 105 (15)

6. *Find the lowest common multiple of each pair.* (Lesson 35)

    a. 18, 12 (36)

    b. 24, 28 (168)

7. *Write these numbers as Hindu-Arabic numerals.* (Lesson 3)

    a. Five hundred nineteen million, two hundred fourteen
    (519,000,214)

    b. Eighty-six billion, two hundred thirty million (86,230,000,000)

Introduction

Begin with a little mental math involving fractions.

a. 20 is $\frac{1}{2}$ of ___ (40)
b. 10 is $\frac{1}{3}$ of ___ (30)
c. 14 is $\frac{2}{3}$ of ___ (21)
d. 21 is $\frac{3}{4}$ of ___ (28)

Finding the base when you know the percentage and the rate works on exactly the same principle.

a. 20 is 50% of ___ (40)
b. 10 is $33\frac{1}{3}$% of ___ (30)
c. 14 is $66\frac{2}{3}$% of ___ (21)
d. 21 is 75% of ___ (28)

Teaching Guide

To find the base when the percentage and the rate are known, divide the percentage by the rate.

Base = Percentage ÷ Rate, or B = $\frac{P}{R}$

(If base is missing in the mnemonic triangle, the rest indicates percentage divided by rate.)

If the rate is less than 100%, the base will be larger than the percentage. If the rate is more than 100%, the base will be less than the percentage. Special care must be exercised in changing the percent to a decimal when it includes a fraction.

a. 12 is 40% of ___ (12 ÷ 0.4 = 30)
b. 14 is 35% of ___ (14 ÷ 0.35 = 40)

Rates greater than 100%

c. 30 is 125% of ___ (30 ÷ 1.25 = 24)
d. 36 is 240% of ___ (36 ÷ 2.4 = 15)

Rates less than 1%

e. 12 is $\frac{1}{2}$% of ___
   (12 ÷ 0.005 = 2,400)
f. 36 is $1\frac{4}{5}$% of ___
   (36 ÷ 0.018 = 2,000)

Solutions for Part B

15. 675 = 0.002 × ___; 675 ÷ 0.002
16. 720 = 0.02 × ___; 720 ÷ 0.02
17. $2,515.60 = 0.93 × ___; $2,515.60 ÷ 0.93
18. $325.50 = 0.05 × ___; $325.50 ÷ 0.05

Lesson 67

WRITTEN EXERCISES

A. *Find the base in each problem. All answers are whole numbers.*

1. 40 is 20% of ___   200
2. 50 is 25% of ___   200
3. 36 is 75% of ___   48
4. 45 is 36% of ___   125
5. 19 is 76% of ___   25
6. 24 is 32% of ___   75
7. 63 is 15% of ___   420
8. 42 is 70% of ___   60
9. 42 is $\frac{3}{4}$% of ___   5,600
10. 36 is $\frac{4}{5}$% of ___   4,500
11. 14 is $3\frac{1}{2}$% of ___   400
12. 9 is $1\frac{1}{4}$% of ___   720
13. 40 is 125% of ___   32
14. 36 is 240% of ___   15

B. *Solve these reading problems.*

15. When the children of Israel warred against the Midianites (Numbers 31), God directed that the booty be divided. One half would go to the 12,000 soldiers and the other half to the other Israelites. Of the soldiers' booty, 0.2% was to be given to the Lord. If the soldiers gave 675 sheep to the Lord, how many sheep were in their portion?   337,500 sheep

And levy a tribute unto the LORD.
Numbers 31:28

16. The other Israelites were required to give 2% of their portion to the Lord. If they gave 720 head of cattle, how many had they received as their portion?   36,000 head

17. Oak Valley Furniture Store received a 7% discount for buying a large quantity of furniture from their supplier. This means the store paid 93% of the regular price. If the amount paid was $2,515.60, what was the regular price of the furniture, to the nearest cent? $2,704.95

18. Uncle Douglas paid a sales tax of $325.50 when he registered a car he had purchased. If the tax rate was 5%, what was the price of the car? $6,510

224   Chapter 5   *Mastering Percents*

19. Job said a man could not answer one of a thousand questions that God might ask (Job 9:3). What percent is 1 of 1,000?   0.1%

20. Noah was 600 years old when the Flood came upon the earth, and he died at the age of 950. To the nearest whole percent, what part of Noah's life was lived before the Flood?   63%

REVIEW EXERCISES

C. *Find the rate of each change, to the nearest whole percent. Include the label* increase *or* decrease. *(Lesson 66)*

21. From $6.50 to $5.95   8% decrease   **22.** From $2.80 to $3.20   14% increase

D. *Find each rate to the nearest whole percent. (Lesson 65)*

23. 9 is 64% of 14   **24.** $52 is 149% of $35

E. *Find these percentages. (Lesson 62)*

25. 125% of 18   22.5   **26.** $\frac{4}{5}$% of 215   1.72

F. *Solve mentally by changing the decimals to fractions. (Lesson 51)*

27. $0.33\frac{1}{3}$ of 15   5   **28.** 0.8 of 35   28

G. *Find the greatest common factor of each pair. (Lesson 35)*

29. 16, 48   16   **30.** 24, 42   6

H. *Find the lowest common multiple of each pair. (Lesson 35)*

31. 9, 15   45   **32.** 12, 14   84

I. *Write these numbers as Hindu–Arabic numerals. (Lesson 1)*

33. Two hundred fifteen million, four hundred four thousand   215,404,000

34. Nine billion, thirty-eight   9,000,000,038

19. $\frac{1}{1,000} = 1 \div 1,000$

20. $\frac{600}{950} = \frac{12}{19} = 12 \div 19$

LESSON 68

Objectives

- To review finding each of the three parts in percent problems.
- To apply the percent formula practically by working with commissions.

Review

1. Give Lesson 68 Quiz (Finding Percents).

2. *Find the base in each problem. All answers are whole numbers.* (Lesson 67)

    a. 36 is 60% of ___ (60)

    b. 42 is $\frac{1}{4}$% of ___ (16,800)

3. *Find the rate of each change, to the nearest whole percent. Label your answer as an increase or a decrease.* (Lesson 66)

    a. From $8.29 to $7.75 (7% decrease)

    b. From $8.65 to $9.35 (8% increase)

4. *Find the new price after each change, to the nearest cent.* (Lesson 64)

    a. $62.50 decreased by 12% ($55.00)

    b. $11.42 increased by 25% ($14.28)

5. *Express these ratios as percents.* (Lesson 60)

    a. 11:25 (44%)

    b. 4:5 (80%)

6. *Write a ratio in lowest terms for each fact.* (Lesson 52)

    a. Ruth Ann ironed 8 shirts in 32 minutes. (1:4)

    b. The Kimble family traveled 45 miles in 51 minutes. (15:17)

7. *Reduce these fractions to lowest terms.* (Lesson 36)

    a. $\frac{42}{70}$ ($\frac{3}{5}$)   b. $\frac{48}{64}$ ($\frac{3}{4}$)

68. Working With Commission

In this chapter you have used the same basic formula to do many different things. In Lesson 61, you found a percentage by multiplying the base times the rate. In Lesson 65, you found the rate by dividing the percentage by the base. In Lesson 67, you found the base by dividing the percentage by the rate.

The **commission** is one practical way to use the three variations of the percent formula. Many salesmen receive a percentage of their sales as a reward for their work and as an incentive to sell more goods. This percentage is their commission. The same formula is used to solve commission problems as other percent problems, but the terms have new names that relate to commission.

Sales—This is the same as the **base**. It is the selling price.

Rate—This is the percent. It is sometimes called the **rate of commission**.

Commission—This is the same as the **percentage**. It is the part of the sale that the salesman receives.

Finding a commission when the sales and the rate of commission are known is the same as finding a percentage of a number.

$$\text{Percentage} = \text{Base} \times \text{Rate} \quad \text{or} \quad \text{commission} = \text{sales} \times \text{rate}$$

Example A

Sales = \$550; rate = $7\frac{3}{4}\%$; commission = ___

$\$550 \times 7\frac{3}{4}\% = \$550 \times 0.0775 = \$42.6250 = \$42.63$ (rounded)

Finding the rate of commission when the sales and the amount of commission are known is the same as finding the rate in a percent problem.

$$\text{Rate} = \frac{\text{Percentage}}{\text{Base}} \quad \text{or} \quad \text{rate of commission} = \frac{\text{commission}}{\text{sales}}$$

Example B

Sales = \$850; commission = \$80.75; rate = ___

$\frac{80.75}{850} = \$80.75 \div \$850 = 0.095 = 9\frac{1}{2}\%$

Chapter 5 Mastering Percents

Finding the sales when the commission and the rate of commission are known is the same as finding the base in a percent problem.

$$\text{Base} = \frac{\text{Percentage}}{\text{Rate}} \quad \text{or} \quad \text{sales} = \frac{\text{commission}}{\text{rate}}$$

Example C
 commission = \$34; rate = $8\frac{1}{2}\%$; sales = ___

$$\frac{34}{8\frac{1}{2}\%} = 34 \div 0.085 = \$400$$

In working with sales and commissions, round any answer to the nearest cent if it contains a fraction of a cent. All rates in this lesson are either whole or half percents.

CLASS PRACTICE

Find the missing parts. If an answer does not work out evenly, round it to the nearest cent (money amounts), or express the remainder as a fraction in lowest terms (percents).

	Sales	Rate	Commission		Sales	Rate	Commission
a.	\$4,250	6%	$255.00	b.	\$1,980.50	$9\frac{1}{2}\%$	$188.15
c.	\$2,340	$6\frac{1}{2}\%$	\$152.10	d.	\$6,750	9%	\$607.50
e.	$1,401.25	4%	\$56.05	f.	$4,000	$10\frac{1}{2}\%$	\$420.00
g.	\$7,100	$8\frac{1}{2}\%$	\$603.50	h.	$7,406	$12\frac{1}{2}\%$	\$925.75

WRITTEN EXERCISES

A. *Find the missing parts. If an answer does not work out evenly, round it to the nearest cent (money amounts), or express the remainder as a fraction in lowest terms (percents).*

	Sales	Rate	Commission		Sales	Rate	Commission
1.	\$600	7%	$42.00	2.	\$900	9%	$81.00
3.	\$4,166.37	$7\frac{1}{2}\%$	$312.48	4.	\$2,657.25	$6\frac{1}{2}\%$	$172.72
5.	\$275	4%	\$11.00	6.	\$360	8%	\$28.80
7.	\$2,180	$5\frac{1}{2}\%$	\$119.90	8.	\$3,146	$8\frac{1}{2}\%$	\$267.41
9.	$49.00	4%	\$1.96	10.	$66.00	6%	\$3.96
11.	$910.00	$6\frac{1}{2}\%$	\$59.15	12.	$3,120	$3\frac{1}{2}\%$	\$109.20
13.	\$3,190.00	7%	$223.30	14.	\$2,350.50	8%	$188.04
15.	$887.00	3%	\$26.61	16.	$3,987.00	8%	\$318.96

Lesson 68 T–226

Introduction

Review the concept of commissions. Ask the students to give examples they know of people working on commissions. (This includes people who sell real estate, fertilizer, and seed, and those who sell livestock and other things at auctions.)

Review the principle of how it works to sell on commission. If a person sells $100 worth of goods at a 7% rate of commission, he receives 7% of $100 or $7.00.

Teaching Guide

1. **The commission is one practical way to use the three variations of the percent formula.** The same formula is used as in other percent problems, but the terms have new names that relate to commission.

   Sales—This is the same as the base. It is the selling price.

   Rate—This is the percent. It is sometimes called the rate of commission.

   Commission—This is the same as the percentage. It is the part of the sale that the salesman receives.

2. **Finding a commission when the sales and the rate of commission are known is the same as finding a percentage of a number.**

   Percentage = Base × Rate *or*
   commission = sales × rate

   a. Sales = $225.00; rate = 9%
   (commission = $20.25)

   b. Sales = $775.60; rate = 7%
   (commission = $54.29)

   c. Sales = $855.60; rate = $5\frac{1}{2}$%
   (commission = $47.06)

   d. Sales = $3,245.71; rate = $6\frac{1}{2}$%
   (commission = $210.97)

3. **Finding the rate of commission when the sales and the amount of commission are known is the same as finding the rate in a percent problem.**

   Rate = $\frac{\text{Percentage}}{\text{Base}}$ *or*

   rate of commission = $\frac{\text{commission}}{\text{sales}}$

   a. Sales = $275.00
   commission = $22.00 (rate = 8%)

   b. Sales = $325.00
   commission = $42.25 (rate = 13%)

T-227   Chapter 5 Mastering Percents

   c. Sales = $270.00
      commission = $55.35 (rate = $20\frac{1}{2}$%)
   d. Sales = $3,166.00
      commission = $237.45 (rate = $7\frac{1}{2}$%)
4. **Finding the sales when the commission and the rate of commission are known is the same as finding the base in a percent problem.**

   $$\text{Base} = \frac{\text{Percentage}}{\text{Rate}} \quad or$$

   $$\text{sales} = \frac{\text{commission}}{\text{rate}}$$

   a. Commission = $25.75
      rate = 5%   (sales = $515.00)
   b. Commission = $60.75
      rate = 9%   (sales = $675.00)
   c. Commission = $316.92
      rate = 6%   (sales = $5,282.00)
   d. Commission = $635.25
      rate = $5\frac{1}{2}$%   (sales = $11,550.00)

Solutions for Part B

17. 0.07 × $615.23
18. $45.36 ÷ 0.07
19. $125.00 ÷ $1,562.50
20. 1,250 × $0.66 × 2% = $16.50
21. 69 − 66 = 3; 3 ÷ 66
22. 66 − 62 = 4; 4 ÷ 66

B. Solve these reading problems.

17. Dale Martin sells furniture on 7% commission. At one furniture store that he contacts on a monthly basis, he made a sale of $615.23. What was his commission? $43.07

18. The next month, Dale's commission from the same store was $45.36. At a 7% commission, what were his sales that month? $648.00

19. If an auctioneer receives a commission of $125.00 on a sale of $1,562.50, what is the rate of commission?   8%

20. Father sent a 1,250-pound steer to market, and it sold for $0.66 per pound. What was the commission on the steer if it was 2% of the selling price? $16.50

21. If cattle prices rise from $0.66 to $0.69 per pound, what is the rate of increase to the nearest whole percent?   5%

22. If cattle prices drop from $0.66 to $0.62 per pound, what is the rate of decrease to the nearest whole percent?   6%

REVIEW EXERCISES

C. Find the base. All answers are whole numbers. (Lesson 67)

23. 104 is $1\frac{5}{8}$% of ___   6,400
24. 88 is 275% of ___   32

D. Find the rate of each change, to the nearest whole percent. Include the label increase or decrease. (Lesson 66)

25. From $9.95 to $9.30   7% decrease
26. From $8.20 to $9.00   10% increase

E. Find the new price after each change, to the nearest cent. (Lesson 64)

27. $36.25 increased by 12% $40.60
28. $12.05 decreased by 19% $9.76

F. Express these ratios as percents. (Lesson 60)

29. 24:25   96%
30. 9:20   45%

G. Write a ratio in lowest terms for each fact. (Lesson 52)

31. In Henry's classroom, 6 of the 16 students are in his grade.   3:8
32. After visiting Canaan, 2 of the 12 spies brought back a good report.   1:6

H. Reduce these fractions to lowest terms. (Lesson 36)

33. $\frac{15}{25}$ $\frac{3}{5}$
34. $\frac{27}{81}$ $\frac{1}{3}$

I. Solve these addition problems. (Lesson 4)

35. 28,952
    + 2,058
    31,010

36. $419,025.75
    + 159,843.63
    $578,869.38

69. Solving Percent Problems Mentally

Sometimes a percent problem can be solved mentally by changing the percent to a common fraction. This is true especially when the common fraction is a unit fraction such as $\frac{1}{2}$ or $\frac{1}{3}$, or when the numerator and the denominator are both small, as in $\frac{2}{3}$ and $\frac{3}{4}$.

Finding the Percentage (P = B × R)

To find the percentage mentally, follow these steps.

1. Change the rate to a fraction.
2. Divide the base by the denominator of the fraction.
3. Multiply the quotient found in step 2 by the numerator. (This step is not necessary if the numerator is 1.)

Example A
$33\frac{1}{3}\%$ of 27 = ___   Think: $\frac{1}{3} \times 27 = 27 \div 3 = 9$

Example B
75% of 24 = ___   Think: $\frac{3}{4} \times 24 = 24 \div 4 \times 3 = 18$

Finding the Rate (R = $\frac{P}{B}$)

To find the rate mentally, follow these steps.

1. Write a fraction with the percentage as the numerator and the base as the denominator.
2. Reduce the fraction to lowest terms.
3. Change the fraction to a percent mentally. In this lesson, all rates can be calculated mentally.

Example C
6 is ___ of 24   Think: $6 \div 24 = \frac{6}{24} = \frac{1}{4} = 25\%$

Example D
12 is ___ of 32   Think: $12 \div 32 = \frac{12}{32} = \frac{3}{8} = 37\frac{1}{2}\%$

LESSON 69

Objectives

- To review finding the rate, the base, and the percentage mentally.
- To teach *finding the percentage mentally when the rate is a fractional percent.

Review

1. Find each commission, to the nearest cent. (Lesson 68)

   a. sales: $6,945.50; rate: 4%

   ($277.82)

   b. sales: $1,184.95; rate: 9%

   ($106.65)

2. Find the missing parts. Round rates to the nearest whole percent. (Lessons 61, 65, 67)

	Base	Rate	Percentage	
a.	___	40%	86	(215)
b.	___	24%	6	(25)
c.	26	___	19	(73%)
d.	84	___	12	(14%)
e.	26	49%	___	(12.74)
f.	122	15%	___	(18.3)

3. Solve these addition and subtraction problems involving fractions. (Lesson 37)

   a. $\frac{2}{9}$
      $+\frac{5}{18}$
      $(\frac{1}{2})$

   b. 12
      $-\frac{7}{8}$
      $(11\frac{1}{8})$

4. Do these additions mentally. (Lesson 5)

   a. 62 + 81 (143)

   b. 55 + 26 (81)

5. Solve by using direct or inverse proportions. (Lesson 55)

   a. Last year Sister Jenson planted peas in rows measuring a total of 400 feet, and the yield was 160 pounds of peas. This year she wants to harvest 200 pounds of peas. If the garden produces at the same rate as last year, what total length will her rows of peas need to be? (500 feet)

   $$\begin{array}{c} \text{feet} \\ \text{pounds} \end{array} \quad \frac{400}{160} = \frac{n}{200} \quad \begin{array}{c} \text{feet} \\ \text{pounds} \end{array}$$

   b. It normally takes the Martin family 8 minutes to make the $4\frac{1}{2}$-mile drive to school, which works out to an average of $33\frac{3}{4}$ miles per hour. One day the Martin boys rode their bicycles to school in 18 minutes. What was their average rate of speed? (15 m.p.h.)

   $$\begin{array}{c} \text{minutes (1)} \\ \text{minutes (2)} \end{array} \quad \frac{8}{18} = \frac{n}{33\frac{3}{4}} \quad \begin{array}{c} \text{m.p.h. (2)} \\ \text{m.p.h. (1)} \end{array}$$

Introduction

Drill the fraction–percent equivalents on the chart below.

Review the three variations of the percent formula, and calculate some simple ones mentally.

Percentage = base × rate

a. 50% of 60 = ___ ($\frac{1}{2}$ of 60 = 30)

b. 25% of 40 = ___ ($\frac{1}{4}$ of 40 = 10)

Rate = percentage ÷ base

c. 10 is ___ of 20 ($\frac{10}{20} = \frac{1}{2} = 50\%$)

d. 6 is ___ of 18 ($\frac{6}{18} = \frac{1}{3} = 33\frac{1}{3}\%$)

Base = percentage ÷ rate

e. 18 is 50% of ___
   ($50\% = \frac{1}{2}$; $18 \div \frac{1}{2} = 18 \times 2 = 36$)

f. 12 is 25% of ___
   ($25\% = \frac{1}{4}$; $12 \div \frac{1}{4} = 12 \times 4 = 48$)

Teaching Guide

1. To find the percentage mentally, follow the steps below.

(1) Change the rate to a fraction.

(2) Divide the base by the denominator of the fraction.

(3) Multiply the quotient found in step 2 by the numerator. (This step is not necessary if the numerator is 1.)

a. $33\frac{1}{3}\%$ of 66 = ___
   ($\frac{1}{3}$ of 66 = 66 ÷ 3 = 22)

b. $12\frac{1}{2}\%$ of 64 = ___
   ($\frac{1}{8}$ of 64 = 64 ÷ 8 = 8)

c. $87\frac{1}{2}\%$ of 32 = ___
   ($\frac{7}{8}$ of 32 = 32 ÷ 8 × 7 = 28)

d. 10% of 36 = ___
   ($\frac{1}{10}$ of 36 = 36 ÷ 10 = 3.6)

2. To find the rate mentally, follow the steps below.

(1) Write a fraction with the percentage as the numerator and the base as the denominator.

(2) Reduce the fraction to lowest terms.

10%	20%	30%	40%	50%	60%	70%	80%	90%	100%
$\frac{1}{2} = 50\%$									
$\frac{1}{3} = 33\frac{1}{3}\%$					$\frac{2}{3} = 66\frac{2}{3}\%$			
$\frac{1}{4} = 25\%$				$\frac{3}{4} = 75\%$			
$\frac{1}{5} = 20\%$		$\frac{2}{5} = 40\%$		$\frac{3}{5} = 60\%$		$\frac{4}{5} = 80\%$		
$\frac{1}{6} = 16\frac{2}{3}\%$						$\frac{5}{6} = 83\frac{1}{3}\%$		
$\frac{1}{8} = 12\frac{1}{2}\%$		$\frac{3}{8} = 37\frac{1}{2}\%$			$\frac{5}{8} = 62\frac{1}{2}\%$			$\frac{7}{8} = 87\frac{1}{2}\%$	
$\frac{1}{10} = 10\%$	$\frac{3}{10} = 30\%$					$\frac{7}{10} = 70\%$		$\frac{9}{10} = 90\%$	

Finding the Base (B = $\frac{P}{R}$)

To find the base mentally, follow these steps.

1. Change the percent to a fraction.
2. Divide the percentage by the numerator of the fraction. (This step is not necessary if the numerator is 1.)
3. Multiply the quotient found in step 2 by the denominator of the fraction.

Example E
8 is 10% of ___   Think: $8 \div 10\% = 8 \div \frac{1}{10} = 8 \times 10 = 80$

Example F
21 is $87\frac{1}{2}\%$ of ___   Think: $21 \div 87\frac{1}{2}\% = 21 \div \frac{7}{8} = 21 \div 7 \times 8 = 24$

Finding 1% or a Fraction of 1%

To find 1% of a number mentally, divide the number by 100. (Move the decimal point two places to the left.) To find a fractional percent of a number, first find 1% and then multiply by the fraction.

Example G
1% of 325 = ___   Think: $\frac{1}{100}$ of $325 = 325 \div 100 = 3.25$

Example H
$\frac{1}{2}\%$ of 400 = ___   Think: 1% of 400 = 4; $\frac{1}{2} \times 4 = 2$

Example I
$\frac{5}{6}\%$ of 1,200 = ___   Think: 1% of 1,200 = 12; $\frac{5}{6}$ of $12 = 12 \div 6 \times 5 = 10$

CLASS PRACTICE

Mentally calculate each percentage.

a. 25% of 36 = __9__
b. $83\frac{1}{3}\%$ of 18 = __15__
c. $12\frac{1}{2}\%$ of 24 = __3__
d. 60% of 25 = __15__

Mentally calculate each percent (rate).

e. 2 is __25%__ of 8
f. 5 is __$33\frac{1}{3}\%$__ of 15
g. 15 is __$62\frac{1}{2}\%$__ of 24
h. 12 is __$66\frac{2}{3}\%$__ of 18

Mentally calculate each base.

i. 4 is 40% of __10__
j. 3 is 25% of __12__
k. 7 is $33\frac{1}{3}\%$ of __21__
l. 10 is $62\frac{1}{2}\%$ of __16__

230 Chapter 5 *Mastering Percents*

Find 10% of each number.
m. 210   21   n. 800   80   o. 258   25.8   p. 1,064   106.4

Find 1% of each number.
q. 200   2   r. 390   3.9   s. 3,850   38.5   t. 4,195   41.95

Find $\frac{1}{2}$% of each number.
u. 800   4   v. 600   3   w. 2,600   13   x. 3,000   15

WRITTEN EXERCISES

A. *Mentally calculate each percentage.*
1. 50% of 12 = __6__
2. 20% of 15 = __3__
3. $33\frac{1}{3}$% of 9 = __3__
4. 10% of 12 = __1.2__
5. 75% of 20 = __15__
6. $66\frac{2}{3}$% of 15 = __10__

B. *Mentally calculate each percent (rate).*
7. 10 is __$83\frac{1}{3}$%__ of 12
8. 5 is __20%__ of 25
9. 9 is __$33\frac{1}{3}$%__ of 27
10. 6 is __$66\frac{2}{3}$%__ of 9
11. 15 is __60%__ of 25
12. 12 is __80%__ of 15

C. *Mentally calculate each base.*
13. 3 is 50% of __6__
14. 8 is 25% of __32__
15. 6 is $33\frac{1}{3}$% of __18__
16. 9 is $12\frac{1}{2}$% of __72__
17. 15 is 60% of __25__
18. 8 is $66\frac{2}{3}$% of __12__

D. *Find 10% of each number.*
19. 120   12   20. 140   14   21. 1,323   132.3   22. 3,559   355.9

E. *Find 1% of each number.*
23. 500   5   24. 600   6   25. 2,746   27.46   26. 3,987   39.87

F. *Find $\frac{1}{2}$% of each number.*
27. 200   1   28. 1,600   8   29. 2,800   14   30. 3,400   17

G. *Solve these reading problems. You should be able to do numbers 31–34 mentally. Write proportions for numbers 35 and 36.*

31. When the children of Israel left Egypt, their number included over 600,000 men. What is $\frac{1}{2}$% of 600,000?   3,000

32. After the death of Solomon, the 10 northern tribes rebelled against King David's dynasty, but his descendants continued to reign over Judah and Benjamin. What percent of the 12 tribes were the 2 tribes? $16\frac{2}{3}$%

(3) Change the fraction to a percent mentally.

   a. 15 is ___ of 24
      ($\frac{15}{24} = \frac{5}{8} = 62\frac{1}{2}\%$)

   b. 12 is ___ of 18
      ($\frac{12}{18} = \frac{2}{3} = 66\frac{2}{3}\%$)

   c. 36 is ___ of 60 ($\frac{36}{60} = \frac{3}{5} = 60\%$)

   d. 30 is ___ of 75 ($\frac{30}{75} = \frac{2}{5} = 40\%$)

3. **To find the base mentally, follow the steps below.**

  (1) Change the percent to a fraction.

  (2) Divide the percentage by the numerator of the fraction. (This step is not necessary if the numerator is 1.)

  (3) Multiply the quotient found in step 2 by the denominator of the fraction.

   a. 12 is 75% of ___
      (75% = $\frac{3}{4}$; 12 ÷ 3 × 4 = 16)

   b. 26 is $66\frac{2}{3}\%$ of ___
      ($66\frac{2}{3}\% = \frac{2}{3}$; 26 ÷ 2 × 3 = 39)

   c. 15 is 60% of ___
      (60% = $\frac{3}{5}$; 15 ÷ 3 × 5 = 25)

   d. 24 is 80% of ___
      (80% = $\frac{4}{5}$; 24 ÷ 4 × 5 = 30)

4. **To find 1% of a number mentally, divide the number by 100. To find a fractional percent of a number, first find 1% and then multiply by the fraction.**

   a. 1% of 450 = ___
      ($\frac{1}{100}$ of 450 = 450 ÷ 100 = 4.5)

   b. $\frac{2}{3}$% of 900 = ___
      ($\frac{1}{100}$ of 900 = 9; 9 ÷ 3 × 2 = 6)

   c. $\frac{3}{4}$% of 400 = ___
      ($\frac{1}{100}$ of 400 = 4; 4 ÷ 4 × 3 = 3)

   d. $\frac{5}{8}$% of 1,600 = ___ ($\frac{1}{100}$ of 1,600 = 16; 16 ÷ 8 × 5 = 10)

Solutions for Part G

31. 1% of 600,000 = 6,000; 6,000 ÷ 2

32. $\frac{2}{12} = \frac{1}{6} = 16\frac{2}{3}\%$

T–231   Chapter 5   Mastering Percents

33. $2{,}000 \div 4$

34. 1% of 120,000 = 1,200; $\frac{3}{4} \times 1{,}200$

35. $\dfrac{\text{yellow}}{\text{total}} \ \dfrac{2}{5} = \dfrac{n}{15} \ \dfrac{\text{yellow}}{\text{total}}$ $\qquad n = 6$ rows

36. $\dfrac{\text{pages}}{\text{minutes}} \ \dfrac{12}{9} = \dfrac{n}{15} \ \dfrac{\text{pages}}{\text{minutes}}$ $\qquad n = 20$ pages

33. The Dark Ages included about $\frac{1}{4}$ of the 2,000 years since the birth of Christ. What was the approximate length of the Dark Ages? 500 years

34. The odometer in the Wengers' van shows 120,000 miles. They drove $\frac{3}{4}$% of that distance on a recent trip. How far did they travel on the trip? 900 miles

35. Mother planted 2 rows of yellow beans for every 3 rows of green beans. In all she planted 15 rows of beans. How many rows of yellow beans did she plant? 6 rows

36. One afternoon, Thelma read 12 pages in 9 minutes. At that rate, how many pages could she read in 15 minutes? 20 pages

REVIEW EXERCISES

H. Find each commission, to the nearest cent. *(Lesson 68)*

37. Sales: $4,652.35
Rate: 6%
Commission: $279.14

38. Sales: $2,155.90
Rate: 5%
Commission: $107.80

I. Find the missing parts. Round rates to the nearest whole percent. *(Lessons 61, 65, 67)*

	Base	Rate	Percentage		Base	Rate	Percentage
39.	260	25%	65	**40.**	375	12%	45
41.	35	51%	18	**42.**	72	64%	46
43.	64	50%	32	**44.**	48	75%	36

J. Solve these addition and subtraction problems involving fractions. *(Lesson 37)*

45. $\frac{1}{4} + \frac{7}{12} = \frac{5}{6}$

46. $\frac{8}{15} - \frac{2}{15} = \frac{2}{5}$

K. Do these additions mentally. *(Lesson 5)*

47. 51 + 12   63

48. 88 + 45   133

232   Chapter 5   *Mastering Percents*

70. Reading Problems: Using Sketches

Reading problem skills include much more than being able to calculate with the numbers in the problems. Especially when a reading problem involves an area, a distance, or a sequence, it is important that you form a clear picture of it in your mind. Consider the following example.

Example
Before planting tomatoes, Mother covered the ground with plastic. She planted 3 rows of tomatoes 3 feet apart and 3 feet from all edges of the plastic. The rows were 50 feet long. How many square feet did she cover with plastic?

How can we get a clear mental picture of all these details? The best way is to draw a sketch that puts them together in an understandable way.

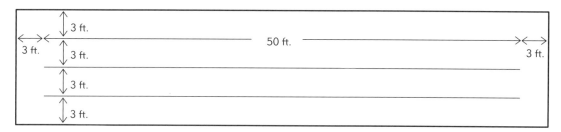

Now the problem is simple. The surface covered with plastic is
  56 feet long (3 + 50 + 3) and
  12 feet wide (4 × 3).
Its area is 56 × 12 or 672 square feet.

A sketch is valuable in the following ways.

a. A sketch simplifies the problem by showing how the facts relate to each other. Remember the saying, "One picture is worth a thousand words."

b. A sketch helps to determine which facts are necessary and which are unnecessary.

c. A sketch shows the steps that are needed to find the solution.

d. A sketch helps to show whether a solution is logical.

LESSON 70

Objective

- To review using sketches as a practical help to solve reading problems.

Review

1. Give Lesson 70 Quiz (Commissions).

2. *Solve these percent problems mentally.* (Lesson 69)

   a. 50% of 42 = ___   (21)

   b. 18 is ___ of 27   ($66\frac{2}{3}$%)

3. *Find the missing parts in these commission problems.* (Lesson 68)

   a. $s = \$9,520; c = \$571.20$ ($r = 6\%$)

   b. $c = \$638.91; r = 9\%$ ($s = \$7,099$)

4. *Find the rate of each change, to the nearest whole percent. Label your answer as an increase or a decrease.* (Lesson 66)

   a. From $2.65 to $3.40
      (28% increase)

   b. From $7.49 to $6.59
      (12% decrease)

5. *Find the amount of each change, to the nearest cent. Also find the new price.* (Lesson 63)

   a. $45.75 increased by 26%
      ($11.90; $57.65)

   b. $32.45 decreased by 18%
      ($5.84; $26.61)

6. *Solve these subtraction and multiplication problems.* (Lessons 6, 38)

   a. $9,504.09
      − 785.32
      ($8,718.77)

   b. $831.56 − $521.69   ($309.87)

   c. $\frac{5}{6} \times \frac{4}{5}$ ($\frac{2}{3}$)

   d. $3\frac{1}{9} \times 2\frac{1}{2}$ ($7\frac{7}{9}$)

7. *Solve by inverse proportions.* (Lesson 54)

   a. Jesse figured that he would need 14 2-by-4s 8 feet long to build a doghouse. If the lumberyard is out of 8-foot 2-by-4s, how many 10-foot pieces should he buy? Round up to the next whole board. (12 10-foot pieces)

   $$\frac{\text{boards (1)}}{\text{boards (2)}} \quad \frac{14}{n} = \frac{10}{8} \quad \frac{\text{feet (2)}}{\text{feet (1)}}$$

   b. Bethany, who types at an average of 45 words per minute, can type an article in 15 minutes. If her older sister Brenda can type the article in $11\frac{1}{2}$ minutes, what is her average typing speed? Round to the nearest whole number. (59 words per minute)

   $$\frac{\text{words per minute (1)}}{\text{words per minute (2)}} \quad \frac{45}{n} = \frac{11\frac{1}{2}}{15} \quad \frac{\text{minutes (2)}}{\text{minutes (1)}}$$

Introduction

Brother Clifford and Brother Carl both left Flat Rock Church at the same time. Brother Clifford traveled west at an average of 45 miles per hour, and Brother Carl traveled east at an average of 48 miles per hour. How far apart were the two after traveling for 20 minutes?

The sketch below shows that the distances traveled by the men must be added to find the answer. Calculation shows that Brother Clifford traveled 15 miles, and Brother Carl traveled 16 miles. After 20 minutes, the brethren were 31 miles apart.

Brother Clifford	Brother Carl
←	→
$\frac{1}{3}$ hr. × 45 m.p.h.	$\frac{1}{3}$ hr. × 48 m.p.h.

Sketches for CLASS PRACTICE

a.

b.
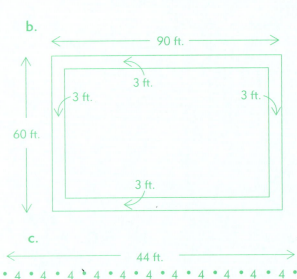

c.

Teaching Guide

1. **A sketch is valuable in the following ways.**

    (1) A sketch simplifies the problem by showing how the facts relate to each other. Remember the saying, "One picture is worth a thousand words."

    (2) A sketch helps to determine which facts are necessary and which are unnecessary.

    (3) A sketch shows the steps that are needed to find the solution.

    (4) A sketch helps to show whether a solution is logical.

2. **To produce a useful sketch, follow these steps.**

    (1) Picture the problem and sketch it, using simple but neat shapes. Be sure to include all the necessary information from the problem. If the problem relates to distance, it is helpful to draw the sketch approximately to scale.

    (2) Write the information given in the problem where it belongs on the sketch.

    (3) Solve the problem. Be sure your answer agrees with the sketch.

    Apply the steps above to *Class Practice*. First discuss the Example sketch in the lesson. Next draw a sketch for *Class Practice* problem *a*. Then have the students draw sketches for problems *b–d*, perhaps having several do each problem on the board.

d.
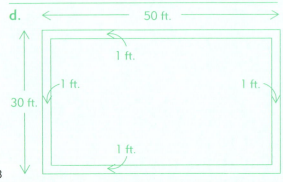

To produce a useful sketch, follow these steps.

(1) Picture the problem and sketch it, using simple but neat shapes. Be sure to include all the necessary information from the problem. If the problem relates to distance, it is helpful to draw the sketch approximately to scale.

(2) Write the information given in the problem where it belongs on the sketch.

(3) Solve the problem. Be sure your answer agrees with the sketch.

CLASS PRACTICE

Draw a sketch for each problem, and use it to find the solution. Show all your work. (See facing page for sketches.)

a. Father is putting up posts for a fence to enclose a rectangular meadow. The meadow is 180 feet long and 90 feet wide. If the fence posts are 10 feet apart, how many posts will it take for the entire fence? 54 posts

b. The Newswangers have a garden that measures 90 feet by 60 feet. If they leave a 3-foot border around the garden unplanted, how many square feet are planted in all? 4,536 square feet

c. Marvin is planting blueberry bushes at a distance of 4 feet apart. If his row of blueberries is 44 feet long, how many bushes did he plant? 12 bushes

d. The foundation of the Metzler house is a 12-inch poured concrete wall. The outside measurements of the wall are 30 feet by 50 feet. What are the inside measurements of the wall? 28 feet by 48 feet

WRITTEN EXERCISES

A. *Draw a sketch for each problem, and use it to find the solution. Show all your work.* (See page T-234 for sketches.)

1. Darla hung 5 towels edge to edge on the clothesline. She fastened each with two clothespins. Wherever the towels overlapped, she used one clothespin for both towels. How many clothespins did she need? 6 clothespins

2. Brother David built 40 feet of chainlink fence along the west side of his property and 56 feet along the south side of the property. The two sides met at the corner post. If he set the posts at 8-foot intervals, how many posts did he use? 13 posts

3. Lewis is planting a row of 20 bushes. If he plants the bushes 2 feet apart, how far apart will he plant the first bush and the last one? 38 feet

4. A farm has 6 silos standing in a row with 4 feet between each one and the next. All the silos are 20 feet in diameter. A conveyor runs beside the silos from the midpoint of the second one to the midpoint of the sixth one. How long is the conveyor? 96 feet

5. The Nolts planted sweet corn on a plot 24 feet wide by 100 feet long. The rows run lengthwise and are spaced 3 feet apart and 3 feet in from all edges of the patch. How many rows of sweet corn did they plant? 7 rows

234   Chapter 5   Mastering Percents

6. Before planting melons, Mother covered the ground with plastic. She planted 2 rows of melons 5 feet apart and 3 feet from all edges of the plastic. The rows are 40 feet long. How many square feet did she cover with plastic?   506 square feet

7. A section of cabinets in the Schrocks' kitchen is 82 inches long and has 4 doors. There is a 2-inch vertical strip of wood at each end of the section and separating each door from the next. How wide are the openings for the doors?   18 inches

8. Carol is making a get-well card for her cousin. The front of the card measures $5\frac{1}{2}$ inches by $4\frac{1}{4}$ inches, and there is a $\frac{3}{4}$-inch border along all four edges. What are the dimensions inside the border?   4 inches by $2\frac{3}{4}$ inches

REVIEW EXERCISES

B. Solve these reading problems. Use inverse proportions for numbers 13 and 14. *(Lessons 54, 66, 68)*   (See page T–235 for solutions.)

9. Merlin Albright sells fence posts at a 7% commission for extra income. If the commission for one week was $74.97, what were his sales that week? $1,071

10. Merlin calculated that his operating expenses were $30.00 per week. At 7% commission, what must his sales amount to each week to cover his operating expenses? (Answer to the nearest cent.) $428.57

11. Merlin's total sales amounted to $2,250.00 the first month and $2,800.00 the second month. To the nearest whole percent, what was the rate of increase in sales?   24%

12. During the winter, Merlin's monthly sales are below average. One month his total sales were $1,650. The next month's total sales were $375. What was the rate of decrease, to the nearest whole percent?   77%

13. On the school playground, an upper-grade boy weighing 120 pounds is sitting 5 feet from the fulcrum, seesawing with a lower-grade boy weighing 75 pounds. How far must the smaller boy sit from the fulcrum for the seesaw to balance?   8 feet

14. A bicycle has a drive sprocket with 42 teeth and a driven sprocket with 16 teeth. When the drive sprocket is pedaled at 48 revolutions per minute, how fast does the driven sprocket turn?   126 r.p.m.

C. Solve these percent problems mentally. *(Lesson 69)*

15. 6 is $66\frac{2}{3}$% of   9   16. $\frac{1}{2}$% of 600 =   3

D. Find the missing parts in these commission problems. *(Lesson 68)*

	Sales	Rate	Commission		Sales	Rate	Commission
17.	$8,112.50	6%	$486.75	18.	$2,684	12%	$322.08

E. Find the rate of each change, to the nearest whole percent. Include the label *increase* or *decrease*. *(Lesson 66)*

19. From $4.50 to $5.04   12% increase   20. From $9.75 to $8.00   18% decrease

Sketches for WRITTEN EXERCISES

1.
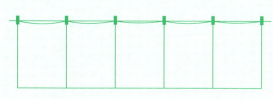

2.
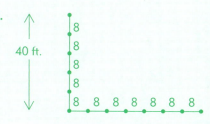

3.

4.

5.
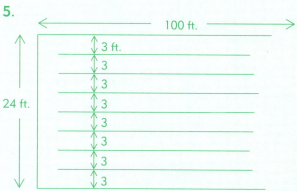

6.
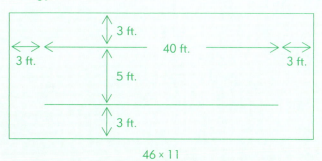

46 × 11

7.
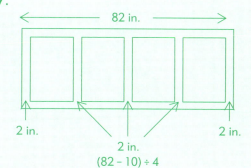

8.
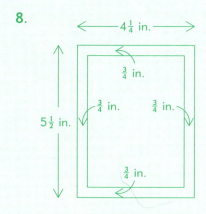

T–235   Chapter 5   Mastering Percents

Solutions for Part B

9. $74.97 ÷ 0.07

10. $30 ÷ 0.07

11. $2,800 − $2,250 = $550; $550 ÷ $2,250

12. $1,650 − $375 = $1,275; $1,275 ÷ $1,650

13. $\dfrac{\text{pounds on side 1}}{\text{pounds on side 2}}\quad \dfrac{120}{75} = \dfrac{n}{5}\quad \dfrac{\text{feet from fulcrum (2)}}{\text{feet from fulcrum (1)}}\qquad n = 8 \text{ feet}$

14. $\dfrac{\text{teeth (1)}}{\text{teeth (2)}}\quad \dfrac{42}{16} = \dfrac{n}{48}\quad \dfrac{\text{r.p.m. (2)}}{\text{r.p.m. (1)}}\qquad n = 126 \text{ r.p.m.}$

F. Find the amount of each change, to the nearest cent. Also find the new price. *(Lesson 63)*

　　　　　　　　　　　　　　　　　　change　　　new price

21. $86.09 decreased by 85%　　$73.18　　$12.91
22. $174.16 increased by 36%　　$62.70　　$236.86

G. Solve these subtraction and multiplication problems. *(Lessons 6, 38)*

23. $1,593.48
　　 − 861.36
　　　$732.12

24. $853.12 − $214.75　　$638.37

25. $\frac{3}{5} \times \frac{2}{9}$　　$\frac{2}{15}$

26. $5\frac{1}{2} \times 3\frac{3}{4}$　　$20\frac{5}{8}$

This is a faithful saying, and these things I will that thou affirm constantly, that they which have believed in God might be careful to maintain good works. These things are good and profitable unto men.

Titus 3:8

236   Chapter 5   Mastering Percents

71. Chapter 5 Review

A. Do these percent exercises. *(Lesson 60)*

1. Express 58% as a decimal.   0.58
2. Express 30% as a decimal.   0.3
3. Express 45% as a fraction in lowest terms. $\frac{9}{20}$
4. Express 59% as a fraction in lowest terms. $\frac{59}{100}$
5. Express 19% as a ratio in which the consequent is 100.   19:100
6. Express 23% as a ratio in which the consequent is 100.   23:100
7. Express 78% as a ratio in lowest terms.   39:50
8. Express 64% as a ratio in lowest terms.   16:25
9. Express $\frac{4}{5}$ as a percent.   80%
10. Express 0.7 as a percent.   70%
11. Express $\frac{3}{4}$ as a percent. (You will need to know the fraction–percent equivalents in Lesson 60 by memory for the test.)   75%
12. Express $\frac{5}{8}$ as a percent. (You will need to know the fraction–percent equivalents in Lesson 60 by memory for the test.) $62\frac{1}{2}\%$

B. Answer with base, rate, **or** percentage. *(Lesson 61)*

13. In "30% of 33 = 9.9," 30% is the ____.   rate
14. In "50% of 26 = 13," 13 is the ____.   percentage

C. Add or subtract these percents as indicated. *(Lesson 61)*

15. 100% + 38%   138%   16. 100% − 89%   11%

D. Find these percentages. *(Lessons 61–62)*

17. 35% of 48   16.8   18. 19 × 41%   7.79
19. 55 × 7%   3.85   20. 130% of 80   104
21. $2\frac{1}{2}\%$ of 450   11.25   22. $5\frac{3}{4}\%$ of 250   14.375

E. Express these percents as decimals. *(Lesson 62)*

23. 328%   3.28   24. 4,369%   43.69
25. $\frac{4}{5}\%$ 0.008   26. $\frac{3}{8}\%$ 0.00375

LESSON 71

Objective

- To review the material taught in Chapter 5 (Lessons 60–70).

Teaching Guide

1. Lesson 71 reviews the material taught in Lessons 60–70. For pointers on using review lessons, see *Teaching Guide* for Lesson 15.

2. Be sure to review the new concepts introduced in this chapter. They are listed below along with the exercises in this lesson that review those concepts.

Lesson number and new concept	Exercises in Lesson 71
60—Changing percents to ratios.	5, 6
65—Finding what percent one number is of another, rounded to nearest whole percent.	35–38
66—Finding the rate of increase when the rate is greater than 100%.	40
67—Finding the base when the rate is a fractional part of a percent.	45, 46
67—Finding the base when the rate is more than 100%.	47, 48
69—Finding percentage mentally when the rate is a fractional percent.	57, 58

T–237 Chapter 5 Mastering Percents

Sketches for Part M

59.

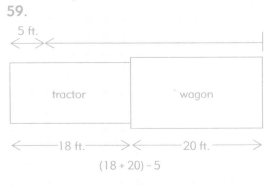

$(18 + 20) - 5$

60.

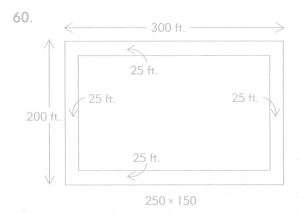

250×150

61.

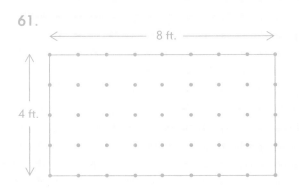

62.

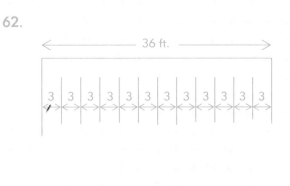

63.

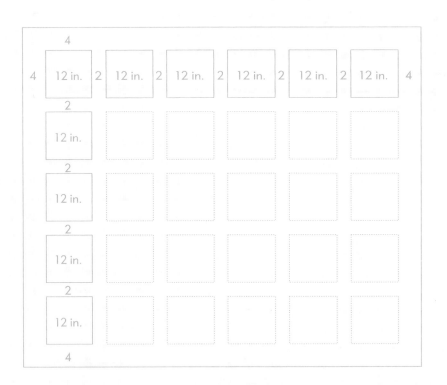

F. Find the amount of each change, to the nearest cent. Also find the new price. *(Lesson 63)*

		change	new price
27.	$6.27 increased by 28%	$1.76	$8.03
28.	$25.76 increased by 32%	$8.24	$34.00
29.	$6.75 increased by 31%	$2.09	$8.84
30.	$16.89 decreased by 15%	$2.53	$14.36

G. Find the new amount after each change, to the nearest cent. *(Lesson 64)*

31. $22.75 increased by 24% $28.21
32. $15.65 decreased by 38% $9.70
33. $122.75 decreased by 8% $112.93
34. $157.60 increased by 7% $168.63

H. Find the missing rates, to the nearest whole percent. *(Lesson 65)*

35. 9 is 60% of 15   36. 7 is 47% of 15
37. 13 is 76% of 17   38. 16 is 73% of 22

I. Find the rate of each change, to the nearest whole percent. Include the label *increase* **or** *decrease*. *(Lesson 66)*

39. week 1 production, 21   43%   40. week 3 production, 15   140%
    week 2 production, 12   decrease   week 4 production, 36   increase
41. year 1 weight, 24 lb.   29%   42. week 1 weight, 3,500 g   6%
    year 2 weight, 31 lb.   increase   week 2 weight, 3,300 g   decrease

J. Find the base. All answers are whole numbers. *(Lesson 67)*

43. 14 is 25% of 56   44. 45 is 30% of 150
45. 9 is $3\frac{3}{5}$% of 250   46. 12 is $1\frac{1}{4}$% of 960
47. 30 is 150% of 20   48. 48 is 160% of 30

K. Find the missing parts in these commission problems. *(Lesson 68)*

	Sales	Rate	Commission		Sales	Rate	Commission
49.	$1,231	8%	$98.48	50.	$1,525	9%	$137.25
51.	$4,100	7%	$287	52.	$3,223	4%	$128.92

L. Solve these percentage problems mentally. *(Lesson 69)*

53. 9 is $37\frac{1}{2}$% of 24   54. 15 is 60% of 25
55. 13 is $33\frac{1}{3}$% of 39   56. 44 is 80% of 55
57. $\frac{1}{2}$% of 1,400 = 7   58. $\frac{3}{4}$% of 1,600 = 12

238 Chapter 5 Mastering Percents

M. Draw a sketch for each problem, and use it to help you find the solution.
(Lesson 70) (See page T–237 for sketches.)

59. Darvin drove the tractor and wagon forward until the back of the wagon was 5 feet behind where the front of the tractor had been before. If the tractor was 18 feet long and the wagon was 20 feet long, how far did Darvin drive the tractor?   33 feet

60. The Witmers have a plot of land measuring 200 feet by 300 feet. Around the edge of the land is a 25-foot strip of lawn. The remainder of the land is garden. How many square feet are in the garden?   37,500 square feet

61. When David was nailing plywood, he put in one nail every foot, lengthwise and crosswise. How many nails did David use for one sheet of 4-foot by 8-foot plywood? (Count the first and last nails in a row as if they were right at the edge.)   45 nails

62. Mother's garden is 36 feet wide, and she plants the rows 3 feet apart. How many rows can she have in the garden if the first and last rows are 3 feet in from the edge?   11 rows

63. Mabel sewed a quilt top together, using 30 patches arranged in 6 rows with 5 in each row. After the seams were sewn, the patches measured 12 inches square. She has 2-inch-wide strips of fabric between all patches and a 4-inch border all around the quilt. What are the dimensions of Mabel's quilt top?   76 inches by 90 inches

64. Loretta is making curtains for a window that is 34 inches wide. When the curtains are closed, she wants the two halves to overlap 6 inches in the center of the window. How wide must each half be?   20 inches

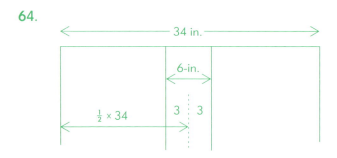

64.

72. Chapter 5 Test

LESSON 72

Objective

- To test the students' mastery of the concepts in Chapter 5.

Teaching Guide

1. Correct Lesson 71.
2. Review any areas of special difficulty.
3. Administer the test. For pointers on giving tests, see *Teaching Guide* for Lesson 16.

It was the opinion of Rehum the chancellor and Shimshai the scribe that Jerusalem was "a rebellious city and hurtful unto kings and provinces" (Ezra 4:15). Their letter to King Artaxerxes appealed for support of this opinion from past records.

On a later occasion, King Darius was asked to have records searched for a decree by Cyrus that the temple should indeed be built. (Ezra 5:17)

Statistics and graphs often depend on the availability and accuracy of past records.

Chapter 6
Statistics and Graphs

Statistics is a system used to draw conclusions about groups of numbers. For example, the fact that it was 0° one morning does not mean that the whole winter was especially cold. But if the morning temperature was 0° or lower for half of the winter, you may well conclude that the winter was quite cold. Gathering information and drawing conclusions like this is the purpose of statistics.

Graphs are used to give a picture of varying amounts so that a person can quickly compare them and draw conclusions about them. For example, by looking at a double-line graph of winter temperatures for this year and last year, you can easily tell whether this winter was colder, milder, or about the same as last year. Four common graphs that you will study in this chapter are the picture graph, the line graph, the bar graph, and the circle graph.

That search may be made in the book of the records of thy fathers: so shalt thou find in the book of the records, and know.
Ezra 4:15

73. The Arithmetic Mean

Suppose you were asked to compare the heights of 6,000 fourteen-year-old students and the heights of 6,000 twelve-year-old students. How could you compare 12,000 measurements?

Statistics is a system for dealing with large groups of numbers such as these. One tool of comparison for such groups is the **arithmetic mean** (ăr′ ĭth mĕt′ ĭk). This is the same as the average, and it is the most common tool used in statistics. For typical American students, the average height would be about 58 inches for twelve-year-olds and about 60 inches for fourteen-year-olds.

To calculate the arithmetic mean, use the following steps.

1. Find the sum of the group of numbers being considered.
2. Divide the sum by the total number of items in the group.

Often the arithmetic mean is not a whole number. Then you should use one of the following options.

(1) Express the remainder as a fraction or decimal. Do this if an exact amount is needed and if it is logical to have a fractional part of an item. (See Examples A and B.)

(2) Round to the nearest whole number. Do this if a fraction is not logical or practical, such as for persons or percent scores. (See Example C.)

Example A On the past four Saturdays, Alvin and Earl picked the following numbers of bushels of apples: 198, 169, 179, 188. What was the mean number of bushels per day? Express any remainder as a fraction.	198 169 179 + 188 ——— 734	$183\frac{2}{4} = 183\frac{1}{2}$ $4\overline{)734}$
Example B Mr. Fields delivered the following numbers of gallons of heating oil to six households: 278, 286, 417, 350, 366, 378. Find the mean, to the nearest tenth.	278 286 417 350 366 + 378 ——— 2,075	$345.83 = 345.8$ gal. $6\overline{)2,075.00}$ (rounded)

LESSON 73

Objectives

- To review the procedures for finding the arithmetic mean of a set of data.
- To introduce the term *arithmetic mean*. Previous experience: Using *average* as the main term, with *arithmetic mean* briefly mentioned.
- To review expressing the remainder as a fraction and rounding the arithmetic mean to the nearest whole number.
- To teach *rounding the arithmetic mean to the nearest tenth. (This does not involve any new mathematical concept.)

Review

1. *Calculate these rates mentally.* (Lesson 69)

    a. 4 is ___ of 12 ($33\frac{1}{3}\%$)

    b. 15 is ___ of 18 ($83\frac{1}{3}\%$)

2. *Find the missing parts. Express any partial percent as a fraction.* (Lessons 41, 65)

    a. 6 is $\frac{1}{9}$ of ___ (54)

    b. 16 is $\frac{3}{4}$ of ___ ($21\frac{1}{3}$)

    c. 6 is what percent of 11? ($54\frac{6}{11}\%$)

    d. 3 is what percent of 21? ($14\frac{2}{7}\%$)

3. *Do these exercises with scale drawings.* (Lesson 57)

    a. Scale: $\frac{1}{8}$ in. = 1 ft.; scale distance: $3\frac{1}{8}$ in.; actual distance: ___ (25 ft.)

    b. Scale: 1:80; actual distance: 4.2 m; scale distance: ___ cm (5.25 cm)

4. *Solve by horizontal multiplication.* (Lesson 9)

    a. 9 × 382 (3,438)

    b. 3 × 483 (1,449)

Introduction

Place the following table of daily temperatures on the board.

Mean Daily Temperatures

	S	M	T	W	T	F	S
Week 1:	55°	60°	65°	40°	45°	47°	55°
Week 2:	55°	52°	53°	52°	53°	52°	50°

What is meant by mean daily temperature? (average daily temperatures)

Which week was cooler? How could you find out? The best way is to find the mean temperature for each week. The mean temperature for both weeks is the same—52.4°.

Teaching Guide

1. **One tool of comparison for groups of numbers is the arithmetic mean.** (Notice that *arithmetic* is accented on the third syllable.) The arithmetic mean is the same as the average.

2. **Following are the steps for calculating the arithmetic mean.**

    (1) Find the sum of the group of numbers being considered.

    (2) Divide the sum by the total number of items in the group.

    Find the arithmetic mean of each set below.
    a. 45, 51 (48)
    b. 34, 45, 53 (44)
    c. 56, 64, 76, 72 (67)
    d. 38, 32, 35, 41, 34 (36)

3. If the arithmetic mean is not a whole number, one of the following options is generally used.

    (1) Express the remainder as a fraction or decimal. Do this if an exact amount is needed and if it is logical to have a fractional part of an item.

    (2) Round to the nearest whole number. Do this if a fraction is not logical or practical, such as for persons or percent grades.

    Find the arithmetic mean of each set, and express the remainder as indicated.
    a. 58, 23, 48, 61, 59, 53; remainder as a fraction $(50\frac{1}{3})$
    b. 43, 44, 47, 51, 53; remainder as a decimal (47.6)

    Find the arithmetic mean of each set, to the nearest whole number.
    c. 45, 53, 56, 57 (53)
    d. 75, 73, 68, 49, 62 (65)
    e. 95%, 98%, 86%, 99%, 100%, 89% (95%)
    f. 88%, 87%, 90%, 93%, 80%, 85% (87%)

Lesson 73

Example C

Jean had the following math scores: 95%, 97%, 89%, 96%, 100%, 100%, 94%, 92%, 100%. Find her mean score, to the nearest whole percent.

$$95 + 97 + 89 + 96 + 100 + 100 + 94 + 92 + 100 = 863$$

$$9\overline{)863.0} = 95.8 = 96\% \text{ (rounded)}$$

CLASS PRACTICE

Find the arithmetic mean of each set. Answers are whole numbers.

a. 19, 22, 24, 20, 25   22

b. 61, 71, 70, 64, 66, 64   66

Find the arithmetic mean of each set. Deal with remainders as indicated.

Express any remainder as a fraction.

c. 515, 499, 428, 501 $485\frac{3}{4}$

d. 853, 785, 812, 834   821

Round to the nearest tenth.

e. 96, 78, 85, 88, 94, 91   88.7

f. 181, 168, 175, 162, 177   172.6

Round to the nearest whole percent.

g. 100%, 86%, 89%, 97%, 100%   94%

h. 79%, 92%, 100%, 88%, 92%, 85%, 96%   90%

WRITTEN EXERCISES

A. *Find the arithmetic mean of each set. Answers are whole numbers.*

1. 12, 16, 19, 13, 15   15
2. 27, 23, 25, 28, 32   27
3. 45, 47, 68, 59, 51, 48   53
4. 79, 75, 68, 91, 87, 74   79

B. *Find the arithmetic mean of each set. Deal with remainders as indicated.*

Express any remainder as a fraction.

5. 195, 271, 215, 265 $236\frac{1}{2}$
6. 319, 249, 356, 321 $311\frac{1}{4}$
7. 874, 775, 689, 795 $783\frac{1}{4}$
8. 974, 915, 993, 910   948

Round to the nearest tenth.

9. 95, 89, 77, 98, 92, 87   89.7
10. 79, 86, 83, 98, 91, 69   84.3
11. 125, 158, 171, 161, 141   151.2
12. 271, 289, 251, 271, 291   274.6

Round to the nearest whole percent.

13. 95%, 97%, 92%, 99%, 100%   97%
14. 89%, 91%, 100%, 85%, 86%   90%
15. 100%, 100%, 100%, 70%, 100%   94%
16. 88%, 74%, 91%, 88%, 81%   84%

242 Chapter 6 *Statistics and Graphs*

C. Solve these reading problems.

17. Numbers 1–3 describes a census in which 603,550 men of war were numbered. All but 2 of these men died during the 40 years of wilderness wanderings. What was the average number of men that died each year, to the nearest whole number?
15,089 men

18. The Psalms of Degrees are short psalms that were sung by the Israelites as they traveled to feasts in Jerusalem. The following chart shows the numbers of verses in these psalms. Find the average number of verses in each psalm, to the nearest whole number.
7 verses

Psalms	120	121	122	123	124	125	126	127	128	129	130	131	132	133	134
Verses	7	8	9	4	8	5	6	5	6	8	8	3	18	3	3

19. When Neil and his brothers dug potatoes, the rows yielded the following numbers of bags: $4\frac{1}{2}$, 4, $3\frac{1}{2}$, 5, 4, $5\frac{1}{2}$, $4\frac{1}{2}$, 6. What was the average yield per row? (Express any remainder as a fraction.)
$4\frac{5}{8}$ bags

20. Lisa is usually able to complete the weekly ironing in $1\frac{1}{4}$ hours, of which she spends 45 minutes on shirts. If she irons 11 shirts, how much time does it take for each one? (Answer to the nearest whole number.)
4 minutes

21. Virgil noted that of the 50 cows in his father's herd, 43 were being milked and 7 were dry. What percent of the herd was dry cows? (See if you can find the answer mentally.)
14%

22. Martha's math lesson includes 25 exercises. She has already completed 19 of them. What percent of her math assignment is finished? (See if you can find the answer mentally.)
76%

REVIEW EXERCISES

D. Calculate these rates mentally. *(Lesson 69)*

23. 7 is 25% of 28

24. 10 is $66\frac{2}{3}$% of 15

E. Find the missing parts. Express any partial percent as a fraction. *(Lessons 41, 65)*

25. 11 is $\frac{1}{4}$ of 44

26. 3 is $\frac{2}{5}$ of $7\frac{1}{2}$

27. 5 is $55\frac{5}{9}$ % of 9

28. 18 is $85\frac{5}{7}$ % of 21

F. Do these exercises with scale drawings. *(Lesson 57)*

29. Scale: $\frac{1}{4}$ in. = 1 ft.; actual distance: 19 ft.; scale distance: $4\frac{3}{4}$ in.

30. Scale: 8:3; scale distance: $3\frac{1}{2}$ in.; actual distance: $1\frac{5}{16}$ in.

G. Solve by horizontal multiplication. *(Lesson 9)*

31. 6 × 254   1,524

32. 7 × 683   4,781

Lesson 73 T–242

Solutions for Part C

17. 603,548 ÷ 40
18. 101 (sum of verses) ÷ 15
19. 37 (sum of bags) ÷ 8
20. 45 ÷ 11
21. $\frac{7}{50} = \frac{14}{100}$
22. $\frac{19}{25} = \frac{76}{100}$

T-243 Chapter 6 Statistics and Graphs

LESSON 74

Objectives

- To teach *finding the median of a set of data.
- To teach *finding the mode of a set of data.

Review

1. Give Lesson 74 Quiz (Practice With Commissions).

2. *Find the mean of each set, to the nearest tenth.* (Lesson 73)
   a. 95, 97, 96, 84, 82, 91 (90.8)
   b. 41, 35, 46, 38, 49, 35 (40.7)

3. *Find the rate of each change, to the nearest whole percent. Label your answer as an increase or a decrease.* (Lesson 66)
   a. Week 1 production, 72; week 2 production, 85 (18% increase)
   b. Week 3 production, 34; week 4 production, 30 (12 % decrease)

4. *Do these divisions mentally.* (Lesson 42)
   a. $8 \div \frac{2}{3}$ (12)
   b. $18 \div \frac{3}{4}$ (24)

5. *Multiply mentally by applying the distributive law.* (Lesson 10)
   a. 11×35 (385)
   b. 18×12 (216)

6. *Draw a sketch for each problem, and use it to find the solution.* (Lesson 70)
   a. Justin mowed alfalfa in a field measuring 270 by 400 feet. He began by cutting a swath 9 feet wide around the perimeter of the field. After he mowed 3 rounds, how many square feet of alfalfa did he still have to mow?

   (74,736 sq. ft.)

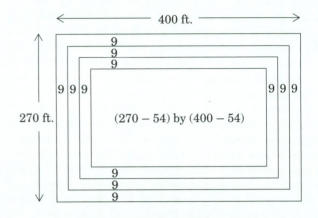

b. Geraldine is piecing a quilt with patches that measure 9 inches square. She is arranging the patches in 5 rows with 7 in each row and with 2-inch strips between the patches. She will put a 19-inch border along both sides (lengthwise) and one end (crosswise). What will be the dimensions of the finished quilt? (94 in. by 91 in.)

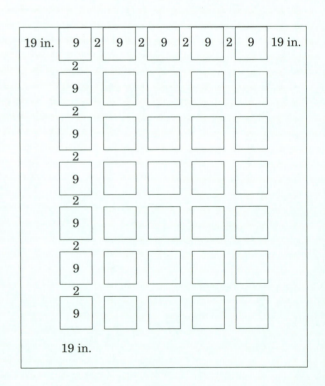

74. The Median and the Mode

The following chart is a record of the books seven students read in one month. The row of totals is a set of data that can be used to summarize the amount of reading done by the students.

Library Books Read in November	Alta	Edith	Henry	Jason	Leon	Marlin	Verna
A Home for Grandma	X				X		
All on a Mountain Day					X		
Coals of Fire		X		X	X	X	X
Coon Tree Summer	X				X		
Fisherman's Daughter, The	X	X					X
Home Fires Beneath the Northern Lights				X			
Home Fires at the Foot of the Rockies				X			
Ice Slide Winter	X				X		
Marita	X				X		X
Mary Jones and Her Bible			X			X	
Mohan in the Jungle				X			
Stand By, Boys!					X		
Tip Lewis and His Lamp					X		
Trapped by the Mountain Storm					X		
War-torn Valley	X				X		
Total Number Read	6	2	2	3	10	2	3

The mean number of books read is 4. Is that a good generalization of the amount of reading done by this class? The mean can give a wrong impression if one number in a set is far different from the rest. Only two students read as many as the mean, and one of them read more than twice that many. In such a case, the **median** gives a more realistic picture.

Finding the Median

The median is the middle value. It is most useful for comparing sets of numbers when a few numbers vary widely from the others. To find the median of a set of numbers, use the following steps.

1. Arrange the numbers in order from smallest to largest. This is known as ranking the data.

2. Find the middle number. If the number of values is even, find the mean (average) of the two middle numbers. Express any remainder as a fraction. See Example B.

Example A
Find the median of these numbers.
6, 2, 2, 3, 10, 2, 3

(1) 2, 2, 2, <u>3</u>, 3, 6, 10
(2) Median = 3

Example B
Find the median of these numbers.
75, 77, 78, 70, 0, 88, 69, 66

(1) 0, 66, 69, <u>70, 75,</u> 77, 78, 88
(2) Median = mean of 70 and 75
= $72\frac{1}{2}$.

Finding the Mode

Another statistical tool is the **mode**. This is the value that occurs most often in a set of numbers. To find the mode of a set of numbers, use the following steps.

1. Rank the data.

Example C
Find the mode of these numbers.
6, 2, 2, 3, 10, 2, 3

(1) <u>2, 2, 2,</u> 3, 3, 6, 10
(2) Mode = 2

2. Find the number that occurs most often. If more than one value occurs the same number of times, these numbers are all modes (Example D). If no number occurs more than once, there is no mode (Example E).

Example D
Find the mode(s) of these numbers.
9, 5, 2, 8, 2, 9, 6, 9, 5, 2, 8

(1) <u>2, 2, 2,</u> 5, 5, 6, 8, 8, <u>9, 9, 9</u>
(2) Mode = 2, 9

Example E
Find the mode of these numbers.
22, 18, 19, 24, 23, 21, 26, 17, 11

(1) 11, 17, 18, 19, 21, 22, 23, 24, 26
(2) No mode

A related application of mode is to find the item that occurs most often in a group of observations. For example, which is the most-read title in the list of books at the beginning of the lesson? The title read by the most students is *Coals of Fire*; that is the mode of this set of books.

CLASS PRACTICE

Find the median of each set.

a. 25, 62, 14, 28, 20   25
b. 64, 57, 45, 68, 80   64
c. 65, 83, 68, 61, 9, 68 $66\frac{1}{2}$
d. 88, 82, 91, 54, 98, 95 $89\frac{1}{2}$

Find the mode(s) of each set.

e. 2, 8, 4, 5, 2, 4, 3, 2   2
f. 3, 5, 8, 7, 2, 4, 3, 6, 8   3, 8
g. 16, 15, 18, 8, 9, 14, 16, 14, 11, 9, 14, 6, 15, 19   14
h. 30, 36, 33, 37, 33, 32, 36, 34, 37, 32, 31, 33, 35, 34, 31, 30   33

Lesson 74 T–244

Introduction

Find the mean of the following set.
23, 26, 24, 21, 29, 862 (mean = $164\frac{1}{6}$)
Ask the students, "Is this mean useful? Why or why not?"

The mean is of little value because one of the numbers is much larger than the other five. This causes the mean to be about six times as great as most of the numbers.

When numbers vary greatly, the mean has little significance. The median is a better statistical tool in such a case.

Teaching Guide

1. **The median is the middle value in a set of numbers.** It is most useful when a few numbers vary widely from the others.

2. **Following are the steps for finding the median of a set of numbers.**

   (1) Arrange the numbers in order from smallest to largest. This is known as ranking the data.

   (2) Find the middle number. If you are working with an even number of values, find the mean (average) of the two middle numbers. Express any remainder as a fraction.

   a. 47, 53, 45, 56, 98
      (45, 47, <u>53</u>, 56, 98
         median = 53)

   b. 85, 83, 0, 88, 89, 84
      (0, 83, <u>84</u>, <u>85</u>, 88, 89
         median = $84\frac{1}{2}$)

3. **The mode is the value that occurs most often in a set of numbers.** To find the mode, use the following steps.

   (1) Rank the data.

   (2) Find the number that occurs most often. If more than one value occurs the same number of times, these numbers are all modes. If no number occurs more than once, there is no mode.

   a. 15, 16, 17, 14, 15, 14, 13, 13, 18, 14
      (13, 13, <u>14</u>, <u>14</u>, <u>14</u>, 15, 15, 16, 17, 18; mode = 14)

   b. 1, 3, 5, 1, 3, 6, 3, 6, 3, 6, 7, 9, 2, 1, 2, 3
      (1, 1, 1, 2, 2, <u>3</u>, <u>3</u>, <u>3</u>, <u>3</u>, <u>3</u>, 5, 6, 6, 6, 7, 9; mode = 3)

   c. 1, 8, 5, 1, 3, 6, 3, 6, 7, 6, 7, 9, 2, 1, 2, 3
      (<u>1</u>, <u>1</u>, <u>1</u>, 2, 2, <u>3</u>, <u>3</u>, <u>3</u>, 5, <u>6</u>, <u>6</u>, <u>6</u>, 7, 7, 8, 9; mode = 1, 3, 6)

d. 3, 1, 2, 18, 5, 4, 7, 8, 17, 6, 9, 13, 19, 15
(1, 2, 3, 4, 5, 6, 7, 8, 9, 13, 15, 17, 18, 19; no mode)

Further Study

The tools discussed in Lessons 73 and 74 are only a few of the methods used in statistics. *Frequency distribution* deals with the frequency at which various values occur; this tool is presented in Lesson 75 with the study of histograms. *Probability* deals in mathematical terms with how likely it is that certain things will happen, as based on past records. Sampling, graphs, and polls of public opinion are other statistical tools that are widely used. Encyclopedias contain articles discussing these and other methods used in statistics.

Solutions for Part D

21. 7,625 (sum of years) ÷ 9

Find the mean, the median, and the mode(s) of each set. Express any remainder as a fraction.

i. 18, 24, 2, 8, 16, 14, 16, 20 mean: $14\frac{3}{4}$ median: 16 mode: 16
j. 84, 86, 83, 19, 89, 142, 86 mean: $84\frac{1}{7}$ median: 86 mode: 86
k. 51, 58, 56, 45, 321, 49, 52, 45 mean: $84\frac{5}{8}$ median: $51\frac{1}{2}$ mode: 45
l. 92, 58, 753, 125, 88, 99, 78, 94 mean $173\frac{3}{8}$ median 93 no mode

WRITTEN EXERCISES

A. *Find the median of each set.*

1. 55, 57, 89, 49, 63   57
2. 75, 76, 63, 94, 79   76
3. 98, 94, 99, 81, 0, 79 $87\frac{1}{2}$
4. 95, 93, 95, 96, 97, 42   95
5. 79, 76, 62, 79, 70, 80, 73, 71 $74\frac{1}{2}$
6. 95, 26, 77, 79, 82, 87, 93, 99 $84\frac{1}{2}$
7. 105, 109, 257, 116, 92, 115, 117, 123, 131, 115 $115\frac{1}{2}$
8. 215, 226, 210, 316, 205, 399, 275, 268, 289, 225, 278, 291 $271\frac{1}{2}$

B. *Find the mode(s) of each set. If there is none, write* no mode.

9. 3, 5, 6, 8, 3, 6, 8, 9   3, 6, 8
10. 2, 7, 9, 6, 5, 2, 1, 3, 9   2, 9
11. 12, 11, 6, 9, 14, 13, 15, 17, 10, 7, 16, 4, 18, 19   no mode
12. 25, 23, 24, 23, 24, 25, 27, 21, 27, 22, 21, 23, 24, 26, 21, 25   21, 23, 24, 25
13. 45, 47, 45, 47, 43, 46, 49, 47, 42, 49, 41, 45, 43, 43   43, 45, 47
14. 95, 92, 97, 93, 95, 97, 97, 97, 97, 92, 91, 99, 92, 97, 91, 90   97
15. 215, 214, 213, 214, 216, 215, 214, 213, 217, 214, 213, 213   213, 214
16. 125, 123, 131, 129, 135, 136, 118, 126, 124, 122, 117, 115   no mode

C. *Find the mean, the median, and the mode(s) of each set. Express any remainder as a fraction. Write* no mode *if there is none.*

17. 25, 23, 0, 2, 29, 27, 22, 21   mean: $18\frac{5}{8}$ median: $22\frac{1}{2}$ no mode
18. 46, 48, 3, 5, 47, 46, 47, 46   mean: 36   median: 46   mode: 46
19. 39, 38, 32, 39, 999, 31, 31, 37   mean: $155\frac{3}{4}$ median: $37\frac{1}{2}$ mode: 31 and 39
20. 78, 72, 877, 918, 72, 75, 73, 71   mean: $279\frac{1}{2}$ median: 74   mode: 72

D. *Solve these reading problems. Draw sketches for numbers 25 and 26.*

21. Find the median age of these nine men who lived before the Flood.   910 (years)

Name	Age	Name	Age	Name	Age
Adam	930	Cainan	910	Enoch	365
Seth	912	Mahalaleel	895	Methuselah	969
Enos	905	Jared	962	Lamech	777

246   Chapter 6   *Statistics and Graphs*

22. Find the median and the mode(s) of the number of verses in Psalms 115–125.
    median: 8 verses   mode: 8 verses

Psalms	115	116	117	118	119	120	121	122	123	124	125
Verses	18	19	2	29	176	7	8	9	4	8	5

23. When the Witmers harvested their 40 acres of corn, they tested the moisture of each truckload before unloading it into the corn-drying bin. Find the median and the mode(s) of the following test results from the 11 loads: 27.6%, 26.1%, 28.3%, 28.2%, 25.0%, 27.6%, 28.4%, 28.4%, 28.4%, 33.6%, 27.6%.
    median: 28.2%   modes: 27.6%, 28.4%

24. The Lapps raise tomatoes in two greenhouses and pick them three times each week. During the first three weeks of June, the number of pounds of tomatoes picked from both houses was as follows: 115, 180, 155, 160, 225, 175, 190, 300, 180. Find the median number of pounds.
    180 pounds

25. The Masts are planning an extension to their cherry orchard. They have prepared a plot 100 feet wide and 125 feet long beside the existing orchard. The trees will be spaced 25 feet apart, and there will be a $12\frac{1}{2}$-foot border on all four sides of the new part of the orchard. How many trees may be planted in this area?
    20 trees

26. Brother Leroy is posting his students' art pictures between two windows in his classroom. There are 14 pictures each 10 inches square, and he will mount them in 2 rows. If the pictures are separated from the windows by a 6-inch space at each end of a row, and a 3-inch space separates each picture from the next, how far apart are the two windows?
    100 inches

REVIEW EXERCISES

E. Find the mean of each set, to the nearest tenth. *(Lesson 73)*

27. 85, 69, 73, 71, 76, 77   75.2   28. 44, 48, 46, 42, 38, 52   45

F. Find the rate of each change, to the nearest whole percent. Include the label *increase* **or** *decrease.* *(Lesson 66)*

29. year 1 enrollment, 75
    year 2 enrollment, 81   8% increase

30. year 1 enrollment, 45
    year 2 enrollment, 39   13% decrease

G. Do these divisions mentally. *(Lesson 42)*

31. $9 \div \frac{3}{5}$ 15   32. $20 \div \frac{5}{6}$ 24

H. Multiply mentally by applying the distributive law. *(Lesson 10)*

33. 21 × 13   273   34. 32 × 17   544

22. 2, 4, 5, 7, 8, 8, 9, 18, 19, 29, 176

23. 25.0, 26.1, 27.6, 27.6, 27.6, 28.2, 28.3, 28.4, 28.4, 28.4, 33.6

24. 115, 155, 160, 175, 180, 180, 190, 225, 300

25.

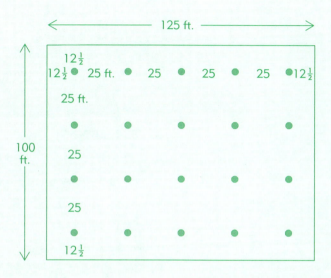

26.

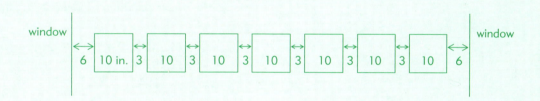

LESSON 75

Objectives

- To review reading and drawing histograms.
- To review that *range of data* refers to the distance between the highest and lowest values in a set of data.

Review

1. *Find the mean of each set, to the nearest tenth.* (Lesson 73)

   a. 32, 38, 41, 45, 37, 35 (38)

   b. 98, 87, 86, 92, 95, 94 (92)

2. *Find the base in each problem. All answers are whole numbers.* (Lesson 67)

   a. 21 is 30% of ___ (70)

   b. 36 is 45% of ___ (80)

   c. 75 is $2\frac{1}{2}$% of ___ (3,000)

   d. 52 is $6\frac{1}{2}$% of ___ (800)

   e. 81 is 135% of ___ (60)

   f. 96 is 480% of ___ (20)

3. *Solve by long division, and check by casting out nines.* (Lesson 11)

   a. $26\overline{)4,536}$ (174 R 12)

   b. $287\overline{)91,284}$ (318 R 18)

4. *Solve these problems by short division.* (Lesson 11)

   a. $9\overline{)37,281}$ (4,142 R 3)

   b. $6\overline{)37,284}$ (6,214)

75. The Histogram

Brother Kreider gave his seventh and eighth grade students a music test. Ranked from lowest to highest, the scores were as follows: 66%, 72%, 79%, 83%, 85%, 85%, 88%, 90%, 90%, 91%, 92%, 92%, 92%, 95%, 96%, 98%, 100%.

The mean score is 88%, the median is 90%, and the mode is 92%. However, the mean, median, and mode do not indicate the range from the lowest to the highest scores. Neither do they show how many students received good, fair, and poor scores.

The histogram shows both the **range of data** (distance between the lowest and highest values) and the distribution of data within the range. In this case, the range was from 66% to 100%, but most of the students' scores were 81% or higher.

The data above can be organized on a **frequency distribution table**. Each time a value occurs, it is recorded in the proper interval with a tally mark (/). Then the number of marks is recorded, and the table is used to make a histogram.

Frequency Distribution Table

Intervals		Frequency
More than 65% but not more than 70%	/	1
More than 70% but not more than 75%	/	1
More than 75% but not more than 80%	/	1
More than 80% but not more than 85%	///	3
More than 85% but not more than 90%	///	3
More than 90% but not more than 95%	/////	5
More than 95% but not more than 100%	///	3

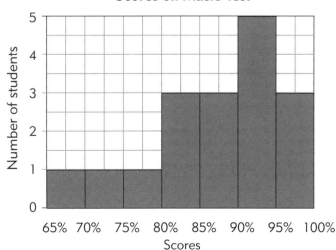

248   *Chapter 6   Statistics and Graphs*

A histogram shows the approximate mean, median, and mode in graph form. The tallest bar shows the mode, which is the range from 91% to 95%. The median is the point at which half of the students are below and half are above. The median in the histogram on page 247 is the range from 86% to 90%.

A histogram looks much like a bar graph. On a histogram, however, each bar shows the number of items within a given interval, whereas each bar on a bar graph represents one specific value. You will study bar graphs in a later lesson.

Some values represented by a histogram are exactly at the dividing point between intervals (for example, a score of 90% in the example above). Such a value could be included in either the next lower interval or the next higher interval. **In this course, the value at the dividing point between two intervals is counted in the lower interval.** (The tally for the interval of 95%–100% does not count the 95% score, but it does count the 100%.) Compare the frequency distribution table and the histogram above, and be sure to follow the same pattern whenever you work with histograms.

Constructing a Histogram

To make a histogram, you first need to make a distribution table like the one in the lesson. Use the following steps.

1. Decide on a suitable interval for grouping your data, and arrange the data according to those intervals on a frequency distribution table. (For the lesson example, the data was divided into intervals of 5 percent each.)

2. Mark off the horizontal scale of the histogram, starting with the lowest value and showing the intervals that you chose in step 1. Label the horizontal scale.

3. Mark off the vertical scale of the histogram, with the highest number equal to or slightly greater than the highest frequency on the frequency distribution table. (The highest number on the sample histogram is 5 because the highest frequency on the table is 5.) Label the vertical scale.

4. Draw each bar to the correct height. The bars should all be the same width, with no space between them.

5. Write a title for the histogram. If you obtained the data from a reference source, name that source in the lower right corner of the histogram.

CLASS PRACTICE

Use the histogram in the lesson to answer the following questions.

a. How many students scored between 76% and 80%?   1 student
b. How many students scored between 81% and 85%?   3 students
c. How many students scored between 91% and 95%?   5 students
d. How many students scored between 96% and 100%?   3 students
e. In what percent range did the class mode fall?   91%–95%
f. How many students received a grade higher than 90%? 8 students
g. How many students received a grade 90% or lower?   9 students

Introduction

Call attention to the histogram in the lesson text. How does it compare with a bar graph? (Both kinds of graphs show varying amounts. But a histogram shows the number of items within a given range, whereas each bar on a bar graph represents one specific value.)

Which of the statistical tools—the mean, the median, or the mode—is illustrated by the histogram? (A histogram shows the mode or the number of items in specified ranges. The approximate mean and median can also be found).

Teaching Guide

1. **A histogram shows the number of times that values occur within specified ranges.** Demonstrate this point by using the first set of exercises in *Class Practice*.

2. **A histogram shows the range of data.** In the lesson example, the range is from 66% to 100%. The bars show the distribution of data within the range. Notice that most of the students' scores were 81% or higher.

3. **In this course, the value at the dividing point between two intervals is counted in the lower interval.** Be sure students understand and follow this pattern when working with histograms.

4. **Use the following steps to draw a histogram.**

    (1) Decide on a suitable interval for grouping your data, and arrange the data according to those intervals on a frequency distribution table.

    (2) Mark off the horizontal scale of the histogram, starting with the lowest value and showing the intervals that you chose in step 1. Label the horizontal scale.

    (3) Mark off the vertical scale of the histogram, with the highest number equal to or slightly greater than the highest frequency on the frequency distribution table. Label the vertical scale.

    (4) Draw each bar to the correct height. The bars should all be the same width, with no space between them.

    (5) Write a title for the histogram. If you obtained the data from a reference source, name that source in the lower right corner of the histogram.

T–249   Chapter 6   Statistics and Graphs

An Ounce of Prevention

1. Stress neatness and accuracy in making graphs. Use your discretion in deciding how many graphs to assign in each lesson.

2. For scoring graphs in this chapter, one exercise number (one point) is assigned to each item of data and one for the title and labels. Line graphs have a maximum of 10 points each.

Histogram for CLASS PRACTICE

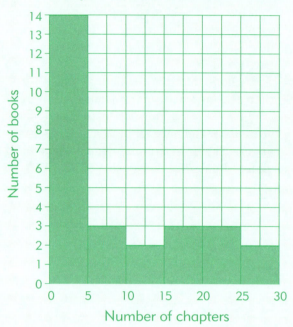

Prepare a histogram from the following data. Begin by completing the frequency distribution table that is shown. (See facing page.)

Chapters in New Testament Books

Matthew 28	2 Corinthians 13	1 Timothy 6	2 Peter................3
Mark................16	Galatians............6	2 Timothy 4	1 John 5
Luke 24	Ephesians 6	Titus 3	2 John 1
John 21	Philippians 4	Philemon 1	3 John 1
Acts..................28	Colossians 4	Hebrews 13	Jude 1
Romans 16	1 Thessalonians ..5	James................5	Revelation.........22
1 Corinthians 16	2 Thessalonians ..3	1 Peter...............5	

Frequency Distribution Table

Intervals	Frequency
More than 0 but not more than 5 chapters	14
More than 5 but not more than 10 chapters	3
More than 10 but not more than 15 chapters	2
More than 15 but not more than 20 chapters	3
More than 20 but not more than 25 chapters	3
More than 25 but not more than 30 chapters	2

WRITTEN EXERCISES

A. Use the following histogram to do the exercises below.

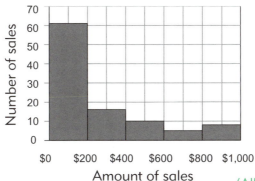

Amount Sold per 100 Sales at Sterling Electric Motors

(Allow reasonable variation.)

1. Estimate the number of sales in the $0–$200 interval.   61 (percent)
2. What number of sales were in the $400–$600 interval?   10 (percent)
3. In all, the sales between $200 and $600 were about ($\frac{1}{3}$, $\frac{1}{4}$, $\frac{1}{5}$) of the total.
4. The sales above $600 were about ($\frac{1}{5}$, $\frac{1}{6}$, $\frac{1}{8}$) of the total.
5. What was the total number of sales in all the intervals?   100 (percent)
6. In what interval was the mode of the sales? $0–$200

B. Prepare a histogram from each set of data. Begin by making a frequency distribution table with the intervals given. (See facing page.)

7–14. Use intervals of 10 years.

Reigns of Rulers in Judah

Rehoboam .... 17 yr.	Joash .... 40 yr.	Amon .... 2 yr.
Abijah .... 3 yr.	Amaziah .... 29 yr.	Josiah .... 31 yr.
Asa .... 41 yr.	Uzziah .... 52 yr.	Jehoahaz .... 3 mo.
Jehoshaphat .... 25 yr.	Jotham .... 16 yr.	Jehoiakim .... 11 yr.
Jehoram .... 8 yr.	Ahaz .... 16 yr.	Jehoiachin .... 3 mo.
Ahaziah .... 1 yr.	Hezekiah .... 29 yr.	Zedekiah .... 11 yr.
Athaliah .... 7 yr.	Manasseh .... 55 yr.	

15–22. Use intervals of 2 years.

Time in Office for Presidents of the United States

Washington .... 8 yr.	Buchanan .... 4 yr.	Coolidge .... $5\frac{1}{2}$ yr.
J. Adams .... 4 yr.	Lincoln .... 4 yr.	Hoover .... 4 yr.
Jefferson .... 8 yr.	A. Johnson .... 4 yr.	F. D. Roosevelt.* $12\frac{1}{4}$ yr.
Madison .... 8 yr.	Grant .... 8 yr.	Truman .... $7\frac{3}{4}$ yr.
Monroe .... 8 yr.	Hayes .... 4 yr.	Eisenhower .... 8 yr.
J. Q. Adams .... 4 yr.	Garfield .... $\frac{1}{2}$ yr.	Kennedy .... $2\frac{5}{6}$ yr.
Jackson .... 8 yr.	Arthur .... $3\frac{1}{2}$ yr.	L. B. Johnson .... $5\frac{1}{6}$ yr.
Van Buren .... 4 yr.	Cleveland .... 8 yr.	Nixon .... $5\frac{1}{2}$ yr.
W. H. Harrison .... 1 mo.	B. Harrison .... 4 yr.	Ford .... $2\frac{1}{2}$ yr.
Tyler .... 4 yr.	McKinley .... $4\frac{1}{2}$ yr.	Carter .... 4 yr.
Polk .... 4 yr.	T. Roosevelt .... $7\frac{1}{2}$ yr.	Reagan .... 8 yr.
Taylor .... $1\frac{1}{3}$ yr.	Taft .... 4 yr.	Bush .... 4 yr.
Fillmore .... $2\frac{2}{3}$ yr.	Wilson .... 8 yr.	Clinton .... 8 yr.
Pierce .... 4 yr.	Harding .... $2\frac{1}{2}$ yr.	*Place in 10–12 yr. interval.

C. Solve these reading problems.

23. Cheryl and her mother were making applesauce. When 75% of the apples were processed, Cheryl counted 44 full quarts on the countertop. To the nearest quart, how many could they expect to have when they were finished? 59 quarts

24. A newly hatched Nile crocodile is about 12 inches long. If that is 5% of its full-grown length, how many feet long may it grow to be? 20 feet

25. Linford's father sent 48 steers to market, of which 39 were Holsteins. Write a ratio in lowest terms to compare the number of Holsteins with the total number of steers. 13:16

26. Of the 75 rows of celery on the Hottenstein farm, 15 rows still need to be cultivated. Write a ratio in lowest terms to compare the celery rows already cultivated with the total number of rows. 4:5

Answers for Part B
7–14.

Frequency Distribution Table

Intervals	Frequency
More than 0 but not more than 10 years	7
More than 10 but not more than 20 years	5
More than 20 but not more than 30 years	3
More than 30 but not more than 40 years	2
More than 40 but not more than 50 years	1
More than 50 but not more than 60 years	2

15–22.

Frequency Distribution Table

Intervals	Frequency
More than 0 but not more than 2 years	3
More than 2 but not more than 4 years	20
More than 4 but not more than 6 years	4
More than 6 but not more than 8 years	13
More than 8 but not more than 10 years	0
More than 10 but not more than 12 years	1

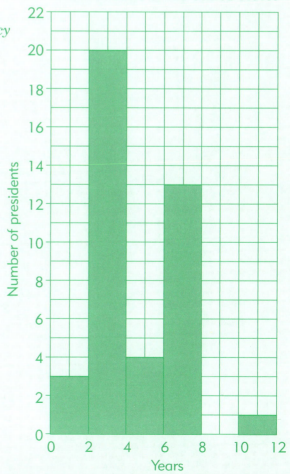

Solutions for Part C
23. $44 = 75\%$ of ___; $44 \div 0.75$
24. $12 = 5\%$ of ___; $12 \div 0.05$
25. $39:48 = 13:16$
26. $60:75 = 4:5$

T–251   Chapter 6   *Statistics and Graphs*

Further Study

Businessmen use distribution tables and histograms to observe trends in sales, expenses, and inventory. This helps them answer management questions: How much inventory should we carry? Which type of advertising has brought the best results? What percent of the balance in Accounts Receivable is 31 to 60 days past due? What percent of our sales have been in the $500 to $1,000 range? What period of the day or year has been our most profitable?

Histograms help statisticians analyze observations and trends so that they can better predict changes in prices, weather, and so forth.

In the classroom, you could check the height, weight, or pulse rate of your students and graph it on a histogram.

27. $(3{,}000 + 1{,}005) \div (40 \times 12)$

28. $(40 \times 12) \div 42$

27. In 1 Kings 4:32, we read that Solomon spoke 3,000 proverbs and wrote 1,005 songs. Apparently he wrote all these during his 40-year reign after he had received special wisdom from God. If that is true, he averaged how many songs and proverbs each month? (Answer to the nearest whole number.)   8 songs and proverbs

28. Numbers 33 lists the encampments of the children of Israel from the time they left Egypt until they pitched by Jordan near Jericho. If they camped at 42 sites in 40 years, what was the average number of months that they stayed at each camping location, to the nearest whole month?   11 months

REVIEW EXERCISES

D. Find the mean of each set, to the nearest tenth. *(Lesson 73)*

29. 64, 75, 68, 71, 67, 73   69.7   **30.** 99, 86, 95, 94, 89, 105   94.7

E. Find the base in each problem. *(Lesson 67)*

31. 90 is 40% of 225   **32.** 14 is 25% of 56

33. 102 is $8\frac{1}{2}$% of 1,200   **34.** 63 is $1\frac{1}{2}$% of 4,200

35. 56 is 175% of 32   **36.** 42 is 350% of 12

F. Solve by long division, and check by casting out nines. Show your check numbers. *(Lesson 11)*   (Check numbers are in parentheses.)

(7) × (5) + (4) = (39) → (3)   (0) × (6) + (1) = (1)

37. 34)7,824   230 R 4   (3)   **38.** 342)63,811   186 R 199   (1)

G. Solve these problems by short division. *(Lesson 11)*

39. 8)3,508   438 R 4   **40.** 4)28,542   7,135 R 2

76. The Picture Graph

Graphs present data in picture form. It is much easier to compare data by looking at a graph than by reading paragraphs of text or lists of numbers. The most common kinds of graphs are the picture graph, the bar graph, the line graph, and the circle graph.

Here is some data showing the total number of sheep in the main sheep-grazing states. Notice how this data was used to make a picture graph.

States With Largest Numbers of Sheep

Texas	1,200,000
California	800,000
Wyoming	570,000
Colorado	440,000
South Dakota	420,000

States With Largest Numbers of Sheep

🐑 = 100,000 sheep

Source: *Funk & Wagnall's New Encyclopedia*

Reading a Picture Graph

To read a picture graph, count the number of symbols shown and multiply by the value that each symbol represents. If a part symbol is shown, estimate what fractional part of the whole it is. For example, Wyoming is shown with $5\frac{1}{2}$ symbols each representing 100,000 sheep. Multiplying $5\frac{1}{2}$ times 100,000 shows that Wyoming has about 550,000 sheep.

A picture graph usually shows approximate amounts. The graph above shows that South Dakota has about 400,000 sheep. The source data indicates that there are 420,000 sheep in that state.

The range of data on the graph above is from about 400,000 to 1,200,000 sheep. This is fairly simple to determine because values on a picture graph are normally arranged from largest to smallest.

LESSON 76

Objectives

- To review reading and drawing picture graphs, including those with partial symbols.
- To teach *choosing the value that each symbol on a picture graph should represent.

Review

1. Give Lesson 76 Quiz (Mean, Median, and Mode).

2. *Find the missing part in each commission problem.* (Lesson 68)

   Sales   Rate   Commission
   a. $4,960 $6\frac{1}{2}$%   ___   ($322.40)
   b. $3,250   ___ $276.25   ($8\frac{1}{2}$%)
   c. ___   2% $186   ($9,300)

3. *Express each number as a percent.* (Lesson 60)

   a. $\frac{3}{4}$ (75%)
   b. 0.54 (54%)
   c. 14:25 (56%)

4. *Tell whether each division will work out evenly.* (Lesson 12)

   a. 489 ÷ 6 (no)
   b. 318 ÷ 3 (yes)

5. *Use the divide-and-divide method to solve these problems mentally.* (Lesson 12)

   a. 168 ÷ 14 (12)
   b. 132 ÷ 44 (3)

6. *Tell what missing information is needed to solve each problem.* (Lesson 44)

   a. Jonathan read 40% of the pages in a 275-page book. At the same rate of reading, how long will it take him to read the entire book? (how long it took to read 40% of the book)

   b. A fuel deliveryman filled Maple Farm's new 1,000-gallon tank. He filled the 250-gallon house tank, which had 75 gallons left from the previous winter, as well as the shop tank, which had 45 gallons left in it. What was the total amount of heating oil delivered to Maple Farm? (capacity of the shop tank)

T–253   Chapter 6   *Statistics and Graphs*

Introduction

Call attention to the table and the first picture graph in the lesson. Ask, "Which state had about half as many sheep as Texas had?" (Wyoming) Point out that the picture graph is much more helpful than the table in considering this question. One can find the answer at a glance, without even calculating.

This brings to mind the saying, "A picture is worth a thousand words." A picture graph is useful for comparing quantities because we can see the relative numbers of the different items.

Teaching Guide

1. **To read a picture graph, count the number of symbols shown and multiply by the value that each symbol represents. If a part symbol is shown, estimate what fractional part of the whole it is.**
   Have the students compare the table and the picture graph, and notice how the picture graph gives about the same information as the table. Then do the first set of *Class Practice* exercises.

2. **Use the following steps to draw a picture graph.**

   (1) Collect the information for the graph. Rank it in descending order (from the highest value to the lowest).

   (2) Choose a scale that allows the lowest value to have at least half of a symbol and the highest value to have no more than twenty symbols.

   (3) Divide each value by the chosen scale number to determine how many symbols it should have. Round each number of symbols to the nearest half symbol.

   (4) Choose a simple symbol with a logical relationship to the subject.

   (5) Write a label for each row of the graph. Draw the symbols carefully and space them evenly. Arranging the symbols in groups of five aids accurate reading and visual comparison.

   (6) Write a title for the graph and make a scale to explain what each symbol represents. If you obtained the data from a reference source, name that source in the lower right corner.

   Use the steps above to draw a picture graph as indicated in *Class Practice*. It is recommended that you work through this procedure with the class.

Constructing a Picture Graph

Drawing a picture graph requires careful planning as well as neatness and accuracy. Following are the steps for drawing a picture graph.

1. Collect the information for the graph. Rank it in descending order (from the highest value to the lowest).

2. Choose a reasonable scale. The lowest value should have at least half of a symbol and the highest value no more than twenty symbols.

3. Divide each value by the chosen scale number to determine how many symbols it should have. Round each number of symbols to the nearest half symbol.

4. Choose a simple symbol with a logical relationship to the subject.

5. Write a label for each row of the graph. Draw the symbols carefully and space them evenly. Arranging the symbols in groups of five aids accurate reading and visual comparison.

6. Write a title for the graph and make a scale to explain what each symbol represents. If you obtained the data from a reference source, name that source in the lower right corner.

In step 3 you are directed to round to the nearest half symbol. The following diagram shows how to do this.

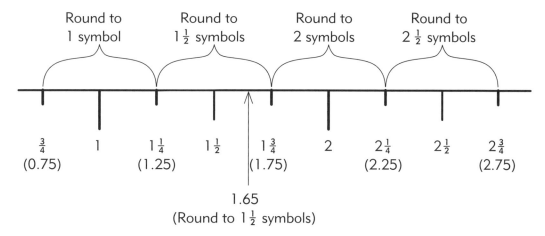

Any value of $\frac{3}{4}$ (0.75) or greater, but less than $1\frac{1}{4}$ (1.25), should be rounded to 1 symbol. If a value is $1\frac{1}{4}$ (1.25) or greater, but less than $1\frac{3}{4}$ (1.75), it should be rounded to $1\frac{1}{2}$ symbols; and so on. (The fourths are the dividing points between the halves.) As shown above, 1.65 would be rounded to $1\frac{1}{2}$ symbols because its value is between 1.25 and 1.75.

254 Chapter 6 Statistics and Graphs

CLASS PRACTICE

Use the following picture graph to do the exercises below.

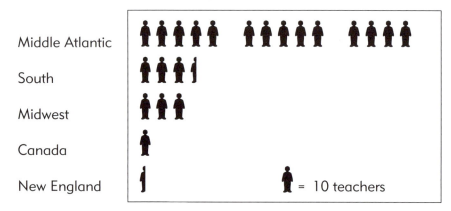

a. Estimate the number of teachers from the Middle Atlantic States. 140 teachers

b. Estimate the number of teachers from Canada. 10 teachers

c. Estimate the number of teachers from the United States other than the Middle Atlantic States. 70 teachers

d. Estimate the number of teachers from the South and the Midwest together. 65 teachers

e. Estimate the range of data. from 5 to 140

f. Estimate how many more teachers were from the Middle Atlantic States than from all the other regions of the United States and Canada combined. 60 teachers

Use the following information to do the exercises below.

The following list shows the 5 generals in Jehoshaphat's army and the number of men under the command of each one (2 Chronicles 17:14–18).

Adnah 300,000
Jehohanan 280,000
Amasiah 200,000
Eliada 200,000
Jehozabad 180,000

g. What symbol would you use for the graph? a man

h. Choose the most reasonable of the following numbers for the scale: One symbol = (10,000; 20,000; 50,000; 100,000) men.

i. Construct a picture graph from this information. (See facing page.)

Picture Graph for CLASS PRACTICE i

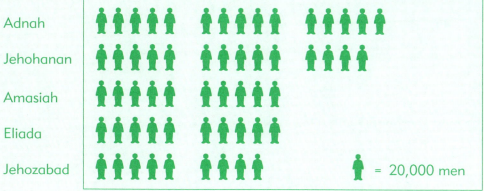

Generals and Men in Jehoshaphat's Army

Source: *2 Chronicles 17:14–18*

T–255 Chapter 6 *Statistics and Graphs*

Picture Graphs for Part B
9–13.

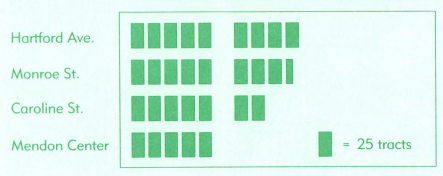

14–19.

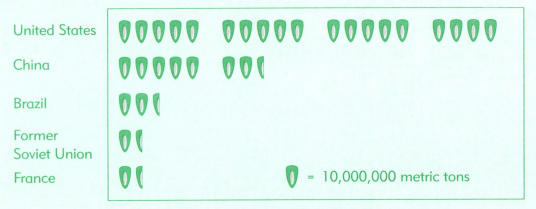

Source: *1993 Information Please Almanac*

Lesson 76

WRITTEN EXERCISES

A. Use the following picture graph to do the exercises below.

States With Greatest Lumber Production

Oregon	▱▱▱▱▱ ▱▱▱▱▱ ▱▱▱▱▱ ▱◿
California	▱▱▱▱▱ ▱▱▱▱▱ ◿
Washington	▱▱▱▱▱ ◿
North Carolina	▱▱▱
Georgia	▱▱▱

▱ = 500,000,000 board feet

Source: *World Book Encyclopedia*

1. How many board feet of lumber were produced in Oregon?   8,250,000,000 board feet

2. How many more board feet were produced in Oregon than in Washington?   5,000,000,000 board feet

3. How many board feet were produced on the west coast of the United States?   16,750,000,000 board feet

4. Which state produced about as much lumber as Georgia and North Carolina together?   Washington

5. Estimate the range of data.   from 1,500,000,000 to 8,250,000,000

6. How many board feet of lumber did all five states produce together?   19,750,000,000 board feet

7. What scale of board feet per symbol would be needed in order to have about 4 symbols for the Oregon production?   1 symbol = 2,000,000,000 board feet

8. What scale of board feet per symbol would be needed in order to have about 8 symbols for the Oregon production?   1 symbol = 1,000,000,000 board feet

B. Draw a picture graph for each set of data.   (See facing page.)

9–13. Following are the numbers of tracts distributed on several tract routes. Choose the most reasonable of the following numbers for the scale: One symbol = (**25**), 50, 100) tracts.

       Hartford Avenue 225
       Monroe Street 215
       Caroline Street 175
       Mendon Center 125

14–19. The following nations have the greatest corn production in the world. Choose the most reasonable of the following numbers for the scale: One symbol = (500,000; 1,000,000; **10,000,000**; 100,000,000) metric tons.

       United States . . . . 191,197,000 metric tons
       China. . . . . 75,840,000 metric tons
       Brazil . . . . 26,508,000 metric tons
       Former Soviet Union. . . 17,000,000 metric tons
       France . . . . 12,926,000 metric tons

Source: *1993 Information Please Almanac*

256   Chapter 6   *Statistics and Graphs*

C. Solve these reading problems. In problems 24 and 25, write what missing information is needed to solve the problem.

20. Noah had 3 sons by the time he was 500 years old. He lived a total of 950 years. To the nearest whole percent, what part of his life was past when he reached the age of 500?
53%

21. After the cows had gotten into the garden, Helen noticed that only 45 of the 250 pepper plants were left. Express 45:250 as a percent.
18%

22. Conrad Nolt works as a salesperson at London Furniture. He receives a commission of 5% in addition to a monthly salary. If he sold $15,000 worth of furniture last month, what was his commission?
$750

23. Herbert took 25 rabbits to Morrison's Country Auction. The auction charges a 9% commission for selling animals brought there. If the commission deducted from Herbert's check was $8.10, what was the selling price of the rabbits?
$90

24. Henry is shoveling a path through the snow from the house to the barn. If he can clear a path 2 feet wide and 5 feet long every minute, how long will it take him to reach the barn?
distance to the barn

25. Mother left home at 8:30 one Saturday morning to go shopping. She spent 20 minutes at Lowe's Hardware and 25 minutes at Slade's Farm Store. After refueling at a service station, she made a final stop at the local grocery store. She arrived home at 11:30. If she had spent $1\frac{1}{2}$ hours driving, how long was she at the grocery store?
time spent at the service station

REVIEW EXERCISES

D. Find the missing part in each commission problem. *(Lesson 68)*

	Sales	Rate	Commission
26.	$5,230	$5\frac{1}{2}\%$	$287.65
27.	$6,420	$3\frac{1}{2}\%$	$224.70

E. Express each number as a percent. *(Lesson 60)*

28. $\frac{2}{5}$ 40%   29. 0.09   9%

F. Write yes or no to tell whether each division will work out evenly. *(Lesson 12)*

30. 162 ÷ 4   no   31. 381 ÷ 9   no

G. Use the divide-and-divide method to solve these problems mentally. *(Lesson 12)*

32. 144 ÷ 18   8   33. 192 ÷ 16   12

Solutions for Part C

20. $\frac{500}{950} = \frac{10}{19} = 10 \div 19$

21. $\frac{45}{250} = \frac{9}{50} = \frac{18}{100}$

22. $0.05 \times \$15{,}000$

23. $\$8.10 = 9\% \times \underline{}$; $\$8.10 \div 0.09$

T-257   Chapter 6   Statistics and Graphs

LESSON 77

Objective
- To review reading and drawing bar graphs.

Review
1. *Find the mean of each set, to the nearest tenth.* (Lesson 73)
   a. 216, 231, 221, 218, 243   (225.8)
   b. 601, 585, 594, 611, 607   (599.6)
2. *Solve these percent problems mentally.* (Lesson 69)
   a. $87\frac{1}{2}\%$ of 16 = ___   (14)
   b. $83\frac{1}{3}\%$ of ___ = 20   (24)
   c. 3 is ___ of 12   (25%)
   d. 9 is ___ of 15   (60%)
3. *Find these percentages.* (Lesson 61)
   a. 20% of 16   (3.2)
   b. 90% of 75   (67.5)
4. *Solve mentally by using the double-and-double method.* (Lesson 13)
   a. 350 ÷ 50   (7)
   b. 6,300 ÷ 50   (126)
5. *Solve mentally by using the shortcut for dividing by 25.* (Lesson 13)
   a. 3,600 ÷ 25   (144)
   b. 8,800 ÷ 25   (352)

Introduction
Discuss the first bar graph in the lesson text. Ask, "Could this information be presented on a picture graph? The answer is yes. Bar graphs and picture graphs show exactly the same types of information. They compare the sizes of different items.

Now ask the students, "How is this bar graph different from the picture graphs in the last lesson?" Following are several differences.

(1) Shaded bars are used instead of pictures.

(2) The value of each bar is determined by the vertical scale of the graph. The value of each picture on a picture graph is determined by the scale in the lower right corner.

(3) Picture graphs show data horizontally, but bar graphs usually show data vertically. Bars on bar graphs can also extend horizontally, but it is often more logical to compare heights of bars than lengths of bars.

Teaching Guide
1. **To read a bar graph, compare the height of each bar with the scale to the left of the graph. If the bar ends between two numbers, estimate the value it represents.** Demonstrate with the first set of *Class Practice* exercises.

2. **The bars on a bar graph may be vertical or horizontal, but usually they are vertical. The numerical scale should always begin at zero.** Illustrate by covering 50 years of the graph with a paper. (The visual impression changes so that it looks as if the life expectancy in Sweden and the United States is more than twice

77. The Bar Graph

Instead of symbols or pictures, a bar graph uses bars of varying lengths to compare different values. A scale indicates the approximate value of each bar on the graph.

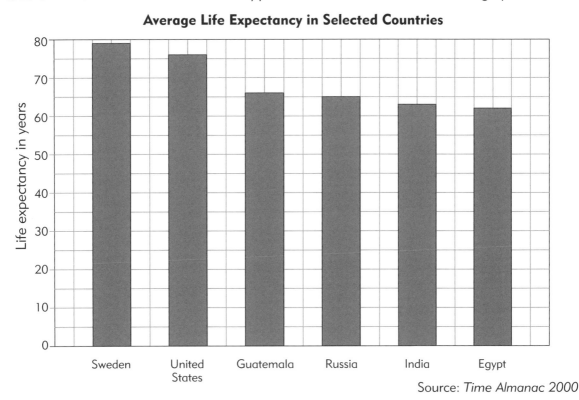

To read a bar graph, compare the height of each bar with the scale to the left of the graph. If the bar ends between two numbers, estimate the value it represents.

The bars on a bar graph may be vertical or horizontal, but usually they are vertical. The numerical scale should always begin at zero. Graph paper is recommended for producing a consistent scale and neat graph.

To draw a bar graph, use the following steps.

1. Rank the data from the highest to the lowest value.

2. Decide on an appropriate interval for the vertical scale. The graph should have between 10 and 20 horizontal lines, with the scale extending a little beyond the highest value. For example, if the largest number is 925, an appropriate scale would reach to 1,000 and have intervals of 50.

3. Mark off the vertical scale, beginning at zero and labeling the lines according to the intervals chosen in step 2. Label the vertical scale.

258   Chapter 6   *Statistics and Graphs*

4. Draw the bars, making them equal in width and leaving a space as wide as a bar between each one. If the correct height of a bar comes between two horizontal lines, estimate where the top of the bar should be.

5. Write a label for each bar and a title for the graph.

6. Write the source of the information in the lower right corner of the graph.

CLASS PRACTICE

Use the graph on page 257 to answer these questions.

a. Estimate the range of data. — from 62 to 79

b. Find the difference in the life expectancies of Swedes and Russians. — 14 years

c. Find the difference in the life expectancies of Swedes and Egyptians. — 17 years

d. The life expectancy in Guatemala is 10 years lower than the life expectancy in what country? — United States

e. The life expectancy in the United States is 13 years higher than the life expectancy in what country? — India

f. To the nearest whole number, find the average of the life expectancies in Guatemala, India, and Egypt. — 64 years

Use the following information to do the exercises below.

This list shows the six longest chapters in the Bible and the number of verses in each one.

Numbers 7 89
1 Chronicles 6 81
Nehemiah 7 73
Psalm 119 176
Matthew 26 75
Luke 1 80

g. The most reasonable interval for the vertical scale is (5, ⑳, 50, 100).

h. Construct a bar graph from this information.   (See facing page.)

WRITTEN EXERCISES

A. *Use the graph on page 259 to do these exercises.* (Allow reasonable variation.)

Write an estimated population of the following cities.

1. Philadelphia   1,580,000   2. Pittsburgh   380,000   3. Erie   110,000
4. Allentown   105,000   5. Scranton   90,000   6. Reading   90,000

Write the answers.

7. Estimate the range of data. — from 90,000 to 1,580,000

8. Estimate how many more people lived in Philadelphia than in all the other cities combined. — 805,000

as great as that in India or Egypt.) The best kind of paper for drawing bar graphs is graph paper.

3. **To draw a bar graph, use the following steps.**

   (1) Rank the data from the highest to the lowest value.

   (2) Decide on an appropriate interval for the vertical scale. The graph should have between 10 and 20 horizontal lines, with the scale extending a little beyond the highest value. For example, if the largest number is 925, the scale should probably reach to 1,000 and have intervals of 50.

   (3) Mark off the vertical scale, beginning at zero and labeling the lines according to the intervals chosen in step 2. Label the vertical scale.

   (4) Draw the bars, making them equal in width and leaving a space as wide as a bar between each one. If the correct height of a bar comes between two horizontal lines, estimate where the top of the bar should be.

   (5) Write a label for each bar and a title for the graph.

   (6) Write the source of the information in the lower right corner of the graph.

Use the second set of *Class Practice* exercises to demonstrate the steps above. It is suggested that you prepare a graph on the board while the students work on the same graph at their desks.

Bar Graph for CLASS PRACTICE h

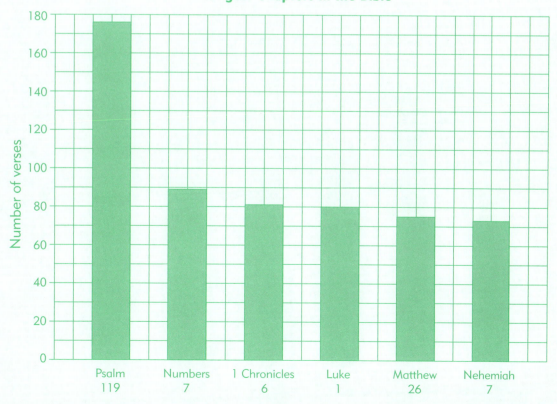

T–259 Chapter 6 Statistics and Graphs

Bar Graphs for Part B

9–15.

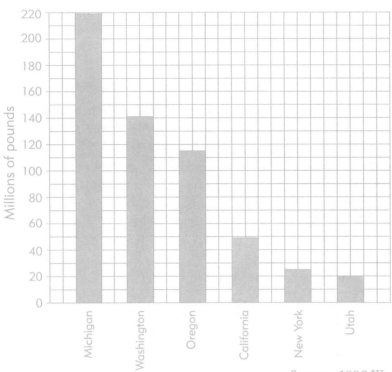

16–20.

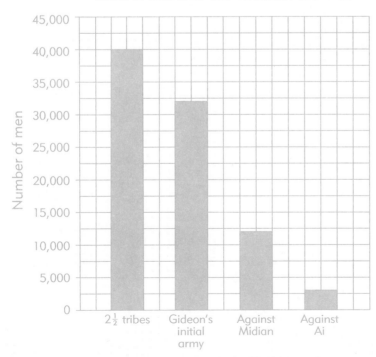

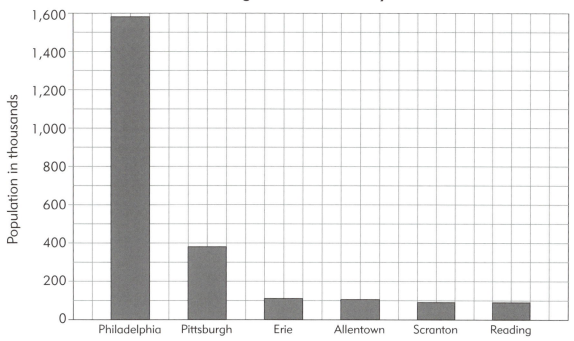

Source: *Census Bureau, 1990*

B. Prepare a bar graph from each set of data. Label every other horizontal line on your graph paper. Make each bar two squares wide, and allow two blank squares between bars. (See facing page.)

9–15. This list shows the production of the leading cherry-growing states. Choose the most reasonable of the following intervals for the vertical scale: 1,000,000; 10,000,000; (20,000,000); 50,000,000.

State	Pounds of cherries
Michigan	219,000,000
Washington	141,000,000
Oregon	115,000,000
California	49,000,000
New York	25,700,000
Utah	20,200,000

Source: *1996 World Book Encyclopedia*

16–20. Following are the sizes of selected armies from the Old Testament. Choose the most reasonable of the following intervals for the vertical scale: 100; 1,000; (5,000); 50,000.

Against Midian (Numbers 31:5)	12,000
$2\frac{1}{2}$ tribes (Joshua 4:13)	40,000
Against Ai (Joshua 7:4)	3,000
Gideon's initial army (Judges 7:3)	32,000

260   Chapter 6   Statistics and Graphs

C. Solve these reading problems. Be careful, for some of them contain extra information.

21. In order to find food, a Siberian tiger occupies a very large territory. One Siberian tiger traveled 620 miles in 22 days in its search for food. How many miles per day did it average? (Express any remainder as a fraction.) $28\frac{2}{11}$ miles

22. Each summer the Groffs buy several trailer loads of rye seed to clean and resell to local farmers for a fall cover crop. One load weighed 46,480 pounds, and they screened 10 bushels of waste out of it. If rye weighs 56 pounds per bushel, how many bushels were they able to resell from this load? (Assume that the waste weighed as much per bushel as the rye.)   820 bushels

23. Gideon had only 300 men with him when he went against the Midianites. His army had originally consisted of 32,000 men. Find what percent of the original army stayed with Gideon, to the nearest whole percent.   1%

24. Kenneth figured that a certain city in Pennsylvania had about $\frac{9}{10}$% of the state's population in 1990. If 108,000 people lived in that city, about how many lived in Pennsylvania?   12,000,000

25. A full-grown grizzly bear weighs 600 to 800 pounds, and its average standing height is 72 inches. A grizzly cub's length at birth is about $\frac{1}{10}$ of its mother's height, whereas the average human baby measures about $\frac{1}{3}$ of its mother's height. About how long is a grizzly cub at birth? $7\frac{1}{5}$ inches

26. One day when Wilmer rode his bicycle to school, he covered the 4 miles in 15 minutes. On the way home he took a shortcut, riding $3\frac{1}{2}$ miles and taking only $\frac{2}{3}$ as much time as it had taken to go to school. How much time did he save by using the shortcut?   5 minutes

REVIEW EXERCISES

D. Find the mean of each set, to the nearest tenth. *(Lesson 73)*

27. 156, 164, 158, 172, 164   162.8   28. 384, 365, 380, 325, 364   363.6

E. Solve these percent problems mentally. *(Lesson 69)*

29. $66\frac{2}{3}$% of 9 = __6__   30. 3 is $16\frac{2}{3}$% of 18

31. 6 is 75% of __8__   32. 5% of 3,000 = 150

F. Find these percentages. *(Lesson 61)*

33. 40% of 14   5.6   34. 80% of 66   52.8

G. Solve mentally by using the double-and-double method. *(Lesson 13)*

35. 750 ÷ 50   15   36. 2,850 ÷ 50   57

H. Solve mentally by using the shortcut for dividing by 25. *(Lesson 13)*

37. 2,800 ÷ 25   112   38. 7,000 ÷ 25   280

Solutions for Part C

21. $620 \div 22$
22. $(46{,}480 \div 56) - 10$
23. $\frac{300}{32{,}000} = \frac{3}{320} = 3 \div 320$
24. $108{,}000 = 0.009 \times \underline{}$; $108{,}000 \div 0.009$
25. $\frac{1}{10} \times 72$
26. $\frac{1}{3} \times 15$

T–261 Chapter 6 Statistics and Graphs

LESSON 78

Objectives

- To review reading and drawing line graphs.
- To review that, whereas picture and bar graphs compare different items, the line graph compares the same item over a period of time.
- To teach *reading and constructing line graphs with more than one line.

Review

1. *Find the median of each set.* (Lesson 74)

   a. 85, 83, 84, 88, 78 (84)

   b. 54, 53, 61, 49, 45, 63 ($53\frac{1}{2}$)

2. *Find the mode of each set.* (Lesson 74)

   a. 5, 9, 8, 4, 2, 5, 5, 8 (5)

   b. 425, 429, 425, 428, 426, 427, 423, 428, 421, 426, 425, 429 (425)

3. *Find the amount of each change, to the nearest whole number. Also find the new amount.* (Lesson 63)

   a. 625 increased by 90% (563; 1,188)

   b. 610 decreased by 45% (275; 335)

4. *Write the operation that would be used to answer each question.* (Lesson 14)

   a. How many eggs will Sandra need to make 6 batches of a recipe?
      (multiplication)

   b. How many times will Sara need to make this recipe to get 10 quarts of fruit salad? (division)

5. *Draw a sketch for each problem, and use it to find the solution.* (Lesson 70)

   a. The living room wall of a new house is 14 feet long. Two feet from the left end of the wall is a window $2\frac{1}{2}$ feet wide. Six inches from the right end of the wall is a door 3 feet wide. If a sofa is 6 feet 3 inches long, will it fit between the window and the door?
   (No; the space is 3 inches too short.)

   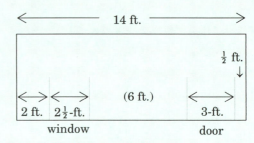

   b. Father plans to build a house on a lot 100 feet square. The town requires the building to be set at least 35 feet in from the road, 20 feet from the back of the lot, and 15 feet from the left and right ends of the lot. What are the maximum dimensions of the space on which Father can build the house? (45 ft. by 70 ft.)

   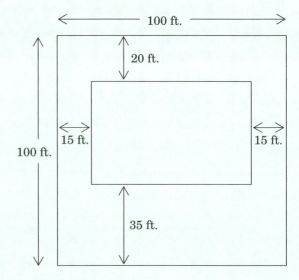

514

78. The Line Graph

Picture graphs and bar graphs usually compare the sizes of several items, whereas line graphs usually show how one item changes over a period of time. A graph may have more than one line, and thereby show comparison of two items as well as changes in each one. The graph below shows the variation in the number of students in two schools over a 10-year period. A heavy line is used for the Glendon School and a lighter line for the Pointview School.

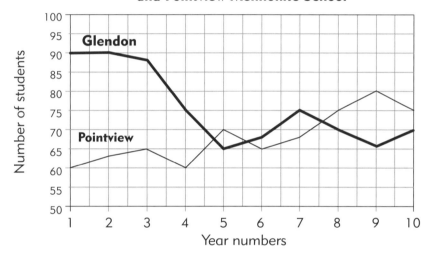

The numerical scale of a line graph is always vertical. However, this scale need not start at zero. A bar graph must start at zero to show the correct proportion of amounts to each other, but a line graph simply shows changes over a period of time.

On the horizontal scale, the data is arranged in order from left to right. Following are the steps for making a line graph.

1. Decide on an appropriate interval for the vertical scale. The graph should have between 10 and 20 horizontal lines, with the scale extending a little below the lowest value and a little above the highest value. Remember that the distance between the lowest and highest values is the range of data.

2. Mark off the vertical scale, labeling the lines according to the intervals chosen in step 1. Label the vertical scale.

3. Mark off and label the lines of the horizontal scale. Space the labels evenly across the bottom of the graph. Also label the horizontal scale.

262   Chapter 6   *Statistics and Graphs*

4. Plot the line of the line graph by placing dots at the correct positions and drawing a line to join them. If the graph has two lines, use pencil for one line and ink for the other, or else use a different color for each line. Label the two lines to tell what each one represents.

5. Write a title for the graph. If you obtained the data from a reference source, name that source in the lower right corner.

CLASS PRACTICE

Use the line graph on page 261 to do the following exercises.

Give the enrollment in the following years.

	Glendon	Pointview
a. Year 3	88	65
b. Year 6	68	65
c. Year 9	66	80

Give the answers.

d. Find Pointview's average enrollment for years 1, 2, 3, and 4.   62 (rounded)

e. Find Glendon's average enrollment for years 7, 8, 9, and 10.   70

f. Estimate the range of data for the Glendon School.   from 65 to 90

Use the following data to make a double-line graph. Choose one of the following intervals for the vertical scale: 20,000; 200,000; 2,000,000. (See facing page.)

Population of California and New York

	California	New York
1940	6,907,387	13,479,142
1950	10,586,223	14,830,195
1960	15,717,204	16,782,304
1970	19,971,069	18,241,391
1980	23,667,764	17,558,165
1990	29,760,021	17,990,455

Source: *1996 Encarta Encyclopedia*

WRITTEN EXERCISES

A. Use the graph on page 263 to do these exercises. (Allow reasonable variation.)

Give the approximate population of Alaska in each of these years.

1. 1900   65,000   2. 1930   60,000   3. 1960   225,000   4. 1990   550,000

Give the approximate population of Wyoming in each of these years.

5. 1900   95,000   6. 1930   225,000   7. 1960   330,000   8. 1990   455,000

Do these exercises.

9. Estimate the range of data for Alaska.   from 65,000 to 550,000

10. Estimate the range of data for Wyoming.   from 95,000 to 470,000

Introduction

Ask the students how the line graph is different from the bar graph and the picture graph. Picture graphs and bar graphs usually compare the sizes of different items, whereas line graphs usually show how the same item changes over a period of time.

Teaching Guide

1. **A line graph usually shows changes over a period of time.** Show the contrast to bar graphs by turning back to Lesson 77. One graph compares the population of different cities in Pennsylvania. If a line graph were used, it would probably show how the population of one city changed over a certain span of years.

2. **The numerical scale of a line graph is always vertical.** However, this scale does not always start at zero. A bar graph must start at zero so that the bars give an accurate visual impression of sizes or amounts in relation to each other. But a line graph simply shows changes over a period of time.

   On the horizontal scale, the data is arranged in order from left to right.

3. **Following are the steps for making a line graph.**

   (1) Decide on an appropriate interval for the vertical scale. The graph should have between 10 and 20 horizontal lines, with the scale extending a little below the lowest value and a little above the highest

Double-line Graph for CLASS PRACTICE

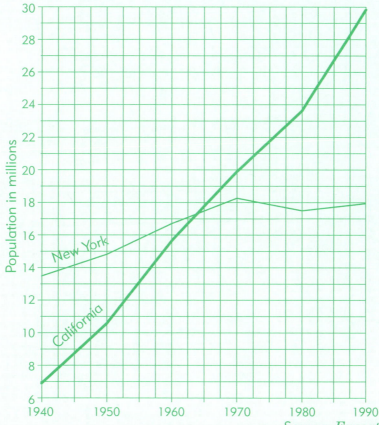

Source: *Encarta Encyclopedia*

value. Remember that the distance between the lowest and highest values is the range of data.

(2) Mark off the vertical scale, labeling the lines according to the intervals chosen in step 2. Label the vertical scale.

(3) Mark off and label the lines of the horizontal scale. Space the labels evenly across the bottom of the

Answers for Part B

11–20.

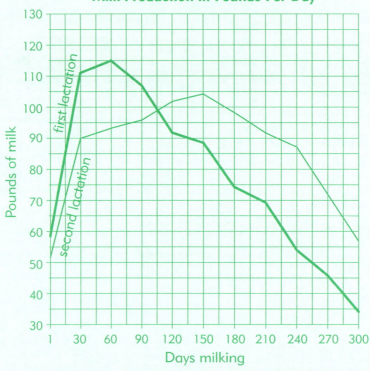

21–30.

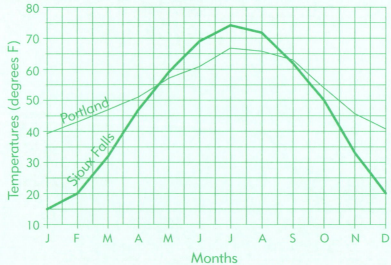

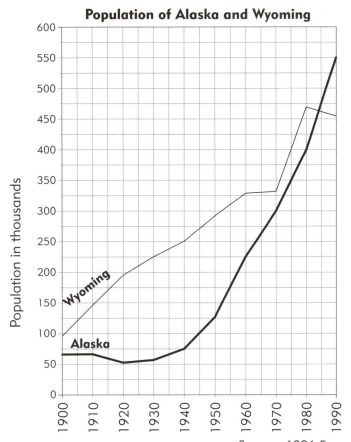

Source: *1996 Encarta Encyclopedia*

B. Construct a double-line graph for each set of data. (See facing page.)

11–20. Following is the milk production record of a certain cow. Choose one of the following intervals for your vertical scale: 10, 20, 80, 100.

Milk Production in Pounds Per Day

Days Milking	1	30	60	90	120	150	180	210	240	270	300
First Lactation	58	111	115	107	92	88	74	69	54	46	34
Second Lactation	52	90	93	96	102	104	98	92	87	72	57

21–30. Portland, Oregon, and Sioux Falls, South Dakota, are about the same distance north of the equator. But because Portland is along a seacoast, the difference between summer and winter temperatures there is not nearly as great as at Sioux Falls. The following table compares the average monthly temperatures of the two cities in degrees Fahrenheit. Choose one of the following intervals for the vertical scale of your graph: 2, 5, 10, 20.

Mean Temperatures at Portland, Oregon, and Sioux Falls, South Dakota

Months	J	F	M	A	M	J	J	A	S	O	N	D
Portland	39	43	47	51	57	61	67	66	63	54	46	41
Sioux Falls	15	20	32	47	59	69	74	72	62	50	33	20

Source: *1992 Weather Almanac*

264   Chapter 6   *Statistics and Graphs*

C. Solve these reading problems. Draw sketches for problems 35 and 36.
(Lesson 70)

31. Last winter, the Millers were able to feed 140 head of cattle in their barn. Since they built an addition to the barn this summer, they have increased the holding capacity by 35%. Their barn can hold how many head of cattle now? 189 head

32. The Millers also improved their hog barn. The building housed 60 sows, but while they were working on it, they needed to find temporary pens for 30% of them. How many sows could remain in the hog barn during the renovation? 42 sows

33. Two bins hold the yearly harvest of corn on the Miller farm. One of them holds 8,500 bushels, and the other has a capacity of 11,000 bushels. This year the Millers had to buy 1,200 bushels of corn so that they would have enough until harvest. If both bins were full after last year's harvest, how many bushels did they use over the past year? 20,700 bushels

34. Matthew grinds about $1\frac{3}{4}$ tons of shelled corn every morning. If each steer eats 19 pounds of corn per day, how many steers will this amount feed? (Round to the nearest whole number.) 184 steers

35. The outer court of the Old Testament tabernacle was a rectangle measuring about 171 feet by 81 feet. A curtain was stretched around this perimeter, held up by pillars. If the pillars stood 9 feet apart, how many were needed? 56 pillars

36. Miriam is placing full quarts of canned fruit on the canning shelves in the basement. Each shelf is 50 inches long and is wide enough for 2 rows of jars. If Miriam places the 4-inch jars with $\frac{1}{2}$ inch of space between them, how many jars can she put on each shelf? 22 jars

REVIEW EXERCISES

D. Find the median of each set. *(Lesson 74)*

37. 153, 164, 121, 174, 134, 147, 182, 133, 98   147
38. 101, 87, 116, 97, 67, 98, 105, 88, 113, 99 $98\frac{1}{2}$

E. Find the mode(s) of each set. *(Lesson 74)*

39. 6, 8, 7, 3, 6, 4, 2, 6   6   40. 15, 19, 21, 16, 15, 22, 21   15, 21

F. Find the amount of each change, to the nearest whole number. Also find the new amount. *(Lesson 63)*

41. 325 increased by 35%   114; 439   42. 420 decreased by 20%   84; 336

G. Write the operation that would be used to answer each question. *(Lesson 14)*

43. What is the difference between Karen's and Carol's ages?   subtraction
44. What is the combined weight of the truck and trailer?   addition

Lesson 78   T-264

graph. Also label the horizontal scale.

(4) Plot the line of the line graph by placing dots at the correct positions and drawing a line to join them. If the graph has two lines, use pencil for one line and ink for the other, or else use a different color for each line. Label the two lines to tell what each one represents.

(5) Write a title for the graph. If you obtained the data from a reference source, name that source in the lower right corner.

Use the steps above to draw line graphs with the students as indicated in *Class Practice*.

An Ounce of Prevention

The students will need compasses and protractors for the next lesson.

Solutions for Part C

31. 1.35×140

32. 0.70×60

33. $8{,}500 + 11{,}000 + 1{,}200$

34. $1\frac{3}{4} \times 2{,}000 \div 19$

35.

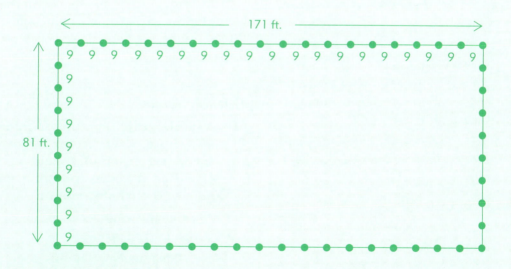

36.

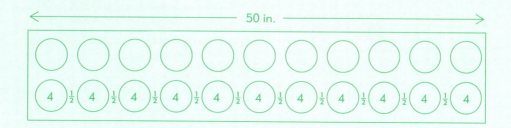

T–265   Chapter 6   *Statistics and Graphs*

LESSON 79

Objective

- To review reading and constructing circle graphs.

Review

1. Give Lesson 79 Quiz (Practice With Mean, Median, and Mode).

2. *Find these percentages. Round money amounts to the nearest cent.* (Lesson 63)

   a. 205% of $4.30   ($8.82)

   b. 195% of $62.15   ($121.19)

   c. $\frac{1}{2}$% of 300   (1.5)

   d. $\frac{3}{5}$% of 610   (3.66)

3. *Write these decimals, using words.* (Lesson 47)

   a. 1.064   (One and sixty-four thousandths)

   b. 2.09818   (Two and nine thousand eight hundred eighteen hundred-thousandths)

4. *Compare each set of decimals, and write < or > between them.* (Lesson 47)

   a. 0.021 (<) 0.199

   b. 0.601 (>) 0.5998

5. *Solve these addition and subtraction problems.* (Lesson 47)

   a. 3.2   b. 9.1274
      + 7.8   − 5.2394
      (11)   (3.888)

Introduction

Discuss the bar graph and the circle graph in the lesson. Point out that both were prepared from the data on the table after the circle graph. Then ask the following questions.

a. Which graph shows the approximate population of each continent? (The bar graph shows this. The circle graph does not show this information at all.)

b. Which graph shows most clearly that well over half of the world population lives in Asia? (The circle graph shows this clearly. This fact could be calculated from the bar graph, but it would take considerably longer.)

Picture graphs and bar graphs are better for comparing approximate amounts. Circle graphs are better for showing what fraction of the whole each part represents.

Teaching Guide

1. **A circle graph contains a number of sectors. Each sector makes up the same fractional part of the circle as its corresponding value does of the whole amount.**

   Ask the following questions about the circle graph in the lesson.

   a. Which region contains over one-half of the world population? (Asia)

   b. Which three regions together contain one-fourth of the world population? (South America, North America, and Africa; *or* Europe, South America, and North America)

79. The Circle Graph

Picture graphs and bar graphs show the sizes of different amounts in relation to each other. A circle graph shows the sizes of different parts that make up a whole.

Compare the bar graph and circle graph below. Both are based on the population figures in the table on the next page. The bar graph shows a comparison of the population in the different regions. The circle graph also shows how the population of each region compares to the total world population. For example, the circle graph shows that well over half the people in the world live in eastern Asia. This fact could be calculated from the bar graph, but the circle graph shows it at a glance.

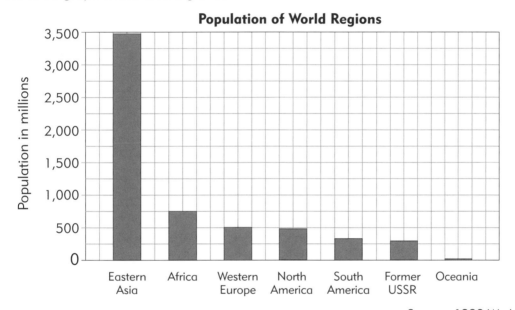

Source: *1998 World Almanac*

Source: *1998 World Almanac*

Chapter 6 Statistics and Graphs

Population of World Regions

Region	Population	Fraction	Decimal	Degrees
Eastern Asia	3,477,000,000	$\frac{3,477}{5,850}$	0.594	214°
Africa	750,000,000	$\frac{750}{5,850}$	0.128	46°
Western Europe	508,000,000	$\frac{508}{5,850}$	0.087	31°
North America	464,000,000	$\frac{464}{5,850}$	0.079	28°
South America	329,000,000	$\frac{329}{5,850}$	0.056	20°
Former USSR	293,000,00	$\frac{293}{5,850}$	0.050	18°
Oceania*	29,000,000	$\frac{29}{5,850}$	0.005	2°
Totals	5,850,000,000		0.999	359°

Source: *1998 World Almanac*
*Includes Australia, New Zealand, and various islands

A circle graph contains a number of **sectors.** Each sector makes up the same fractional part of the circle as its corresponding value does of the whole amount. For example, the data above shows that North America has 8% of the world population; therefore, the sector for North America is 8% of the circle. Because circles are measured in degrees, the size of the sectors is found by calculating with degrees.

Following are the steps to construct a circle graph. Use steps 1–3 to make a chart like the one above.

1. Write a fraction to show what part of the total each sector will be. Reduce the fraction to lowest terms if that is simple to do.

2. Change each fraction found in step 1 to a decimal rounded to the nearest thousandth.

3. Find the number of degrees for each sector by multiplying 360° by the decimal for that sector. Round each answer to the nearest whole degree. The sum of all the degrees should equal 360, but it may vary by 1 or 2 degrees because of rounding.

4. Draw a circle with a 2-inch radius. Then draw a radius in the circle to use as a starting point.

5. Using the degrees calculated in step 3, draw an angle for the first sector. Use the second side of this angle as the base for the next angle. Continue around the circle until all the sectors are drawn.

   The last sector will be what is left when all the other sectors are drawn. Mark it with a small x. Its size should be close to the number of degrees calculated for the last sector.

6. Give each sector a label that includes the name and the corresponding percent. Find the percents by rounding the decimals to the nearest hundredth and changing them to percents. Your teacher may also want you to write the number of degrees beside each sector to simplify checking.

7. Write a title for the graph. If you obtained the data from a reference source, name that source in the lower right corner.

2. Use the following steps to construct a circle graph. For steps 1–3, make a chart like the one in the pupil's lesson.

(1) Write a fraction to show what part of the total each sector will be. Reduce the fraction to lowest terms if that is simple to do.

(2) Change each fraction found in step 1 to a decimal rounded to the nearest thousandth.

(3) Find the number of degrees for each sector by multiplying 360° by the decimal for that sector. Round each answer to the nearest whole degree. The sum of all the degrees should equal 360, but it may vary by 1 or 2 degrees because of rounding.

(4) Draw a circle with a 2-inch radius. Then draw a radius in the circle to use as a starting point.

(5) Using the degrees calculated in step 3, draw an angle for the first sector. Use the second side of this angle as the base for the next angle. Continue around the circle until all the sectors are drawn.

The last sector will be what is left when all the other sectors are drawn. Mark it with a small x. Its size should be close to the number of degrees calculated for the last sector.

(6) Give each sector a label that includes the name and the corresponding percent. Find the percents by rounding the decimals to the nearest hundredth and changing them to percents. (The teacher may also want students to write the number of degrees beside each sector to simplify checking.)

(7) Write a title for the graph. If you obtained the data from a reference source, name that source in the lower right corner.

T–267   Chapter 6   Statistics and Graphs

Circle Graph for CLASS PRACTICE

Chapters in Matthew Devoted to Different Parts of Jesus' Life

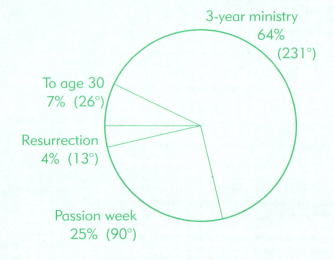

- 3-year ministry 64% (231°)
- To age 30 7% (26°)
- Resurrection 4% (13°)
- Passion week 25% (90°)

Circle Graphs for Part A

1–5.

Chapters in John Devoted to Different Parts of Jesus' Life

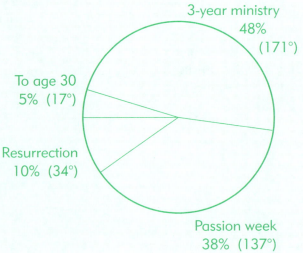

- 3-year ministry 48% (171°)
- To age 30 5% (17°)
- Resurrection 10% (34°)
- Passion week 38% (137°)

6–10.

Surface Area of World's Oceans

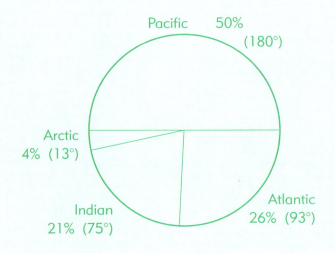

- Pacific 50% (180°)
- Arctic 4% (13°)
- Indian 21% (75°)
- Atlantic 26% (93°)

CLASS PRACTICE

Use the steps in the lesson to complete the following table and make a circle graph.

(See facing page for graphs.)

Chapters in Matthew
Devoted to Different Parts of Jesus' Life

Part	Chapters	Fraction	Decimal	Degrees
To age 30	2	$\frac{1}{14}$	0.071	26°
3-year ministry	18	$\frac{9}{14}$	0.643	231°
Passion Week	7	$\frac{1}{4}$	0.250	90°
Resurrection	1	$\frac{1}{28}$	0.036	13°
Totals	28		1.000	360°

WRITTEN EXERCISES

A. *Complete these tables, and use them to make circle graphs.*

1–5.

Chapters in John
Devoted to Different Parts of Jesus' Life

Part	Chapters	Fraction	Decimal	Degrees
To age 30	1	$\frac{1}{21}$	0.048	17°
3-year ministry	10	$\frac{10}{21}$	0.476	171°
Passion Week	8	$\frac{8}{21}$	0.381	137°
Resurrection	2	$\frac{2}{21}$	0.095	34°
Totals	21		1.000	359°

6–10.

Surface Area of the World's Oceans

Ocean	Millions of Square Miles	Fraction	Decimal	Degrees
Pacific	70	$\frac{1}{2}$	0.500	180°
Atlantic	36	$\frac{9}{35}$	0.257	93°
Indian	29	$\frac{29}{140}$	0.207	75°
Arctic	5	$\frac{1}{28}$	0.036	13°
Totals	140		1.000	361°

B. *Solve these reading problems.* (See page T-268 for solutions.)

11. The direct distance between the Sea of Galilee and the Dead Sea is 70 miles. However, the Jordan River is so winding that water flowing from one sea to the other travels 200 miles. The actual length of the Jordan River is what percent of the direct distance, to the nearest whole percent? 286%

12. The Sea of Galilee is nearly $12\frac{1}{2}$ miles long. Its greatest width is 60% of its length. How wide is it? $7\frac{1}{2}$ miles

13. A steer weighing about 1,250 pounds is adding $\frac{1}{5}$% of that weight each day. What is its daily weight gain? $2\frac{1}{2}$ pounds

268 Chapter 6 *Statistics and Graphs*

14. The Stauffers canned about 300 pounds of applesauce. Sherri figured that if their average daily use was $\frac{1}{4}$% of that amount, their supply would last all year. How many pounds would that allow per day?
$\frac{3}{4}$ pound

15. The electrodes of many spark plugs should have a space of 0.035 inch between them. If a plug has a gap of only 0.0245 inch, how much farther apart should the electrodes be?
0.0105 inch

16. On a certain diesel engine, the camshaft must be able to move back and forth within its bearings at least 0.0508 inch but no more than 0.1524 inch. How great is the range between these limits?
0.1016 inch

REVIEW EXERCISES

C. Find these percentages. Round money amounts to the nearest cent. *(Lesson 62)*

17. 115% of $3.60 $4.14
18. 145% of $88.15 $127.82
19. $\frac{1}{2}$% of 400   2
20. $\frac{2}{5}$% of 520   2.08

D. Write these decimals, using words. *(Lesson 47)* (See facing page.)

21. 3.341
22. 3.06409

E. Compare each set of decimals, and write < or > between them. *(Lesson 47)*

23. 0.0565 < 0.0656
24. 1.0892 < 1.098

F. Solve these addition and subtraction problems. *(Lesson 47)*

25.   6.5
    + 5.9

     12.4

26.   8.174
    − 3.1452

     5.0288

Lesson 79   T–268

An Ounce of Prevention

    Here are some pointers for scoring circle graphs.

(1) Insist on carefulness and neatness. Making circle graphs is meticulous work, but that is no excuse for untidiness and careless mistakes.

(2) Grade the sectors with a tolerance of one degree.

(3) The sector marked with an x is formed by default and cannot be wrong if the others are right. Therefore, do not grade this sector.

(4) It is recommended that you count one point for the title and one for each sector except the one containing the x.

Solutions for Part B

11. $\frac{200}{70} = 20 \div 7$

12. 0.60×12.5

13. $0.002 \times 1,250$

14. 0.0025×300

15. $0.035 - 0.0245$

16. $0.1524 - 0.0508$

Answers for Part D

21. Three and three hundred forty-one thousandths

22. Three and six thousand four hundred nine hundred-thousandths

T-269   Chapter 6   *Statistics and Graphs*

LESSON 80

Objective

- To teach *reading and drawing rectangle graphs.

Review

1. *Find the new price after each increase or decrease, to the nearest cent.* (Lesson 64)

    a. $63.84 increased by 14% ($72.78)

    b. $26.35 decreased by 52% ($12.65)

2. *Find the answers mentally.* (Lesson 48)

    a. 10 × 35.1 (351)

    b. 100 × 15.009 (1,500.9)

3. *Multiply these decimals.* (Lesson 48)

    a. 6.71   b.   12.8
        × 8   × 3.07
        (53.68)   (39.296)

4. *Give these English/metric equivalents by memory.* (Lesson 27)

    a. 1 cm = ___ in. (0.39)
    b. 1 in. = ___ cm (2.54)
    c. 1 m = ___ ft. (3.28)
    d. 1 ft. = ___ m (0.3)
    e. 1 kg = ___ lb. (2.2)
    f. 1 lb. = ___ kg (0.45)
    g. 1 ton = ___ kg (907)
    h. 1 *l* = ___ qt. (1.06)
    i. 1 qt. = ___ *l* (0.95)
    j. 1 ha = ___ a. (2.5)
    k. 1 a. = ___ ha (0.4)

Introduction

Have students look at the rectangle graph in the lesson and tell which other graph it is similar to. Perhaps they will mention the bar graph or the histogram because bars are used. Point out, however, that there is one bar divided into segments according to the fractional part that each segment is of the whole. So the rectangle graph is most like the circle graph.

Teaching Guide

1. **The rectangle graph shows how the parts of an item make up the whole.** Thus the rectangle graph serves the same purpose as the circle graph.

2. **The steps for making a rectangle graph are similar to those for making a circle graph.** However, instead of being measured by degrees, the segments of a rectangle graph are measured by linear units such as inches or centimeters. The graphs in the lesson are measured by millimeters because of the simplicity of expressing percents as millimeters. On a rectangle 100 millimeters long, 1% equals 1 millimeter.

   Using millimeters is simpler than converting decimals to sixteenths of an inch. It is also beneficial for the experience it gives with metric units—especially for students who normally work with inches.

   Here are the steps for making a rectangle graph.

   1. Write a fraction to show what part of the total each sector will be. Reduce the fraction to lowest terms if that is simple to do.

(Teaching guide continues on page T-271.)

80. The Rectangle Graph

The rectangle graph, like the circle graph, shows how the parts of an item make up the whole. The graph below shows what part of the students are in each grade of the upper-grade classroom in the Worcester Christian School. Each segment is the same fractional part of the whole rectangle as that grade is of the whole classroom.

Upper-grade Classroom at Worcester Christian School

Grade 7 28%	Grade 8 33%	Grade 9 17%	Grade 10 22%

The steps for making a rectangle graph are similar to those for making a circle graph. However, instead of measuring by degrees, use linear units such as inches or centimeters to measure the rectangle segments. The millimeter is a convenient unit for working with percents. On a rectangle that measures 100 millimeters long, 1% equals 1 millimeter.

At the Worcester Christian School, the upper-grade classroom has 5 pupils in grade 7, 6 pupils in grade 8, 3 pupils in grade 9, and 4 pupils in grade 10.

Grade	Number of pupils	Fraction	Decimal	Millimeters
Grade 7	5	$\frac{5}{18}$	0.28	28
Grade 8	6	$\frac{1}{3}$	0.33	33
Grade 9	3	$\frac{1}{6}$	0.17	17
Grade 10	4	$\frac{2}{9}$	0.22	22
Totals	18		1.00	100

Here are the steps for making a rectangle graph.

1. Write a fraction to show what part of the total each sector will be. Reduce the fraction to lowest terms if that is simple to do.

2. Change each fraction found in step 1 to a decimal rounded to the nearest hundredth. Multiply this decimal by 100 to find the length of each segment in millimeters.

3. Draw a rectangle 100 millimeters long. Use any convenient width (such as 30 millimeters). Carefully mark off the segments along the length of the rectangle, using the measurements calculated in step 2.

4. Give each segment an appropriate label. Change the decimals to percents, and include the percents with the corresponding labels. (The number for each percent will be the same as the number of millimeters. That is, 24 millimeters will be 24%.)

5. Write a title for the graph. If you obtained the data from a reference source, name that source in the lower right corner.

270 Chapter 6 *Statistics and Graphs*

CLASS PRACTICE

Complete the following tables, and make rectangle graphs by using the steps in the lesson.
(See facing page for graphs.)

a.
Life of Joseph

Part	Years	Fraction	Decimal	Millimeters
Childhood	17	$\frac{17}{110}$	0.15	15
Egyptian bondage	13	$\frac{13}{110}$	0.12	12
Egyptian ruler	80	$\frac{80}{110}$	0.73	73
Totals	110		1.00	100

Source: *Genesis 37:2; 41:46; 50:26*

b.
Gideon's Army

Part	Number of men	Fraction	Decimal	Millimeters
Returned home in fear	22,000	$\frac{11}{16}$	0.69	69
Dismissed after test at water	9,700	$\frac{97}{320}$	0.30	30
Joined Gideon in battle	300	$\frac{3}{320}$	0.01	1
Totals	32,000		1.00	100

Source: *Judges 7:1–7*

WRITTEN EXERCISES

A. Complete each table, and make a rectangle graph based on it.

1–3.
1995 Population of New Jersey

Part	Population	Fraction	Decimal	Millimeters
Urban	7,060,000	$\frac{7,060}{7,931}$	0.89	89
Rural	871,000	$\frac{871}{7,931}$	0.11	11
Totals	7,931,000		1.00	100

Source: *1996 Encarta Encyclopedia*

4–6.
1995 Population of Missouri

Part	Population	Fraction	Decimal	Millimeters
Urban	3,647,000	$\frac{3,647}{5,286}$	0.69	69
Rural	1,639,000	$\frac{1,639}{5,286}$	0.31	31
Totals	5,286,000		1.00	100

Source: *1996 Encarta Encyclopedia*

Lesson 80 T–270

Rectangle Graphs for CLASS PRACTICE

a. **Life of Joseph**

| Childhood 15% | Egyptian bondage 12% | Egyptian ruler 73% |

Source: *Genesis 37:2; 41:46; 50:26*

b. **Gideon's Army**

| Returned home in fear 69% | Dismissed after test at water 30% | ← Joined Gideon in battle 1% |

Source: *Judges 7:1–7*

Rectangle Graphs for Part A

1–3. **1995 Population of New Jersey**

| Urban 89% | Rural 11% |

Source: *1996 Encarta Encyclopedia*

4–6. **1995 Population of Missouri**

| Urban 69% | Rural 31% |

Source: *1996 Encarta Encyclopedia*

T–271 *Chapter 6 Statistics and Graphs*

2. Change each fraction found in step 1 to a decimal rounded to the nearest hundredth. Multiply this decimal by 100 to find the length of each segment in millimeters.
3. Draw a rectangle 100 millimeters long. Use any convenient width (such as 30 millimeters). Carefully mark off the segments along the length of the rectangle, using the measurements calculated in step 2.
4. Give each segment an appropriate label. Change the decimals to percents, and include the percents with the corresponding labels. (The number for each percent will be the same as the number of millimeters. That is, 24 millimeters will be 24%.)

   Sometimes a segment is so small that the label must be written outside the rectangle. See the second graph for *Class Practice*.
5. Write a title for the graph. If you obtained the data from a reference source, name that source in the lower right corner.

7–11. **Provinces in Canada**

| Atlantic provinces 40% | Ontario and Quebec 20% | Prairie provinces 30% | B.C. 10% |

12–20. **States in Regions of the United States**

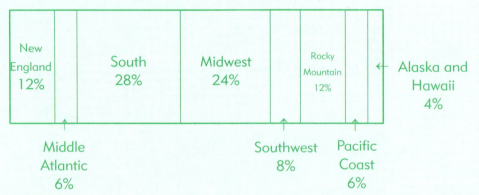

Solutions for Part B

21. 49.5 × 7.48
22. 20 × 30 × 8 × 0.08
23. 1,000 × 0.01212
24. 10,000 × 0.004

7–11. *(Express fractions on the table as tenths.)*

Provinces of Canada

Name or Group	Number	Fraction	Decimal	Millimeters
Atlantic provinces	4	$\frac{4}{10}$	0.40	40
Ontario and Quebec	2	$\frac{2}{10}$	0.20	20
Prairie provinces	3	$\frac{3}{10}$	0.30	30
British Columbia	1	$\frac{1}{10}$	0.10	10
Totals	10		1.00	100

12–20. *(Express fractions on the table as hundredths.)*

States in Regions of the United States

Region	Number	Fraction	Decimal	Millimeters
New England	6	$\frac{12}{100}$	0.12	12
Middle Atlantic	3	$\frac{6}{100}$	0.06	6
South	14	$\frac{28}{100}$	0.28	28
Midwest	12	$\frac{24}{100}$	0.24	24
Southwest	4	$\frac{8}{100}$	0.08	8
Rocky Mountain	6	$\frac{12}{100}$	0.12	12
Pacific Coast	3	$\frac{6}{100}$	0.06	6
Alaska and Hawaii	2	$\frac{4}{100}$	0.04	4
Totals	50		1.00	100

B. Solve these reading problems. Be careful, for some of them contain extra information.

21. One cubic foot equals 7.48 gallons. If a gasoline tank has a capacity of 49.5 cubic feet, how many gallons can it hold, to the nearest tenth of a gallon? 370.3 gallons

22. Air weighs 0.08 pound per cubic foot. If a classroom is 20 feet wide, 30 feet long, and 8 feet high, how many pounds of air does it contain? 384 pounds

23. If one sheet of paper weighs 0.01212 pound, what is the weight of 1,000 sheets? (Calculate mentally.) 12.12 pounds

24. A sheet of 20-pound typing paper measures $8\frac{1}{2}$ by 11 inches and is about 0.004 inch thick. What is the total thickness of 10,000 such papers? (Calculate mentally.) 40 inches

272 Chapter 6 *Statistics and Graphs*

25. The Rohrers have a roadside produce stand. They buy some fruits and vegetables to sell along with what they raise on their 10 acres. One week they bought cherries for $11.49 a case, but the next week their supplier was asking 47% more than that. To the nearest cent, what was the new cost per case? $16.89

Blessed be the LORD, who daily loadeth us with benefits. Psalm 68:19

26. Due to an abundance of green beans from the patch, the Rohrers reduced their price 15%. If the former price was $1.35 a quart, what was the new price? (Round to the nearest cent.) $1.15

REVIEW EXERCISES

C. Find the new price after each increase or decrease, to the nearest cent. *(Lesson 64)*

27. $43.28 increased by 19% $51.50
28. $63.18 decreased by 15% $53.70

D. Find the answers mentally. *(Lesson 48)*

29. 10 × 2.61   26.1
30. 10 × 118.42   1,184.2
31. 1,000 × 2.09   2,090
32. 1,000 × 5.15009   5,150.09

E. Multiply these decimals. *(Lesson 48)*

33. 8.6
    × 2.5
    ────
    21.5

34. 6.12
    × 6.85
    ─────
    41.922

F. Find these English/metric equivalents. *(Lesson 27)*

35. 34 kg = 74.8 lb.
36. 14 in. = 35.56 cm
37. 7 qt. = 6.65 l
38. 27 m = 88.56 ft.
39. 12 cm = 4.68 in.
40. 20 l = 21.2 qt.

25. 1.47 × $11.49
26. 0.85 × $1.35

T-273 Chapter 6 *Statistics and Graphs*

LESSON 81

Objective

- To review the material taught in Chapter 6 (Lessons 73–80).

Teaching Guide

1. Lesson 81 reviews the material taught in Lessons 73–80. For pointers on using review lessons, see *Teaching Guide* for Lesson 15.
2. Be sure to review the new concepts introduced in this chapter. They are listed below along with the exercises in this lesson that review those concepts.

Lesson number and new concept	Exercises in Lesson 81
73—Using *arithmetic mean* instead of *average*.	Directions for parts A, B
74—Finding the median.	9–12
74—Finding the mode.	13–16
76—Choosing the value of each symbol on a picture graph.	None
78—Reading double-line graphs.	29–32
80—Drawing rectangle graphs.	37–42

81. Chapter 6 Review

A. Find the mean of each set, to the nearest whole number or whole percent. *(Lesson 73)*

1. 26, 28, 22, 25, 24   25
2. 90, 94, 81, 83, 92   88
3. 148, 161, 129, 113   138
4. 454, 452, 416, 499   455
5. 92%, 93%, 89%, 93%, 99%   93%
6. 91%, 82%, 100%, 83%, 88%   89%

B. Find the mean of each set, to the nearest tenth. *(Lesson 73)*

7. 37, 38, 42, 41, 47, 46   41.8
8. 57, 59, 48, 62, 55, 53   55.7

C. Find the median of each set. *(Lesson 74)*

9. 36, 39, 32, 19, 79   36
10. 48, 45, 99, 42, 44   45
11. 315, 299, 317, 316, 499, 75, 369, 352, 314, 335 $316\frac{1}{2}$
12. 415, 426, 410, 516, 405, 599, 475, 467, 489, 425, 478, 491   471

D. Find the mode of each set. *(Lesson 74)*

13. 2, 3, 2, 6, 4, 2, 3, 3   2, 3
14. 7, 6, 8, 3, 4, 8, 9, 3, 2   3, 8
15. 177, 176, 175, 179, 177, 172, 173, 175, 174, 177, 169, 188   177
16. 322, 326, 327, 322, 323, 324, 322, 325, 326, 327, 326, 323   322, 326

E. Use the histogram at the right to do these exercises. *(Lesson 75)*

17. How many cows produced between 16,001 and 18,000 pounds of milk?   6 cows
18. How many cows produced between 18,001 and 20,000 pounds of milk?   8 cows
19. How many more cows were in the 19,000–20,000 interval than in the 21,000–22,000 interval?   2 cows
20. How many cows were in the herd?   19 cows
21. What is the range of data?   from 16,001 to 22,000 pounds
22. The mode is from ___ to ___ pounds.   18,000 to 20,000

One Year's Milk Production at Riverside Farm

F. Use this picture graph to do the exercises below. *(Lesson 76)*

Chief Cotton-producing States

Texas	🟰🟰🟰🟰🟰 🟰🟰🟰🟰🟰 🟰🟰
California	🟰🟰🟰🟰🟰 🟰🟰
Mississippi	🟰🟰🟰🟰🟰 🟰
Arkansas	🟰🟰🟰🟰
Louisiana	🟰🟰🟰🟰

🟰 = 400,000 bales

Source: *1993 Information Please Almanac*

23. Tell which state had the greatest cotton production, and how many bales it produced.   **Texas—4,800,000 bales**
24. Tell which state ranked fourth in cotton production, and how many bales it produced.   **Arkansas—1,600,000 bales**
25. Which two states together produced about 3,600,000 bales of cotton?   **Mississippi and Louisiana**
26. What was the total number of bales from the two states with the largest cotton production?   **7,400,000 bales**

G. Use this bar graph to do the exercises below. *(Lesson 77)*

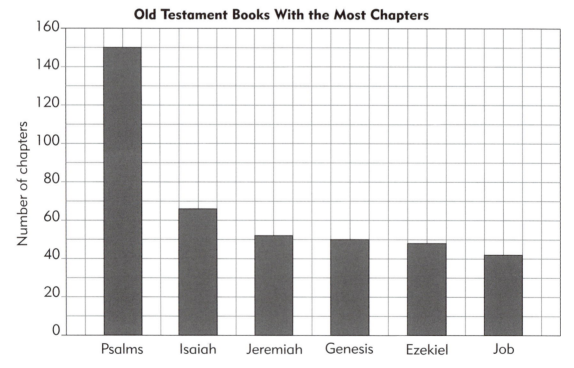

27. Estimate the number of chapters in the three largest books together.   **268 chapters**
28. Estimate the number of chapters in all the books on the graph.   **408 chapters**

T-275   Chapter 6 Statistics and Graphs

Rectangle Graph for Part J

39–44. **Students at Millville Mennonite School**

Grades 1–3	Grades 4–6	Grades 7–9
39%	33%	28%

H. Use the line graph at the right to do these exercises. *(Lesson 78)*
(Accept reasonable variation)

29. Estimate the population of Canada in 1971.   22,200,000

30. Estimate the population of Australia in 1976.   13,600,000

31. In 1991, how many more people lived in Canada than in Australia?   10,000,000

32. In which 5-year span did the population of both Canada and Australia rise more slowly than usual?   1971–1976

33. How many years did it take for the population of Australia to rise from 15 million to 17 million?   10 years

34. What was the approximate increase of population from 1991 to 1996?
    a. in Canada   3 million (3,000,000)
    b. in Australia $1\frac{1}{2}$ million (1,500,000)

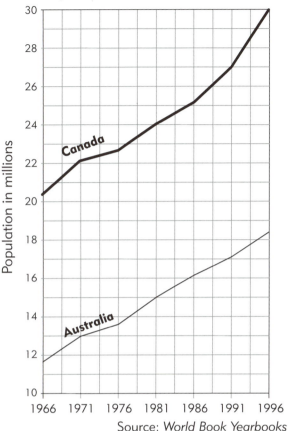

Population of Canada and Australia

Source: *World Book Yearbooks*

I. Complete this table for a circle graph. You do not need to draw the graph. *(Lesson 79)*

Book Divisions of the New Testament

Division	Number of books	Fraction	Decimal	Degrees
35. Gospels and Acts	5	$\frac{5}{27}$	0.185	67°
36. Pauline Epistles	13	$\frac{13}{27}$	0.481	173°
37. General Epistles	8	$\frac{8}{27}$	0.296	107°
38. Revelation	1	$\frac{1}{27}$	0.037	13°
Totals	27		0.999	360°

J. Complete this table, and use it to make a rectangle graph 100 millimeters long and 30 millimeters wide. *(Lesson 80)*   (See facing page.)

39–44.   **Students at Millville Mennonite School**

Room	Number of students	Fraction	Decimal	Millimeters
Grades 1–3	21	$\frac{7}{18}$	0.39	39
Grades 4–6	18	$\frac{1}{3}$	0.33	33
Grades 7–9	15	$\frac{5}{18}$	0.28	28
Totals	54		1.00	100

276 Chapter 6 *Statistics and Graphs*

K. *Solve these reading problems.*

45. Throughout the past spring and summer, Gerald's father sold 2,592 tons of silage and hauled it to other farmers. If he hauled the silage in 134 loads, what was the mean weight of each load? Answer to the nearest tenth of a ton.   19.3 tons

46. One morning in March, the temperatures each hour from 1:00 A.M. to noon were as follows: 34°, 33°, 32°, 32°, 31°, 30°, 30°, 30°, 32°, 38°, 40°, 43°. What was the mean temperature that morning, to the nearest whole degree?   34°

47. One day the Sensenigs did a random survey of 10 of their dairy cows. The pounds of milk produced by these cows were as follows: 72, 56, 89, 62, 66, 55, 90, 58, 52, 40. What was their median milk production?   60 pounds

48. Last week Carolyn recorded how many minutes it took each time she washed the dishes by herself. Find the median of these numbers of minutes: 20, 22, 18, 30, 25, 24, 28, 45. $24\frac{1}{2}$ minutes

49. Brother James sometimes uses the mode to determine which items on a test are the hardest for his 4 eighth graders. On a certain test, they had the following exercise numbers wrong: Joshua—2, 8, 12, 26; Janelle—3, 5, 12; Sharon—2, 12, 18, 21; and Kevin—2, 8, 17, 21, 26. Which number or numbers proved the most difficult?   2 and 12

50. Joyce looked through her typing papers from the past quarter to note her typing speed on timed drills. She jotted down these numbers of words per minute: 32, 34, 32, 35, 33, 32, 33, 38, 34, 32. What is the mode of these numbers?   32

Solutions for Part K

45. 2,592 ÷ 134

46. 405 (sum of degrees) ÷ 12

47. 40, 52, 55, 56, <u>58, 62</u>, 66, 72, 89, 90; (58 + 62) ÷ 2

48. 18, 20, 22, <u>24, 25</u>, 28, 30, 45; (24 + 25) ÷ 2

49. <u>2, 2, 2</u>, 3, 5, 8, 8, <u>12, 12, 12</u>, 17, 18, 21, 21, 26, 26

50. <u>32, 32, 32, 32</u>, 33, 33, 34, 34, 35, 38

82. Chapter 6 Test

LESSON 82

Objective

- To test the students' mastery of the concepts in Chapter 6.

Teaching Guide

1. Correct Lesson 81.
2. Review any areas of special difficulty.
3. Administer the test. For pointers on giving tests, see *Teaching Guide* for Lesson 16.

T-277   Chapter 6   Statistics and Graphs

LESSON 83

Objective

- To review the material taught in Chapters 1–6.

Review

Give Lesson 83 Quiz (General Calculations).

Teaching Guide

Lesson 83 reviews the material taught in the first six chapters. For pointers on how to use review lessons, see *Teaching Guide* for Lesson 15.

83. Semester 1 Review

A. Solve these problems. *(Lessons 37–41, 47–51)*

1. $\frac{3}{4}$
    $+ \frac{5}{6}$
    _____
    $1\frac{7}{12}$

2. $\frac{11}{12}$
    $- \frac{5}{8}$
    _____
    $\frac{7}{24}$

3. $\frac{4}{7}$ of 35   20
4. $1\frac{1}{3} \times 2\frac{1}{5}$ $2\frac{14}{15}$
5. $4\frac{1}{6} \div 1\frac{1}{5}$ $3\frac{17}{36}$
6. $\frac{\frac{2}{3}}{\frac{5}{6}}$ $\frac{4}{5}$
7. $\frac{4}{9}$ is $\frac{2}{3}$ of ___ $\frac{2}{3}$
8. $4\frac{3}{8}$ is $\frac{3}{4}$ of ___ $5\frac{5}{6}$
9. 18 + 12.05 + 19.091   49.141
10. 4.026 + 2.7 + 2.69 + 1.098   10.514
11. 1 − 0.7893   0.2107
12. 0.8 − 0.0978   0.7022
13. 32.6 × 1.109   36.1534
14. 1.45 ÷ 0.05   29

B. Solve these problems mentally. *(Lessons 12, 13, 48–51)*

15. 242 ÷ 22   11
16. 144 ÷ 18   8
17. 750 ÷ 50   15
18. 270 ÷ 45   6
19. 100 × 21.1774   2,117.74
20. 35.76 ÷ 1,000   0.03576
21. 0.4 of 45   18
22. 0.375 of 24   9
23. $5 \div \frac{1}{12}$ = ___   60
24. $12 \div \frac{1}{4}$ = ___   48

C. Do these exercises with numerals. *(Lessons 1–3, 50)*

25. Write 11,070,009,800,000, using words.   Eleven trillion, seventy billion, nine million, eight hundred thousand
26. Use digits to write this number: Twenty-two quadrillion, two million.   22,000,000,002,000,000
27. Round 4,476,555 to the nearest million.   4,000,000
28. Round 685,010,333 to the nearest ten million.   690,000,000
29. Round 3.98189 to the nearest thousandth.   3.982
30. Round 2.414971 to the nearest ten-thousandth.   2.4150
31. Write $\overline{\text{MMCM}}$CLXXII as a Hindu-Arabic numeral.   2,900,172
32. Write 3,157 as a Roman numeral.   MMMCLVII

278   Chapter 6   *Statistics and Graphs*

D. Solve these measurement problems. *(Lessons 25–29)*

33. Subtract 2 lb. 12 oz. from 6 lb. 7 oz.   3 lb. 11 oz.
34. Divide 22 lb. 12 oz. by 4.   5 lb. 11 oz.
35. 74°F = __23__ °C, to nearest whole degree.
36. 26°C = __79__ °F, to nearest whole degree.
37. d = 275 mi.; r = 50 m.p.h.; t = __$5\frac{1}{2}$__ hr.
38. r = 54 m.p.h.; t = $6\frac{1}{2}$ hr.; d = __351__ mi.

E. Do these exercises with factors and multiples. *(Lesson 35)*

39. Find the greatest common factor of 35 and 49.   7
40. Find the greatest common factor of 39 and 65.   13
41. Find the lowest common multiple of 15 and 20.   60
42. Find the lowest common multiple of 18 and 24.   72

F. Solve these percent problems. *(Lessons 62, 68)*

43. In the following problem, 40 is the (base), rate, percentage). 50% of 40 = 20
44. In the following problem, 40% is the (base, (rate), percentage). 40% of 20 = 8
45. Change 238% to a decimal.   2.38
46. Express 950% as a mixed number. $9\frac{1}{2}$
47. Express $\frac{4}{5}$% as a decimal.   0.008
48. Express $1\frac{3}{4}$% as a decimal.   0.0175
49. Find 125% of 63.   78.75
50. Find 275% of 78.   214.5
51. Find $\frac{3}{8}$% of 880.   3.3
52. Find $1\frac{1}{2}$% of 250.   3.75

G. Solve these statistics problems. *(Lessons 73, 74)*

53. Find the mean of the following set, and express any remainder as a fraction.
    48, 45, 46, 78, 32 $49\frac{4}{5}$
54. Find the mean of the following set, to the nearest tenth.
    785, 467, 843, 897, 654, 414   676.7
55. Find the median of this set: 68, 66, 65, 69, 32, 0. $65\frac{1}{2}$
56. Find the median of this set: 212; 256; 234; 202; 3,000; 249. $241\frac{1}{2}$
57. Find the mode(s) of the following set.
    16, 15, 16, 18, 19, 15, 19, 13, 14, 15, 13, 18, 17, 16, 14   15, 16
58. Find the mode(s) of the following set.
    48, 45, 46, 47, 44, 43, 49, 51, 40, 33, 25, 28, 38, 55, 54   no mode

T-279   Chapter 6   Statistics and Graphs

Solutions for Part N
77. 5,000 ÷ 12
78. $\frac{7}{8} \times 64$

H. Find these equivalents. You should be able to do numbers 63–66 without looking at a table of English/metric equivalents. *(Lessons 17–24, 27)*

59. 1 tbsp. = __3__ tsp.
60. 1 fl. oz. = __$\frac{1}{8}$__ cup
61. 25 kg = ____ mg   25,000,000
62. 7 MT = ____ kg   7,000
63. 49 m = ____ ft.   160.72
64. 188 km = ____ mi.   116.56
65. 26 cm = __10.14__ in.
66. 45 kg = __99__ lb.

I. Find the distance represented by each length on a map or scale drawing. *(Lessons 56, 57)*

67. Scale: 1 in. = 40 mi.
    length: $2\frac{1}{4}$ in.   90 mi.
68. Scale: 1 in. = $\frac{1}{4}$ in.
    length: $7\frac{1}{2}$ in. $1\frac{7}{8}$ in.

J. Find the amount of each change, to the nearest cent. Also find the new amount. *(Lesson 63)*

69. $4.26 increased by 20% $0.85 $5.11
70. $15.98 decreased by 35% $5.59 $10.39

K. Find the new amount after each change by first adding or subtracting the percent of change. Round answers to the nearest cent. *(Lesson 64)*

71. $41.25 increased by 15% $47.44
72. $31.10 decreased by 52% $14.93

L. Find the missing rates, to the nearest tenth of a percent. *(Lesson 65)*

73. 13 is __81.3%__ of 16
74. 9 is __52.9%__ of 17

M. Find the rate of each change, to the nearest whole percent. Include the label *increase* **or** *decrease*. *(Lesson 66)*

75. year 1 enrollment, 55
    year 2 enrollment, 57   4% increase
76. year 2 enrollment, 57
    year 3 enrollment, 52   9% decrease

N. Solve these reading problems. *(Lesson 14)*

77. When Jesus fed the 5,000, His disciples distributed the loaves and fishes. What was the average number of men served by each of the 12 disciples?   417 men

78. David and Brian are 1 year apart in age. David is 64 inches tall and weighs 145 pounds. Brian weighs $\frac{5}{6}$ as much as David and is $\frac{7}{8}$ as tall. How tall is Brian?   56 inches

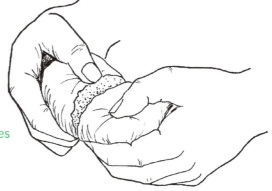

280 Chapter 6 *Statistics and Graphs*

O. Write a proportion to solve each problem. It may be a direct proportion or an indirect proportion, and you may need to calculate one part of a proportion. *(Lessons 53–55)*

79. Kevin figured that there were 3 girls for every 2 boys in his classroom. If the total number of students was 15, how many boys were there? 6 boys

80. Marvin and Daniel hoed in the strawberry patch for 2 hours one morning. That afternoon their sister Janet joined them. How long should it take them to do the same amount of work that they did in the morning? $1\frac{1}{3}$ hours

P. Draw a sketch for each problem, and use it to find the solution. *(Lesson 70)*

81. Sylvia is hanging out the wash. She has 20 shirts and 12 pairs of trousers that must be hung on the three clotheslines behind the house. Up to 12 pieces can be hung on each line. How many clothespins will she need if she uses one pin wherever two items come together? 35 clothespins

82. Brother Alvin is building a small greenhouse for bedding plants. He is using 18 hoops that arch across the width of the greenhouse. If he sets the hoops 3 feet apart, how long will the greenhouse be? 51 feet

81.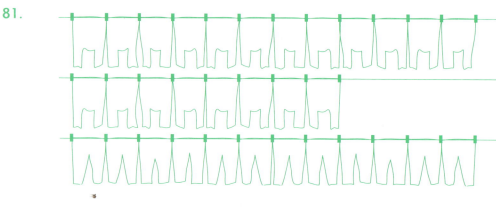

82.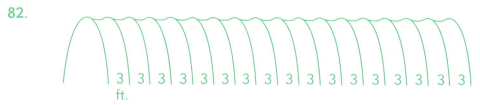

84. Semester 1 Test

Proportions for Part O

79. $\dfrac{\text{boys}}{\text{students}}\quad \dfrac{2}{5}=\dfrac{n}{15}\quad \dfrac{\text{boys}}{\text{students}}\qquad n=6\text{ boys}$

80. $\dfrac{\text{workers (1)}}{\text{workers (2)}}\quad \dfrac{2}{3}=\dfrac{n}{2}\quad \dfrac{\text{hours (2)}}{\text{hours (1)}}\qquad n=1\tfrac{1}{3}\text{ hours}$

LESSON 84

Objective
- To test the students' mastery of the concepts in Chapters 1–6.

Teaching Guide
1. Correct Lesson 83.
2. Review any areas of special difficulty.
3. Administer the test. For pointers on giving tests, see *Teaching Guide* for Lesson 16.

QUIZZES and SPEED TESTS

Answer Key

*He maketh my feet like hinds' feet:
and setteth me upon my high places.
2 Samuel 22:34*

Lesson 3 ~ Speed Test Preparation

Name _____

Date _____

Addition

Score _____

1.	9 +3 = 12	7 +5 = 12	3 +7 = 10	2 +3 = 5	6 +5 = 11	0 +1 = 1	6 +3 = 9	2 +8 = 10	2 +0 = 2	0 +5 = 5
2.	8 +3 = 11	3 +5 = 8	2 +7 = 9	1 +3 = 4	5 +5 = 10	1 +1 = 2	5 +3 = 8	1 +8 = 9	4 +2 = 6	1 +5 = 6
3.	1 +4 = 5	2 +5 = 7	9 +7 = 16	4 +4 = 8	8 +5 = 13	5 +1 = 6	7 +3 = 10	7 +8 = 15	2 +1 = 3	3 +2 = 5
4.	5 +7 = 12	1 +6 = 7	7 +9 = 16	6 +7 = 13	2 +2 = 4	9 +8 = 17	5 +2 = 7	4 +6 = 10	8 +4 = 12	0 +2 = 2
5.	3 +3 = 6	9 +9 = 18	6 +6 = 12	8 +9 = 17	0 +0 = 0	9 +6 = 15	8 +8 = 16	0 +1 = 1	7 +7 = 14	9 +4 = 13
6.	7 +6 = 13	8 +6 = 14	8 +2 = 10	7 +0 = 7	2 +4 = 6	8 +1 = 9	2 +6 = 8	6 +1 = 7	4 +9 = 13	2 +9 = 11
7.	3 +0 = 3	9 +1 = 10	0 +6 = 6	5 +8 = 13	1 +2 = 3	6 +0 = 6	4 +0 = 4	8 +7 = 15	1 +9 = 10	5 +9 = 14
8.	4 +5 = 9	5 +0 = 5	9 +3 = 12	4 +1 = 5	3 +4 = 7	3 +6 = 9	3 +1 = 4	0 +4 = 4	4 +3 = 7	5 +6 = 11
9.	8 +0 = 8	6 +9 = 15	9 +2 = 11	4 +8 = 12	7 +1 = 8	4 +7 = 11	5 +4 = 9	3 +8 = 11	0 +9 = 9	6 +8 = 14
10.	1 +7 = 8	9 +0 = 9	2 +7 = 9	6 +4 = 10	6 +2 = 8	7 +4 = 11	0 +8 = 8	3 +9 = 12	7 +0 = 7	9 +5 = 14

Lesson 4 ~ Speed Test (Time: 1 minute) Addition

1. 9+5=14, 7+0=7, 3+7=10, 2+5=7, 6+4=10, 0+1=1, 6+0=6, 2+4=6, 2+9=11, 0+6=6

2. 3+3=6, 1+9=10, 7+6=13, 6+9=15, 2+0=2, 9+9=18, 5+8=13, 4+0=4, 8+7=15, 0+9=9

3. 7+3=10, 8+5=13, 8+0=8, 7+2=9, 2+6=8, 8+1=9, 2+3=5, 6+8=14, 4+1=5, 3+1=4

4. 4+2=6, 9+3=12, 0+7=7, 5+3=8, 1+5=6, 6+1=7, 4+3=7, 8+8=16, 0+0=0, 7+5=12

5. 1+3=4, 9+4=13, 2+7=9, 8+3=11, 3+5=8, 7+1=8, 3+2=5, 9+6=15, 5+0=5, 9+1=10

6. 8+9=17, 6+5=11, 9+8=17, 4+7=11, 2+2=4, 4+8=12, 5+2=7, 3+9=12, 8+4=12, 6+2=8

7. 4+6=10, 5+5=10, 8+2=10, 4+4=8, 3+8=11, 3+4=7, 1+7=8, 0+1=1, 4+9=13, 5+1=6

8. 3+0=3, 9+2=11, 6+6=12, 6+7=13, 0+3=3, 9+7=16, 5+9=14, 1+2=3, 7+7=14, 5+6=11

9. 1+8=9, 2+1=3, 7+8=15, 1+1=2, 8+6=14, 5+4=9, 0+5=5, 7+9=16, 2+8=10, 2+9=11

10. 9+0=9, 0+8=8, 1+4=5, 5+7=12, 6+3=9, 0+2=2, 7+4=11, 4+5=9, 1+8=9, 0+4=4

Lesson 6 ~ Speed Test Preparation

Name _____

Date _____

Subtraction

Score _____

1.   6 − 3 = 3   10 − 9 = 1   13 − 6 = 7   15 − 9 = 6   2 − 0 = 2   18 − 9 = 9   13 − 8 = 5   4 − 0 = 4   15 − 7 = 8   9 − 9 = 0

2.   10 − 3 = 7   13 − 5 = 8   8 − 0 = 8   9 − 2 = 7   8 − 6 = 2   9 − 1 = 8   5 − 3 = 2   14 − 8 = 6   5 − 1 = 4   4 − 1 = 3

3.   14 − 5 = 9   7 − 0 = 7   10 − 7 = 3   7 − 5 = 2   10 − 4 = 6   1 − 1 = 0   6 − 0 = 6   6 − 4 = 2   11 − 9 = 2   6 − 6 = 0

4.   17 − 9 = 8   11 − 5 = 6   17 − 8 = 9   11 − 7 = 4   4 − 2 = 2   12 − 8 = 4   7 − 2 = 5   12 − 9 = 3   12 − 4 = 8   8 − 2 = 6

5.   6 − 2 = 4   12 − 3 = 9   7 − 7 = 0   8 − 3 = 5   6 − 5 = 1   7 − 1 = 6   7 − 3 = 4   16 − 8 = 8   0 − 0 = 0   12 − 5 = 7

6.   4 − 3 = 1   13 − 4 = 9   9 − 7 = 2   11 − 3 = 8   8 − 5 = 3   8 − 1 = 7   5 − 2 = 3   15 − 6 = 9   5 − 0 = 5   10 − 1 = 9

7.   9 − 0 = 9   8 − 8 = 0   5 − 4 = 1   12 − 7 = 5   9 − 3 = 6   2 − 2 = 0   11 − 4 = 7   9 − 5 = 4   9 − 8 = 1   4 − 4 = 0

8.   3 − 0 = 3   11 − 2 = 9   12 − 6 = 6   13 − 7 = 6   3 − 3 = 0   16 − 7 = 9   14 − 9 = 5   3 − 2 = 1   14 − 7 = 7   11 − 6 = 5

9.   10 − 6 = 4   10 − 5 = 5   10 − 2 = 8   8 − 4 = 4   11 − 8 = 3   7 − 4 = 3   8 − 7 = 1   1 − 1 = 0   13 − 9 = 4   6 − 1 = 5

10.   9 − 8 = 1   3 − 1 = 2   15 − 8 = 7   2 − 1 = 1   14 − 6 = 8   9 − 4 = 5   5 − 5 = 0   16 − 9 = 7   10 − 8 = 2   11 − 9 = 2

Lesson 7 ~ Speed Test (Time: 1 minute) *Subtraction*

1. 5 − 4 = 1; 7 − 5 = 2; 16 − 7 = 9; 8 − 4 = 4; 13 − 5 = 8; 6 − 1 = 5; 10 − 3 = 7; 15 − 8 = 7; 3 − 1 = 2; 5 − 2 = 3

2. 12 − 3 = 9; 12 − 5 = 7; 10 − 7 = 3; 5 − 3 = 2; 11 − 5 = 6; 1 − 1 = 0; 9 − 3 = 6; 10 − 8 = 2; 2 − 0 = 2; 5 − 5 = 0

3. 11 − 3 = 8; 8 − 5 = 3; 9 − 7 = 2; 4 − 3 = 1; 10 − 5 = 5; 2 − 1 = 1; 8 − 3 = 5; 9 − 8 = 1; 6 − 2 = 4; 6 − 5 = 1

4. 12 − 7 = 5; 7 − 6 = 1; 16 − 9 = 7; 13 − 7 = 6; 4 − 2 = 2; 17 − 8 = 9; 7 − 2 = 5; 10 − 6 = 4; 12 − 4 = 8; 2 − 2 = 0

5. 13 − 6 = 7; 14 − 6 = 8; 10 − 2 = 8; 7 − 0 = 7; 6 − 4 = 2; 9 − 1 = 8; 8 − 6 = 2; 7 − 1 = 6; 13 − 9 = 4; 11 − 9 = 2

6. 6 − 3 = 3; 18 − 9 = 9; 12 − 6 = 6; 17 − 9 = 8; 0 − 0 = 0; 15 − 6 = 9; 16 − 8 = 8; 1 − 1 = 0; 14 − 7 = 7; 13 − 4 = 9

7. 8 − 0 = 8; 15 − 9 = 6; 11 − 2 = 9; 12 − 8 = 4; 8 − 1 = 7; 11 − 7 = 4; 9 − 4 = 5; 11 − 8 = 3; 9 − 9 = 0; 14 − 8 = 6

8. 3 − 0 = 3; 10 − 1 = 9; 6 − 6 = 0; 13 − 8 = 5; 3 − 2 = 1; 6 − 0 = 6; 4 − 0 = 4; 15 − 7 = 8; 10 − 9 = 1; 14 − 9 = 5

9. 8 − 7 = 1; 9 − 0 = 9; 9 − 7 = 2; 10 − 4 = 6; 8 − 2 = 6; 11 − 4 = 7; 8 − 8 = 0; 12 − 9 = 3; 7 − 0 = 7; 14 − 5 = 9

10. 9 − 5 = 4; 5 − 0 = 5; 12 − 3 = 9; 5 − 1 = 4; 7 − 4 = 3; 9 − 6 = 3; 4 − 1 = 3; 4 − 4 = 0; 7 − 3 = 4; 11 − 6 = 5

Lesson 9 ~ Speed Test Preparation

Name _____

Date _____

Multiplication Score _____

1. | 12 × 2 = 24 | 7 × 4 = 28 | 11 × 6 = 66 | 7 × 6 = 42 | 12 × 10 = 120 | 12 × 7 = 84 | 6 × 1 = 6 | 9 × 3 = 27 | 12 × 5 = 60 | 10 × 7 = 70 |

2. | 12 × 6 = 72 | 6 × 6 = 36 | 10 × 12 = 120 | 5 × 3 = 15 | 5 × 9 = 45 | 12 × 8 = 96 | 9 × 8 = 72 | 10 × 8 = 80 | 12 × 0 = 0 | 12 × 3 = 36 |

3. | 4 × 5 = 20 | 11 × 10 = 110 | 8 × 7 = 56 | 12 × 4 = 48 | 11 × 3 = 33 | 3 × 6 = 18 | 1 × 0 = 0 | 10 × 11 = 110 | 4 × 4 = 16 | 11 × 5 = 55 |

4. | 7 × 5 = 35 | 8 × 5 = 40 | 9 × 4 = 36 | 7 × 7 = 49 | 12 × 12 = 144 | 8 × 1 = 8 | 10 × 3 = 30 | 6 × 8 = 48 | 4 × 0 = 0 | 10 × 5 = 50 |

5. | 3 × 3 = 9 | 12 × 11 = 132 | 7 × 8 = 56 | 6 × 5 = 30 | 2 × 1 = 2 | 9 × 6 = 54 | 5 × 4 = 20 | 4 × 8 = 32 | 8 × 2 = 16 | 11 × 7 = 77 |

6. | 1 × 5 = 5 | 10 × 2 = 20 | 9 × 7 = 63 | 5 × 5 = 25 | 11 × 2 = 22 | 5 × 7 = 35 | 7 × 0 = 0 | 12 × 1 = 12 | 2 × 5 = 10 | 10 × 10 = 100 |

7. | 9 × 0 = 0 | 11 × 4 = 44 | 6 × 7 = 42 | 9 × 9 = 81 | 10 × 4 = 40 | 9 × 1 = 9 | 8 × 0 = 0 | 0 × 7 = 0 | 11 × 12 = 132 | 10 × 9 = 90 |

8. | 3 × 9 = 27 | 9 × 2 = 18 | 0 × 2 = 0 | 5 × 0 = 0 | 8 × 4 = 32 | 6 × 4 = 24 | 8 × 6 = 48 | 11 × 11 = 121 | 5 × 6 = 30 | 7 × 3 = 21 |

9. | 8 × 3 = 24 | 7 × 9 = 63 | 6 × 2 = 12 | 2 × 9 = 18 | 6 × 0 = 0 | 11 × 9 = 99 | 6 × 9 = 54 | 10 × 0 = 0 | 2 × 7 = 14 | 0 × 9 = 0 |

10. | 4 × 7 = 28 | 11 × 8 = 88 | 8 × 8 = 64 | 12 × 9 = 108 | 3 × 2 = 6 | 3 × 8 = 24 | 1 × 2 = 2 | 10 × 6 = 60 | 4 × 9 = 36 | 12 × 2 = 24 |

Lesson 10 ~ Speed Test   (Time: 1 minute)   *Multiplication*

1. $\;\;4\times2=8\;\;\;\;11\times4=44\;\;\;\;8\times6=48\;\;\;\;12\times4=48\;\;\;\;10\times2=20\;\;\;\;12\times3=36\;\;\;\;1\times1=1\;\;\;\;10\times3=30\;\;\;\;4\times5=20\;\;\;\;12\times7=84$

2. $\;\;12\times5=60\;\;\;\;10\times7=70\;\;\;\;9\times2=18\;\;\;\;2\times5=10\;\;\;\;12\times6=72\;\;\;\;0\times8=0\;\;\;\;6\times0=0\;\;\;\;10\times11=110\;\;\;\;2\times3=6\;\;\;\;0\times6=0$

3. $\;\;3\times3=9\;\;\;\;11\times11=121\;\;\;\;0\times0=0\;\;\;\;5\times5=25\;\;\;\;12\times1=12\;\;\;\;6\times6=36\;\;\;\;4\times4=16\;\;\;\;9\times9=81\;\;\;\;2\times2=4\;\;\;\;7\times7=49$

4. $\;\;9\times6=54\;\;\;\;9\times5=45\;\;\;\;6\times2=12\;\;\;\;9\times0=0\;\;\;\;6\times9=54\;\;\;\;9\times4=36\;\;\;\;8\times7=56\;\;\;\;0\times1=0\;\;\;\;10\times9=90\;\;\;\;11\times3=33$

5. $\;\;1\times3=3\;\;\;\;2\times9=18\;\;\;\;5\times6=30\;\;\;\;5\times9=45\;\;\;\;10\times8=80\;\;\;\;11\times9=99\;\;\;\;7\times9=63\;\;\;\;12\times0=0\;\;\;\;2\times7=14\;\;\;\;10\times6=60$

6. $\;\;3\times7=21\;\;\;\;11\times8=88\;\;\;\;7\times8=56\;\;\;\;6\times7=42\;\;\;\;2\times8=16\;\;\;\;9\times7=63\;\;\;\;5\times2=10\;\;\;\;4\times6=24\;\;\;\;8\times9=72\;\;\;\;12\times2=24$

7. $\;\;7\times0=0\;\;\;\;11\times2=22\;\;\;\;8\times3=24\;\;\;\;7\times4=28\;\;\;\;11\times1=11\;\;\;\;8\times1=8\;\;\;\;10\times0=0\;\;\;\;6\times5=30\;\;\;\;10\times4=40\;\;\;\;11\times6=66$

8. $\;\;6\times4=24\;\;\;\;10\times10=100\;\;\;\;7\times5=35\;\;\;\;11\times5=55\;\;\;\;12\times11=132\;\;\;\;3\times6=18\;\;\;\;1\times0=0\;\;\;\;10\times12=120\;\;\;\;4\times8=32\;\;\;\;12\times12=144$

9. $\;\;6\times3=18\;\;\;\;8\times5=40\;\;\;\;11\times10=110\;\;\;\;5\times3=15\;\;\;\;9\times5=45\;\;\;\;7\times1=7\;\;\;\;9\times3=27\;\;\;\;12\times10=120\;\;\;\;11\times0=0\;\;\;\;10\times1=10$

10. $\;\;12\times8=96\;\;\;\;11\times7=77\;\;\;\;11\times12=132\;\;\;\;7\times3=21\;\;\;\;12\times9=108\;\;\;\;7\times6=42\;\;\;\;6\times8=48\;\;\;\;9\times8=72\;\;\;\;7\times2=14\;\;\;\;10\times5=50$

Lesson 12 ~ Speed Test Preparation   Name _____

  Date _____

Division   Score _____

1. $11\overline{)99}^{\,9}$ $5\overline{)25}^{\,5}$ $4\overline{)24}^{\,6}$ $3\overline{)27}^{\,9}$ $2\overline{)10}^{\,5}$ $7\overline{)56}^{\,8}$ $5\overline{)45}^{\,9}$ $6\overline{)42}^{\,7}$

2. $4\overline{)36}^{\,9}$ $12\overline{)108}^{\,9}$ $6\overline{)12}^{\,2}$ $12\overline{)144}^{\,12}$ $5\overline{)40}^{\,8}$ $10\overline{)80}^{\,8}$ $3\overline{)18}^{\,6}$ $9\overline{)63}^{\,7}$

3. $3\overline{)12}^{\,4}$ $8\overline{)0}^{\,0}$ $2\overline{)18}^{\,9}$ $9\overline{)81}^{\,9}$ $4\overline{)12}^{\,3}$ $11\overline{)110}^{\,10}$ $1\overline{)7}^{\,7}$ $12\overline{)60}^{\,5}$

4. $12\overline{)0}^{\,0}$ $8\overline{)72}^{\,9}$ $3\overline{)15}^{\,5}$ $10\overline{)120}^{\,12}$ $2\overline{)8}^{\,4}$ $4\overline{)32}^{\,8}$ $11\overline{)66}^{\,6}$ $1\overline{)9}^{\,9}$

5. $8\overline{)24}^{\,3}$ $3\overline{)6}^{\,2}$ $10\overline{)70}^{\,7}$ $9\overline{)36}^{\,4}$ $6\overline{)36}^{\,6}$ $4\overline{)16}^{\,4}$ $8\overline{)32}^{\,4}$ $12\overline{)84}^{\,7}$

6. $10\overline{)20}^{\,2}$ $7\overline{)28}^{\,4}$ $11\overline{)88}^{\,8}$ $8\overline{)16}^{\,2}$ $4\overline{)40}^{\,10}$ $6\overline{)24}^{\,4}$ $4\overline{)48}^{\,12}$ $3\overline{)0}^{\,0}$

7. $9\overline{)27}^{\,3}$ $7\overline{)84}^{\,12}$ $5\overline{)35}^{\,7}$ $11\overline{)121}^{\,11}$ $2\overline{)0}^{\,0}$ $8\overline{)96}^{\,12}$ $9\overline{)45}^{\,5}$ $3\overline{)21}^{\,7}$

8. $2\overline{)18}^{\,9}$ $6\overline{)30}^{\,5}$ $12\overline{)132}^{\,11}$ $9\overline{)99}^{\,11}$ $3\overline{)9}^{\,3}$ $11\overline{)132}^{\,12}$ $1\overline{)11}^{\,11}$ $5\overline{)20}^{\,4}$

9. $2\overline{)22}^{\,11}$ $2\overline{)6}^{\,3}$ $4\overline{)16}^{\,4}$ $7\overline{)21}^{\,3}$ $6\overline{)54}^{\,9}$ $9\overline{)72}^{\,8}$ $7\overline{)35}^{\,5}$ $4\overline{)28}^{\,7}$

10. $12\overline{)72}^{\,6}$ $7\overline{)63}^{\,9}$ $5\overline{)30}^{\,6}$ $2\overline{)14}^{\,7}$ $9\overline{)54}^{\,6}$ $7\overline{)49}^{\,7}$ $5\overline{)60}^{\,12}$

11. $6\overline{)48}^{\,8}$ $8\overline{)64}^{\,8}$ $11\overline{)44}^{\,4}$ $4\overline{)0}^{\,0}$ $6\overline{)66}^{\,11}$ $12\overline{)48}^{\,4}$ $7\overline{)77}^{\,11}$

12. $7\overline{)42}^{\,6}$ $8\overline{)48}^{\,6}$ $2\overline{)24}^{\,12}$ $9\overline{)108}^{\,12}$ $11\overline{)132}^{\,12}$ $10\overline{)60}^{\,6}$ $7\overline{)14}^{\,2}$

13. $3\overline{)33}^{\,11}$ $10\overline{)100}^{\,10}$ $2\overline{)6}^{\,3}$ $12\overline{)96}^{\,8}$ $8\overline{)56}^{\,7}$ $10\overline{)110}^{\,11}$ $8\overline{)40}^{\,5}$

Lesson 13 ~ Speed Test (Time: 1 minute) *Division*

1. 7)77̄ = 11   7)42̄ = 6   8)48̄ = 6   2)24̄ = 12   9)108̄ = 12   11)99̄ = 9   5)25̄ = 5   4)24̄ = 6

2. 3)27̄ = 9   2)10̄ = 5   7)63̄ = 9   12)72̄ = 6   9)81̄ = 9   10)110̄ = 11   8)64̄ = 8   6)12̄ = 2

3. 12)144̄ = 12   5)40̄ = 8   10)80̄ = 8   3)18̄ = 6   2)14̄ = 7   9)63̄ = 7   7)49̄ = 7   5)60̄ = 12

4. 6)42̄ = 7   4)12̄ = 3   11)110̄ = 10   1)7̄ = 7   12)60̄ = 5   12)0̄ = 0   2)8̄ = 4   4)36̄ = 9

5. 7)56̄ = 8   6)54̄ = 9   9)72̄ = 8   11)66̄ = 6   1)9̄ = 9   8)72̄ = 9   3)15̄ = 5   10)120̄ = 12

6. 3)21̄ = 7   2)18̄ = 9   6)30̄ = 5   12)132̄ = 11   9)99̄ = 11   10)20̄ = 2   7)28̄ = 4   11)88̄ = 8

7. 8)16̄ = 2   4)40̄ = 10   6)24̄ = 4   4)48̄ = 12   3)0̄ = 0   9)27̄ = 3   7)84̄ = 12   5)35̄ = 7

8. 11)121̄ = 11   2)0̄ = 0   8)96̄ = 12   9)45̄ = 5   9)36̄ = 4   6)36̄ = 6   4)16̄ = 4   8)32̄ = 4

9. 12)108̄ = 9   2)12̄ = 6   11)132̄ = 12   1)11̄ = 11   5)20̄ = 4   2)22̄ = 11   8)56̄ = 7   2)6̄ = 3

10. 12)96̄ = 8   9)54̄ = 6   4)32̄ = 8   7)35̄ = 5   4)28̄ = 7   12)84̄ = 7   2)16̄ = 8

11. 5)30̄ = 6   7)14̄ = 2   3)9̄ = 3   8)0̄ = 0   8)40̄ = 5   10)70̄ = 7   8)88̄ = 11

12. 11)44̄ = 4   4)0̄ = 0   6)66̄ = 11   8)80̄ = 10   7)21̄ = 3   5)45̄ = 9   6)48̄ = 8

13. 4)44̄ = 11   6)60̄ = 10   11)132̄ = 12   10)90̄ = 9   5)15̄ = 3   3)33̄ = 11   6)18̄ = 3

Lesson 21 ~ Quiz

Name _____

Date _____

English Measures

Score _____

A. Write these equivalents by memory.

1. 1 ft. = __12__ in.
2. 1 yd. = __3__ ft.
3. 1 mi. = __5,280__ ft.
4. 1 tbsp. = __3__ tsp.
5. 1 fl. oz. = __2__ tbsp.
6. 1 cup = __8__ fl. oz.
7. 1 qt. = __2__ pt.
8. 1 gal. = __4__ qt.
9. 1 pk. = __8__ qt.
10. 1 bu. = __4__ pk.
11. 1 lb. = __16__ oz.
12. 1 ton = __2,000__ lb.
13. 1 sq. ft. = __144__ sq. in.
14. 1 sq. yd. = __9__ sq. ft.
15. 1 a. = __43,560__ sq. ft.
16. 1 sq. mi. = __640__ a.

B. Find these equivalents.

17. 15 ft. = __180__ in.
18. 80 oz. = __5__ lb.
19. 3 ft. 8 in. = __$3\frac{2}{3}$__ ft.
20. 8 lb. 5 oz. = __133__ oz.

Lesson 23 ~ Quiz　　　　　　　　　　　Name _____

　　　　　　　　　　　　　　　　　　　　　Date _____

English and Metric Measures　　　　　　Score _____

A. Write these equivalents by memory.

1. 1 ft. = __12__ in.　　　　　　2. 1 yd. = __3__ ft.

3. 1 mi. = __5,280__ ft.　　　　　4. 1 tbsp. = __3__ tsp.

5. 1 fl. oz. = __2__ tbsp.　　　　6. 1 cup = __8__ fl. oz.

7. 1 qt. = __2__ pt.　　　　　　　8. 1 gal. = __4__ qt.

9. 1 pk. = __8__ qt.　　　　　　　10. 1 bu. = __4__ pk.

11. 1 lb. = __16__ oz.　　　　　　12. 1 ton = __2,000__ lb.

13. 1 sq. ft. = __144__ sq. in.　　14. 1 sq. yd. = __9__ sq. ft.

15. 1 a. = __43,560__ sq. ft.　　　16. 1 sq. mi. = __640__ a.

17. 1 m = __100__ cm　　　　　　18. 1 cm = __0.01__ m

19. 1 km = __1,000__ m　　　　　20. 1 mm = __0.1__ cm

B. Write the metric prefix for each number.

21. one hundred __hecto-__　　　22. one thousand __kilo-__

23. one tenth __deci-__　　　　　24. one thousandth __milli-__

25. one hundredth __centi-__

Lesson 27 ~ Quiz Name _____

Metric Measures Date _____

Score _____

A. Write these equivalents by memory.

1. 1 m = __100__ cm
2. 1 g = __100__ cg
3. 1 kg = __1,000__ g
4. 1 kg = __10__ hg
5. 1 kl = __100,000__ cl
6. 1 hl = __10,000__ cl
7. 1 m^2 = __10,000__ cm^2
8. 1 ha = __10,000__ m^2
9. 1 km^2 = __100__ ha
10. 1 MT = __1,000__ kg

B. Find these equivalents.

11. 18 cm = __0.18__ m
12. 75 km = __75,000__ m
13. 23 MT = __23,000__ kg
14. 600 kg = __0.6__ MT
15. 25.2 kl = __25,200__ l
16. 750 mm = __0.75__ m
17. 15 m = __49.2__ ft.
18. 18 lb. = __8.1__ kg
19. 28 cm = __10.92__ in.
20. 15 in. = __38.1__ cm

Lesson 31 ~ Quiz

Practice With Measures

A. Write these equivalents by memory.

1. 1 wk. = __7__ days
2. 1 min. = __60__ sec.
3. 1 yr. = __52__ wk.
4. 1 millennium = __1,000__ yr.
5. 1 mi. = __5,280__ ft.
6. 1 yd. = __3__ ft.
7. 1 lb. = __16__ oz.
8. 1 ton = __2,000__ lb.
9. 1 cup = __8__ fl. oz.
10. 1 fl. oz. = __2__ tbsp.
11. 1 pt. = __½__ qt.
12. 1 gal. = __4__ qt.
13. 1 pk. = __8__ qt.
14. 1 bu. = __4__ pk.
15. 1 cm = __10__ mm
16. 1 km = __100__ dkm
17. 1 ha = __10,000__ m²
18. 1 km² = __1,000,000__ m²
19. 1 kl = __1,000__ l
20. 1 m² = __10,000__ cm²

B. Find these equivalents. Give temperatures to the nearest whole degree.

21. 70°F = __21__ °C
22. 80°C = __176__ °F
23. 8 in. = __20.32__ cm
24. 87 kg = __191.4__ lb.
25. 8 ft. = __2.4__ m
26. 9 cm = __3.51__ in.
27. 125 l = __132.5__ qt.
28. 12 ha = __30__ a.

Lesson 35 ~ Quiz

Distance, Rate, and Time

Name _____
Date _____
Score _____

Find the missing facts. Write any remainder as a fraction.

	Distance	Rate	Time
1.	_192_ mi.	48 m.p.h.	4 hr.
2.	_1,980_ mi.	165 m.p.h.	12 hr.
3.	_920_ mi.	46 m.p.h.	20 hr.
4.	630 mi.	_42_ m.p.h.	15 hr.
5.	2,450 mi.	_175_ m.p.h.	14 hr.
6.	2 mi.	$\frac{2}{3}$ m.p.h.	3 hr.
7.	2 mi.	$\frac{1}{2}$ m.p.h.	4 hr.
8.	180 mi.	45 m.p.h.	_4_ hr.
9.	825 mi.	150 m.p.h.	$5\frac{1}{2}$ hr.
10.	357 km	84 km/h	$4\frac{1}{4}$ hr.

Lesson 37 ~ Quiz   Name _____

                         Date _____

Factoring Numbers   Score _____

A. Write P after each prime number and C after each composite number.

1. 59 __P__   2. 63 __C__   3. 65 __C__

B. Find the prime factors of these composite numbers. Use exponents for repeating factors.

4. 38 = __2 × 19__   5. 64 = __2^6__   6. 54 = __$2 × 3^3$__

C. Find the greatest common factor of each pair.

7. 40, 56   8. 63, 90
   g.c.f. = __8__   g.c.f. = __9__

D. Find the lowest common multiple of each pair.

9. 9, 12   10. 16, 20
   l.c.m. = __36__   l.c.m. = __80__

Lesson 39 ~ Quiz   Name _____

                          Date _____

Adding and Subtracting Fractions   Score _____

Solve numbers 1–5 by addition and numbers 6–10 by subtraction.

1. $3\frac{1}{4}$
    $+1\frac{1}{2}$
    $\overline{4\frac{3}{4}}$

2. $4\frac{7}{8}$
    $+5\frac{3}{4}$
    $\overline{10\frac{5}{8}}$

3. $5\frac{7}{9}$
    $+2\frac{5}{6}$
    $\overline{8\frac{11}{18}}$

4. $3\frac{2}{3}$
    $+1\frac{4}{5}$
    $\overline{5\frac{7}{15}}$

5. $2\frac{11}{12}$
    $+1\frac{3}{8}$
    $\overline{4\frac{7}{24}}$

6. $4\frac{3}{4}$
    $-1\frac{5}{8}$
    $\overline{3\frac{1}{8}}$

7. $3\frac{1}{3}$
    $-1\frac{1}{4}$
    $\overline{2\frac{1}{12}}$

8. $4\frac{6}{7}$
    $-2\frac{1}{2}$
    $\overline{2\frac{5}{14}}$

9. $5\frac{4}{9}$
    $-1\frac{5}{6}$
    $\overline{3\frac{11}{18}}$

10. $4\frac{3}{8}$
    $-1\frac{3}{4}$
    $\overline{2\frac{5}{8}}$

Lesson 42 ~ Quiz

Name _____

Date _____

Calculating With Fractions

Score _____

Solve these problems.

1. $5\frac{3}{4}$
   $+2\frac{1}{2}$
   $\overline{8\frac{1}{4}}$

2. $3\frac{3}{5}$
   $+1\frac{3}{4}$
   $\overline{5\frac{7}{20}}$

3. $3\frac{2}{3}$
   $+1\frac{7}{8}$
   $\overline{5\frac{13}{24}}$

4. $4\frac{5}{8}$
   $-1\frac{4}{9}$
   $\overline{3\frac{13}{72}}$

5. $3\frac{1}{4}$
   $-1\frac{1}{3}$
   $\overline{1\frac{11}{12}}$

6. $4\frac{1}{5}$
   $-2\frac{1}{3}$
   $\overline{1\frac{13}{15}}$

7. $1\frac{3}{4} \times 9 =$ __$15\frac{3}{4}$__

8. $3\frac{3}{5} \times \frac{3}{4} =$ __$2\frac{7}{10}$__

9. $6\frac{3}{8} \div 6 =$ __$1\frac{1}{16}$__

10. $4\frac{1}{2} \div \frac{3}{4} =$ __6__

Lesson 44 ~ Quiz

Name _____

Date _____

Using Fractions

Score _____

Solve these problems.

1. $2\frac{3}{5}$
   $+ 3\frac{1}{2}$
   $\overline{6\frac{1}{10}}$

2. $4\frac{5}{6}$
   $+ 3\frac{7}{8}$
   $\overline{8\frac{17}{24}}$

3. $2\frac{9}{10}$
   $+ 1\frac{1}{8}$
   $\overline{4\frac{1}{40}}$

4. $3\frac{3}{4}$
   $- 2\frac{1}{5}$
   $\overline{1\frac{11}{20}}$

5. $3\frac{5}{8}$
   $- 2\frac{7}{12}$
   $\overline{1\frac{1}{24}}$

6. $3\frac{7}{10}$
   $- 1\frac{5}{6}$
   $\overline{1\frac{13}{15}}$

7. $7 \times 1\frac{4}{5} =$ __$12\frac{3}{5}$__

8. $3\frac{1}{8} \times 1\frac{2}{5} =$ __$4\frac{3}{8}$__

9. $4\frac{3}{4} \div 6 =$ __$\frac{19}{24}$__

10. $3\frac{1}{9} \div 3\frac{1}{2} =$ __$\frac{8}{9}$__

Lesson 50 ~ Quiz

Multiplying Decimals

Name _____

Date _____

Score _____

A. Solve these multiplication problems mentally.

1. 10 × 2.35 = __23.5__

2. 1,000 × 5.04 = __5,040__

3. 1,000 × 3.6 = __3,600__

4. 10,000 × 2.0716 = __20,716__

5. 100 × 0.0012 = __0.12__

6. 10 × 0.00139 = __0.0139__

B. Solve these multiplication problems.

7.   5.7
  × 3.2
  18.24

8.   3.9
  × 2.3
  8.97

9.   2.25
  × 3.08
  6.93

10.   2.78
  × 3.15
  8.757

11.   0.165
  × 0.2
  0.033

12.   0.156
  × 0.034
  0.005304

13.   0.017
  × 0.032
  0.000544

14.   0.012
  × 0.35
  0.0042

15. 0.072 × 0.065 = __0.00468__

16. 0.91 × 0.056 = __0.05096__

Lesson 53 ~ Quiz

Name _____

Date _____

Using Decimals

Score _____

1.   2.7
   + 3.8
     6.5

2.   2.1678
   + 3.719
     5.8868

3.   4.46
   − 1.6285
     2.8315

4.   5.5
   − 2.7983
     2.7017

5.   1.645
   × 0.032
    0.05264

6.   4.814
   × 0.53
    2.55142

7. $0.8\overline{)7.12}$ = 8.9

8. $0.05\overline{)1.95}$ = 39

9. 17 + 10.98 + 16.3 = __44.28__

10. 6.3 − 2.7889 = __3.5111__

Lesson 56 ~ Quiz　　　　　　　　　　Name _____

　　　　　　　　　　　　　　　　　　　　Date _____

Proportions　　　　　　　　　　　　　Score _____

A. Write whether the facts in each relationship vary directly (D) or inversely (I).

__I__ 1. The time spent on a project in relation to the number of persons working on the project. (Do more people mean more or less time to complete the project?)

__D__ 2. The amount of berries picked in relation to the number of people picking berries. (Do more pickers mean more or fewer berries picked?)

B. Identify each proportion as direct (D) or inverse (I).

__I__ 3. $\dfrac{\text{m.p.h. (1)}}{\text{m.p.h. (2)}}\ \dfrac{40}{n} = \dfrac{2\ \text{hours (2)}}{3\ \text{hours (1)}}$

__D__ 4. $\dfrac{\text{m.p.h.}}{\text{mi. in 3 hr.}}\ \dfrac{50}{150} = \dfrac{60\ \text{m.p.h.}}{n\ \text{mi. in 3 hr.}}$

C. Find the missing parts of these proportions.

5. $\dfrac{25}{20} = \dfrac{75}{n}$　　6. $\dfrac{3\frac{1}{2}}{4} = \dfrac{n}{5}$　　7. $\dfrac{5.75}{1} = \dfrac{10.35}{n}$　　8. $\dfrac{6}{9} = \dfrac{14}{n}$

　$n = \underline{\ 60\ }$　　　　$n = \underline{\ 4\frac{3}{8}\ }$　　　$n = \underline{\ 1.8\ }$　　　$n = \underline{\ 21\ }$

D. Use proportions to solve these problems.

9. Marcus is mixing mortar cement at the rate of 2 parts sand to 1 part dry mortar. How many pounds of sand should he use to get a total dry mix of 300 pounds?

　　　　　　　　$\dfrac{\text{sand}}{\text{mix}}\ \dfrac{2}{3} = \dfrac{n\ \text{sand}}{300\ \text{mix}}$　　　　$n = \underline{\ 200\ \text{pounds}\ }$

10. Darlene typed a report in 30 minutes at the rate of 40 words per minute (w.p.m.). How long would it take to type the same report at 50 words per minute?

　　　　　　　　$\dfrac{\text{w.p.m. (1)}}{\text{w.p.m. (2)}}\ \dfrac{40}{50} = \dfrac{n\ \text{minutes (2)}}{30\ \text{minutes (1)}}$　　$n = \underline{\ 24\ \text{minutes}\ }$

Lesson 58 ~ Quiz

Name _____

Date _____

Proportions and Scale Drawings

Score _____

A. Find the distance represented by each measurement on a map.

Scale for numbers 1–2: *1 in. = 24 mi.*

1. $3\frac{3}{4}$ in.   __90 mi.__

2. $4\frac{3}{8}$ in.   __105 mi.__

Scale for numbers 3–4: *1 in. = 40 mi.*

3. $1\frac{3}{8}$ in.   __55 mi.__

4. $6\frac{5}{8}$ in.   __265 mi.__

Scale for numbers 5–6: $\frac{1}{2}$ *in. = 1 mi.*

5. $3\frac{3}{4}$ in.   __$7\frac{1}{2}$ mi.__

6. $2\frac{1}{2}$ in.   __5 mi.__

Scale for numbers 7–8: $\frac{3}{8}$ *in. = 1 mi.*

7. $2\frac{1}{16}$ in.   __$5\frac{1}{2}$ mi.__

8. $3\frac{15}{16}$ in.   __$10\frac{1}{2}$ mi.__

B. Find the correct length on a map for each distance. Scale: *1 in. = 4 mi.*

9. 12 mi.   __3 in.__

10. 25 mi.   __$6\frac{1}{4}$ in.__

Lesson 62 ~ Speed Test   (Time: 3 minutes)   *Percents, Fractions, and Decimals*

A. Express these percents as decimals.

1. 45% = __0.45__
2. 53% = __0.53__
3. 87% = __0.87__
4. 60% = __0.6__
5. 9% = __0.09__
6. 4% = __0.04__

B. Express these percents as fractions in lowest terms.

7. 75% = __$\frac{3}{4}$__
8. 40% = __$\frac{2}{5}$__
9. 35% = __$\frac{7}{20}$__
10. 74% = __$\frac{37}{50}$__
11. 38% = __$\frac{19}{50}$__
12. 68% = __$\frac{17}{25}$__

C. Express these decimals as percents.

13. 0.78 = __78%__
14. 0.89 = __89%__
15. 0.3 = __30%__
16. 0.7 = __70%__
17. 0.04 = __4%__
18. 0.09 = __9%__

D. Express these fractions as percents.

19. $\frac{7}{10}$ = __70%__
20. $\frac{9}{10}$ = __90%__
21. $\frac{13}{20}$ = __65%__
22. $\frac{17}{20}$ = __85%__
23. $\frac{9}{50}$ = __18%__
24. $\frac{3}{50}$ = __6%__

Lesson 65 ~ Quiz

Name _____

Date _____

Finding Percentages

Score _____

1. 35% of 89 = _31.15_

2. 46% of 78 = _35.88_

3. 9% of 98 = _8.82_

4. 7% of 712 = _49.84_

5. 124% of 65 = _80.6_

6. 420% of 25 = _105_

7. ½% of 430 = _2.15_

8. ¾% of 280 = _2.1_

9. 1½% of 540 = _8.1_

10. 5½% of 390 = _21.45_

Lesson 68 ~ Quiz

Name _____

Date _____

Finding Percents

Score _____

A. Find each rate to the nearest whole percent.

1. 20 is __80%__ of 25

2. 20 is __40%__ of 50

3. 16 is __27%__ of 60

4. 35 is __44%__ of 80

B. Find the rate of each change, to the nearest whole percent. Include the label *increase* **or** *decrease*.

	Original Price	*New Price*	
5.	$40.00	$42.00	5% increase
6.	$35.00	$38.00	9% increase
7.	$26.00	$24.00	8% decrease
8.	$34.00	$26.00	24% decrease
9.	$12.00	$55.00	358% increase
10.	$18.00	$2.00	89% decrease

Lesson 70 ~ Quiz

Name _____

Date _____

Commissions

Score _____

Find the missing parts. Round money amounts to the nearest cent and rates to the nearest whole percent.

	sales	rate	commission		sales	rate	commission
1.	$456.00	3%	$13.68	2.	$578.00	8%	$46.24
3.	$257.23	4%	$10.29	4.	$279.95	9%	$25.20
5.	$620.00	4%	$24.80	6.	$230.90	8%	$18.47
7.	$128.60	16%	$20.58	8.	$314.23	9%	$28.28
9.	$1,399.00	5%	$69.95	10.	$568.29	7%	$39.78

Lesson 74 ~ Quiz

Name _____

Date _____

Practice With Commissions

Score _____

Find the missing parts. See how many you can solve mentally.

	sales	rate	commission		sales	rate	commission
1.	$100.00	5%	$5.00	2.	$200.00	8%	$16.00
3.	$300.00	12%	$36.00	4.	$400.00	11%	$44.00
5.	$100.00	4%	$4.00	6.	$200.00	7%	$14.00
7.	$500.00	5%	$25.00	8.	$400.00	7%	$28.00
9.	$200.00	5%	$10.00	10.	$750.00	4%	$30.00

Lesson 76 ~ Quiz   Name _____
                             Date _____
Mean, Median, and Mode   Score _____

A. Find the mean and the median of each set. Express remainders as fractions.

45, 43, 46, 49, 53

1. mean = __$47\frac{1}{5}$__   2. median = __46__

29, 46, 22, 25, 27, 28

3. mean = __$29\frac{1}{2}$__   4. median = __$27\frac{1}{2}$__

25, 29, 32, 24, 21, 26

5. mean = __$26\frac{1}{6}$__   6. median = __$25\frac{1}{2}$__

34, 35, 32, 36, 29, 28

7. mean = __$32\frac{1}{3}$__   8. median = __33__

B. Find the mode(s) of each set.

9. 12, 11, 13, 15, 12, 13, 16, 17, 12, 13, 15, 11, 11, 16, 17, 19

   mode(s) = __11, 12, 13__

10. 36, 34, 38, 37, 35, 36, 39, 31, 32, 35, 36, 37, 34, 35, 38, 39, 31

    mode(s) = __35, 36__

Lesson 79 ~ Quiz

Name _____

Date _____

Practice With Mean, Median, and Mode   Score _____

A. Find the mean and the median of each set. Express remainders as fractions.

225, 218, 232, 226, 278

1. mean = $235\frac{4}{5}$ 2. median = 226

316, 323, 315, 361, 356, 301

3. mean = $328\frac{2}{3}$ 4. median = $319\frac{1}{2}$

414, 200, 495, 434, 446, 395

5. mean = $397\frac{1}{3}$ 6. median = 424

534, 527, 516, 426, 329, 528

7. mean = $476\frac{2}{3}$ 8. median = $521\frac{1}{2}$

B. Find the mode(s) of each set.

9. 32, 31, 39, 36, 38, 37, 33, 35, 34, 40, 29, 31, 43, 45, 31, 34, 40

   mode(s) = 31

10. 11, 12, 11, 12, 12, 11, 11, 10, 16, 12, 11, 12, 12, 11, 15, 14, 16

    mode(s) = 11, 12

Lesson 83 ~ Quiz

Name _____

Date _____

General Calculations

Score _____

1. 34,521
  + 27,831
  _____
   62,352

2. $1,356.78
   - 958.97
   _____
    $397.81

3.   251
   × 169
   _____
   42,419

4. 35)14,151 = 404 R 11

5. 0.5 + 0.205 = 0.705

6. 0.4 - 0.17 = 0.23

7. 0.19 × 2.017 = 0.38323

8. 1.2)3.24 = 2.7

9. 124% of 45 = 55.8

10. 10 is 40 % of 25

Chapter Tests

Answer Key

Grade 8   Applying Mathematics   Chapter 1 Test   3

16. Chapter 1 Test

Name _____ Date _____ Score _____

A. Write the labels for the parts of these problems.

1. 45   addend   2.   67   minuend
   + 37 addend   − 46   subtrahend
    82   sum   21   difference

3. 23   multiplicand   4. _quotient_   23 R 12   remainder
   × 17 multiplier   divisor   15)357   dividend
   391   product

B. Match the shortcuts to the times when they are most useful. You will not use all the choices.

_e_ 5. When dividing by 25.

_c_ 6. When multiplying by 5, 50, or 500.

_b_ 7. When the divisor and the dividend are divisible by the same factor.

_d_ 8. When dividing by 5, 50, or 500

a. distributive method

b. divide-and-divide method

c. double-and-divide method

d. double-and-double method

e. multiplying both numbers by 4

C. Solve these problems. Check multiplication and division by casting out nines.

9. $2,818.81   10.   238,324,124
    + 4,821.88   − 143,780,234
     $7,640.69   94,543,890

11.   567   201 R 39
     × 436   12. 125)25,164
     247,212

4   Chapter 1 Test

D. Solve by horizontal calculation or short division.

13. 32,765 + 15,681 =   48,446

14. 18,653 − 9,482 =   9,171

15. 8 × 5,682 =   45,456

16. $8\overline{)25,288}$ = 3,161

E. Solve these problems mentally.

17. 356 + 346 =   702   18. 20 × 23 =   460

19. 5 × 326 =   1,630   20. 288 ÷ 36 =   8

F. Do these exercises.

21. Write the number 14,000,600,000,000, using words.   fourteen trillion, six hundred million

22. Use digits to write "sixteen trillion, five hundred billion."   16,500,000,000,000

23. Write the value of the 2 in 120,000,000,000,000,000.   20,000,000,000,000,000

24. Round 3,578,000,034 to the nearest billion.   4,000,000,000

25. Write CDXLVIII as an Arabic numeral.   448

26. The equation 7(4 × 3) = (7 × 4)3 illustrates the   associative   law.

27. Complete this equation: (10 × 12) + (8 × 12) = 18 × 12.

G. Write a number sentence for each reading problem, and find the answer.

28. One of the beauties that God created is waterfalls. The highest waterfall in the world is Angel Falls in Venezuela, whose water drops a total of 3,212 feet. The highest falls in the United States is Ribbon Falls in California, where the water drops a total of 1,612 feet. How much greater is the height (h) of Angel Falls than Ribbon Falls?

  Number sentence: $h = 3,212 - 1,612$

  Answer: 1,600 feet

Grade 8   Applying Mathematics

29. Yosemite Falls in California is a series of falls and cascades. The water drops 1,430 feet in the Upper Yosemite Falls, 675 feet in the cascade below these falls, and 320 feet at the Lower Yosemite Falls. What is the total descent (d) of the water at Yosemite Falls?

   Number sentence: $d = 1{,}430 + 675 + 320$

   Answer: 2,425 feet

30. Niagara Falls is noted for its huge volume of water. An average of about 212,000 cubic feet of water passes over the falls each second. Water weighs 62.4 pounds per cubic foot. What is the weight (w) in pounds of the water passing over Niagara Falls in one second?

   Number sentence: $w = 62.4 \times 212{,}000$

   Answer: 13,228,800 pounds

31. The largest desert in the world is the Sahara in Africa; it covers an area of 3,500,000 square miles. This is 7 times larger than the Gobi Desert in Mongolia and China. What is the area (a) of the Gobi Desert?

   Number sentence: $a = 3{,}500{,}000 \div 7$

   Answer: 500,000 square miles

32. Greenland, the largest island in the world, has an area of 840,000 square miles and a population of 53,000. The second largest island in the world, New Guinea, has an area 534,000 square miles less than that of Greenland but a population 81 times as great as that of Greenland. What is the area (a) of New Guinea?

   Number sentence: $a = 840{,}000 - 534{,}000$

   Answer: 306,000 square miles

33. Another large island, Taiwan, has an area of 13,823 square miles and a population density of 1,510 persons per square mile. What is the population (p) of Taiwan? (Round your answer to the nearest million.)

   Number sentence: $p = 1{,}510 \times 13{,}823$

   Answer: 21,000,000 people

Grade 8   Applying Mathematics   Chapter 2 Test   7

33. Chapter 2 Test

Name _____   Date _____   Score _____

A. Write the abbreviations for these units.

1. square mile   sq. mi.
2. month   mo.
3. dekameter   dkm
4. square kilometer km^2

B. Write these equivalents by memory.

5. 1 bu. = 4 pk.
6. 1 cup = 8 fl. oz.
7. 1 l = 1.06 qt.
8. 1 qt. = 0.95 l

C. Find these equivalents.

9. 47,000 lb. = $23\frac{1}{2}$ tons
10. 15 lb. = 240 oz.
11. 20 hr. 20 min. = $20\frac{1}{3}$ hr.
12. 15 yr. 5 mo. = 185 mo.
13. 4.2 kl = 4,200 l
14. 51 l = 51,000 ml
15. 95°C = 203 °F
16. 40°F = 4 °C

D. Solve these problems involving compound measures.

17. 7 ft. 11 in.
    + 6 ft. 10 in.
    ―――――――――
     14 ft. 9 in.

18. 7 bu. 1 pk.
    − 2 bu. 3 pk.
    ―――――――――
     4 bu. 2 pk.

19. 6 hr. 45 min.
    ×   6
    ―――――――――
    40 hr. 30 min.

20. 8)$\overline{62 \text{ lb. }10 \text{ oz.}}$ = 7 lb. $13\frac{1}{4}$ oz.

21. 2.2 kg − 450 g = 1,750 g
22. 9 l 456 ml ÷ 6 = 1.576 l

Chapter 2 Test

E. Answer these questions.

23. If an airplane flies at 300 kilometers per hour for 4 hours, what is the total distance traveled?

    _____1,200 kilometers_____

24. How long will it take a man to walk 10 miles at a rate of 3 miles per hour?

    _____$3\frac{1}{3}$ hours (3 hr. 20 min.)_____

25. If it is 5:00 P.M. Eastern Standard Time, what time is it Pacific Standard Time?

    _____2:00 P.M._____

26. Write the formula for finding time when distance and rate are known.

    $$\text{time} = \frac{\text{distance}}{\text{rate}}$$

F. Solve these reading problems.

27. The Roseville Industrial Park paid $224,000 for 15 acres of land to build a 40,000-square-foot industrial center. To the nearest cent, what was the price per hectare?

    _____$37,333.33_____

28. The Masts canned 175 quarts of peaches and 150 quarts of pears. How many liters of pears did they can?

    _____142.5 liters_____

29. The World Trade Center in New York City was built in 1972 and destroyed in 2001. It had twin towers each 110 stories high and was the work site for more than 100,000 people. To the nearest whole number, how many workers was that for each floor of one tower?

    _____455 people_____

30. The Sears Tower in Chicago, built in 1974, also has 110 stories. With a height of 1,454 feet, however, this building is 104 feet taller than the World Trade Center was. To the nearest foot, what is the average height per story of the Sears Tower?

_____ 13 feet _____

31. The West family lives 1.25 kilometers from their school and 6 kilometers from their church. How many meters do the Wests live from their school?

_____ 1,250 meters _____

32. A water reservoir measuring 50 feet in diameter and 40 feet high contains 1,950 kiloliters of water. How many liters of water does it contain?

_____ 1,950,000 liters _____

Grade 8 Applying Mathematics   Chapter 3 Test 11

46. Chapter 3 Test

Name _____ Date _____ Score _____

A. Solve these problems.

1. $\frac{5}{8}$
   $+ \frac{3}{10}$
   _____
   $\frac{37}{40}$

2. $2\frac{1}{9}$
   $- 1\frac{5}{6}$
   _____
   $\frac{5}{18}$

3. $\frac{3}{5}$ of 40 =   __24__

4. $54 \times \frac{5}{9}$ =   __30__

5. $16 \div \frac{2}{5}$ =   __40__

6. $2\frac{4}{7} \div 7\frac{1}{2}$ =   __$\frac{12}{35}$__

7. $15 \div 2\frac{1}{4}$ =   __$6\frac{2}{3}$__

8. $\frac{3}{8}$ is $\frac{4}{5}$ of   __$\frac{15}{32}$__

B. Solve by vertical multiplication.

9. 32
   $\times 3\frac{1}{4}$
   _____
   104

10. 22
    $\times 2\frac{2}{3}$
    _____
    $58\frac{2}{3}$

C. Simplify these complex fractions.

11. $\frac{\frac{3}{4}}{\frac{7}{8}}$ = __$\frac{6}{7}$__

12. $\frac{4\frac{2}{3}}{\frac{5}{6}}$ = __$5\frac{3}{5}$__

D. Find these answers mentally.

13. $\frac{3}{4}$ of 28 = __21__

14. $8 \div \frac{1}{7}$ = __56__

15. $10 \div \frac{1}{8}$ = __80__

16. $12 \div \frac{3}{8}$ = __32__

12   Chapter 3 Test

E. Write the answers.

17. Is the number 19 prime or composite?   _prime_

18. Find the prime factors of 45.
    Use exponents to show repeating factors. $45 = 3^2 \times 5$

19. Find the greatest common factor of 40 and 64.   _8_

20. Find the lowest common multiple of 15 and 33.   _165_

21. Compare the fractions at the right, and place < or > between them. $\frac{2}{5} < \frac{7}{16}$

22. What kind of fraction is $\frac{41}{2}$?   _improper fraction_

23. Reduce $\frac{54}{90}$ to lowest terms. $\frac{3}{5}$

24. What is the reciprocal of 27? $\frac{1}{27}$

F. Solve these reading problems.

25. Father's chain-reference Bible is $1\frac{7}{8}$ inches thick. Sharon's Bible is $1\frac{1}{8}$ inches thick. How many times thicker is Father's Bible than Sharon's?

    $1\frac{2}{3}$ times

26. Father has a Bible dictionary with 1,152 pages. The pages have a combined thickness of $1\frac{1}{2}$ inches. How many pages are there per inch of thickness?

    768 pages

Grade 8   *Applying Mathematics*　　　　　　　　　　　　　　　　Chapter 3 Test   13

27. One day it rained for $3\frac{1}{2}$ hours at an average rate of $\frac{3}{4}$ inch per hour. What was the total rainfall that day?

$\underline{2\frac{5}{8} \text{ inches}}$

28. One winter day it snowed 12 inches in $4\frac{1}{2}$ hours. What was the average snowfall per hour?

$\underline{2\frac{2}{3} \text{ inches}}$

29. A printing press can print 13,000 sheets per hour. How many sheets can it print in an $8\frac{1}{2}$-hour work day?

$\underline{110{,}500 \text{ sheets}}$

30. Mr. Davis is planting wheat in a $21\frac{1}{3}$-acre field at the rate of $2\frac{1}{8}$ bushels per acre. How many bushels of wheat will Mr. Davis plant in the field?

$\underline{45\frac{1}{3} \text{ bushels}}$

Grade 8 Applying Mathematics   Chapter 4 Test   15

59. Chapter 4 Test

Name _____   Date _____   Score _____

A. Solve these problems.

1.   4.6   2.   8.52   3.   0.065   60.2
   + 1.7849   − 7.7284   × 0.064   4. 0.05)3.01
    6.3849   0.7916   0.00416

5. 6.161 + 5.2 + 6.89 =   18.251

6. 3.1 − 2.095 =   1.005

7. 15.8 × 2.015 =   31.837

8. 172.5 ÷ 0.24 =   718.75

B. Solve these problems mentally. Change decimals to fractions if it makes calculation easier.

9. 3.17 × 1,000 =   3,170

10. 3,516.2 ÷ 10,000 =   0.35162

11. 0.7 of 90 =   63

12. 0.625 of 40 =   25

C. Write the answers.

13. Change 0.28 to a common fraction in lowest terms. $\frac{7}{25}$

14. Name the place farthest to the right in the number 0.98765.   hundred-thousandths

15. Write 1.000008, using words.   one and eight millionths

16 Chapter 4 Test

16. Compare the decimals at the right, and place < or > between them.   0.023 __>__ 0.0024

17. Round 2.08176 to the nearest ten-thousandth.   __2.0818__

18. Express $\frac{15}{32}$ as a decimal.   __0.46875__

19. Is the following a direct or an inverse proportion?   __inverse__

$$\frac{\text{pounds on side 1}}{\text{pounds on side 2}} \quad \frac{60}{80} = \frac{40}{n} \quad \frac{\text{inches from fulcrum (2)}}{\text{inches from fulcrum (1)}}$$

20. Does the scale 3:2 indicate an enlargement or a reduction?   __enlargement__

21. Use a proportion to find the distance represented by $2\frac{5}{8}$ inches on a map with a scale of 1 inch = 16 miles.   __42 miles__

$$\frac{\text{scale in.}}{\text{actual mi.}} \quad \frac{1}{16} = \frac{2\frac{5}{8}}{n} \quad \frac{\text{scale in.}}{\text{actual mi.}}$$

22. A scale drawing has a scale of 3 inches = 2 inches. Use a proportion to find how long a line should be to represent $2\frac{1}{2}$ inches. $3\frac{3}{4}$ inches

$$\frac{\text{scale in.}}{\text{actual mi.}} \quad \frac{3}{2} = \frac{n}{2\frac{1}{2}} \quad \frac{\text{scale in.}}{\text{actual mi.}}$$

D. Write a ratio in lowest terms for each statement.

23. In the first inning of a softball game, 6 of the 9 team members were up to bat.   __2 to 3__

24. Randall has memorized 10 of the 12 verses in Isaiah 53.   __5 to 6__

E. Use direct or inverse proportions to solve these problems. Be careful, for one part of a proportion may need to be calculated.

25. The ark that Noah built was 6 cubits long for every cubit of its width. The length was 300 cubits. What was the width? $\frac{\text{length}}{\text{width}} \quad \frac{6}{1} = \frac{300}{n} \quad \frac{\text{length}}{\text{width}}$

$n = $ __50 cubits__

Grade 8 *Applying Mathematics* Chapter 4 Test

26. Dale weighs 100 pounds and Levi weighs 120 pounds. If Dale sits 7 feet from the fulcrum of a seesaw, how far from the fulcrum must Levi sit for the seesaw to balance?

    $n = $ _____ $5\frac{5}{6}$ feet _____

    $$\frac{\text{pounds on side 1}}{\text{pounds on side 2}} \quad \frac{100}{120} = \frac{n}{7} \quad \frac{\text{feet from fulcrum (2)}}{\text{feet from fulcrum (1)}}$$

27. Last week it took 3 of the Masts $1\frac{1}{3}$ hours to do the cleaning at church. This week one of their cousins is helping them. How long should it take to do the cleaning this week?

    $n = $ _____ 1 hour _____

    $$\frac{\text{workers (1)}}{\text{workers (2)}} \quad \frac{3}{4} = \frac{n}{1\frac{1}{3}} \quad \frac{\text{hours (2)}}{\text{hours (1)}}$$

28. One afternoon during rush hour, it took the Sensenigs 3 hours at an average rate of 48 miles per hour to travel to a distant church service. When they returned home late that evening, their average speed was 54 miles per hour. How long did it take for the return trip?

    $n = $ _____ $2\frac{2}{3}$ hours _____

    $$\frac{\text{hours (1)}}{\text{hours (2)}} \quad \frac{3}{n} = \frac{54}{48} \quad \frac{\text{m.p.h. (2)}}{\text{m.p.h. (1)}}$$

29. A post 6 feet tall casts a 5-foot shadow. At the same time, a nearby utility pole casts a 40-foot shadow. How tall is the telephone pole?

    $n = $ _____ 48 feet _____

    $$\frac{\text{height}}{\text{shadow}} \quad \frac{6}{5} = \frac{n}{40} \quad \frac{\text{height}}{\text{shadow}}$$

30. One week, Woodvale Furniture finished 3 sheafback chairs for every 5 of their regular style of chairs. They finished 120 sheafback chairs. How many chairs did they finish in all?

    $n = $ _____ 320 chairs _____

    $$\frac{\text{sheafback}}{\text{total}} \quad \frac{3}{8} = \frac{120}{n} \quad \frac{\text{sheafback}}{\text{total}}$$

Grade 8   Applying Mathematics   Chapter 5 Test   19

72. Chapter 5 Test

Name _____   Date _____   Score _____

A. Write the answers.

1. Change 24% to a fraction in lowest terms. $\frac{6}{25}$

2. Change 91% to a ratio in which the consequent is 100.   91:100

3. Change $\frac{3}{4}$ to a percent.   75%

4. Change 0.01 to a percent.   1%

5. In the problem 25% of 32 = 8, 32 is the (rate, percentage, base).   base

6. Write 3,121% in decimal form.   31.21

7. Express $\frac{3}{4}$% as a decimal.   0.0075

8. sales = $1,357; rate = 8%; commission = $108.56

9. commission = $125; rate = 5%; sales = $2,500.00

10. base = 60; percentage = 21; rate =   35%

11. Find 3% of 67. Express the answer in decimal form.   2.01

12. Add 100% and 28%.   128%

Chapter 5 Test

B. Find these percentages.

13. 135% of 80 = _____108_____   14. $3\frac{1}{4}$% of 280 = _____9.1_____

C. Find these percentages mentally.

15. 40% of 35 = _____14_____   16. $83\frac{1}{3}$% of 48 = _____40_____

17. $66\frac{2}{3}$% of 270 = _____180_____   18. $62\frac{1}{2}$% of 560 = _____350_____

D. Find the new price after each change, to the nearest cent.

19. $280 increased by 48% = _$414.40_   20. $27.93 decreased by 38% = _$17.32_

E. Find these rates. Express remainders as fractions.

21. 6 is __$33\frac{1}{3}$%__ of 18   22. 9 is __$34\frac{8}{13}$%__ of 26

F. Find the rate of each change, to the nearest whole percent. Include the label *increase* **or** *decrease.*

23. Term 1 enrollment, 95; term 2 enrollment, 101   _____6% increase_____

24. Term 3 enrollment, 86; term 4 enrollment, 79   _____8% decrease_____

G. Find the base in each problem.

25. 33 is $1\frac{1}{2}$% of _____2,200_____   26. 99 is 180% of _____55_____

Grade 8   *Applying Mathematics*   Chapter 5 Test   21

H. Solve these reading problems. Draw sketches for numbers 27 and 28.

27. Marvin planted sweet corn in a row 20 feet long. He began planting 2 feet from each end of the row, and he dropped 2 seeds every $\frac{1}{2}$ foot. How many seeds did Marvin plant in the row?

_____ 66 seeds _____

28. Leonard is setting fence posts to enclose a rectangular area measuring 108 feet by 84 feet. If he sets the posts 12 feet apart, how many posts will he need?

_____ 32 posts _____

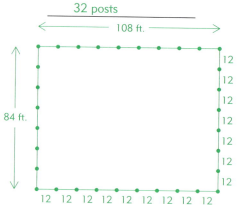

29. Melody is planting flowers. She has planted 5 flowers, or 20% of the number that she will plant in the flower bed. How many plants does she have in all?

_____ 25 plants _____

30. A salesman received a 5% commission on a sale of $225.00. What was the amount of his commission?

$$\underline{\$11.25}$$

31. Of the 48 bundles of shingles at the job site, Melvin has carried 23 bundles onto the roof. What percent of the shingles is that, to the nearest whole percent?

$$\underline{48\%}$$

32. Delbert read 60 pages or 48% of a book. How many pages does the book contain?

$$\underline{125 \text{ pages}}$$

Grade 8   *Applying Mathematics*   *Chapter 6 Test*   23

82. Chapter 6 Test

Name _____ Date _____ Score _____

A. *Do these exercises.*

1. Find the mean of this set: 36, 38, 32, 35, 34.   <u> 35 </u>

2. Find the mean of the following percent scores, to the nearest whole percent. 93%, 84%, 100%, 77%, 99%   <u> 91% </u>

3. Find the median of the set of numbers below.   <u> $924\frac{1}{2}$ </u>

4. Find the mode(s) of the set of numbers below.   <u> 922, 926 </u>

   922, 926, 927, 922, 923, 924, 922, 925, 926, 927, 926, 923

B. *Answer the questions about this histogram.*

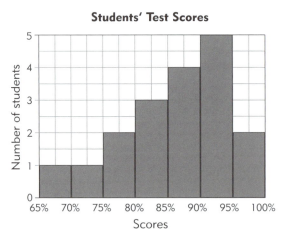

Students' Test Scores

5. How many students scored between 76% and 80%?   <u> 2 students </u>

6. How many students received a grade higher than 90%?   <u> 7 students </u>

7. How many students were in the entire room?   <u> 18 students </u>

24   Chapter 6 Test

C. Answer the questions about this bar graph.

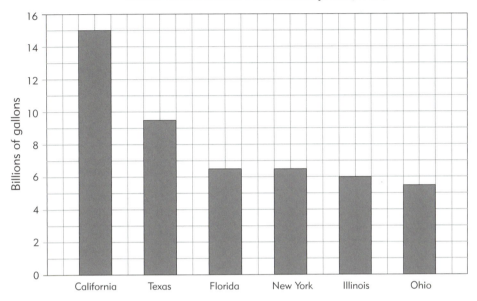

(Accept reasonable variation.)

8. What was the fuel consumption in California?   _15,000,000,000 gallons_

9. What was the fuel consumption in Texas?   _9,500,000,000 gallons_

10. How much more fuel was used in California than in Ohio?   _9,500,000,000 gallons_

11. How many gallons of fuel were used in all six states?   _49,000,000,000 gallons_

D. Complete the following chart. (You do not need to prepare a circle graph.) Express decimals to the nearest thousandth.

Topical Outline of the Book of Romans

Topic	Number of Chapters	Fraction	Decimal	Degrees
God's Righteousness Revealed	8	12. $\frac{1}{2}$	0.500	180°
God's Righteousness Vindicated	3	13. $\frac{3}{16}$	0.188	68°
God's Righteousness Applied	5	14. $\frac{5}{16}$	0.313	113°
Totals	16		1.001	361°

Grade 8 *Applying Mathematics* Chapter 6 Test 25

E. Make a line graph as indicated.

15–24. Use the data on this table to complete the double-line graph below.

Yards Quilted by Sharon and Doris

Week	1	2	3	4	5	6	7	8	9	10
Sharon	28	32	35	37	42	51	68	71	78	88
Doris	18	22	27	32	35	40	45	50	60	68

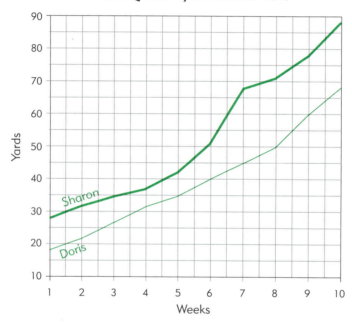

F. Solve these reading problems.

25. On seven consecutive math lessons, Benjamin's scores were as follows: 87%, 96%, 85%, 83%, 90%, 88%, 92%. What was his average for these lessons, to the nearest whole percent?

_____89%_____

26. The tributaries of the Dead Sea pour about 6,500,000 tons of water daily into it. How many tons is that per hour? (Answer to the nearest whole ton.)

_____270,833 tons_____

26 *Chapter 6 Test*

27. In 1990, the state of New York had a population of 17,990,455. This state has 34 representatives in Congress. To the nearest whole number, how many people did each member represent that year?

$$\text{529,131 people}$$

28. One day in March, the temperatures from 9:00 A.M. to 7:00 P.M. were as follows: 32°, 38°, 40°, 43°, 44°, 44°, 44°, 45°, 44°, 42°, 40°. What was the median temperature?

$$43°$$

29. Spelling test scores of the seventh and eighth graders were as follows: 100%, 92%, 96%, 100%, 96%, 84%, 92%, 88%, 96%, 100%. Find the median and the mode of these scores.

median = __96%__

mode(s) = __96%, 100%__

30. Randy wanted to find the average age of the students in his classroom. He compiled this list: 13, 15, 14, 14, 13, 15, 13, 16, 14, 13, 15, 13. Find the mean and the mode of these ages.

mean = __14__

mode(s) = __13__

Grade 8   Applying Mathematics   Semester 1 Test   27

84. Semester 1 Test

Name _____   Date _____   Score _____

A. Solve these problems.

1. $25,467.32
    + 18,693.59
    $44,160.91

2. $59,369.96
    − 27,499.96
    $31,870.00

3.   963
    × 369
    355,347

4. 178)22,191   124 R 119

5. $\frac{3}{5}$
    $+ \frac{5}{6}$
    $1\frac{13}{30}$

6. $\frac{5}{9}$
    $- \frac{3}{10}$
    $\frac{23}{90}$

7. $\frac{3}{8}$ of 26 = $9\frac{3}{4}$

8. $3\frac{1}{6} \times 2\frac{1}{4}$ = $7\frac{1}{8}$

9. $5\frac{1}{4} \div 1\frac{1}{6}$ = $4\frac{1}{2}$

10. $3\frac{3}{5}$ is $\frac{2}{5}$ of   9

11. $\frac{\frac{3}{5}}{\frac{2}{5}}$ = $1\frac{1}{2}$

12. 22 + 14.61 + 18.0282 =   54.6382

13. 5.05 − 0.0009 =   5.0491

14. 35.8 × 2.035 =   72.853

B. Do these exercises.

15. Express 4,669 as a Roman numeral. $\overline{IV}$DCLXIX

16. Solve mentally: $11 \div \frac{1}{7}$ = ___.   77

17. d = 140 mi.; r = ___ m.p.h.; t = $2\frac{1}{2}$ hr.   56

28 Semester 1 Test

18. Multiply 12 × 4 lb. 6 oz. 52 lb. 8 oz.

19. 250 kg × 20 = ___ MT 5

20. 1 cup = ___ pt. $\frac{1}{2}$

21. 4.75 m = ___ cm 475

22. Find the greatest common factor of 54 and 90. 18

23. Round 3.00945 to the nearest thousandth. 3.009

24. The following is a/an (direct, inverse) proportion. direct

$$\frac{\text{height}}{\text{shadow}} \quad \frac{40}{60} = \frac{n}{90} \quad \frac{\text{height}}{\text{shadow}}$$

25. The base in the following problem is ___. 20
200% of 20 = 40

26. Express 121% as a decimal. 1.21

27. Find the mean of this set, and express any remainder as a fraction.
151, 181, 292, 161, 111, 222 $186\frac{1}{3}$

28. Find the median of this set.
399, 366, 322, 417, 999, 15 $382\frac{1}{2}$

29. Find the mode(s) of this set.
22, 21, 26, 22, 13, 22, 29, 23, 22, 15, 17, 23, 22, 19, 23, 25 22

30. Find 225% of 70. 157.5

31. Find $2\frac{1}{2}$% of 300. 7.5

Grade 8   Applying Mathematics   Semester 1 Test   29

32. Write a proportion to find the actual length of an object that measures $2\frac{3}{4}$ inches on a scale drawing with a ratio of 1:15.

   proportion: $\dfrac{\text{scale}}{\text{actual}} \ \dfrac{1}{15} = \dfrac{2\frac{3}{4}}{n} \ \dfrac{\text{scale in.}}{\text{actual in.}}$ solution: $n = 41\frac{1}{4}$ inches

33. Find the new price, to the nearest cent. $22.70

   $25.79 decreased by 12% = ___

34. Find the percent of change, to the nearest whole percent.   13% increase
   Label the answer as an increase or a decrease.

   Original price, $15.00; new price, $17.00

C. *Solve these reading problems.*

35. The New Jerusalem described in Revelation 21 is a cube 12,000 furlongs broad and long and high. Its wall measures 144 cubits high and has 3 gates and 12 foundations garnished with precious stones. If a cubit is $1\frac{1}{2}$ feet, how high is the wall?

   216 feet

36. Maria folded 48 pieces of laundry. If she worked a total of 36 minutes, how much time did she spend on each article?

   $\frac{3}{4}$ minute (or 45 seconds)

D. *Write proportions to solve these problems. Be sure to use the right kind each time.*

37. Darlene and her younger sister Jean are picking strawberries. For every 2 quarts Jean picks, Darlene picks 5. If Jean picks 12 quarts, how many will Darlene have picked?

   $\dfrac{\text{Jean}}{\text{Darlene}} \ \dfrac{2}{5} = \dfrac{12}{n} \ \dfrac{\text{Jean}}{\text{Darlene}}$

   $n =$ 30 quarts

38. If a car travels at 40 miles per hour to reach a certain place in 3 hours, at what rate would it need to travel to go the same distance in 2 hours?

   $\dfrac{\text{m.p.h. (1)}}{\text{m.p.h. (2)}} \ \dfrac{40}{n} = \dfrac{2}{3} \ \dfrac{\text{hours (2)}}{\text{hours (1)}}$

   $n =$ 60 miles per hour

30 Semester 1 Test

E. Draw a sketch for each problem, and use it to find the solution. Show all your work.

39. Luke is building a bookcase having 3 shelves. There will be 12 inches between the top of one shelf and the bottom of next higher shelf. The shelves are $\frac{3}{4}$ inch thick, and the first shelf is 8 inches above the floor. How high will the top of the highest shelf be above the floor?

$34\frac{1}{4}$ inches

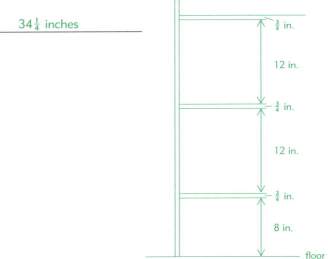

40. Mother is cutting old towels into smaller rags. One towel is 45 inches long and 29 inches wide. If the rags are $9\frac{1}{2}$ inches square, how many full squares can she make out of that towel?

12 squares

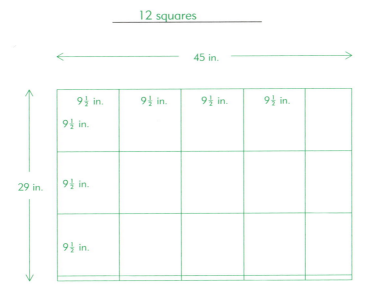

INDEX

absolute value 483
account balance 401
acre 66, 84
acute angle 286
acute triangle 291
addition
    associative law 21
    axiom 458
    commutative law 21
    compound measure 87
    decimals 153
    fractions 124
    mental 24, 437
    monomials 455
    signed numbers 486
algebra 445–497
    equations 458–466
    evaluating 452
    exponents 470–476
    expressions 446, 447
    literal numbers 446
    order of operations 449
    signed numbers 482–497, 507
    terms 446
angle 282, 286
    acute 286
    complementary 286
    obtuse 286
    reflex 286
    right 282, 286
    sides 282
    straight 286
    supplementary 286
    vertex 282
antecedent 169, 191
Arabic number system 12, 15, 19

arc 308
area 66, 83
    circle 323
    compound figures 326
    metric 83
    parallelogram 315
    rectangle 312
    square 312
    trapezoid 319
    triangle 315
arithmetic mean 240
arithmetic notation 12
associative law 21
average 240
balancing accounts 401
bank number 395
bank statement 405
bar graph 257
base (of exponent) 470
base in numeration 527
    twelve 530, 537
    two 533, 537
base, percentage 204, 222
bath 102
Bible measure 102, 103
binary system 533, 537
binomial 447
bisecting
    angles 288, 300
    line segments 300
blueprint 191
bushel 103
cab 103

cancellation   127
capacity   63, 80
capitalism   426
caret   159
casting out nines   33, 44
Celsius   96
Centigram   77
centiliter   80
centimeter   73
century   69
check register   401, 405
checking
   casting out nines   33, 44
   division   44
   multiplication   33
checking account   394–406
chord   308
choosing correct operation   52
circle   283, 308
   area   323
   circumference   308
circle graph   265
circumference   308
coefficient   446
commission   225
common fraction   120
commutative law   21, 33
comparing
   common fractions   121
   decimal fractions   153
compass   297, 300
complementary angle   286
complex fraction   120
composite number   114
compound figures   326
consequent   169
cost of goods   426

common denominator   121
commutative law   21
complex fractions   134
compound interest   414, 417
compound interest table   418
compound measure
   adding, subtracting   87
   multiplying, dividing   90
cone   340
   volume   360
congruent figures   292
consequent   169, 191
constructing
   angles   287
   equilateral triangles   301
   hexagons   301
   graph from formula   503
   perpendicular lines   301
   table from formula   500
   triangles   296, 301
cost of goods   426
counting numbers   482
cube   340
   surface area   340
   volume   357
cubit   102
cylinder   340
   surface area   347
   volume   360
decade   69
decagon   283
decigram   77
deciliter   80
decimals   152, 200
   adding and subtracting   153
   dividing   159
   multiplying   156, 166
   nonterminating   163
   repeating   163
   terminating   162

decimal system   15
decimeter   73
decrease by percent   210, 213
dekagram   77
dekaliter   80
dekameter   73
deposit ticket   394
destination   490
diameter   308
direct proportion   173
directly proportionate   503
distance, rate, time   99
distributive law   39
divide-and-divide method   47
divisibility rules   47
division   43
   axiom   461
   checking   44
   compound measure   90
   decimals   159
   exponents   476
   fractions   131, 134
   mental   47–50, 139, 438
   signed numbers   494
double-and-divide-method   40
double-and-double method   50
dry measure   63
duodecimal system   530, 537
electric meter   518
English measure   60–69
English/metric conversion   93, 96
enlargement   191
ephah   103
equation   458
   solving   458, 461
   writing   465

equiangular triangle   291
equilateral triangle   291
estimation
   products   33
evaluating expressions   452
expanding fractions   120
expense   426
exponent   114
   in division   476
   in multiplication   473
   with literal numbers   470
exponentially proportionate   504
expressions, algebraic   446
extracting square root   373, 376
extremes   173
factors   114, 117, 446
factoring   114, 117, 118
Fahrenheit   96
fathom   60
finger   102
formulas   582, 583
fractional part   137
fractions   120, 200
   adding and subtracting   124
   comparing   121
   decimal equivalents   162, 163
   dividing   131, 134
   multiplying   127, 166
   percent equivalents   200, 201
   reciprocal   131
frequency distribution table   247
furlong   103
geometry   281
   plane   282–326
   solid   340–366
   terms   282–283, 340
gerah   102

gram   77

graph
   bar   257
   circle   265
   from formula   503
   histogram   247
   line   261
   picture   252
   rectangle   269

greatest common factor   117

gross profit   426

handbreadth   102

hectare   84

hectogram   77

hectoliter   80

hectometer   73

heptagon   283

hexagon   283

hin   102

Hindu-Arabic system   12

histogram   247

homer   102, 103

hypotenuse   379

improper fraction   120

income   426

increase by percent   210, 213

integers   482

interest   408–418
   compound   414, 417
   part-year   412
   simple   408

International System of Units   73

intersecting lines   283

inverse proportion   178

irrational number   308

irregular polygon   283

isosceles triangle   291

kilogram   77

kiloliter   80

kilometer   73

kilowatt-hour   518

lateral surface   341

league   60

leap year   69

like fractions   121

like terms   455

line   282
   intersecting   283
   parallel   283
   perpendicular   283
   segment   282

linear measure   60, 73

line graph   261

line segment   282

liquid measure   63

liter   80

literal coefficient   446

literal number   446
   with exponent   470

log   102

long division   43

long ton   60

lowest common multiple   118

maps   187

mathematical law
   associative   21
   commutative   21, 33
   distributive   39

means   173

measures
    adding, subtracting   87
    Bible   102, 103
    capacity   63, 80
    dry   63
    English   60–66
    linear   60, 73
    liquid   63
    metric   73–84
    metric/English conversion   93, 96
    multiplying, dividing   90
    square   66
    temperature conversion   96
    time   69
    weight   60, 77
median   243
mental math   437–438
    addition   24, 437
    dividing by proper fractions   139
    division   47–50, 438
    multiplication   36, 39, 437
    multiplying by fractions   127
    percent   228
    subtraction   30, 437
meter   73
metric/English conversion   93, 96
metric measure   73–84
metric ton   77
millennium   69
milligram   77
milliliter   80
millimeter   73
mixed number   120
mode   244
monomial   447

multiplication   32
    associative law   21
    axiom   461
    checking   33
    commutative law   21, 33
    compound measure   90
    decimals   156
    distributive law   39
    exponents   473
    fractions   127, 128
    mental   36, 437–438
    signed numbers   493
natural numbers   482
negative numbers   482
net profit   426
nonagon   283
nonterminating decimal   163
numbers
    Arabic   12
    counting   482
    integers   482
    natural   482
    negative   482
    place value   15
    positive   482
    real   482
    Roman   18
    rounding   16
    signed   482
numerical coefficient   446
obtuse angle   286
obtuse triangle   291
octagon   283
omer   103
order of operations   449
origin   490

578 Index

outstanding check/deposit 405
overhead 426
parallel lines 283
parallelogram 284
    area 315
pentagon 283
percent 200
    finding base 222
    finding percentage 204
    finding rate 216, 219
    fraction equivalents 201
    increase and decrease 210, 213, 219
percentage 204, 225
perfect square 369
perimeter 305
perpendicular lines 283
pi 308
picture graph 252
place value 15
plane 283
plane geometry 282
point 282
polygon 283
    regular/irregular 283
polynomial 447, 507
positive numbers 482
prime factor 114
prime number 114
principal, rate, time 408–414
profit 426, 430
proper fraction 120
property tax 422
proportion 173–191
    direct 173
    inverse 178
protractor 286, 287

pyramid 340
    surface area 351
    volume 363
Pythagorean rule 379, 382
quadrilateral 283
quadruple-and-quadruple method 50
radical sign 369
radicand 369
radius 308
ranking data 243
range of data 247
rate 204, 216, 219
greater than 100% or less than 1% 207
rate of commission 225
ratio 169, 200
ray 282
reading problems
    equations 465
    finances 434
    fractions in 141
    multistep 330
    missing information 145
    necessary information 106
    operation 52
    parallel problems 386
    proportions 183
    sketches 232
real numbers 482
reciprocal 131
reconciling account 405
rectangle 284
    area 312
    perimeter 305
rectangle graph 269
rectangular solid 340
    surface area 344
    volume 357
reducing fractions 120

reduction (drawing)   191
reed   102
reflex angle   286
regular polygon   283
repeating decimal   163
rhombus   284
right angle   282, 286
right triangle   291
rod   60
Roman numerals   18
rounding
   decimals   162
   picture graph symbols   253
   whole numbers   16
sales   225
sales tax   422
savings account   408
scale drawing   187, 191
scalene triangle   291
scientific notation   523
sector   266, 308
semicircle   308
shekel   102
short cut
   addition   24
   divide-and-divide   47
   double-and-divide   40
   double-and-double   50
   quadruple-and-quadruple   50
short division   44
signed numbers   482–497, 507
   adding   486
   dividing   494
   evaluating expressions   497
   multiplying   493
   polynomials   507
   subtracting   490
similar figures   292

simple interest   408
simplifying expressions   449
sketches   232
solid geometry   340
span   102
sphere   340
   surface area   354
   volume   366
square   284
   area   312
   perfect   369
   perimeter   305
square root   369
   extracting   373, 376
statistics   240, 247
   mean   240
   median   243
   mode   244
straight angle   286
substitution   452
subtraction   27
   axiom   458
   compound measures   87
   decimals   153
   fractions   124
   mental   30, 437
   monomials   455
   signed numbers   490
supplementary angles   286
surface area   340
   cube   340
   cylinder   347
   pyramid   351
   rectangular solid   344
   sphere   354
tables for formulas   500
talent   102
tax   422
temperature   96

terminating decimal 162
terms 120
time 69
time zones 69
trapezoid 284
    area 319
triangle 283
    area 315
    acute 291
    construction 296, 301
    equiangular 291
    equilateral 291
    isosceles 291
    obtuse 291
    right 291
    scalene 291
trinomial 447

unit fraction 120
unlike fractions 124
unlike terms 455
vertex 282
vinculum 18, 163
volume
    cone 360
    cube 357
    cylinder 360
    pyramid 363
    rectangular solid 357
    sphere 366
weight 60, 77
whole number 120
writing checks 397

SYMBOLS

Symbol	Name	Example	Meaning
>	greater than	7 > 3	"7 is greater than 3"
<	less than	2 < 5	"2 is less than 5"
·	multiplication	6 · 8	"6 times 8"
√	radical	√9	"the square root of 9"
∧	caret	2.05∧	insertion of decimal point
⟷	line	$\overleftrightarrow{AB}$	"line AB"
→	ray	$\overrightarrow{EF}$	"ray EF"
∠	angle	∠G	"angle G"
△	triangle	△JKL	"triangle JKL"
⌐	right angle	⌐	indication that an angle is a right angle
⊥	perpendicular	$\overleftrightarrow{ST} \perp \overleftrightarrow{UV}$	"line ST is perpendicular to line UV"
∥	parallel	$\overleftrightarrow{WX} \parallel \overleftrightarrow{YZ}$	"line WX is parallel to line YZ"
~	similar	△DEF ~ △GHI	"triangle DEF is similar to triangle GHI"
≅	congruent	△UVW ≅ △XYZ	"triangle UVW is congruent to triangle XYZ"
π	pi		relation of circumference to diameter of a circle; value near 3.14 or $3\frac{1}{7}$
—	vinculum	$\overline{FG}$	"line segment FG"
			$\overline{XXVI}$	thousands in Roman numerals
			$1.0\overline{298}$	non-terminating decimal repetition
			$\frac{a}{b+c}$	grouping and/or division in algebra
+	positive	+8	"positive 8"
−	negative	−8	"negative 8"
| |	absolute value	|−8|	"the absolute value of negative 8"

FORMULAS

Distance, rate, time

distance	$d = rt$
rate	$r = \frac{d}{t}$
time	$t = \frac{d}{r}$

Percent

percentage	$P = BR$
rate	$R = \frac{P}{B}$
base	$B = \frac{P}{R}$

Commission

commission	$c = sr$
rate	$r = \frac{c}{s}$
sales	$s = \frac{c}{r}$

Interest

interest	$i = prt$
rate	$r = \frac{i}{p}$
principal	$p = \frac{i}{r}$
compound interest	$a = p(1 + r)^n$

Temperature

Fahrenheit to Celsius	$C = \frac{5}{9}(F - 32)$
Celsius to Fahrenheit	$F = \frac{9}{5}C + 32$

FORMULAS

Geometric

PLANE GEOMETRY

	perimeter	*area*	
square	$p = 4s$	$a = s^2$	
rectangle	$p = 2(l + w)$	$a = lw$	
parallelogram		$a = bh$	
triangle	$p = a + b + c$	$a = \frac{1}{2}bh$	*hypotenuse*
right triangle			$c^2 = a^2 + b^2$
trapezoid		$a = \frac{1}{2}h(b_1 + b_2)$	
circle	$c = \pi d$	$a = \pi r^2$	*diameter*
	$c = 2\pi r$		$d = \frac{c}{\pi}$

SOLID GEOMETRY

	surface area	*volume*
cube	$a_s = 6e^2$	$v = e^3$
rectangular solid	$a_s = 2lw + 2wh + 2lh$	$v = lwh$
cylinder	$a_s = 2\pi r^2 + 2\pi rh$	$v = \pi r^2 h$ or $v = Bh$
cone		$v = \frac{1}{3}\pi r^2 h$ or $v = \frac{1}{3}Bh$
square pyramid	$a_s = 4(\frac{1}{2}b\ell) + b^2$	$v = \frac{1}{3}lwh$
sphere	$a_s = 4\pi r^2$	$v = \frac{4}{3}\pi r^3 = 4.18 \times r^3$ (for decimals)
		$v = \frac{4}{3}\pi r^3 = \frac{88}{21} \times r^3$ (for fractions and multiples of 7)

TABLE OF SQUARE ROOTS

Number	Square root	Number	Square root	Number	Square root	Number	Square root
1	1.000	51	7.141	101	10.050	151	12.288
2	1.414	52	7.211	102	10.100	152	12.329
3	1.732	53	7.280	103	10.149	153	12.369
4	2.000	54	7.348	104	10.198	154	12.410
5	2.236	55	7.416	105	10.247	155	12.450
6	2.449	56	7.483	106	10.296	156	12.490
7	2.646	57	7.550	107	10.344	157	12.530
8	2.828	58	7.616	108	10.392	158	12.570
9	3.000	59	7.681	109	10.440	159	12.610
10	3.162	60	7.746	110	10.488	160	12.649
11	3.317	61	7.810	111	10.536	161	12.689
12	3.464	62	7.874	112	10.583	162	12.728
13	3.606	63	7.937	113	10.630	163	12.767
14	3.742	64	8.000	114	10.677	164	12.806
15	3.873	65	8.062	115	10.724	165	12.845
16	4.000	66	8.124	116	10.770	166	12.884
17	4.123	67	8.185	117	10.817	167	12.923
18	4.243	68	8.246	118	10.863	168	12.961
19	4.359	69	8.307	119	10.909	169	13.000
20	4.472	70	8.367	120	10.954	170	13.038
21	4.583	71	8.426	121	11.000	171	13.077
22	4.690	72	8.485	122	11.045	172	13.115
23	4.796	73	8.544	123	11.091	173	13.153
24	4.899	74	8.602	124	11.136	174	13.191
25	5.000	75	8.660	125	11.180	175	13.229
26	5.099	76	8.718	126	11.225	176	13.266
27	5.196	77	8.775	127	11.269	177	13.304
28	5.292	78	8.832	128	11.314	178	13.342
29	5.385	79	8.888	129	11.358	179	13.379
30	5.477	80	8.944	130	11.402	180	13.416
31	5.568	81	9.000	131	11.446	181	13.454
32	5.657	82	9.055	132	11.489	182	13.491
33	5.745	83	9.110	133	11.533	183	13.528
34	5.831	84	9.165	134	11.576	184	13.565
35	5.916	85	9.220	135	11.619	185	13.601
36	6.000	86	9.274	136	11.662	186	13.638
37	6.083	87	9.327	137	11.705	187	13.675
38	6.164	88	9.381	138	11.747	188	13.711
39	6.245	89	9.434	139	11.790	189	13.748
40	6.325	90	9.487	140	11.832	190	13.784
41	6.403	91	9.539	141	11.874	191	13.820
42	6.481	92	9.592	142	11.916	192	13.856
43	6.557	93	9.644	143	11.958	193	13.892
44	6.633	94	9.695	144	12.000	194	13.928
45	6.708	95	9.747	145	12.042	195	13.964
46	6.782	96	9.798	146	12.083	196	14.000
47	6.856	97	9.849	147	12.124	197	14.036
48	6.928	98	9.899	148	12.166	198	14.071
49	7.000	99	9.950	149	12.207	199	14.107
50	7.071	100	10.000	150	12.247	200	14.142

Tables of Measure

Bible Measure

	Approximate English Equivalent	Approximate Metric Equivalent

Length

1 finger	$\frac{3}{4}$ inch	1.9 centimeters
1 handbreadth = 4 fingers	3 inches	7.6 centimeters
1 span = 3 handbreadths	9 inches	23 centimeters
1 cubit = 2 spans	18 in. or $1\frac{1}{2}$ feet	46 centimeters or 0.46 meters
1 fathom = 4 cubits	6 feet	1.8 meters
1 reed = 7 cubits	$10\frac{1}{2}$ feet	3.25 meters
1 furlong	606 feet or $\frac{1}{9}$ mile	184.7 meters or 0.18 kilometers

Weight

1 gerah	$\frac{1}{50}$ ounce	0.6 gram
1 bekah = 10 gerahs	$\frac{1}{5}$ ounce	$5\frac{2}{3}$ grams
1 shekel = 2 bekahs	$\frac{2}{5}$ ounce	11.3 grams
1 pound (maneh) = 50 shekels	20 ounces	566 grams
1 talent = 60 manehs	75 pounds	34 kilograms

Liquid Measure

1 log	almost 1 pint	$\frac{1}{3}$ liter
1 hin = 12 logs	$5\frac{1}{3}$ quarts	5 liters
1 bath = 6 hins	8 gallons	30.3 liters
1 firkin	9 gallons	34 liters
1 homer = 10 baths	80 gallons	303 liters

Dry Measure

1 cab	$2\frac{3}{4}$ pints	1.5 liters
1 omer = almost 2 cabs	5 pints	2.8 liters
1 seah = 6 cabs or $3\frac{1}{3}$ omers	1 peck	9.3 liters
1 ephah = 3 seahs or 10 omers	$3\frac{1}{4}$ pecks	28.2 liters
1 homer = 10 ephahs	8 bushels	282 liters

Tables of Measure

Metric Measure

Length

basic unit: **meter**

1 millimeter = 0.001 meter
1 centimeter = 0.01 meter
1 decimeter = 0.1 meter
1 dekameter = 10 meters
1 hectometer = 100 meters
1 kilometer = 1,000 meters

Area

1 hectare = 10,000 square meters

Capacity

basic unit: **liter**

1 milliliter = 0.001 liter
1 centiliter = 0.01 liter
1 deciliter = 0.1 liter
1 dekaliter = 10 liters
1 hectoliter = 100 liters
1 kiloliter = 1,000 liters

Weight

basic unit: **gram**

1 milligram = 0.001 gram
1 centigram = 0.01 gram
1 decigram = 0.1 gram
1 dekagram = 10 grams
1 hectogram = 100 grams
1 kilogram = 1,000 grams
1 metric ton = 1,000 kilograms

Temperature, Celsius scale

0° = freezing point
100° = boiling point

Metric-to-English Conversion

Linear Measure

1 centimeter = 0.39 inch
1 meter = 39.4 inches
1 meter = 3.28 feet
1 kilometer = 0.62 mile

Weight

1 gram = 0.035 ounce
1 kilogram = 2.2 pounds

Capacity

1 liter = 1.06 quarts (liquid)

Area

1 hectare = 2.5 acres
1 square kilometer = 0.39 square mile

Temperature

$\frac{9}{5}$ degrees Celsius + 32 = degrees F

English-to-Metric Conversion

Linear Measure

1 inch = 2.54 centimeters
1 foot = 0.3 meter
1 mile = 1.61 kilometers

Weight

1 ounce = 28.3 grams
1 pound = 0.45 kilogram

Capacity

1 quart (liquid) = 0.95 liter
1 tablespoon = 15 milliliters

Area

1 acre = 0.4 hectare
1 square mile = 2.59 square kilometers

Temperature

$\frac{5}{9}$ (degrees Fahrenheit − 32) = degrees C